Advances in Spatial Science

Titles in the Series

H. Eskelinen and *F. Snickars* (Eds.)
Competitive European Peripheries
VIII, 271 pages. 1995. ISBN 3-540-60211-9

C. S. Bertuglia, S. Lombardo and *P. Nijkamp* (Eds.)
Innovative Behaviour in Space and Time
X, 437 pages. 1997. ISBN 3-540-62542-9

A. Nagurney and *S. Siokos*
Financial Networks
XVI, 492 pages. 1997. ISBN 3-540-63116-X

M. M. Fischer and *A. Getis* (Eds.)
Recent Developments in Spatial Analysis
X, 434 pages. 1997. ISBN 3-540-63180-1

P. McCann
The Economics of Industrial Location
XII, 228 pages. 1998. ISBN 3-540-64586-1

R. Capello, P. Nijkamp and *G. Pepping* (Eds.)
Sustainable Cities and Energy Policies
XI, 282 pages. 1999. ISBN 3-540-64805-4

M. M. Fischer, L. Suarez-Villa and *M. Steiner* (Eds.)
Innovation, Networks and Localities
XI, 336 pages. 1999. ISBN 3-540-65853-X

J. Stillwell, S. Geertman and *S. Openshaw* (Eds.)
Geographical Information and Planning
X, 454 pages. 1999. ISBN 3-540-65902-1

G. J. D. Hewings, M. Sonis, M. Madden and *Y. Kimura* (Eds.)
Understanding and Interpreting Economic Structure
X, 365 pages. 1999. ISBN 3-540-66045-3

D. G. Janelle and *D. C. Hodge* (Eds.)
Information, Place, and Cyberspace
XII, 381 pages. 2000. ISBN 3-540-67492-6

G. Clarke and *M. Madden* (Eds.)
Regional Science in Business
VIII, 363 pages. 2001. ISBN 3-540-41780-X

M. M. Fischer and *Y. Leung* (Eds.)
GeoComputational Modelling
XII, 279 pages. 2001. ISBN 3-540-41968-3

M. M. Fischer and *J. Fröhlich* (Eds.)
Knowledge, Complexity and Innovation Systems
XII, 477 pages. 2001. ISBN 3-540-41969-1

M. M. Fischer, J. Revilla Diez and *F. Snickars*
Metropolitan Innovation Systems
VIII, 270 pages. 2001. ISBN 3-540-41967-5

L. Lundqvist and *L.-G. Mattsson* (Eds.)
National Transport Models
VIII, 202 pages. 2002. ISBN 3-540-42426-1

J. R. Cuadrado-Roura and *M. Parellada* (Eds.)
Regional Convergence in the European Union
VIII, 368 pages. 2002. ISBN 3-540-43242-6

G. J. D. Hewings, M. Sonis and *D. Boyce* (Eds.)
Trade, Networks and Hierarchies
XI, 467 pages. 2002. ISBN 3-540-43087-3

G. Atalik and *M. M. Fischer* (Eds.)
Regional Development Reconsidered
X, 220 pages. 2002. ISBN 3-540-43610-3

Z. J. Acs, H. L. F. de Groot and *P. Nijkamp* (Eds.)
The Emergence of the Knowledge Economy
VII, 388 pages. 2002. ISBN 3-540-43722-3

R. J. Stimson, R. R. Stough and *B. H. Roberts*
Regional Economic Development
X, 397 pages. 2002. ISBN 3-540-43731-2

S. Geertman and *J. Stillwell* (Eds.)
Planning Support Systems in Practice
XII, 578 pages. 2003. ISBN 3-540-43719-3

B. Fingleton (Ed.)
European Regional Growth
VIII, 435 pages. 2003. ISBN 3-540-00366-5

T. Puu
Mathematical Location and Land Use Theory, 2nd Edition
X, 362 pages. 2003. ISBN 3-540-00931-0

J. Bröcker, D. Dohse and *R. Soltwedel* (Eds.)
Innovation Clusters and Interregional Competition
VIII, 409 pages. 2003. ISBN 3-540-00999-X

D. A. Griffith
Spatial Autocorrelation and Spatial Filtering
XIV, 247 pages. 2003. ISBN 3-540-00932-9

J. R. Roy
Spatial Interaction Modelling
X, 239 pages. 2004. ISBN 3-540-20528-4

M. Beuthe, V. Himanen
A. Reggiani and *L. Zamparini* (Eds.)
Transport Developments and Innovations in an Evolving World
XIV, 346 pages. 2004. ISBN 3-540-00961-2

Y. Okuyama and *S. E. Chang* (Eds.)
Modeling Spatial and Economic Impacts of Disasters
X, 323 pages. 2004. ISBN 3-540-21449-6

Luc Anselin · Raymond J. G. M. Florax
Sergio J. Rey (Editors)

Advances in Spatial Econometrics

Methodology, Tools and Applications

With 41 Figures
and 83 Tables

Springer

Dr. Luc Anselin
Regional Economics Applications Laboratory
Dept. of Agricultural and Consumer Economics
University of Illinois, Urbana-Champaign
1301 Gregory Drive
Urbana, IL 61801
USA
E-mail: anselin@uiuc.edu

Dr. Sergio J. Rey
Dept. of Geography
San Diego State University
San Diego, CA 92182-4493
USA
E-mail: rey@typhoon.sdsu.edu

Dr. Raymond J. G. M. Florax
Dept. of Spatial Economics
Free University
De Boelelaan 1105
1081 HV Amsterdam
The Netherlands
E-mail: rflorax@feweb.vu.nl

Cataloging-in-Publication Data applied for
A catalog record for this book is available from the Library of Congress.
Bibliographic information published by Die Deutsche Bibliothek
Die Deutsche Bibliothek lists this publication in the Deutsche Nationalbibliografie; detailed bibliographic data available in the internet at *http://dnb.ddb.de*

ISBN 3-540-43729-0 Springer Berlin Heidelberg New York

Springer is a part of Springer Science+Business Media
springeronline.com

Printed in Germany

Cover design: Erich Kirchner, Heidelberg
Production: Helmut Petri
Printing: betz-druck

Printed on acid-free paper – 42/3130 – 5 4 3 2 1 0

To Jean Paelinck

Preface

The volume on *New Directions in Spatial Econometrics* appeared in 1995 as one of the first in the then new Springer series on *Advances in Spatial Sciences*. It very quickly became evident that the book satisfied a pent up demand for a collection of advanced papers dealing with the methodology and application of spatial econometrics. This emerging subfield of applied econometrics focuses on the incorporation of location and spatial interaction in the specification, estimation and diagnostic testing of regression models.

The current effort is a follow up to the *New Directions* volume. Even though the number of empirical and theoretical journal articles dealing with various aspects of spatial econometrics has grown tremendously in the recent past, the need remained to bring together an advanced collection on methodology, tools and applications. This volume contains several papers that were presented at special sessions on spatial econometrics organized as part of a number of conferences of the Regional Science Association International. In addition, a few papers were invited for submission. All papers were refereed.

The focus in the volume reflects the advances made in the field in recent years. In terms of methodology, attention has moved to models for discrete dependent variables, endogeneity in systems of equations and advanced diagnostic tests for multiple sources of misspecification. In addition, the Bayesian and non-parametric perspectives on spatial analysis are becoming increasingly important parts of the methodological toolbox. Applications reflect topical interests in regional science and the new economic geography, centered around the concepts of externalities, agglomeration economies, and economic growth and convergence. New software tools have been developed as well, facilitating the dissemination of existing methods and the stimulation of new ones.

The growing appreciation for the role of a spatial perspective in social science research is evidenced in the United States by the establishment of the Center for Spatially Integrated Social Science, funded by the U.S. National Science Foundation under grant BCS-9978058. CSISS has supported the editorial efforts behind this volume and has included it as a part of its *best practices* program. Prof. Michael Goodchild, the Director of CSISS, authored the Foreword.

A volume such as this could not have come to be without the assistance of many individuals. We gratefully acknowledge the time (patience) and effort spent by all authors and referees, and the editorial guidance provided by Marianne Bopp at Springer Verlag. We particularly appreciate the technical typesetting prowess of Mark Janikas of the Geography Department at San Diego State University, who served as the LaTeX guru on the project, and without whose tremendous effort and dedication this volume would not have existed. We also thank students in the Spatial Econometrics course at the University of Illinois, Urbana-Champaign, who reviewed and commented on draft copies of various chapters. We are extremely grateful to Carolyn (Dong) Guo of REAL at the University of Illinois, who proof-read the complete manuscript and suggested several useful corrections.

The Bruton Center at the University of Texas at Dallas provided institutional support in the early stages of the editorial project. In addition, we are grateful for the open source software movement, which has given us tools such as TeX, LaTeX, Vim and Python that were instrumental in facilitating the technical aspects of typesetting and indexing.

Finally, we would like to dedicate this volume to Jean Paelinck, who coined the term *spatial econometrics* in the early 1970s and has remained a strong and active force behind the growth of the ˇ eld throughout the years.

Urbana, IL, USA — *Luc Anselin*
Amsterdam, The Netherlands — *Raymond J.G.M. Florax*
San Diego, CA, USA — *Sergio Rey*

March 2004

Foreword

Space is an essential part of human experience: along with time it frames events, since everything that happens happens *somewhere* in space and time. The power of science lies in its ability to discover general truths that are independent of space and time, and can therefore be expressed economically, and applied anywhere, at any time, to solve problems of human importance. So it is not at all obvious that space is important to science, except as a complication to be removed during the process of generalization.

This book is about advances in spatial econometrics, a discipline founded on the principle that space *is* important to our understanding of economic and other social processes operating in human societies, distributed over the surface of the Earth. It has strong links with the older disciplines of geography and regional science, and of course economics. It takes a quantitative approach, modeling the interactions that occur across space and that influence economies, labor markets, housing markets, and a myriad of forms of economic and social activity. Spatial variables such as distance appear explicitly in spatial econometric models, to capture these interactions and their response to location. Space is thus an inherent part of the scientific generalizations that result from spatial econometric analysis, but in an abstracted form, typically as a matrix of interactions W, rather than as locations *per se*. Such models are therefore invariant under a range of spatial operations, including rotation, translation, and inversion. The interaction matrix captures *relative* location only, *absolute* location being irrelevant to most spatial econometric theory.

Two arguments underlie this approach, the first behavioral and the second artifactual. Human societies interact in numerous ways, through migration, journeys to work, telephone and mail communication, transportation of goods, and flows of information. In all of these forms interaction tends to react to distance, because interaction cost is a function of distance, or because human acquaintance networks depend in part on face-to-face contact, or because it takes time to overcome distance. Thus space, in the form of distance, becomes a direct causal factor in processes that are impacted by interaction. Recently, of course, there has been much speculation over the distance-conquering effects of the Internet on flows of information.

The second argument results from the tendency of human societies to impose largely arbitrary boundaries on what is in many respects a continuous surface, in part to preserve confidentiality, and in part for economy. Statistical reporting agencies assemble data for bounded zones, masking within-zone variation, and limiting social scientists to the study of between-zone variation. This would be fine if zones behaved as independent social aggregates, but of course they do not; if there are such things as independent social aggregates on the Earth's surface, they are almost certainly cut frequently by zone boundaries. Thus models must include space, again in the form of a matrix of interactions, to deal with what is in essence an inability of data-gathering practice to provide data in a theoretically coherent form.

Over the past three decades spatial econometrics has advanced from a fringe scientific activity to the status of a fledgling discipline. Many of its leaders are represented in the pages of this book, and almost all are cited. The book comes at a time when space is more important than ever in social science, not only for the reasons cited above, but also because of the dramatic increase in recent years in the supply of spatially referenced data; the widespread adoption of geographic information systems (GIS) and other software for handling spatial data and for performing spatial analysis and modeling; and the increasing pressure on science to deliver results that are readily incorporated into policy. The book is a welcome addition to the literature, providing a single source for the most important recent work in the field.

The Center for Spatially Integrated Social Science (CSISS) was funded in 1999 by the U.S. National Science Foundation to improve the research infrastructure for spatial analysis and modeling in the social and behavioral sciences. The arguments for CSISS, including those already outlined above, are elaborated by Goodchild *et al.* (2000). CSISS sponsors seven programs, including the development of tools for analysis and modeling; full descriptions can be found on the Center's website, `http://www.csiss.org`. As Director of CSISS, I am honored to contribute this Foreword, and I welcome the book as an important product of the Center's work and as a significant contribution to the field.

Santa Barbara, CA, USA *Michael F. Goodchild*

March 2004

Contents

List of Tables

List of Figures

1 Econometrics for Spatial Models: Recent Advances

Luc Anselin[1], Raymond J.G.M. Florax[2], and Sergio J. Rey[3]

[1] University of Illinois
[2] Free University Amsterdam
[3] San Diego State University

1.1 Introduction

In the introduction to *New Directions in Spatial Econometrics* (Anselin and Florax, 1995b), the precursor to the current volume, we set out by arguing that "it would be an overstatement to suggest that spatial econometrics has become accepted practice in current empirical research in regional science and regional economics." However, we also pointed out that "there is evidence of an increased awareness of the importance of space in recent empirical work in 'mainstream' economics" (Anselin and Florax, 1995a, p. 3). In the few years since *New Directions* appeared, the latter observation has been confirmed by a tremendous growth in the number of publications in which spatial econometric techniques are applied, not only within regional science and economic geography, but also increasingly in the leading journals of economics, sociology and political science. This has not gone unnoticed, and the wealth of new publications has resulted in a separate classification in the *Journal of Economic Literature* devoted solely to cross-sectional and spatial models.[1] Parallelling the growth in applications, several new methods have been introduced as well, yielding a spatial econometric toolbox that is becoming ever more sophisticated.

Arguably, the renewed interest in a spatial perspective in social science research was also behind the establishment of the Center for Spatially Integrated Social Science (CSISS), funded by the U.S. National Science Foundation (Goodchild *et al.*, 2000). As part of its activities, CSISS has organized several workshops and specialist meetings dealing with the incorporation of spatial analysis concepts and methods in the social sciences. Of direct relevance to spatial econometrics were the workshops on modeling spatial externalities (Anselin, 2003b), on the development of spatial software tools (Anselin and Rey, 2002), and, most recently, on the importance of spatial and social interactions in economics.[2]

Given these developments, we felt it would be timely to bring together a number of papers that reflect the advances made in recent years, both in terms of new methodological approaches as well as in the application of spatial econometrics to

[1] JEL C21, Econometric Methods, Cross-Sectional Models; Spatial Models.

[2] The full set of materials on this meeting can be found on the CSISS web site at: http://www.csiss.org/events/meetings/spatial-interactions/agenda.htm

a broad range of fields in applied economics and regional science. The current volume is the result of this compilation.[3] The nineteen chapters are organized into five parts, two dealing primarily with methodological issues, and three geared to applications. These five parts are, respectively, Specification, Testing and Estimation; Discrete Choice, Nonparametric and Bayesian Approaches; Spatial Externalities; Urban Growth and Agglomeration Economies; and Trade and Economic Growth. Before providing a brief summary of the different chapters, we review recent advances in spatial econometrics, as reflected in the literature that appeared since the publication of the *New Directions* volume. We close this introductory chapter with some speculations about future directions.

1.2 Recent Advances

Since the *New Directions* volume was published, several other extensive reviews of the state of the art in spatial econometrics appeared, such as Anselin and Bera (1998), LeSage (1999), Anselin (2001b, 2002), and, most recently, Florax and van der Vlist (2003). In addition, the review article by Dubin *et al.* (1999) dealt specifically with the application of spatial econometrics in real estate analysis. Also, since 1995, a number of special journal issues were devoted to spatial econometrics. In contrast to the period before 1995, these did not only appear in the traditionally hospitable regional science journals, such as the two special issues of the *International Regional Science Review* (Anselin and Rey, 1997; Florax and van der Vlist, 2003). Specialized "field" journals in economics published special issues on spatial analysis and spatial econometrics as well. This includes, in real estate and housing economics, the *Journal of Real Estate Finance and Economics* (Pace *et al.*, 1998b), and the *Journal of Housing Research* (Can, 1998), and, in agricultural and natural resource economics, a recent issue of *Agricultural Economics* (Nelson, 2002). Also, a main methods journal in criminology, *The Journal of Quantitative Criminology* (Cohen and Tita, 1999), and two political science journals, *Political Analysis* (Ward and O'Loughlin, 2002), and *Political Geography* (Ward, 2002) published recent special issues that dealt with the application of spatial analysis, including spatial regression methods. On the downside, the notion of spatial correlation as an equivalent form of serial correlation is still mostly absent in mainstream econometrics textbooks, with only a few exceptions, such as Johnston and DiNardo (1997). Refreshing in this respect is the inclusion of a section on spatial panels in the second edition of Baltagi's well known panel data econometrics text (Baltagi, 2001, pp. 195–197).

In their recent review article, Florax and van der Vlist (2003) surveyed examples of applications of spatial econometrics based on the contents of the subject and author index of regional science journals (broadly defined), as published by the *International Regional Science Review*.[4] Since their review centered on the adoption of spatial econometrics in regional science, here we provide some complementary

[3] Parenthetically, the current volume was supported by CSISS as part of its *best practices* program.

[4] For details on the scope and methodology used for this index, see Anselin *et al.* (2000).

Table 1.1. Spatial Econometrics in Econometric Methods Journals

Journal	Articles
Econometrica	Pinkse *et al.* (2002)
Econometric Reviews	Baltagi and Li (2001a)
Econometric Theory	Lee (2002)
Journal of Applied Econometrics	Conley and Topa (2002)
Journal of Business and Economic Statistics	Gelfand (1998)
Journal of Econometrics	Blommestein and Koper (1998)
	Pinkse and Slade (1998)
	Conley (1999)
	Kelejian and Prucha (2001)
	Chen and Conley (2001)
	Baltagi *et al.* (2003)
	Kelejian and Prucha (2003)
	Giacomini and Granger (2003)
The Review of Economics and Statistics	Driscoll and Kraay (1998)
	Bell and Bockstael (2000)
	Beron *et al.* (2003)

insight into the current state of diffusion of spatial techniques by focusing specifically on publications in economics journals, and only for the period since 1995.

We find that, in contrast to an almost total absence before 1995, the latter part of the nineties and especially the beginning of the twenty-first century has seen spatial econometrics become a constant (though sparse) presence in the mainstream econometric literature, as illustrated in Table 1.1. The seven journals listed in the table include the main publications in theoretical econometrics, such as *Econometrica*, the *Journal of Econometrics*, and *Econometric Theory*, as well as the leading journals in applied econometrics. In the period surveyed, they contained sixteen articles dealing specifically with spatial econometric topics, but it is notable that eleven of those only appeared after 2000 (including four in 2003).

A similar pattern emerges when considering "field" journals in economics during the same period, but excluding the contents of the special issues mentioned earlier (specifically, the 6 articles contained in the 1998 special issue of the *Journal of Real Estate Finance and Economics* and the 14 articles in the 2002 special issue of *Agricultural Economics*). Table 1.2 lists twenty such publications that contained a total of 43 articles dealing with spatial econometric topics (either methodological or empirical). Of those, 30 appeared since 2000, including 10 in the year 2003.[5] This near exponential growth constitutes a sea change in the acceptance of spatial econometric methods in mainstream empirical economic research, and represents a significant advance relative to the state of the field reviewed in 1995.

[5] This figure is a potential undercount, since it includes only articles that appeared in the first six months of 2003, or were included as in press on journal web sites.

Table 1.2. Spatial Econometric Applications in Economic Field Journals

Journal	Articles
American Journal of Agricultural Economics	Bockstael (1996)
	Nelson and Hellerstein (1997)
	Irwin and Bockstael (2001)
	Anselin (2001c)
	Roe *et al.* (2002)
Applied Economics	Revelli (2001)
	Revelli (2002b)
Ecological Economics	Geoghegan *et al.* (1997)
	Bastian *et al.* (2002)
Economics Letters	Bivand and Szymanski (1997)
	Pace (1997)
	Lahatte (2003)
Economica	Murdoch *et al.* (1997)
International Economic Review	Kelejian and Prucha (1999)
Journal of Economic Behavior and Organization	Hautsch and Klotz (2003)
Journal of Economic Geography	Irwin and Bockstael (2002)
Journal of Economic Growth	Moreno and Trehan (1997)
	Conley and Ligon (2002)
Journal of Economics and Management Strategy	Kalnins (2003)
Journal of Environmental Economics and Management	Kim *et al.* (2003a)
Journal of Public Economics	Murdoch *et al.* (2003)
Journal of Real Estate Finance and Economics	Can and Megbolugbe (1997)
	Pace and Gilley (1997)
	Gillen *et al.* (2001)
	Cano-Guervós *et al.* (2003)
Journal of Urban Economics	Anselin *et al.* (1997)
	Brueckner (1998)
	Saavedra (2000)
	Boarnet and Glazer (2002)
	Plantinga *et al.* (2002)
	Buettner (2003)
	Revelli (2003)
Land Economics	Nelson *et al.* (2001)
	Irwin (2002)
	Paterson and Boyle (2002)
	Lynch and Lovell (2003)
National Tax Journal	Brueckner and Saavedra (2001)
Real Estate Economics	Pace and Gilley (1998)
	Clapp *et al.* (2002)

continued on next page

Table 1.2. Continued

Journal	Articles
	Thibodeau (2003)
Research Policy	Acs *et al.* (2002)
Review of Economic Studies	Topa (2001)
Structural Change and Economic Dynamics	Agnihotri *et al.* (2002)

In *New Directions*, we suggested three major reasons for (then) future growth in the importance and relevance of spatial methods: a renewed interest in the role of space and spatial interactions in social science theory; the increased availability of large socio-economic data sets with geo-referenced observations; and the existence of low cost geographic information systems to manipulate spatial data (Anselin and Florax, 1995a, pp. 4–5). Since 1995, both the use of georeferenced data and GIS technology have become common in empirical social science research. From a theoretical perspective, there have been several exciting developments, strengthening the importance of the first argument made in *New Directions*. In addition, two other significant factors may be suggested that hightened the attention to and acceptance of spatial modeling techniques in the social sciences. One is the tremendous activity (relative to earlier periods) in methodological research to deal with spatially correlated data. The other is the ready availability of software to estimate and test these models, mimicking but also extending the functionality of the legacy *SpaceStat* software (Anselin, 1992). In the following sections, we briefly review some highlights of recent advances (since 1995) along the three dimensions of spatial theory, methodology and software.

1.2.1 Spatial Theory

Perhaps the most visible form of an explicit spatial approach in modern economic theory is the *new economic geography*, typically identified with the publications of Krugman, Fujita, Henderson, Glaeser and co-workers (e.g., Fujita and Krugman, 2004). The theoretical focus on imperfect competition and increasing returns to scale led to an growing attention to the identification and measurement of spatial externalities (Anselin, 2003c). In the specific context of public economics, a recently formulated model for *strategic interaction* (Brueckner, 1998, 2003) forms the theoretical basis for the specification of a so-called spatial lag model, well known in spatial econometrics. Similarly, the notion of a *social multiplier*, popularized in the work of Glaeser *et al.* (1996, 2002) is for all practical purposes identical to the familiar concept of a spatial multiplier in spatial econometric models (Anselin and Bera, 1998). Several chapters in Parts III–V of this volume deal with applications of these concepts to empirical studies related to urban growth and agglomeration economies, international trade, and growth and convergence.

Maybe even more important as a driver of theoretical interest in a spatial perspective is the explicit introduction of social interaction in mainstream economic models dealing with the behavior of individual agents. This has led to a proliferation of models for various forms of spatial interaction, peer influence, neighbor and network effects (Dietz, 2002). The multiple equilibria typically associated with such models require an explicit consideration of *spatial heterogeneity*, whereas spatial interaction brings the role of *spatial dependence* to the fore.

The interplay between social and spatial interaction follows from a formal model of individual decision making that incorporates the role of "context." This yields intricate patterns of interrelations that are conceptualized using notions such as socio-economic distance and spatial correlation (e.g., Akerlof, 1997; Brock and Durlauf, 2001; Conley and Topa, 2002). The modeling of the resulting complex network and neighborhood effects (e.g., Topa, 2001; Aizer and Currie, 2002) requires considerable attention to identification issues, maybe best known from the work of Manski on the "reflection problem" (e.g., Manski, 2000). These theoretical developments have focused considerable attention on the specification and estimation of discrete choice models with spatial correlation, a topic dealt with in several chapters of Part II.

The tremendous recent growth in interest in spatial and social interaction has not been confined to economics. In sociology, building upon the distinguished tradition of the Chicago school, an explicit consideration of neighborhood and context has re-emerged as a central focus in recent work in criminology and urban sociology (Abbot, 1997; Sampson *et al.*, 2002). An increasing number of applications deal with specifications that incorporate externalities, diffusion and contagion in spatial analyses of crime, violence and neighborhood transition (e.g., Morenoff and Sampson, 1997; Sampson *et al.*, 1999; Morenoff *et al.*, 2001; Baller *et al.*, 2001; Baller and Richardson, 2002; Messner and Anselin, 2004). In addition, there are many formal similarities between the treatment of spatial correlation in spatial econometrics and the conceptualization of network correlation in social network analysis (Leenders, 2002).

In political science, explicit spatial models have seen recent application in studies of elections and American politics, for example, in the the work of Gimpel (1999), Gimpel and Schuknecht (2003), Revelli (2002a), Cho (2003), and Kim *et al.* (2003b). The link between social networks and individual voting behavior and the resulting spatial networks are analyzed in Baybeck and Huckfeldt (2002). Also, the formal expression of contagion and spatial externalities continues to be included in studies of international relations and conflict analysis (e.g., Gleditsch and Ward, 2000; Starr, 2001).

Most of the theoretical models of spatial effects turn out to be implemented as standard linear spatial regressions, either of the lag or error form. However, increasingly, the complex specifications resulting from the social and spatial interaction literature require more advanced methods, several of which were only developed in the past few years. We turn to this second driving force next.

1.2.2 Spatial Econometric Methods

Recent years have seen a level of activity in the development of new methods for spatial econometrics that is well above anything experienced prior to 1995. Many new model specifications have been considered, different test statistics proposed, novel estimation methods developed and their computational aspects assessed. In this respect, the current state of the art in spatial econometric methodology has moved significantly beyond the consideration of maximum likelihood estimation in the spatial lag and spatial error model, popularized in Ord (1975), Cliff and Ord (1981), and Anselin (1988b), which was still prevalent at the time the *New Directions* volume appeared.

It should be noted that this recent pattern in spatial econometrics has an arguably even more pronounced counterpart in spatial statistics. We will not consider this aspect in depth, but it is useful to acknowledge the prominent presence of spatial work in the modern statistical literature, with extensive applications in the natural sciences, environmental analysis and epidemiology. For example, the importance of contributions in spatial statistics is highlighted in several of the "vignettes" that appeared in the year 2000 issues of the *Journal of the American Statistical Association*, including those reviewing environmental statistics (Guttorp, 2000), environmental epidemiology (Thomas, 2000), and atmospheric sciences (Nychka, 2000).[6] The recent spatial statistical literature is characterized by a predominant Bayesian perspective, used to model complex space-time interactions by employing hierarchical specifications and simulation estimators, such as Markov Chain Monte Carlo (MCMC) and the Gibbs sampler. Reviews of some of the salient issues can be found in, among others, Wikle *et al.* (1998), Wolpert and Ickstadt (1998), Best *et al.* (1999), and Royle and Berliner (1999). It is worth noting that, to date, the adoption of the Bayesian hierarchical modeling paradigm in spatial econometrics has been limited.

We now turn to a brief review of recent (post 1995) results in the spatial econometric literature that pertain to model specification, testing, estimation and computation. This review is not intended to be comprehensive, but rather to be representative of the range of results that appeared in the literature.

Model Specification. The traditional specification of cross-sectional spatial correlation in the form of a linear regression model with a spatial lag or spatial error term is fairly constraining when it comes to expressing the full range of spatial externalities and spatial multipliers suggested in the theoretical literature. However, while more flexible specifications have been outlined (Anselin, 2003c; Lahatte, 2003), their estimation remains largely unexplored and they have (to date) seen no empirical application. In addition, standard concerns from the time series literature pertaining to unit roots and cointegration in models with lagged variables (or lagged error

[6] Statistical methods for social network analysis are referred to in the vignette on sociology (Raftery, 2000). See also Hoff *et al.* (2002) and Leenders (2002) for a recent review and examples.

terms) are only starting to receive some attention in spatial econometrics, although with mixed results (Fingleton, 1999c; Mur and Trívez, 2003). For example, such concerns are still absent from the treatment of spatial filtering, as exemplified in the recent paper of Getis and Griffith (2002). Some novel specifications have been introduced, primarily in the literature dealing with economic growth and convergence, such as spatial Markov models and models for spatial inequality (Rey, 2001, 2004).

The bulk of recent papers dealing with model specification remains focused on the linear regression model. Examples are closer scrutiny of the implications of the use of various formulations for the spatial correlation structure, as in Anselin (2002), Lee (2002), Dubin (2003) and Wall (2003). Also, the specification of spatial weights continues to receive attention (Bavaud, 1998; Tiefelsdorf *et al.*, 1999). More recently, the linear model has also been more frequently applied in the space-time domain, for example, in Gelfand (1998), Pace *et al.* (1998a), Elhorst (2001, 2003), and Giacomini and Granger (2003).

Finally, an interesting development, also receiving considerable attention in the chapters by Fleming, and Beron and Vijverberg in Part II of this volume, is the incorporation of spatial correlation in models with limited dependent variables, such as specifications used in discrete choice analysis. The spatial probit model in particular has been the focus of several recent papers, e.g., Pinkse and Slade (1998), LeSage (2000), Beron *et al.* (2003), and Murdoch *et al.* (2003).

Specification Testing. Several new test statistics for spatial correlation were developed since the *New Directions* volume appeared, and specification testing continues to be a very active area of research. The Moran's *I* test statistic remains an important focus of investigation. Further insight has been gained into its finite sample distribution (Tiefelsdorf, 2002), and it has been extended to new models, such as the residuals in a 2SLS estimation (Anselin and Kelejian, 1997). More importantly, the Moran's *I* statistic and its Lagrange Multiplier form have been generalized to apply to probit and tobit models by Pinkse and Slade (1998) and Kelejian and Prucha (2001). Other applications of the Lagrange Multiplier principle include tests for additional types of spatial error autocorrelation, such as direct representation (geostatistical model) and spatial error components (Anselin, 2001a; Anselin and Moreno, 2003). It has also been extended to a more general panel data setting (Baltagi *et al.*, 2003).

Recent findings include tests to deal with more complex alternative hypotheses, such as moving average or autoregressive spatial error processes (Mur, 1999), the combination of spatial correlation and heteroskedasticity (Kelejian and Robinson, 1998), as well as spatial correlation and functional misspecification (Baltagi and Li, 2001b). de Graaff *et al.* (2001) outline a general misspecification test against spatial correlation, heteroskedasticity and nonlinearity.

While most of these approaches rely on the Moran statistic and its Lagrange Multiplier counterpart (couched in a maximum likelihood estimation framework), other test strategies have been implemented as well. For example, a general nonparametric test against spatial dependence is suggested by Brett and Pinkse (1997), and spatial test statistics based on the results of method of moments estimation are considered by Kelejian and Robinson (1997) and Saavedra (2003). Baltagi and Li (2001a) extend the principle of double length artificial regression to testing for spatial lag and spatial error autocorrelation. Finally, Florax *et al.* (2003) consider the relative merits of forward and backward specification searches in spatial regression models.

The chapters by Florax and de Graaff, Pinkse, and Kelejian and Robinson in Part I of this volume elaborate on these themes.

Estimation. Some research efforts in recent years continued the tradition of applying the maximum likelihood estimation framework to spatial models. For example, Elhorst (2001, 2003) outlines ML estimation in a range of spatial panel data specifications. However, perhaps the most exciting developments in spatial econometrics involved the application of estimation paradigms other than ML to models with spatial dependence. Foremost among these is the general method of moments approach (including instrumental variables and generalized moments estimators) exemplified in the work of Kelejian and Robinson (1997), Kelejian and Prucha (1998, 1999), and Conley (1999). The derivation of the asymptotic properties of these estimators required the use of novel laws of large numbers and central limit theorems, based on the notion of triangular arrays, as demonstrated by Kelejian and Prucha (1999). GMM and generalized moments estimators also saw application to the spatial probit model by Pinkse and Slade (1998), and to systems of equations by Kelejian and Prucha (2003).

A second approach applies insights from Bayesian statistics. This is evident in work on developing spatial priors for space-time (vector autoregressive) forecasting models, for example, by Dowd and LeSage (1997) and LeSage and Krivelyova (1999). However, the most extensive use of Bayesian techniques in spatial econometrics is in the estimation of spatial autoregressive models, including the spatial probit model (LeSage, 1997a, 2000; Holloway *et al.*, 2002). In practice, this requires the application of simulation estimators, such as the Gibbs sampler.

Non-Bayesian simulation estimators, such as the recursive importance sampler (RIS) are evident in alternative approaches to estimating the spatial probit model. For example, Beron *et al.* (2003) and Murdoch *et al.* (2003) apply the RIS procedure to a spatial probit specification. Both Bayesian and non-Bayesian methods to estimate spatial discrete choice models are treated in the chapters by Fleming, and Beron and Vijverberg in Part II of this volume.

A totally different approach to the estimation problem is based on the use of semi-parametric methods, recently suggested by Driscoll and Kraay (1998), Chen and Conley (2001), and Pace and LeSage (2002).

In addition to the derivation and application of new estimators, the recent literature also includes several comparative studies. These contain both theoretical as well as empirical evaluations of alternative estimation procedures. Examples are Kelejian and Prucha (1997, 2002), Lee (2002), and Das *et al.* (2003).

Finally, it is worthwhile to point out considerable research effort in dealing with spatial heterogeneity in the form of spatially varying parameters. This is probably best known from the work of Fotheringham and colleagues on the geographically weighted regression, or GWR (for a recent comprehensive overview, see Fotheringham *et al.*, 2002, and the references contained therein). An alternative approach is outlined in the chapter by McMillen and McDonald in Part II of this volume. Yet a different perspective is offered in the recent literature on Bayesian spatially varying coef˜ cients, such as Gelfand *et al.* (2003) and Gamerman *et al.* (2003), as well as the chapter by LeSage in Part II of the volume.

Computation. An important practical issue related to the maximum likelihood estimation of spatial autoregressive models is the need to compute the determinant of the Jacobian of the spatial transformation, involving a matrix of dimension equal to the number of observations. For small and medium sized data sets, an eigenvalue decomposition suggested by Ord (1975) provides a satisfactory solution to this problem. However, this procedure breaks down for data sets larger than 1000 observations, due to the numerical instability of eigenvalue routines. The period since 1995 saw considerable activity dealing with approaches to address these computational issues. A number of different methods have been proposed, including the application of Choleski or LU decomposition for sparse matrices (Pace, 1997; Pace and Barry, 1997b,c), simulation approximations to the determinant (Barry and Pace, 1999), a characteristic polynomial approach (Smirnov and Anselin, 2001), and a Chebyshev approximation (Pace and LeSage, 2003a). Slight reformulations of the traditional likelihood in order to make the problem numerically more tractable have been suggested by Pace and Zou (2000) and Pace and LeSage (2003b). These new methods accomplish ML estimation of spatial autoregressive models for data sets with over a million observations in a few minutes, removing most impediments to their application in practice.

1.2.3 Software Tools

A third factor that helped promote the dissemination of spatial econometric methods to empirical practice was undeniably the availability of a growing number of software tools for spatial data analysis. In 1995 only *SpaceStat* (Anselin, 1992) was available as a freestanding program, followed in 1996 by the *S+SpatialStats* extension to the *S-PLUS* statistical package (Kaluzny *et al.*, 1997). While commercial econometric software packages still lack the built-in functionality to carry out spatial econometric analyses, a wide range of toolboxes now exists that overcome this limitation. Many of these implement exploratory spatial data analysis as well as the "core" functionality for linear spatial regression (for recent reviews, see Anselin, 2000; Anselin and Rey, 2002).

Perhaps the best known among the toolboxes are the spatial statistical toolbox of Pace and Barry (1998) and James LeSage's spatial econometrics toolbox.[7] Both

[7] http://www.spatial-econometrics.com/

of these are implemented as modules within the *Matlab* environment. They contain maximum likelihood estimation routines for spatial autoregressive models, as well as specialized sparse matrix procedures to handle large data sets. LeSage's toolbox also includes the Gibbs sampler as the foundation for Bayesian procedures to estimate spatial models, including spatial probit. A similar toolbox for *Stata*, containing regression diagnostics and maximum likelihood estimation is described by Pisati (2001). *Stata* functions that implement the Conley (1999) GMM estimator are available as well.[8] In addition, several more specialized functions have been developed by various individuals and posted on the internet. For example, an extension to the *Rats* time series package (available from the *Rats* support pages) implements the Driscoll and Kraay (1998) spatial correlation consistent covariance matrix estimator for panel data.[9]

As an increasingly attractive alternative to the use of toolboxes that operate as extensions to commercial software, there is a very active community involved in developing statistical sofware in the open source *R* environment.[10] This has led to an extensive collection of functions to analyze spatial data, including descriptive spatial autocorrelation statistics and the full range of spatial regression analyses in Roger Bivand's *spdep* package (see Bivand and Gebhardt, 2000; Bivand, 2002b, as well as the Bivand-Portnov chapter in Part I of this volume). Most recently, the various efforts related to spatial data analysis in *R* have been coordinated through the *R-Geo* initiative.[11]

Finally, it is worth mentioning the spatial software tools development program that is being carried out under the auspices of CSISS. This involves several ongoing activities, including a spatial software tools clearing house, as well as the development of a user-friendly freestanding software package for spatial data analysis, *GeoDa*. *GeoDa* implements mapping, geovisualization and exploratory spatial data analysis using dynamic linking and brushing, and contains functions for global and local spatial autocorrelation indices, as well as rudimentary spatial regression methods (Anselin, 2003a). A comprehensive collection of modules for spatial econometric analysis, referred to as *PySpace*, is being implemented in the open source Python language. This library currently contains all the standard estimation procedures and test statistics for linear spatial regression specifications, as well as methods to analyze spatial panel data models (Anselin and Le Gallo, 2003).[12]

1.3 Specification, Testing and Estimation

Part I of this volume contains five chapters dealing with the specification, testing and estimation of spatial econometric models. The first three chapters, by Florax

[8] http://www.faculty.econ.nwu.edu/faculty/conley/statacode.html

[9] http://www.estima.com/procs_panel.shtml

[10] http://www.r-project.org/

[11] http://sal.agecon.uiuc.edu/csiss/Rgeo/

[12] All the software tools developed as part of the CSISS initiative can be freely downloaded from http://sal.agecon.uiuc.edu/csiss/.

and de Graaff, Pinkse, and Kelejian and Robinson, extend and evaluate test statistics for spatial autocorrelation in regression models. Rey and Boarnet propose a framework of models and estimators to combine simultaneity across equations with spatial dependence, and Bivand and Portnov focus on the implementation of spatial econometric methods in open source sofware.

In "The performance of diagnostics for spatial dependence in regression models: a meta-analytical approach," Raymond Florax and Thomas de Graaff set out to assess and summarize the literature that uses experimental Monte Carlo simulation techniques to document the small sample properties of tests for spatial correlation in the residuals of a linear regression model. They present a taxonomy of the various tests, and review the experimental literature as it came about over the last twenty-five years. In doing so, they bring together numerous reported quantitative results. More precisely, they apply a technique known as *meta-analysis* to obtain general conclusions from the evidence presented in the literature.

The meta-analysis boils down to a regression of the experimentally derived rejection probabilities (of the null hypothesis of no spatial correlation) on various characteristics of the simulation design, such as the sample size, error distribution, spatial weights characteristics, strength of the induced correlation, and the presence of other misspecifications. They find that, unlike what is suggested by accepted wisdom, the Moran's I test is not uniformly more powerful than the Kelejian-Robinson test. They also find support for the "classical" forward specification search using the results from the Lagrange Multiplier tests. The analysis by Florax and de Graaff makes clear that there is a real need for continued work using experimental simulation to further investigate the properties of test statistics for spatial effects.

Joris Pinkse takes a closer look at the limiting distribution of a class of diagnostics for spatial dependence in "Moran-flavored tests with nuisance parameters: examples." He defines Moran-flavored tests as those that are either based on the well known Moran's I statistic, or that can be rewritten in the form of a Moran test. He builds on his earlier theoretical findings to introduce an approach based on a set of formal conditions to obtain a limiting normal distribution. More precisely, when these conditions are satisfied, Moran-flavored tests statistics reach a normal limiting distribution under the null hypothesis of no spatial dependence.

The conditions formulated by Pinkse pertain to the convergence rate of the parameter estimates and/or moment conditions on the variables in the model. Pinkse argues that checking these conditions provides an attractive alternative to having to prove the asymptotic validity for each test statistic from scratch. Moreover, this approach can be used for newly suggested tests in models where the asymptotic properties of the statistic have not yet been established in a rigorous manner. The utility of the approach is demonstrated in an empirical application involving six different spatial econometric specifications. In addition to tests against the standard linear regression spatial error and lag alternatives, he considers models estimated by nonlinear least squares and GMM, a probit and a spatial probit specification.

In the chapter on "The influence of spatially correlated heteroskedasticity on tests for spatial correlation," Harry Kelejian and Dennis Robinson expand on their recent work on tests against multiple sources of misspecification in the linear regression model. They examine the effects of heteroskedasticity on the properties of Moran's I and the Lagrange Multiplier tests against spatial correlation. A fundamental result is the formal demonstration of the role of spatial correlation in the heteroskedasticity itself. They show how not only the presence of this form of spatial correlation matters, but also the sign. Positive spatially correlated heteroskedasticity leads to a higher probability of rejecting the null, while the reverse holds when the heteroskedasticity is negatively correlated. In both instances the large sample properties of the classic tests no longer hold. However, Kelejian and Robinson also show that when the heteroskedasticity is not spatially correlated, there is no effect on the asymptotic properties of the tests for spatial correlation.

This important contribution provides a basis for extending current model specification strategies to consider spatial heteroskedasticity as well as spatial correlation. In addition, it emphasizes the relevance of acknowledging the effect of multiple sources for misspecification on the properties of the test statistics.

Sergio Rey and Marlon Boarnet move beyond the classical linear regression model in "A taxonomy of spatial econometric models for simultaneous equations systems." Their chapter is the first comprehensive discussion of the interrelation between simultaneity among multiple endogenous variables and spatial correlation, with specific attention to estimation issues. Rey and Boarnet start by reviewing some of the empirical literature in which systems of simultaneous equations are employed in models of regional employment and population change, typified by the Carlino-Mills tradition. They use this as a motivation to develop a taxonomy of models that embody both spatially as well as simultaneous endogenous variables.

They demonstrate how a formulation with both types of endogeneity yields a general specification as a "two sided reduced form." Interestingly, this form does not lend itself to the standard rank and order conditions for identification. The framework encompasses no less than 35 special cases, illustrated for a two equation system. Rey and Boarnet point to three important issues to consider in the estimation of such models: feedback simultaneity, spatial autoregressive lag simultaneity and spatial crossregressive lag simultaneity.

They next move to a close scrutiny of estimation issues and consider the properties of four estimators in a series of Monte Carlo simulation experiments. Specifically, ordinary least squares, spatial two stage least squares and two versions of the Kelejian-Robinson-Prucha instrumental variables estimators are compared in terms of bias and root mean squared error (RMSE). Their results demonstrate the importance of taking into account the *spatial* nature of the endogeneity by using spatially explicit instruments. Those estimators turn out to have lower bias and generally lower RMSE than estimators that do not include spatial instruments. This chapter provides a useful point of departure for future work to combine more realistic economic models, including complex endogenous effects, with specifications for spatial dependence.

In "Exploring spatial data analysis techniques using R: the case of observations with no neighbors," Roger Bivand and Boris Portnov demonstrate the flexibility and great potential of spatial data analysis implemented in the open source interactive software environment R. They focus in particular on conceptual and practical issues associated with the speciˇ cation of a spatial weights matrix, and how this affects the computation of spatial correlation statistics when "islands" occur.

Bivand and Portnov start by outlining the different ways in which spatial weights objects are implemented in the R package *spdep*. This includes weights where the neighbor relation is deˇ ned by common boundary, distance band, nearest neighbors, and Delaunay triangulation, as well as cases where they are derived from graph-theoretic concepts such as Gabriel graphs. This is illustrated with various code snippets. They next proceed to discuss the problem of how to deˇ ne a spatially lagged variable for observations that have no neighbors, and whether this should be accommodated by a missing value code or an explicit assignment of zero. They compare the two approaches in terms of their impact on a spatial autocorrelation statistic both for Cressie's well known North Carolina SIDS data set as well as in a study of clustering in the Israeli urban system.

Using data on 157 urban localities, Bivand and Portnov compare the connectedness characteristics of different spatial weights and provide illustrative R code to demonstrate the practical implementation of these concepts. They use the weights in an analysis of spatial autocorrelation in the percentage population change during the second half of the 1990s. The results illustrate how one can explore the spatial dependence in "realistic but challenging" distributions using the R programming environment.

Bivand and Portnov close with a strong argument in favor of an open source software development community for spatial data analysis. This allows users to access and modify the source code of interpreted and compiled functions. It also widens the range of potential contributors for further package development.

1.4 Discrete Choice, Nonparametric and Bayesian Approaches

Part II continues the discussion of model speciˇ caton and estimation, but the attention focuses speciˇ cally on models for discrete choice (with limited dependent variables) and on the application of nonparametric and Bayesian techniques. The chapters by Fleming and by Beron and Vijverberg deal with estimation in the spatial probit model, Pace *et al.*, and McMillen and McDonald introduce nonparametric methods. Finally, LeSage considers a Bayesian approach to estimating a family of geographically weighted regression models.

In "Techniques for estimating spatially dependent discrete choice models," Mark Fleming reviews several solutions that have been suggested in the literature to deal with the estimation of probit models that incorporate spatial correlation. The correlation is speciˇ ed in the form of the usual spatial lag and spatial autoregressive error processes. However, these models do not pertain to the observed dependent variable, which is only measured as 0 or 1, but rather to a latent or unobserved variable, that is assumed to follow a continuous distribution. He sets out by outlining two aspects of the complications caused by the presence of spatial correlation. First, it induces heteroskedasticity, which makes the standard probit estimator inconsistent. More importantly, maximum likelihood estimation that accounts for the spatial correlation structure requires the evaluation of an n-dimensional integral, which imposes a computational burden that cannot be handled in practice.

Fleming goes on to classify solutions to the estimation problem into three categories, which he reviews in turn. The first category tackles the heteroskedasiticy induced by the spatial autoregressive processes, but ignores the spatial correlation structure. A GMM estimator can be derived that incorporates the heteroskedastic variances. While it achieves consistency, it is not efficient relative to estimators that do take the correlation structure into account. This is the case for the second category, which Fleming refers to as "full spatial information estimators." This class consists of simulation estimators, where the parameters are obtained by estimating the spatial model for a simulated sample of "observations" on the latent variable or from draws from the simulated distribution of the error terms. This includes an estimation-maximization (EM) estimator and the recursive importance sampling (RIS) estimator, which are both formulated in a classical framework. A third example is the Bayesian Gibbs sampler.

Fleming also suggests a third category of estimators, based on weighted nonlinear least squares applied to the linear probability model. These estimators can be formulated as GMM estimators, but also turn out to be weighted nonlinear forms of familiar spatial two stage least squares and feasible generalized least squares estimators. He concludes his review with a very useful summary table. Here, he evaluates the different estimators in terms of the degree to which they address and/or solve various critical computational and methodological issues, such as the induced heteroskedasticity, the computation of a n-dimensional determinant, the evaluation of n-dimensional integrals, and the derivation of asymptotic standard errors.

Kurt Beron and Wim Vijverberg elaborate on the properties of the RIS estimator for the spatial probit model in "Probit in a spatial context: a Monte Carlo analysis." They start by outlining the implications of the specification of spatial lag and spatial error probit models for the interpretation of the parameters of the model, such as the marginal impact. In the presence of spatial correlation, the usual expression for the effect of a change in one of the explanatory variables on the probability of observing an outcome is no longer valid, and this "spatial multiplier" effect must be accounted for.

Beron and Vijverberg next spell out the principle behind the recursive importance sampling or RIS simulator. The application of this procedure to the spatially correlated case depends on the Cholesky decomposition of the inverse variance matrix. The resulting triangular structure lends itself well to a recursive approach, which simplifies the computation of the joint multivariate normal probability.

The properties of a Likelihood Ratio test derived by using the RIS simulator are evaluated in a Monte Carlo simulation exercise. The LR test is used on a number of artificial data sets with the spatial structure based on both the contiguity for the US states as well as randomly generated spatial weights. The power of the LR test turns out to be rather weak in the small data sets employed in the experiment, suggesting that much larger samples may be needed before the asymptotic properties apply. Also, it is difficult to distinguish between the error and lag alternatives, especially when the models are misspecified.

Beron and Vijverberg also briefly consider the properties of a spatial linear probability model, which ignores the dichotomous nature of the dependent variable. Overall, however, the spatial probit model was found to be superior to both this linear model as well as to the standard probit model. The simulation study considered here is a beginning, but clearly further work is needed to gain better insight into the finite sample properties of the spatial probit estimators.

In "Simultaneous spatial and functional form transformations," Kelley Pace, Ronald Barry, Carlos Slawson and C.F. Sirmans consider a complex transformation of variables in a spatial regression specification. The transformation takes into account both functional form and spatial dependence and is intended to deal with a number of issues that plague applied spatial data analysis, such as the influence of outliers, heteroskedasticity and non-normality.

Pace *et al.* employ B-splines to implement the functional and spatial transformation. These are piecewise polynomials with conditions enforced among the pieces, in terms of where each local polynomial begins and ends (knots), and the amount of smoothness among the pieces (degree). Relative to the familiar Box-Cox transformation, the B-splines can assume more complicated shapes and can handle more severe transformations of extreme values. The resulting log-likelihood contains three important components, the spatial Jacobian (for the spatial transformation), the functional form Jacobian (for the functional transformation) and the log of the sum of squared errors. Pace *et al.* employ sparse matrix techniques in the computational implementation of the estimation technique.

The new approach is applied to a study of housing values in Baton Rouge, Louisiana, using a data set with 11,000 observations. Spatial dependence is incorporated by means of spatial weights based on four nearest neighbors. The full model contains 113 parameters. Pace *et al.* compare the model to simpler forms using a likelihood ratio test for inference. Relative to a traditional approach, they conclude that the joint transformation leads to an improvement in overall model efficacy. Specifically, the degree of spatial autocorrelation in the residuals is greatly reduced and the interquartile range for the residuals is also lowered dramatically.

Daniel McMillen and John McDonald also take a nonparametric approach in "Locally weighted maximum likelihood estimation: Monte Carlo evidence and an application." McMillen and McDonald introduce a nonparametric estimator to account for spatial heterogeneity in the form of local parameter variation in a probit model. This variant of a geographically weighted regression consists of computing local probit estimates that only use a subset of the data. They include the computational steps in an appendix, which facilitates the implementation of this method in econometric software packages that allow do-loops and have built-in maximization routines. Evidence from Monte Carlo simulation experiments suggests that the locally weighted probit provides accurate estimates, even when the base model is misspecified. McMillen and McDonald therefore conclude that there is little cost and potentially much to benefit from using this approach as an alternative to the standard probit estimator.

They apply the technique to a study of the first Chicago zoning ordinance, employing an original data set on city blocks in 1923. Specifically, they compute both standard as well as local probit estimates for the probability that a city block was zoned for high, medium, or low building heights. The locally weighted ordinal probit results turn out to be very similar to the standard ordinal probit results, and the prediction of the nonparametric estimator is slightly more accurate. The results provided by McMillen and McDonald provide promise for the application of locally weighted discrete choice estimators to visualize potential problems with standard discrete choice methods. Further work is needed, however, to obtain a better understanding of the statistical properties of the estimator and to establish a formal basis (in the form of useful regularity conditions) for the derivation of these results.

In the final chapter of Part II, James LeSage suggests an altemative approach to estimation in local spatial regression analysis in "A family of geographically weighted regression models." He starts out by outlining some methodological concerns associated with a local linear spatial regression approach, such as as geographically weighted regression (GWR). The essence of GWR consists of a series of local estimations where only a subset of the data is used. This subset is determined by a "kernel," a general spatial distance decay function which crucially depends on a range or bandwidth parameter. LeSage lists three important problems pertaining to this approach.

First, since the GWR estimates are conditional upon the selection of a bandwidth parameter, but the distance-decay weights are not adjusted for outliers or aberrant observations, the local linear estimates may be unduly influenced by these outliers. This is important in the interpretation of local variation, since the outliers may spuriously suggest the presence of spatial heterogeneity where in fact there is none. Second, the locally linear estimates derived from a distance weighted subsample of observations may display "weak data" problems, in the sense that insufficient degrees of freedom are available to obtain reliable estimates. Third, inference in GWR based on traditional concepts derived from least squares fit are inappropriate, due to the reuse of the sample for multiple estimations and the resulting spatial correlation between results.

As an alternative to the traditional GWR approach, LeSage suggests a Bayesian approach, referred to as BGWR. The BGWR uses robust estimates that are insensitive to aberrant observations by detecting such observations and downweighting their influence on the estimates. Also, subjective prior information may be introduced to address the weak data problem. Finally, the Bayesian formulation encompasses a range of parameter smoothing relationships. Well known models to deal with spatial heterogeneity, such as the spatial expansion method and GWR are shown to be special cases of LeSage's general parameter smoothing model. This smoothing relationship stochastically restricts the estimates based on spatial (local) relationships.

LeSage goes on to outline the formal structure of the model and its estimation by means of Markov Chain Monte Carlo (MCMC) methods. He compares the results of BGWR to GWR in three sample data sets. First, he uses a generated set of 100 observations to illustrate the main features of the model. He next uses the familiar crime data for 49 Columbus (OH) neighborhoods, as well as a more extensive data set consisting of employment, payroll earnings and establishments for all 50 zip codes in Cuyahoga county in Ohio for 1989. These examples underscore the advantages of an approach that subsumes the GWR as a special case of the Bayesian model.

1.5 Spatial Externalities

In Parts III to V, attention shifts from mostly methodological concerns to a primary attention to empirical applications. Part III contains chapters where the main interest is an explicit incorporation of notions of spatial externalities. Both Beron *et al.*, and Baltagi and Li formulate demand models with spatial spillovers leading to spatially correlated error terms. Moreno *et al.* consider the role of spatial externalities in models of sectoral productivity.

Kurt Beron, Yaw Hanson, James Murdoch and Mark Thayer explore some econometric issues associated with the estimation of spatial hedonic models in "Hedonic price functions and spatial dependence: implications for the demand for urban air quality." An indirect measure of the willingness to pay for air quality may be derived from the parameters of hedonic models, in which the price (or value) of a house is regressed on its characteristics, including neighborhood characteristics and measures of air quality. A major concern in this respect is the proper speciˇ cation of spatial externalities, or neighborhood effects, in the form of a model that incorporates spatially correlated errors or a spatial lag term. The chapter by Beron *et al.* explores these issues in an analysis of an extensive data set on housing transactions in the Los Angeles (CA) basin, spanning six time periods. The ˇ nal set of 60,000 observations is obtained by sampling from a much larger original data set.

Beron *et al.* start by reviewing the salient theoretical and methodological features associated with the estimation of willingness to pay from hedonic models. They next brie y consider econometric issues, such as the implications for the willingness to pay estimate of including a spatial lag or error term in the hedonic model. They implement three sets of nested speciˇ cations, one including all the usual site-speciˇ c characteristics, including air quality measures as well as all neighborhood variables (county dummies, and variables pertaining to the city, school district or census tract containing the individual properties). The other two are "restricted" speciˇ cations, one without county dummy variables, and one without the dummies and all other regional variables.

Each of the three models is estimated by means of ordinary least squares. Spatial heterogeneity is accounted for by including a spatial trend, as a second order trend surface. Diagnostics for spatial effects suggest a spatial error speciˇ cation, which is estimated by means of maximum likelihood. A main ˇ nding of this empirical study is that the estimates of the site-specific characteristics remain relatively invariant between the non-spatial and the spatial model. The estimates of the spatial model are used to estimate demand functions for air quality, providing some evidence that the restricted models are not statistically justified. Moreover, the incorporation of the spatial trend term turns out to be an effective way to deal with spatial heterogeneity. The sensitivity of the benefit estimates to the specification of the spatial models is a cause for concern, and Beron *et al.* close with a call for more in-depth investigation of the associated trade-offs.

In "Prediction in the panel data model with spatial correlation," Badi Baltagi and Dong Li consider the prediction of demand for cigarettes based on a panel of observations for 46 U.S. states over the period 1963–1992. Cross-state spatial heterogeneity as well as spatial externalities in the form of spatially correlated error terms are incorporated in a number of different specifications. These include both fixed effects as well as random effects models.

Baltagi and Li briefly review the estimation issues associated with the different ways of embedding space- and time-wise heterogeneity in combination with spatial correlation. They consider eight different estimates: pooled OLS, pooled spatial error model (ML), the average of year-specific OLS estimates, the average of year-specific ML-Error estimates, a fixed effects model, a fixed effects model with spatial error autocorrelation, a random effects model, and a random effects model with spatial error autocorrelation. The empirical results vary considerably, leading to the assessment of the consequences for predicted values.

A best linear unbiased predictor (BLUP) is obtained by taking into account the covariance structure between current errors and future errors. In the spatial panels, this structure takes on a more complex form, which Baltagi and Li outline for both the fixed effects as well as the random effects specification. The predictions are carried out for one to five year ahead forecasts, and compared in terms of root mean squared error (RMSE) to actual values observed for the years left out of the estimation exercise. The best forecast performance for all five years is obtained by the fixed effects estimator with spatial autocorrelation, closely followed by the spatial random effects model. This illustrates the value of incorporating both spatial heterogeneity as well as spatial correlation in panel data models.

In "External effects and cost of production," Rosina Moreno, Enrique López-Bazo, Esther Vayá and Manuel Artís provide an innovative spatial econometric perspective on the treatment of regional and industrial externalities. This differs from the standard approach in the literature, not only by the explicit consideration of spatial autocorrelation, but Moreno *et al.* also introduce two other innovations. First, they proxy cross-industry spillovers by a measure accounting for both forward and backward linkages across sectors. Second, they use a cost function to model the externalities, rather than the customary production function. In the cost function, particular attention is paid to the cost saving effects of public capital, by including both a region's own stock of public capital as well as that available in the other regions of the spatial system.

Moreno *et al.* start out with a review of the theoretical and empirical literature pertaining to the treatment of industrial and spatial externalities and the inclusion of external effects in cost functions. They consider the incorporation of sectoral and spatial externalities in an econometric specification through a careful selection of spatial weights. In particular, the use of input-output linkages as the basis for the weights matrix that reflects sectoral externalities is innovative. In addition to the usual factors, their cost function also contains both "external input" (the stock of publicly provided capital) as well as "cross-economy spillovers" (the output of neighboring economies).

In the empirical application, Moreno *et al.* estimate a spatial lag model with additional cross-regressive terms in a flexible translog specification. The model is nonlinear in the parameters, and the authors demonstrate the necessary changes that need to be made to apply Lagrange Multiplier tests against spatial effects in a model estimated by nonlinear least squares. The study uses data for 12 manufacturing industries in 15 Spanish regions (at the NUTS II level) during the period 1980–1991. The results suggest that sectoral spillovers yield significant cost reductions. The effect of spatial externalities, however, is found to be opposite in sign (suggesting higher cost). As is the case in much of the literature, the role of public capital remains ambiguous. The chapter clearly demonstrates that the omission of explicitly modeled spatial externalities in the traditional studies of returns to scale may have led to biased parameter estimates.

1.6 Urban Growth and Agglomeration Economies

Part IV contains three papers dealing with the specification of spatial effects in models for urban growth and development, where agglomeration economies are a central focus of interest. Bao *et al.*, and Irwin and Bockstael study growth at the urban fringe, whereas Ioannides deals with the evolution of the urban system as a whole.

Shuming Bao, Mark Henry and David Barkley study the role of spatial interaction relative to local amenities in the rural development process in "Identifying urban-rural linkages, tests for spatial effects in the Carlino-Mills model." They consider the familiar two-equation simultaneous system for population and employment change, popularized in the research of Carlino-Mills-Boarnet. However, in contrast to earlier work, they focus on the explicit incorporation of spatially lagged variables in this specification. This is applied to a study of rural development in South Carolina, parts of Georgia and parts of North Carolina, using the concept of functional economic areas (FEA). Eight such FEA are identified, using a creative application of GIS techniques. In these areas, the development process is modeled for rural tracts. Spread or backwash effects of the existing urban area are incorporated by means of a spatial interaction term. This distinguishes between the effect of the urban core and the suburban fringe. In all, 268 observations are used at the tract level, for a spatially consistent geography for both 1980 and 1990 U.S. census data.

Central to the specification of the spatial lag models for employment and population change is the choice of a spatial weights matrix. In addition to the traditional contiguity and distance based weights, Bao *et al.* also consider spatial weights derived from detailed commuter flow information, allowing for directional effects.

The results of this spatial econometric analysis suggest a mix of spillover and backwash effects from urban core and fringe areas onto their rural hinterlands. Importantly, the coefficients of the spatial lag term were highly significant in all models, illustrating the value of an explicit spatial econometric approach. This also suggests that other studies of the rural development process that ignored these spatial effects may need to be reinterpreted.

In "Endogenous spatial externalities: empirical evidence and implications for the evolution of exurban residential land use patterns," Elena Irwin and Nancy Bockstael investigate the validity of the "interacting agents" hypothesis from the recent literature on social and spatial interaction. They consider this in the context of changes in residential land use patterns at the urban fringe. The point of departure is that spatial externalities will create interdependence among neighboring agents, such that land use conversion decisions become partially driven by a process of endogenous change.

Irwin and Bockstael outline a micro-economic model of land use conversion in which exogenous features of the landscape are incorporated as well as endogenous interactions. Interest focuses on the interaction parameter and the extent to which it is negative, suggesting repelling effects, compatible with scattered development and landscape fragmentation. The theoretical model is viewed as the solution to a problem of optimal timing of development, and yields an intertemporal formulation of the agent's conversion decision. The model is estimated in the form of a proportional hazards specification. A detailed data set of land use conversions in the exurban area of Washington, D.C. is used in the empirical exercise. This data set contains all parcels that were convertible in a six year period, starting in 1991, and was constructed from the geocoded tax assessment rolls obtained from the Maryland Office of Planning.

Three nested specifications are considered, including an expanding set of explanatory variables. Considerable attention is paid to identification issues. The estimation results reveal that in all three specifications, the effect of an outer neighborhood measure is negative and significant, but there was no effect of inner neighborhood. The estimated parameters were then used in a number of simulation exercises, to gauge the robustness of the models in predicting future patterns of land use. The results suggest that scattered residential land use patterns are more likely to emerge when there is a sufficiently strong centrifugal force from the central city. This itself is a reflection of the spatial externalities induced through interacting agents.

In "Economic geography and the spatial evolution of wages in the United States," Yannis Ioannides takes an innovative approach to modeling the urban growth process. In a novel theoretical framework, he brings together two different strands of literature dealing with the spatial evolution of wages. One emphasizes specialization effects, conceptualizing a system of cities with varying agglomeration economies across sectors. This is a key factor in explaining *intra*-metropolitan specialization. The other, formulated in writings on the new economic geography, stresses the role of “historical accidents” and geographical features. The resulting dynamics of city size play an important role in explaining the *inter*-metropolitan distribution of cities across space and time.

Ioannides formulates a theoretical model, ˇ tting in the new economic geography tradition, that includes city-speciˇ c human capital and Romer-type pecuniary externalities. These cause agglomeration effects to determine marginal labor productivity. The key empirical implication of this model is that the dynamic evolution of wages will mimic spatial characteristics, such as geographical distance and proximity. He estimates the model using a unique data set, combining U.S. census data for metropolitan area populations from 1900 to 1990, with data sources for earnings and schooling.

Ioannides empirically compares the explanatory power of different measures of spatial proximity to test several theories of U.S. urban spatial evolution. He employs an econometric speciˇ cation that resembles a spatial lag model, although it is different from the usual formulation in that it involves a switching regression framework and a varying spatial proximity matrix. The basic model is estimated using both a panel data setup as well as a repeated cross-section perspective. The empirical ˇ ndings are generally supportive of recent theories of urban agglomeration in the Krugman-style new economic geography. This chapter constitutes a ˇ rst attempt to stage formal new economic geography models in a spatial econometric setting.

1.7 Trade and Economic Growth

The ˇ nal part of the volume contains three chapters, dealing with spatial models of international trade (Eliste and Fredriksson), and economic growth and convergence (Fingleton, and Vayá *et al.*).

In “Does trade liberalization cause a race-to-the-bottom in environmental policies? A spatial econometric perspective,” Paavo Eliste and Per Fredriksson use data on agricultural trade ows and environmental standards to assess whether countries strategically interact in setting their environmental regulations. This strategic interaction can take different forms, such as a “race to the bottom,” where countries undercut the regulatory stringency of their neighbors' rules, or refrain from implementing strict regulations (“regulatory chill”). Other phenomena compatible with strategic interaction are “ecological dumping” (lax environmental standards) and “pollution havens” (providing a competitive advantage to polluting industries). Although such phenomena are inherently spatial, they have so far escaped analysis from an explicit spatial analytical perspective.

Eliste and Fredriksson use a combination of exploratory spatial data analysis (ESDA) and spatial econonometrics to study the spatial pattern of agricultural environmental regulations. In this, they consider different formulations for spatial weights, based both on the usual geographic criteria (contiguity, great circle distance, and *k* nearest neighbors), as well as derived from aggregate trade ows between countries. An index of the stringency of environmental regulations was constructed for 62 countries from information compiled for the 1992 United Nations Conference on Environment and Development in Rio.

Eliste and Fredriksson are concerned with the extent to which the legislation implemented by trade partners affects the stringency of the country's own regulations, and the direction of this (potential) influence. They also consider the role of a country's openness of trade as a potential intervening factor. Their results, based on the estimation of a spatial lag model, do not provide support for the notion of a race to the bottom. Instead, they find that the strategic interaction between countries is of a complementary nature, suggesting a "race to the top." In addition, the results indicate the importance of political variables, such as freedom of information and political freedom, suggesting an interaction and threshold effect. This further confirms the importance of taking into account spatial effects in econometric models of strategic interaction. Ignoring the spatial lag term, as in the case in most studies to date, may lead to spurious inference.

Bernard Fingleton revisits a well studied topic in "Regional economic growth and convergence: insights from a spatial econometric perspective." After an extensive review of the literature on economic growth theory (covering the role of returns to scale, externalities, catch up mechanisms and exogenous shocks), he focuses on the familiar Verdoorn law as a model for regional productivity growth.

Fingleton goes beyond the traditional specification, and outlines ways to explicitly include spatial processes into this mechanism. This leads to specifications that incorporate both increasing returns to scale, as well as innovation diffusion, catch up and spatial externalities. They are approached as single equation equations, but also as one element in a simultaneous system. Specifically, Fingleton introduces an augmented spatial lag Verdoorn law, an augmented spatial error Verdoorn law, and a reduced unrestricted spatial effects Verdoorn law. These models incorporate the role of spatial effects through spatially lagged terms for the dependent variable, the error term, or the explanatory variables.

Fingleton goes on to discuss in some detail the implications of these specifications for equilibrium and steady state, which follow from different ways to model the connection between productivity growth and the level of productivity. He also carries out an empirical investigation, estimating the augmented spatial lag Verdoorn law (as well as other specifications) for a data set on manufacturing productivity and output for 178 NUTS regions of the European Union (EU), over a period of twenty years (1975–1995). The results provide strong support for increasing returns, and significant coefficients for catch up, peripherality and urbanization effects. More importantly, the spatial autoregressive (lag) coefficient is highly significant, indicating the existence of cross-region spatial externalities.

Fingleton employs the estimated coefficients in a simulation exercise to track the path towards deterministic and stochastic equilbrium in a regional system. The use of an explicit spatial econometric model underlying this simulation allows for the movement of one region to simultaneously influence and be influenced by that of other regions. This constitutes a significant advance in the modeling of regional growth dynamics.

Esther Vayá, Enrique López-Bazo, Rosina Moreno and Jordi Suriñach consider the role of spatial external effects in the accumulation of factors of production in "Growth and externalities across economies: an empirical analysis using spatial econometrics." They develop a theoretical growth model that allows for externalities due to the accumulation of capital within the regional economy. Furthermore, spatial externalities are introduced and related to the aggregate level of technology of neighboring regions, which in turn are linked to their capital stock. Consequently, innovations and new ideas that follow from investment in new capital can flow across economies.

The theoretical model is operationalized in the form of two regression specifications of the mixed regressive-spatial autoregressive type, one for a production function, the other for a growth equation. These are illustrated with two different data sets. The production function is estimated for data on 17 Spanish regions during 15 time slices drawn from the period 1964-1993. The growth equation is estimated for 108 regions in the European Union during the period 1975-1992. Vayá *et al.* consider spatial weights specifications derived from geographical factors, such as contiguity and distance, as well as from economic indicators, such as trade flows. They outline a specialized Maximum Likelihood estimation procedure that imposes constraints, such that parameters remain in the acceptable range (e.g., avoiding negative spatial spillovers or external effects greater than within-economy returns).

The results of the empirical exercise yield highly significant and positive spatial externality effects. This implies that the usual estimates for the rate of convergence, which ignore these spatial effects, are likely to be biased. The findings also illustrate how the prevalence of interregional externalities can create a "poverty trap," based on geographic location. The efforts required to surmount such a trap position may be substantially less if neighbors simultaneously invest resources. Isolated regional efforts are likely to be suboptimal, illustrating the importance of taking into account spatial multiplier effects.

1.8 Future Directions

At the end of the introductory chapter of the *New Directions* volume, we spelled out an agenda for future work along three broad directions: new specifications for spatial weights; spatial effects in nonlinear and limited dependent variable models; and the treatment of spatial heterogeneity and structural change, primarily through the development of a Bayesian perspective (Anselin and Florax, 1995a, p. 15). The recent explosion in the literature, illustrated earlier in this chapter, as well as the chapters included in the current volume constitute a significant advance along these three dimensions, such that at this point, perhaps a new set of directions needs to be formulated.

We can fairly state that today there is an established body of work (a toolbox) to deal with a wide range of spatial effects in linear regression models and their panel data extensions. However, much remains to be done to incorporate spatial effects in more realistic data settings, such as models of counts, rates, and variously truncated and censored variables with spatial dependence. In addition, data-related concerns that receive a lot of attention in spatial statistics, such as spatial sampling issues, missing values and misaligned spatial units have yet to appear in spatial econometric practice. Similarly, while we include some examples of "economic" spatial weights in the current volume, the integration of spatial and social network analysis and their application in econometric model specification is only in its infancy. Finally, much more is needed in terms of comparative studies of competing paradigms and modeling "philosophies." For example, little is known about the relative advantages of Bayesian and non-Bayesian simulation estimators, the use of varying coefficients vs multilevel models to address heterogeneity, or the relative merits of GMM and Maximum Likelihood estimators.

We hope that the current volume will provide a useful background, stimulus and point of departure for future advances in spatial econometrics.

Part I

Speciˇ cation, Testing and Estimation

2 The Performance of Diagnostic Tests for Spatial Dependence in Linear Regression Models: A Meta-Analysis of Simulation Studies

Raymond J.G.M. Florax and Thomas de Graaff

Free University Amsterdam

2.1 Introduction

One of the reasons for A.D. Cliff and J.K. Ord's 1973 book "Spatial Autocorrelation" achieving the status of a seminal work on spatial statistics and econometrics lies in their careful and lucid treatment of the autocorrelation problem in spatial data series. Cliff and Ord present test statistics for univariate spatial series of categorical (nominal and ordinal) and continuous (interval or ratio scale) data. They extend the use of autocorrelation statistics, specifically Moran's I (Moran, 1948), to the analysis of regression residuals (see also Cliff and Ord, 1972). The detection of spatial autocorrelation among regression residuals implies either a nonlinear relationship between the dependent and independent variables, the omission of one or more spatially correlated regressors, or the appropriateness of an autoregressive error structure. Ignoring the presence of spatial autocorrelation among the population errors causes ordinary least squares (OLS) to be a biased variance estimator and an inefficient regression coefficient estimator. Anselin (1988b) shows that erroneously omitting the spatially lagged dependent variable from the set of explanatory variables causes the OLS estimator to be biased and inconsistent. Cliff and Ord (1981, p. 197) therefore urge the applied researcher to always apply "some check for autocorrelation," and take remedial action when necessary.

Over a decade later, Anselin and Griffith (1988) raise the question "[d]o spatial effects really matter in regression analysis?" They conclude that traditional diagnostics and test statistics should not be taken at face value when spatial effects are present, not even as a first approximation. Their conclusion is substantiated by simulation experiments considering the effect of interactions between heteroskedasticity and spatial dependence.

The term "spatial effects" refers to both spatial dependence and spatial heterogeneity Anselin (1988b). Spatial heterogeneity can be satisfactorily dealt with utilizing concurrent standard techniques from mainstream econometrics. Spatially induced heteroskedasticity can be handled using a generalized least squares (GLS) estimator, or White-adjusted variances. Substantive spatial heterogeneity can be incorporated through specifications allowing for spatial regimes. For spatial dependence, however, there are neither standard econometric tests nor standard estimators that adequately account for the specific nature of spatial dependence (Anselin and Bera, 1998; Anselin, 2001b). Consequently, the development of adequate tests for spatial autocorrelation in linear regression models becomes a key focus of the spatial econometric literature.[1]

Spatial dependence or autocorrelation tests are invariably concerned with the null hypothesis of no spatial correlation, but they typically differ in the specification of the alternative hypothesis. We refer to Moran's I as a "diffuse test," because the alternative hypothesis merely implies spatial autocorrelation among a residual data series. The underlying cause for the autocorrelation (nonlinearity, spatially correlated population errors, or an erroneously omitted spatially lagged dependent variable) is unclear. Burridge (1980) shows that a Lagrange Multiplier (LM) test with a spatial autoregressive error model as the alternative is equivalent to a scaled squared Moran coefficient. This marks the turning point to developing spatial misspecification tests with a clear alternative hypothesis in a Maximum Likelihood framework. Nowadays, practitioners are supplied with an extensive toolbox of diagnostic tests, containing unidirectional, multidirectional as well as robust tests for spatial dependence (Anselin *et al.*, 1996). In practice, most tests are formulated and applied as LM tests, rather than Likelihood Ratio or Wald tests which, although they are asymptotically equivalent, are much more cumbersome to estimate because they require the estimation of the alternative model. Recent additions to the misspecification toolbox include tests for simultaneous equation models (Anselin and Kelejian, 1997), the combination of heteroskedasticity and spatial autocorrelation (Kelejian and Robinson, 1998), and spatial error component models (Anselin and Moreno, 2003; see also Kelejian and Yuzefovic, 2001).[2]

Given the analytical intractability of the small sample distribution of the test statistics, extensive simulation experiments are performed to assess the size and the power of tests for spatial dependence in finite samples. Cliff and Ord (1971) perform Monte Carlo simulation experiments with Moran's I for univariate raw data series (see also Haining, 1977). We do not consider spatial series of raw data, but focus on regression models instead. Bartels and Hordijk (1977) are the first to study the small

[1] A formal definition of spatial autocorrelation is:

$$\mathrm{Cov}\left(y_i, y_j\right) = \mathrm{E}\left(y_i y_j\right) - \mathrm{E}\left(y_i\right) \cdot \mathrm{E}\left(y_j\right) \neq 0, \text{ for } i \neq j,$$

pointing to the coincidence of attribute similarity expressed in y and location similarity for locations i and j. The terms "spatial dependence" and "spatial autocorrelation" are used interchangeably from here on, although strictly speaking spatial dependence requires the complete specification of the joint density (and, as such, is unverifiable except under extremely simplifying conditions, such as normality), while spatial autocorrelation is simply a moment of that joint distribution (Anselin, 2001b). It should also be noted that spatial correlation in a spatial process model induces spatial heteroskedasticity (see Brett and Pinkse (1997); and Kelejian and Robinson in Chapter 4 of this volume).

[2] In this chapter, we discard the growing literature on misspecification testing in spatial discrete choice models (see, for instance, McMillen (1995b); Pinkse and Slade (1998); Kelejian and Prucha (2001); Fleming in Chapter 7 of this volume; and Beron and Vijverberg in Chapter 8 of this volume). Recent state-of-the-art reviews of the spatial econometric literature are provided in, for instance, Chapter 1 of this volume, and in Anselin (2002) and Florax and van der Vlist (2003).

sample behavior of Moran's *I* for regression residuals in a Monte Carlo setting, and by now some 30 simulation studies exist. Anselin and Rey (1991) present a qualitative survey of the early simulation studies of spatial effects in linear regression models.

As a complement to a literature survey, a quantitative analysis of simulation results of different studies provides additional insights. A quantitative multivariate approach across studies has three distinct advantages. First, in a multivariate regression framework it is feasible to control for conditioning factors while assessing marginal effects of pivotal features related to the performance of the test statistics (such as, the weights matrix, the distribution of the error term, or the data generating process; see Florax *et al.*, 2002a). Second, a multivariate approach combining the results of different studies provides information about the effects on the small sample behavior of tests of changing salient aspects of the research design. The research design is oftentimes fixed within studies, but it varies between studies (Hedges, 1997). Finally, simulation results depend on the experimental design used in a Monte Carlo study. Results can therefore in a strict sense not be generalized to a broader population. A multivariate quantitative analysis can reduce this, what Hendry (1984) calls, "specificity" of results of simulation experiments.

A quantitative analysis of research results of previous studies is called "meta-analysis." Meta-analysis is akin to the response surface technique developed in mainstream econometrics (see Hendry, 1984, for a discussion). Although Anselin (1980) does not use the terminology, he does employ the technique to summarize his experimental findings regarding spatial estimators. Kelejian and Robinson (1998), and Florax *et al.* (1998) also use response surface analysis to summarize the abundant output of their simulation experiments (see also Anselin and Moreno, 2003; Kelejian and Yuzefovich, 2001). In this chapter, we perform a meta-analysis on the experimental simulation studies that have been conducted in spatial econometrics over the last twenty years. Several restrictions with respect to sampling studies and outcomes are necessary in order to safeguard that the indicator studied in the meta-analysis is sufficiently homogeneous. Sample selection issues as well as a more detailed comparison of the techniques of response surface analysis and meta-analysis is discussed in more detail below.

The remainder of this chapter appears as follows. Section 2.2 presents the essentials of the meta-analysis and response surface analysis techniques, and discusses their appropriateness for the comparative analysis we undertake. In Sect. 2.3, we briefly review the spatial models and test statistics for spatial dependence that have been studied in Monte Carlo experiments. Section 2.5 presents a narrative overview of the available experimental simulation studies, and addresses the issue of sample selection for the meta-analysis. Section 2.6 explains the specification of the meta-regression, and presents the results of the meta-analysis. Finally, Sect. 2.7 contains conclusions, and delivers useful practical guidelines for the selection and interpretation of test statistics for spatial dependence in specific research contexts.

2.2 Meta-Analysis and Response Surfaces

In our analysis, we use the conventional statistical technique of multivariate regression analysis to synthesize the results of previous studies dealing with Monte Carlo simulation of spatial dependence testing in spatial econometrics. This type of analysis of statistical summary indicators (i.e., "effect sizes," such as standardized regression coefficients, odds ratios, and rejection frequencies) is labeled "meta-analysis" (Hedges and Olkin, 1985). The specific variant centering on a multivariate regression analysis of a series of effect sizes is called "meta-regression" (Sutton *et al.*, 2001).

A related technique, common in mainstream econometrics, is concerned with the estimation of a response surface. Response surfaces can be used to summarize the abundant output of Monte Carlo experiments (Davidson and MacKinnon, 1993). The technique has been employed by, among others, Hendry (1979) and MacKinnon (1991). The response surface technique boils down to the estimation of an auxiliary regression, in which some estimated output quantity of the experiments is treated as the dependent variable, and the experiments' parameters set by the experimenter as the independent variables. The technique is applied to a series of experiments of a specific study and, given the experimental context, the analyst has perfect knowledge about the exogenous variables to be included in the response surface specification.

Davidson and MacKinnon (1993) observe that the response surface technique has much to recommend it. The technique facilitates presenting a succinct and concise account of, for instance, the small sample behavior of an estimator – as opposed to the usual abundance of tabulations and graphs. It also alleviates the problem of "specificity" (Hendry, 1984). The outcome of one experiment merely reflects the characteristics of one specific underlying data generating process (DGP). The combination of various experiments in a response surface warrants the generalization of results to a larger population of DGPs.

Meta-analysis is very similar to response surface analysis. The main difference is that empirical results are compared across different studies using (largely) non-overlapping datasets.[3] The technique emerged in the context of replicated experiments in agronomy, and gradually diffused to experimental sciences, such as medicine and psychology (Rosenthal, 1991). It took much longer for meta-analysis to proliferate to economics. The largely non-experimental character of economics may be a reason, but also the lack of a replication tradition. Instead of replication, the "competition of ideas" (Smith and Pattanayak, 2002) triggers creativity in economists. This results in each paper taking a slightly different perspective, with concurrent differences in operationalization of variables, specifications, and data (Heckman, 2001). Comparing and combining results across studies is then correspondingly more difficult. Nevertheless, during the 1990s, meta-analysis gained

[3] A sort of in-between position is possible as well. Florax *et al.* (2002b) analyze cross-country growth regressions, generating empirical results from one database in a quasi-experimental fashion.

ground in economics, at first in environmental economics, but very rapidly also in labor economics, industrial organization, transport economics, and macroeconomics (see Florax *et al.*, 2002a, for references).

Proponents of the technique maintain that meta-analysis provides a more formal and objective framework for reviewing the literature. It avoids the rather fuzzy sample selection procedures of narrative reviews, and it improves on the practice of simply tallying negative, zero, and positive results of statistical significance testing (Stanley, 2001). This so-called vote-counting procedure is considered statistically flawed and obsolete (see Hedges and Olkin, 1985, for details). In addition, we argue that one of the distinctive advantages of meta-analysis, in particular of multivariate meta-regression, is the possibility to investigate the variability of an "effect size" while controlling for intervening factors.

The comparison across studies evokes specific caveats in meta-analysis as compared to response surface analysis. First, the selection of studies included in the meta-analysis is biased if there is a systematic variation between the sampling decision and the magnitude of the effect and/or its associated variance. When a systematic relationship exists between the statistical significance of an effect and the decision to publish a study, the inferences from a meta-analysis are invalidated by publication bias. We do not pursue the assessment of publication bias in the current analysis, because the number of studies is limited, the sources are well known, and we include both published and unpublished results in the meta-analysis.[4]

Second, even when between studies a uniformly defined and standardized effect size is available, it is imperative to account for heterogeneity between studies. The simplest case, not accounting for heterogeneity, is to combine the effect sizes across studies in an average with associated standard error. This is of course equivalent to an OLS regression with a constant term only. The sampled effects are a priori assumed to come from one population distribution. One step ahead is to hypothesize that the effect sizes are drawn from population distributions that differ between studies. The differences in population distributions can be modeled by means of fixed or random effects, depending on the applicability of the specific assumptions of the different models, and/or the results of statistical inference Hedges and Olkin (1985). The heterogeneity of effect sizes is by definition not restricted to differences in population means. A meta-regression is inherently heteroskedastic, because the estimated standard errors of the effect sizes are different.

Finally, in most meta-analyses in economics multiple measurements from the same study are sampled. This leads to a panel data setup, implying that heterogeneity across studies as well as dependence among measurements of the same study become an issue. Effect sizes sampled from the same study are typically generated

[4] There is an extensive methodological and empirical literature about publication bias. See, for instance, Sutton *et al.* (2001), and Florax (2002) for a discussion of methods, and Card and Krueger (1995), and Ashenfelter *et al.* (1999) for empirical examples. Publication bias is likely to be less of an issue in spatial econometric Monte Carlo studies: there is comparably little orthodoxy for a set of results to challenge and therefore less of an incentive for a journal editor to reject a paper because it does not line up with the status quo.

using the same data and identical or similar speciˇ cations, causing the estimated effect sizes of the same study to be correlated.

We address the issues of heterogeneity and dependence in the meta-regression speciˇ cation in Sect. 2.6, after giving a qualitative review of the setup and the main outcomes of the simulation papers in spatial econometrics published during the last two decades in Sect. 2.5. First, however, we present a concise overview of various spatial dependence tests and the respective data generating processes in the next section.

2.3 Spatial Dependence Tests and Data Generating Processes

In terms of data generating processes, three different types of processes are commonly used in the literature. The ˇ rst and second are familiar. One is the spatial autoregressive or moving average error model, and the other is a model containing a spatially lagged dependent variable. Eventually, both models can be combined in the spatial autoregressive moving average model. The third type of process is less well known, and is introduced as the spatial error component model in Kelejian and Robinson (1995). We discuss the respective data generating processes and their associated tests in Sect. 2.3.1 and 2.3.2. In Sect. 2.4, we provide a taxonomy of misspeciˇ cation tests against spatial dependence.

2.3.1 The Spatial Error, the Spatial Lag, and the ARMA Model

We start from the following linear model that adequately represents a data generating process in a spatial context:

$$y = \xi W y + X\beta + \varepsilon, \tag{2.1}$$

where y is a n by 1 stochastic variate, X a n by k matrix of non-stochastic exogenous variables, a k by 1 vector of parameters, ξ the spatial lag parameter, and W a n by n spatial weights matrix specifying the interconnections between different locations. The speciˇ cation in (2.1) contains a spatially lagged dependent variable and is therefore referred to as the spatial (autoregressive) lag model, assuming the error process is white noise.

Alternatively, we can start from the simple model $y = X\beta + \varepsilon$, and allow for alternative speciˇ cations of the error process. Specifying a ˇ rst order autoregressive error process:

$$\begin{aligned} \varepsilon &= \lambda W \varepsilon + \mu, \\ \mu &\sim \mathrm{N}\left(0, \sigma^2 I\right), \end{aligned} \tag{2.2}$$

where λ is the spatial autoregressive error parameter, leads to a spatial autoregressive error or AR(1) model. Specifying:

$$\begin{aligned} \varepsilon &= \lambda W \mu + \mu, \\ \mu &\sim \mathrm{N}\left(0, \sigma^2 I\right), \end{aligned} \tag{2.3}$$

leads to a spatial moving average or MA(1) process. The moving average process is different from the autoregressive process, among other things, because the spatial effects extend to all locations in the spatial system for the autoregressive error process, but are limited to ˇ rst and second order neighbors in the moving average model (see Anselin, 2003c).

The speciˇ cations in (2.1)–(2.3) can easily be extended to include higher order processes (see, for instance, Anselin and Florax, 1995c). A more general model arises from the combination of (2.1) and (2.3), and is referred to as a spatial autoregressive moving average or ARMA(1,1) model.[5] Four types of spatial dependence tests can be distinguished in the context of the ARMA(1,1) model:

1. *Unidirectional* tests, in particular $H_0 : \xi = 0$ under the assumption that $\lambda = 0$, or $H_0 : \lambda = 0$ under the assumption that $\xi = 0$
2. *Multidirectional* tests, in particular $H_0 : \xi = 0$ and $\lambda = 0$
3. *Robust* tests, in particular $H_0 : \xi = 0$ under the assumption that $\lambda \neq 0$, or $H_0 : \lambda = 0$ under the assumption that $\xi \neq 0$, which can be assessed on the basis of OLS estimation of the simple linear model without spatial effects
4. *Sequential unidirectional* tests, in particular $H_0 : \xi = 0$ under the assumption that $\lambda \neq 0$, or $H_0 : \lambda = 0$ under the assumption that $\xi \neq 0$, which can be attained by means of Maximum Likelihood (ML) or Instrumental Variables (IV) estimation of a speciˇ cation where one of the spatial parameters is set unequal to zero.

We do not investigate sequential test procedures in this chapter, because the prime interest would be the power of the speciˇ cation strategies rather than the power of individual tests, and an assessment of the power of speciˇ cation strategies is generally difˇ cult because of multiple comparisons (Anselin and Grifˇ th, 1988; Florax *et al.*, 2003). We present an overview of the other types of tests below.[6]

Moran's I is a *unidirectional* test against a linear additive spatial dependence pattern among the estimated OLS residuals. It reads as:

$$I = \frac{n}{S_0} \cdot \frac{\hat{\varepsilon}' W \hat{\varepsilon}}{\hat{\varepsilon}' \hat{\varepsilon}}, \tag{2.4}$$

where n is the number of observations, S_0 the sum of the elements of the spatial weights matrix W, and $\hat{\varepsilon}$ the n by 1 vector of OLS residuals of the speciˇ cation $y = X\beta + \varepsilon$.[7] Statistical inference can be based on the assumption of asymptotic

[5] For ease of notation, we do not distinguish between different weights matrices in speciˇ cations containing more than one spatial process, although this may be necessary for particular models to be identiˇ ed.

[6] For more details see, among others, Cliff and Ord (1973, 1981); Burridge (1981); Anselin (1988b); Anselin and Rey (1991); Kelejian and Robinson (1992, 1995); Anselin and Florax (1995c); Anselin *et al.* (1996); Anselin and Moreno (2003).

[7] The ˇ rst term on the right hand side of (2.4) is redundant when the weights matrix is standardized, i.e., the elements of each row are summed to one.

normality, or alternatively, when the distribution is unknown, on a theoretical randomization or empirical permutation approach, eventually using BLUS residuals (Cliff and Ord, 1981, chapter 8). Kelejian and Prucha (2001) show that identical large sample results can be derived without using the normality assumption. Tiefelsdorf and Boots (1995) present an exact approach that depends on the matrix X, and King (1981) shows that Moran's I is a locally best invariant test. Moments and estimation details under various assumptions are given in Cliff and Ord (1972, 1973, 1981), and Anselin (1988b). In the case of the presence of endogenous regressors, Moran's I can be used with IV residuals, but the test needs to be adjusted with appropriately defined moments (Anselin and Kelejian, 1997). The test is applicable in the presence of systems endogeneity and/or a spatially lagged dependent variable, and we label the test I^{IV}.

Kelejian and Robinson (1992) develop an alternative unidirectional large sample test that does not depend on the assumption of normality of the distribution of the error term either, nor on linearity. The test is based on an auxiliary OLS regression of the cross products h of potentially spatially correlated residuals i and j, against the cross-products of the exogenous variables, x_i and x_j:

$$KR = \frac{\hat{\gamma}' Z' Z \hat{\gamma}}{\hat{\sigma}^4}, \tag{2.5}$$

where $\hat{\gamma}$ is the estimated parameter vector of the auxiliary regression, and Z the matrix containing the cross-products of the exogenous variables. A consistent estimator for $\hat{\sigma}^4$ is $\hat{\sigma}'\hat{\sigma}/h_n$, where $\hat{\sigma}$ is the vector of residual cross-products, and h_n the number of observations in the auxiliary residual vector.[8] The KR test is asymptotically distributed as χ^2_k, where k represents the number of variables in Z.

The pairs of cross-products are selected to correspond to the covariance of the spatial units i and j assumed or suspected to be non-zero, presupposing that only a limited number of non-zero correlations is specified. This does not require the specification of a weights matrix (Kelejian and Robinson, 1992). When the selection of pairs of spatial units with non-zero covariances is determined by the criterion of sharing a common border, the information about the "ordering" is straightforwardly represented in a first order contiguity weights matrix. The two approaches are then equivalent, except that the KR test is based on comparing unique pairs of residuals, in effect using only half the information (i.e., the upper or lower triangle of the weights matrix) as compared to tests based on the spatial weights concept.[9]

[8] See Kelejian and Robinson (1992) for an alternative, asymptotically equivalent, estimator.

[9] The KR test is not applicable if a distance decay process is hypothesized, unless an appropriate set of distance-based exogenous variables is defined, and the number of non-zero correlations is limited to, for instance, k neighbors in order to comply to the sparseness requirement. In that case, the claim that the KR test does not require full knowledge of the weighting matrix (see, e.g., Kelejian and Yuzefovich, 2001) is no longer valid. In the first order contiguity case, this claim can be made because only information regarding regions sharing a common border is required. Note that the KR test cannot be applied in cases where the number of interactions is not bounded, and/or the interaction cannot reasonably

This may have implications for the small sample power of the test (see Kelejian and Yuzefovich, 2001). Anselin and Moreno (2003) point out that it is not correct to only account for first order neighbors, because most spatial processes induce non-zero covariances beyond first order neighbors. For instance, a spatial autoregressive error model implies non-zero covariances throughout the spatial system, and a spatial moving error process induces non-zero covariances for first and second order neighbors.[10] Neglecting higher order non-zero covariances may have a negative impact on the power of the *KR* test, and alternative definitions of the "weights" are therefore suggested in Anselin and Moreno (2003), and Kelejian and Yuzefovich (2001).

Moran's *I* as well as the Kelejian-Robinson test are diffuse tests, implying they are indicative of spatial dependence, but they do not point to a specific alternative. The alternative hypotheses of the test statistics are general, and comply with the DGP being, for instance, the spatial autoregressive error or moving average model, or the spatial lag model. This is not without practical relevance, in particular if the power of the tests is high, but at the same time it is indicative of the need for focused tests with a more restricted alternative hypothesis. Focused tests for spatial dependence are developed in a maximum likelihood framework, and usually take the LM rather than the asymptotically equivalent Wald or LR form, because of ease of computation.

Burridge (1980) shows that the LM test for spatially autoregressive errors is proportional to a squared Moran's statistic. The test cannot be used to distinguish between spatial autoregressive and spatial moving average errors, because tests for either form are identical (see, for instance, Bera and Ullah, 1991). The LM test for spatial autoregressive or moving average errors is asymptotically distributed as χ_1^2, and reads as:

$$LM_\lambda = \frac{(\hat{\varepsilon}' W \hat{\varepsilon} / \hat{\sigma}^2)^2}{T_1}, \tag{2.6}$$

where T_1 is the matrix trace expression $\mathrm{tr}((W'W + W)W)$. Anselin and Kelejian (1997) show that (2.6) based on IV residuals, denoted LM_λ^{IV}, is appropriate in a model with endogenous regressors, where the endogeneity is caused by the usual systems feedbacks or by spatial interaction of an endogenous variable.[11]

be assumed symmetric. Both conditions would be violated in, for instance, the approach taken in Moreno *et al.* (Chapter 18 of this volume), where coefficients of an input-output table are used to define the elements of the weights matrix.

10 This follows directly from the difference in the error variance-covariance matrices:

$$\sigma^2(I - \lambda W)^{-1}(I - \lambda W')^{-1}, \text{ and } \sigma^2(I + \lambda W)^{-1}(I + \lambda W')^{-1},$$

for the spatial AR and MA process, respectively. The processes can be seen as "locally equivalent alternatives" (see Godfrey, 1988, for the terminology).

11 Use of the OLS-based tests in (2.4) and (2.6) in the presence of endogenous regressors would be "clearly ad hoc," since the endogeneity of some of the regressors is ignored (Anselin and Kelejian, 1997).

Anselin (1988a) develops an LM test for an erroneously omitted spatially lagged dependent variable:

$$LM_{\xi} = \frac{(\hat{\varepsilon}' Wy/\hat{\sigma}^2)^2}{n\hat{J}_{\xi.\beta}}, \tag{2.7}$$

with,

$$\hat{J}_{\xi.\beta} = \frac{1}{n\hat{\sigma}^2}[(WX\hat{\beta})' M(WX\hat{\beta}) + T_1\hat{\sigma}^2],$$

where $M = I - X(X'X)^{-1}X'$, $\hat{J}_{\xi.\beta}$ is the relevant part of the information matrix. The test statistic again follows a χ^2_1 distribution.

It is easy to see that the spatial lag model with *iid*-distributed errors, given in (2.1), can be restated in "reduced form" as $y = (I - \xi W)^{-1}(X\beta + \varepsilon)$, showing that the spatial lag model is equivalent to a model with spatially lagged exogenous variables and spatially autoregressive errors. It is obvious therefore that the respective LM tests for the spatial error and the spatial lag model, exhibit power against both alternatives (Anselin, 2001b). Several solutions to this problem exist. One is, to rely on the ad hoc decision rule that whichever test statistic is greater and signiˇ cantly different from zero, points to the right alternative. This is the decision rule advocated in Anselin and Rey (1991), and assessed in a Monte Carlo setting in Florax *et al.* (2003). An alternative solution is pointed out in Bera and Yoon (1992; see also Anselin *et al.*, 1996), where misspeciˇ cation tests for the error and the lag model robust to local misspeciˇ cation are derived.

The *robust* unidirectional tests for a spatial error process or an erroneously omitted spatially lagged dependent variable are obviously similar to the tests in (2.6) and (2.7). The latter are extended with a correction factor to account for the local misspeciˇ cation (Anselin *et al.*, 1996). The test for the presence of a spatial AR or MA error process, when the speciˇ cation contains a spatially lagged dependent variable, reads as:

$$LM^*_{\lambda} = \frac{[\hat{\varepsilon}' W\hat{\varepsilon}/\hat{\sigma}^2 - T_1\left(n\hat{J}_{\xi.\beta}\right)^{-1}\hat{\varepsilon}' Wy/\hat{\sigma}^2]^2}{T_1\left[1 - T_1\left(n\hat{J}_{\xi.\beta}\right)\right]^{-1}}. \tag{2.8}$$

Alternatively, the test for a spatially lagged dependent variable in the presence of a spatial error process is given by:

$$LM^*_{\xi} = \frac{[\hat{\varepsilon}' Wy/\hat{\sigma}^2 - \hat{\varepsilon}' W\hat{\varepsilon}/\hat{\sigma}^2]^2}{n\hat{J}_{\xi.\beta} - T_1}. \tag{2.9}$$

Several *multidirectional* Lagrange Multiplier tests are available. They are concerned with higher order processes, spatial ARMA models, and combinations of heteroskedasticity and spatial dependence. The LM tests for higher order spatial processes, pertaining to either the spatial error or the spatial lag, are simply the sum of the respective unidirectional tests given in (2.6) or (2.7) above. These tests follow a χ^2 distribution with the number of degrees of freedom equal to the order of the spatial process. We add a subscript i to the test, as in LM_{λ_i}, to signal that the test is concerned with higher order processes. An LM test with the spatial ARMA model as the alternative follows a χ^2_2 distribution, and can be attained as the sum of the unidirectional tests given in equations (2.6) and (2.9), or alternatively (2.7) and (2.8) (see Anselin *et al.*, 1996, for details). Finally, a multidirectional LM test for the combination of heteroskedasticity and spatial autoregressive errors is simply equal to the sum of a Breusch-Pagan statistic and the LM statistic against autoregressive errors (Anselin, 1988b):

$$LM_{\eta\lambda} = \frac{1}{2} f' Z(Z'Z)^{-1} Z' f + \frac{(\hat{\varepsilon}' W \hat{\varepsilon} / \hat{\sigma}^2)^2}{T_1}, \tag{2.10}$$

where $f_i(\sigma^{-1}\hat{\varepsilon}_i)^2 - 1$ are stacked in the vector f, and Z is an n by $p+1$ matrix containing a constant term and the p variables causing heteroskedasticity. The test asymptotically follows a χ^2_{p+1} distribution. There are many ways to specify the heteroskedasticity, including additive, multiplicative and random coefˇ cients speciˇ cations, usually involving more than one variable determining the heteroskedasticity. The test assumes that both the functional form and the in uencing variables are known. For ease of notation we only add the subscript to the symbol referring to the test.

In addition to the multidirectional LM test involving heteroskedasticity, Kelejian and Robinson (1998) extend the KR formulation in (2.5) to a multidirectional test for the absence of spatial autocorrelation and/or heteroskedasticity by using White's heteroskedasticity robust variance-covariance estimator. The test does require knowledge about the variable(s) relating to the heteroskedasticity, but does not require the functional form to be known. We therefore view the test as a diffuse misspeciˇ cation test, both with respect to spatial autocorrelation and heteroskedasticity, and use the symbol KR^{H} (rather than KR_η) to refer to the test.

2.3.2 The Spatial Error Component Model

A slightly different speciˇ cation of a spatial error model is suggested in Kelejian and Robinson (1995). It combines a local error component and a spillover component, in:

$$\begin{cases} y = X\beta + \varepsilon \\ \varepsilon = W\psi + \mu \\ \psi \sim \mathrm{N}(0, \sigma^2_\psi I),\ \mu \sim \mathrm{N}(0, \sigma^2_\mu I),\ \mathrm{E}(\psi_i \mu_j) = 0, \forall i, j, \end{cases} \tag{2.11}$$

where ψ is a n by 1 vector of spillovers across spatially connected units, as speciˇ ed through the weights matrix, and μ is the familiar unit-speciˇ c disturbance term. Anselin and Moreno (2003) show that this so-called spatial error component model is similar to the spatial moving average model. The respective variance-covariance matrices are nearly identical, and both models induce localized spatial spillovers as opposed to the spatial AR model in which the autocorrelation extends to all units in the spatial system.[12] Assuming uncorrelatedness of the spillover component and the

[12] See Anselin (2003c) for this important distinction, to which he refers as "local" and "global" spatial autocorrelation.

unit-speciˇ c component, the variance-covariance matrix of the spatial error component models is $\sigma_\mu^2(I + \theta WW')$, where $\theta = \sigma_\psi^2/\sigma_\mu^2$ is the ratio of the variances of the two error components (Anselin and Moreno, 2003).

Kelejian and Robinson (1995) point out that the usual *KR* test will exhibit power against the spatial error component model, presuming the selection of pairs forming the cross-products are based on the contiguity criterion, and the number of neighbors considered is bounded. Habitually, ˇ rst order neighbors are considered. Anselin and Moreno (2003) provide a variant that considers ˇ rst as well as second order neighbors, because the error variance-covariance matrix shows that non-zero covariances are not present for ˇ rst order neighbors, but rather for second order neighbors. Kelejian and Yuzefovich (2001) suggest using second order neighbors only.

Anselin (2001a) develops a unidirectional LM test against the spatial error component model, which is again asymptotically distributed as χ_1^2, and reads as:

$$LM_\theta = \frac{[(\hat{\varepsilon}'WW'/\hat{\varepsilon}\hat{\sigma}^2) - T_2]^2}{2[T_3 - (T_2)^2/n]}, \tag{2.12}$$

where $T_2 = \mathrm{tr}(WW')$, and $T_3 = \mathrm{tr}(WW'WW')$. The null hypothesis of the test is $H_0 : \theta = 0$, and the test cannot be straightforwardly expressed as a LR or Wald test because the regularity conditions for spatial ML estimation are not met (see Anselin, 2001a, for details).[13] We note that the null hypothesis differs from the typical tests, because the test is concerned with a ratio of two variance components instead of a ratio of covariances to the variance, considered in the other tests.

2.4 A Taxonomy of Spatial Dependence Tests

In the preceding subsections, we distinguish two general types of tests, "diffuse" and "focused" tests. Diffuse tests are capable of signaling a misspeciˇ cation problem (for instance, through autocorrelated residuals), but the alternative hypothesis does not point to a speciˇ c alternative model. Focused tests have a clear alternative hypothesis, suggesting the researcher in which direction to search for a proper re-speciˇ cation.

In Sect. 2.3.1, we distinguish unidirectional, multidirectional, robust, and sequential unidirectional tests. We do not consider the latter type of tests, because they are in fact a series of tests and should be viewed as a speciˇ cation search strategy. However, the distinction between the former three types of tests applies to both diffuse and focused tests, and leads to the taxonomy of spatial dependence tests given in Table 2.1.

The taxonomy in Table 2.1 is in no sense complete, because we only classify tests used in the meta-analysis of Monte Carlo simulation studies. Most other tests,

[13] Kelejian and Robinson (1993, 1997) suggest a focused unidirectional test for the spatial error component model based on general methods of moments (GMM) estimation, which is easily implemented as a one-sided t-test in an OLS regression (see Anselin and Moreno, 2003). This test is, however, based on estimation of the alternative model.

Table 2.1. A taxonomy of spatial dependence tests

Tests	Unidirectional	Multidirectional	Robust
Diffuse	I, I^{IV}, KR	KR^{H}	
Focused	LM_{ξ}, LM_{λ}, LM_{λ}^{IV}, LM_{θ}	LM_{λ_i}, $LM_{\eta\lambda}$, $LM_{\xi\lambda}$	LM_{ξ}^{*}, LM_{λ}^{*}

however, easily fit the scheme. For instance, the heteroskedasticity-robust test for residual spatial dependence derived in Anselin (1988b, pp. 112–115), and the test for heteroskedasticity given that the error terms are spatially correlated, presented in Kelejian and Robinson (1998, p. 395), can be straightforwardly classified.

2.5 Review of the Simulation Literature on Spatial Dependence Tests

It is imperative that the sample selection process for a meta-analysis is carefully documented. Through a literature search, we attain an exhaustive overview of simulation studies in spatial econometrics, categorized in Table 2.2.

The early simulation studies deal with the small sample performance of dependence tests for "raw data" (Category 1). Subsequently, attention focuses on the investigation of tests for regression residuals. Initially, the studies on regression residuals deal primarily with different statistical inference procedures (Category 2), but afterward a series of studies investigates the small sample properties of tests under various experimental setups (Category 3). A limited number of simulation studies is concerned with the small sample behavior of estimators for spatial models (Category 4). Pertinent problems in spatial data analysis, such as the specification of weights (Category 5), boundary and aggregation effects (Category 6), and missing data (Category 7), generate attention in the simulation literature as well. Finally, a growing number of studies deals with the investigation of specification strategies (Category 8).

We center the meta-analysis on simulation experiments dealing with tests for spatial dependence. Consequently, we sample the studies from Category 2 and 3, although with the exception of Anselin's 1990 study on the effect of spatial error autocorrelation on Chow tests for structural stability, because it is the only study considering spatial heterogeneity. Although it would be interesting to also include studies (or relevant parts of studies) dealing with the impact of misspecification of the weights matrix (Category 5), we exclude those for right now because the differences in the design of these studies cannot be easily accounted for in the specification of the meta-regression. Differences in distributional assumptions can be straightforwardly incorporated in a meta-regression by means of fixed effects.

We provide an annotated chronological listing of the studies included in the meta-analysis in Table 2.3. A number of obvious trends can be deduced from this overview. The vast increase in availability and computational abilities of the personal computer makes that the more recent studies are much more accurate, using a substantially larger number of replications. The table also shows that by now a large number of Lagrange Multiplier tests has been developed and investigated, in addition to Moran's I and, the more recently developed, Kelejian-Robinson test. Over time, the attention for irregular lattice structures increases as well as for alternative error distributions. Although initially very small sample sizes are considered $(n < 25)$, recent studies also occasionally include large sample sizes $(n > 1000)$.

A detailed reading of Table 2.3, including the comments, shows that still more choices are needed as to the exact sampling of measurements from the studies. We concentrate the meta-analysis on misspeciˇ cation tests for spatial dependence that can be computed under the null hypothesis of no spatial dependence, because this resembles current practice best. This implies that Moran's I, the Kelejian-Robinson test, and several Lagrange Multiplier tests are considered. Results referring to Wald and LR tests, such as several heteroskedasticity tests in Anselin and Grifˇ th (1988), the LR test in Brandsma and Ketellapper (1979), and the GMM based test for the spatial error component model in Anselin and Moreno (2003), are not included. We also exclude tests that are not common or not strictly concerned with spatial dependence testing, such as the na¨ve test in Brandsma and Ketellapper (1979), and the RESET test in Florax (1992). Finally, we omit the results for the cross-regressive model in Florax (1992) because an erroneous omission of autocorrelated exogenous variables is an omitted variable problem rather than a spatial dependence problem.[14] The results for unstandardized weights matrices in Florax (1992) are also discarded, because they imply different bounds on the spatial autoregressive parameters and are therefore difˇ cult to compare to concurrent results for standardized weight matrices. Under the above restrictions with regard to sampling, we retrieve 8.460 rejection probabilities (or rejection rates) from 11 studies, of which 980 refer to the size and 7480 to the power of spatial dependence tests.

[14] Consider a simple example, $y = X\beta + \rho WX + \varepsilon$, where ε is the usual *iid* error term with mean zero. If the autocorrelated exogenous variables are ignored, the actual regression becomes, $y = X\beta + \mu$, where $\mu = \varepsilon + \rho WX$, but now $\mathrm{E}(\mu) = W \cdot \mathrm{E}(X) = m \neq 0$, representing the omitted variable bias. If we consider the covariance between the "errors" at locations i and j, where i and j are not ˇ rst or second order neighbors, then:

$$\mathrm{Cov}\left(\mu_i, \mu_j\right) = \mathrm{E}\left([\mu_i - m][\mu_j - m]\right) = \mathrm{E}\left(\mu_i \mu_j\right) - m^2,$$

where,

$$\mathrm{E}\left(\mu_i \mu_j\right) = \mathrm{E}\left([\varepsilon_i + \rho(WX)_i][\varepsilon_j + \rho(WX)_j]\right) = \rho^2 \, (WX)_i \, (WX)_j,$$

so that the "error terms" containing the omitted variable tend to be correlated, irrespective of their spatial arrangement. As a result, it is not fruitful to consider omitted spatially autocorrelated exogenous variables with the typical set of spatial misspeciˇ cation tests. We would like to thank a reviewer for pointing this out. See Anselin (2003c) for the empirical relevance of including spatially correlated exogenous variables in spatial regression models.

2.6 Experimental Design and Meta-Regression Results

The meta-regression specification is similar to the response surface specifications used in, for instance, Kelejian and Robinson (1998), and Anselin and Moreno (2003). We model the experimental probabilities of rejecting the null hypothesis of no spatial dependence as a function of characteristics of the DGP, the test statistics, and the experimental design of the underlying simulation studies. We use a logit transform for the rejection probability in order to avoid the double-sided truncation of p-values, and apply a small correction suggested by Cox (1970, as discussed in Maddala 1983, p. 30) to ensure that the logit is defined even when the rejection probability is 0 or 1. A straightforward meta-regression specification then reads as:

$$\log\left(\frac{p_i+(2n_i)^{-1}}{1-p_i+(2n_i)^{-1}}\right)=p_i^*=\alpha+X\beta+\varepsilon, \tag{2.13}$$

where p_i is the rejection probability from experiment i, n_i the number of replications on which the rejection probability is based, α a constant term, β a vector of parameters, X the design matrix, and ε a vector of error terms. We refer to the dependent variable p_i^* as the "logit," which is the adjusted log of the odds ratio of rejecting the null hypothesis of no spatial dependence. We discuss various assumptions regarding the error term and the specification of the design matrix, below.

In recent response surface analyses, (2.13) is estimated presupposing the experiments are independent, and potential heteroskedasticity can be remedied through a heteroskedasticity-robust variance estimator (see Anselin and Moreno, 2003). The population logit is estimated with some random error, and the variation in the population logit is perfectly predictable by means of the variables included in the design matrix. In formal terms, $p_i^*=\pi_i+\varepsilon_i=x_i'\beta+\varepsilon_i$, where π_i is the population logit, and the error term is independently and identically distributed. We can improve on this specification, because in large samples the variance of the estimated logits can be estimated by $(p_i^*(1-p_i^*)n_i)^{-1}$ (Maddala, 1983). Subsequently, we can use weighted least squares (WLS) defining the weights as the inverse of the estimated variance. Somewhat confusingly, this is called a fixed effects model in the meta-analysis literature, because the variation in the estimated logits is not due to randomness but to a number of fixed exogenous effects represented in the design matrix (see Hedges and Olkin, 1985; Sutton *et al.*, 2001, for details).

The fixed effects model presupposes the experiments in the underlying simulation studies are independent. For a response surface analysis concerning a series of experiments within one study, this may be a reasonable assumption, even although the possibility of autocorrelation among the experiments is ignored. In a meta-analysis covering a series of studies with multiple sampling from each study, we prefer an alternative specification that takes into account the nested error structure.

Table 2.2. Overview of the simulation literature

Focus	Study
1. Tests for "raw data"	Cliff and Ord (1975), see also Cliff and Ord (1973, 1981)
	Haining (1977, 1978)
2. Tests for regression residuals, inference procedures	Bartels and Hordijk (1977)
	Brandsma and Ketellapper (1979)
	Florax (1992)
3. Small sample properties of tests for spatial effects	Anselin and Grif˘ th (1988)
	Anselin (1990)
	Anselin and Rey (1991)
	Florax (1992), see also Florax and Folmer (1992)
	Anselin and Florax (1995c), see also Anselin *et al.* (1996)
	Florax and Rey (1995)
	Anselin and Kelejian (1997)
	Kelejian and Robinson (1998)
	Anselin and Moreno (2003),
	see also Anselin (2001a), and Kelejian and Yuzefovich (2001)
4. Small sample properties of estimators	Anselin (1980)
	Anselin (1981)
	Sneek and Rietveld (1997)
	Das *et al.* (2003)
5. Speci˘ cation of weights	Stetzer (1982)
	Anselin (1986)
	Anselin and Rey (1991)
	Florax and Rey (1995)

continued on next page

Table 2.2. Continued

Focus	Study
	Kelejian and Robinson (1998)
6. Missing data	Haining *et al.* (1983)
	Griffith (1988)
7. Boundary effects and MAUP	Griffith and Amrhein (1982, 1983)
	Griffith (1985), see also Griffith (1988)
	Anselin and Rey (1991)
8. Specification strategies	Anselin (1986)
	Anselin and Griffith (1988)
	Anselin (1990)
	Florax and Folmer (1992), see also Florax (1992)
	Florax *et al.* (2003)

Table 2.3. Annotated chronological listing of Monte Carlo simulation studies of spatial dependence tests in linear regression models

Study: Comments	Type tests	Sample size simulation study	Weights[a]	Error distribution[b]	Replications[c] Meta-sample[d]
Bartels and Hordijk (1977): Compares linear unbiased scalar covariance estimators to traditional inference. The DGP in Examples 3 and 4 contains spatially autocorrelated exogenous variables in addition to spatially autoregressive errors. Brandsma and Ketellapper (1979) note a mistake in their computer program and replicate part of their work.	I, I^{LUS}	26, 39	I(q)	N	100 252 (42)
Brandsma and Ketellapper (1979): Compares linear unbiased scalar covariance estimators to traditional inference. The DGP in Model 3 contains spatially autocorrelated exogenous variables in addition to spatially autoregressive errors. Results for a so-called na¨ve test and the Likelihood Ratio test are omitted.	I, I^{LUS}	24, 39	I(q,m)	N, E	100 240 (60)
Anselin and Grif˜ th (1988): Investigates the joint occurrence of heteroskedasticity and spatial correlation. The heteroskedasticity tests as well as the results for a sequential test procedure are excluded from the meta-analysis.	$LM_{\eta\lambda}$	25, 50, 75	R(q)	N	1000 84 (12)

continued on next page

Table 2.3. Continued

Study: Comments	Type tests	Sample size simulation study	Weights[a]	Error distribution[b]	Replications[c] Meta-sample[d]
Anselin and Rey (1991): The results regarding misspeciˇ cation of the weights matrix, and boundary effects are excluded from the meta-analysis. Very comprehensive study, although unfortunately only the size of the tests is recorded in tables. The power results are given in graphs, and are therefore not included.	I, LM_{λ}, LM_{ξ}	25, 49, 81, 121, 169, 225	R(q,r,k)	N	5000 126 (126)
Florax (1992), see also Florax and Folmer (1992): Compares bootstrapping for Moran's I to the traditional inference procedure based on normality. The results for the cross-regressive model, the RESET test, and the unstandardized weights matrices are discarded. One of the DGPs contains spatially autocorrelated exogenous variables in addition to a spatially lagged dependent variable.	I, LM_{λ}, LM_{ξ}	26	I(q,g)	N	500, 5000 261 (11)

continued on next page

Table 2.3. Continued

Study: Comments	Type tests	Sample size simulation study	Weights[a]	Error distribution[b]	Replications[c] Meta-sample[d]
Anselin and Florax (1995c), see also Anselin *et al.* (1996): Includes robust tests, higher order models, and the ARMA speciˇ cation.	I, LM_{λ}^{i}, LM_{ξ}, LM_{λ}^{*}, LM_{ξ}^{*}, $LM_{\xi\lambda}$, KR	40, 81, 127	R(q,r), I(q)	N, L	5000 5536 (64)
Florax and Rey (1995): Study focusing on misspeciˇ cation of the weights matrix, but it also presents the size (in tabular format) and power (in graphs) of test results when the correct weights are used. Only tabular information is included in the meta-analysis. The study also presents characteristics of pre-test estimators.	I, LM_{λ}, LM_{ξ}	16-49	R(q,r,k)	N	1000 36 (36)
Anselin and Kelejian (1997): The study deals with the performance of tests in models with endogenous regressors. The use of the traditional tests is ad hoc.	I, LM_{λ}, I^{IV}, LM_{λ}^{IV}	48, 81, 121, 900, 1600	R(r), I(q)	N, L, U, χ^2	10000, 20000 308 (200)

continued on next page

Table 2.3. Continued

Study: Comments	Type tests	Sample size simulation study	Weights[a]	Error distribution[b]	Replications[c] Meta-sample[d]
Kelejian and Robinson (1998): Investigates the joint occurrence of heteroskedasticity and spatial dependence. The joint LM test is also applied in a robust version, but the results are similar to the non-robust version and not explicitly recorded. The study also investigates the impact of misspeciˇ cation of the weights.	I, LM_{λ}, $LM_{\eta\lambda}$, KR^{H}	36, 81, 169	R(q,r)	N, L, U, χ^2	5000 816 (240)
Anselin and Moreno (2003)[f], see also Anselin (2001b): Study with the spatial error component model as DGP. Includes a higher order variant of the *KR* test. Inclusion of spatially correlated exogenous variables does not substantially affect the results, and they are therefore not reported. The *KR* test based on general methods of moments estimation is not included in the meta-analysis.	I, LM_{λ}, LM_{θ}, KR	49, 81, 121, 256, 400, 1024[e]	R(r), I(r)	N, L, N^{m}	10000 720 (180)

continued on next page

Table 2.3. Continued

Study: Comments	Type tests	Sample size simulation study	Weights[a]	Error distribution[b]	Replications[c] Meta-sample[d]
Kelejian and Yuzefovich (2001): Partly replicates the Anselin and Moreno (2003) study, and deals with heteroskedasticity, the deˇ nition of the spatial ordering, and induced changes in the R^2 across experiments.	KR, LM_θ	49, 81, 121	R(r)	N	5000 81 (9)

[a] The abbreviations point to a regular (R), or irregular lattice structure (I). Within those categories contiguities are determined using the rook criterion (r), the queen criterion (q), a binary measure for *k* nearest neighbors (k), general weights based on the distance between the geographical centers of the spatial units and the length of the common border (g), or interregional migration ows (m).

[b] The categories for the error distributions are: normal (N), lognormal (L), exponential (E), uniform (U), chi-square (χ^2), and mixed normal (N^m).

[c] The number of replications in the simulation study.

[d] The number of observations sampled for the meta-analysis, with in parentheses, the number of meta-observations referring to the size of the spatial dependence tests (assuming no heteroskedasticity).

[e] Sample sizes are slightly different for the irregular matrices.

[f] The original working paper was published in 2001, and Kelejian and Yuzefovich (2001) react to this working paper, which is the reason for the "reverse" ordering.

Speciˇ cally, we use the following standard random effects model, with the subscripts referring to a speciˇ c measurement m sampled from study s:

$$p^*_{ms} = x'_{ms}\beta + \mu_s + \varepsilon_{ms}, \tag{2.14}$$

where the population effect sizes are assumed to vary between studies, and they are considered random draws from a normal distribution. As indicated above, inverse variance weighting applies in order to account for the difference in precision with which the effect sizes have been measured. The random effects model has a non-diagonal variance-covariance matrix, but the non-zero off-diagonal elements re ect heterogeneity between studies rather than dependence within studies. Given the large sample size in the meta-analysis, we ignore the latter type of autocorrelation. If the random error term is not signiˇ cantly different from zero, weighted least squares is applied.

We use an additional set of weights to account for the unbalanced panel data setup of the meta-sample. Failing to do so would imply that studies for which a larger number of experimental results is reported in print, automatically have a greater in uence in determining the results of the meta-analysis. Hence, on the one hand we correct for differences in precision with which the effect sizes have been measured (see above), and on the other, we want to assign the same importance to each study so that in effect each study contributes equally to the meta-analysis. The latter is achieved by simply weighting the observations with weights deˇ ned by:

$$w_{ms} = \frac{n}{n_s S}, \tag{2.15}$$

where w_{ms} is the weight applied to measurement m (= 1, 2, ..., n_s) from study s (= 1, 2, ..., S), n_s is the number of measurements in study s, and:

$$n = \sum_{s=1}^{S} \sum_{n=1}^{n_s} w_{ms},$$

is the total number of observations in the meta-sample (see Bijmolt and Pieters, 2001). The ultimate set of weights is therefore obtained as:

$$w^*_{ms} = 1/\left(\sqrt{\frac{1}{p^*_{ms}(1-p^*_{ms})n_{ms}}} \cdot \sqrt{\frac{n}{n_s S}}\right), \tag{2.16}$$

where n_{ms} is the number of replications with which each individual rejection probability has been evaluated.

The design matrix for the meta-analysis contains six groups of explanatory variables. First, we specify ˇ xed effects for the different tests. Second, we include dummy variables representing the error distribution, with the normal distribution as the omitted category, and the sample size of the underlying Monte Carlo experiments. Third, the characteristics of the weights matrix are measured by means of the density (i.e., the number of non-zero links as a percentage of the n by $(n-1)$

off-diagonal elements, which is the complement of sparseness), and the connectedness (i.e., the average number of non-zero links) of the weights matrices used in the experiments. A dummy variable accounts for weights derived from irregular lattices. We account for the *KR* test using half the information as compared to tests using a weights matrix, by adjusting the density and the connectedness measure accordingly. Fourth, the strength of the spatial interaction is accounted for by λ_1 for the first order spatial autoregressive error coefficient, λ_2 for the second order spatial autoregressive error coefficient, θ_1 and θ_2 for the first and second order moving average parameters, ξ for the coefficient of a spatially lagged dependent variable, and θ for the variance ratio in the spatial error component model. Fifth, the presence of other "misspecifications" is incorporated through a dummy variable for heteroskedasticity (eventually distinguishing low, medium, and high) when in the experiments heteroskedasticity is added in the generation process in addition to the "normal" heteroskedasticity inherent in spatial models. We also identify the presence of spatially correlated exogenous variables, and the presence of systems endogeneity through fixed effects. Finally, several differences in statistical inference are taken into account. We include the variance of the error distribution, which is usually unity, except in two studies having a greater error variance (Florax, 1992; Florax and Rey, 1995), and for the spatial error component model, where the error variance may vary between experiments (see Kelejian and Yuzefovich, 2001). The use of BLUS or RELUS residuals as well as bootstrap confidence intervals are included as fixed effects. With respect to the latter, the bias-corrected percentile method (BPCM), the percentile method (PM), and the permutation percentile method (PPM) are distinguished (see Florax, 1992, for details).

The meta-regressions pertain to the power of the tests for positive values of the spatial parameters. We omit negative values of the spatial parameters. The results for negative values are difficult to compare to their positive counterparts, because the definition of the boundary space for negative autocorrelation is not uniform across different weight matrices, regardless of whether they are standardized or not.

In order for a meta-analysis to provide more insight than individual simulation studies, test statistics need to be investigated in more than one study. If not, a meta-analysis reduces to a response surface analysis. This would, for instance, be the case for the simultaneous equation results in Anselin and Kelejian (1997), and the robust tests in Anselin and Florax (1995c). In view of the limited number of studies establishing overlap in terms of test statistics considered, the usefulness of meta-analysis is still confined. We identify three specific topics for which meta-analysis provides additional knowledge about the small sample power of spatial misspecification tests. We compare the (relative) performance of the two most important diffuse tests, Moran's *I* and the Kelejian-Robinson test, in Sect. 2.6.1. In Sect. 2.6.2, we compare the performance of focused unidirectional tests among each other as well as to diffuse tests, for various data generating processes. We assess the performance of diffuse and focused multidirectional tests against spatial dependence and heteroskedasticity in Sect. 2.6.3.

2.6.1 Moran's *I* and the Kelejian-Robinson Test

We derive results for meta-regressions with the log of the odds ratios for Moran's *I*, the Kelejian-Robinson test, and the two tests combined, as the dependent variable. The results for the Lagrange Multiplier test of the weighted random effects speciˇ - cation against the simple linear weighted least squares model show that the latter is generally the preferred alternative.

Table 2.4 shows that the *KR* test is sensitive to departures from a normal error distribution, whereas Moran's *I* is not. This result is at odds with Kelejian and Robinson's (1998, pp. 414–415) inference from their response surface analysis. The effect of sample size is not signiˇ cantly different from zero in two cases, and signiˇ cantly positive in one case. One should note that this may be partly a result of including the density and connectedness features of the weights matrix, because these are related to sample size (see below).

The effects of different characteristics of the weights matrix are signiˇ cantly different from zero. As expected, greater connectedness increases the small sample power, but increasing density of the weights matrix seems to lower the power of the test. The bivariate correlation of the two indicators is 0.33, suggesting that both indicators measure something different, and that multicollinearity is not a problem. However, the density and the connectedness measure are related through sample size: the same connectedness with a larger sample size results in a lower density. The nexus of the interrelated variables sample size, and density and connectedness of the weights matrix needs further attention. The use of weights derived for irregular lattices, as compared to regular lattice structures, has a positive effect on the small sample power.

The magnitude of the spatial autocorrelation parameter is the most important determinant of the small sample power distribution. The statistical tests are most responsive to spatial autoregressive correlation, of the spatially lagged dependent variable or a spatial autoregressive error term. The tests are substantially less responsive to higher order auto-correlation. These results are not comparable to the effect of a spatial error component, because θ is a ratio of error variances. The magnitude of the spatial correlation in the spatial error component model is therefore measured by the variable θ as well as by the variable representing the variance of the error distribution.

Moran's *I* is not speciˇ cally designed to have power against heteroskedasticity, and Table 2.4 shows that it does not have power against this alternative. The *KR* test should by design be responsive against heteroskedasticity, because the test contains the cross-products of x-variables that are suspected to in uence the spatial dependence, at the same time inducing heteroskedasticity. Other misspeciˇ cations, such as spatially correlated exogenous variables (in addition to a spatially lagged dependent variable or a spatial autoregressive error), and systems endogeneity increase or decrease the power of the tests, respectively.

Table 2.4. Weighted least squares results for diffuse spatial dependence tests under all data generating processes[a]

Variable	I	KR	Both
Constant	−1.874*	−4.192*	0.389
	(0.245)	(1.551)	(0.326)
Tests			
KR			−1.166*
			(0.036)
Distribution and sample size			
Lognormal	0.038	0.669*	−0.150
	(0.121)	(0.243)	(0.162)
Exponential	0.367		0.511
	(0.535)		(1.016)
Mixed normal	0.021	0.767*	0.095
	(0.348)	(0.243)	(0.217)
Monte Carlo sample size	$-4.9E-6$	−0.001	0.003*
	(0.001)	$(4.4E-4)$	$(3.4E-4)$
Weights			
Density	−0.226*	−0.466*	−0.211*
	(0.008)	(0.016)	(0.007)
Connectedness	0.109*	0.303*	0.073*
	(0.013)	(0.027)	(0.013)
Irregular lattice	0.320*	0.264*	0.135*
	(0.029)	(0.034)	(0.024)
Spatial parameters			
λ_1	8.040*	7.206*	7.346*
	(0.110)	(0.132)	(0.099)
λ_2	2.427*	2.772*	2.525*
	(0.077)	(0.101)	(0.072)
θ_1	6.021*	5.008*	5.253*
	(0.061)	(0.069)	(0.052)
θ_2	0.447*	0.806*	0.639*
	(0.046)	(0.057)	(0.042)
ξ	9.214*	8.286*	8.312*
	(0.146)	(0.153)	(0.120)
θ	0.212*	0.321*	0.290*
	(0.038)	(0.029)	(0.025)
Misspeci˘ cations			
Heteroskedasticity	0.046	6.057*	0.422◇
	(0.661)	(1.396)	(0.168)
Spatially correlated x	1.028*		2.077*
	(0.346)		(0.640)

continued on next page

Table 2.4. Continued[a]

Variable	*I*	*KR*	Both
Systems endogeneity	−2.826*		−2.709*
	(0.198)		(0.366)
Inference			
Variance error distribution	0.377	2.313	−1.730*
	(0.241)	(1.549)	(0.319)
One-sided test Moran's *I*	−0.167		−0.448
	(0.739)		(1.399)
BLUS residuals	−0.852		−0.697
	(0.777)		(1.475)
RELUS residuals	−1.030		−0.879
	(0.775)		(1.472)
Bootstrap, BCPM	1.401◇		3.651*
	(0.626)		(1.147)
Bootstrap, PM	0.830		3.113*
	(0.611)		(1.118)
Bootstrap, PPM	$3.3E-4$		2.337◇
	(0.628)		(1.152)
n	1664	1164	2828
R^2-adjusted	0.88	0.86	0.82
F	524.56*	508.17*	548.02*
Log-likelihood	−1013.33	−768.58	−2235.75
LM(REM)[c]	0.52	0.42	[b]

[a] Estimated standard errors are in parentheses. Signiˇ cance is indicated by *, ◇ and ° for the 0.01, 0.05 and 0.10 level, respectively.

[b] The test is not available because the random effects model cannot be estimated due to a negative residual variance (see Greene, 1997, pp. 333–338, for details).

[c] Test of the model with random study effects vs the model without random study effects, both weighted as indicated in the main text.

The variables related to statistical inference procedures are for the most part not signiˇ cantly different from zero. There are a few exceptions. The higher the variance of the error distribution, the lower the power. This is as expected, because the importance of the systematic part of the DGP is correspondingly lower when the variance of the error distribution is higher. The use of BLUS or RELUS residuals does not have a signiˇ cant impact on the power of the tests. In a sense, this contradicts the early simulation experiments of Bartels and Hordijk (1977), and Brandsma and Ketellapper (1979). The bootstrap results suggest that the use of resampling procedures increases the power of the tests. It is important to note, however, that the size of the tests with bootstrapped conˇ dence intervals is signiˇ cantly higher than the nominal Type-I error (see Florax, 1992).

The results for the two tests combined, are very similar. The marginal effect of increasing the sample size with one observation is approximately one percent

$(= e^{0.003/(1-0.003)})$, implying that the asymptotic characteristics are attainable in medium sized samples with approximately 100 observations, if the magnitude if the autocorrelation is small.[15] The most important result is, however, that the power of the *KR*-test is signiˇ cantly lower than Moran's *I*. This result has been reported in previous studies (for instance Anselin and Florax, 1995c), but our claim is stronger because we account for the precision with which the rejection probabilities are estimated, and we control for the fact that the *KR* test uses less information. The *KR* test also has power against heteroskedasticity, which makes that the optimal test strategy for practitioners is to use Moran's *I* when spatial autocorrelation is expected, and to use the *KR* test when there is suspicion of both substantial heteroskedasticity and spatial dependence.

2.6.2 Focused Tests under Different DGPs

In Table 2.5, we present results for the unidirectional focused tests for single known data generating processes, and for all data generating processes combined. We distinguish the AR(1), MA(1), the spatial lag, and the spatial error component model as DGPs. For each DGP we omit the test that has the speciˇ c DGP as the alternative hypothesis. The results are based on weighted least squares regression, because as a rule the random effects model cannot be estimated due to a negative residual variance estimate.[16]

Table 2.5 shows that overall the *KR* test has lower power than Moran's *I*. However, if we treat the DGPs as known, then the *KR* test has lower power than Moran's *I* for the spatial AR(1) and MA(1) models, but higher power as compared to Moran's *I* against the spatial lag model and the spatial error component model.

Almost uniformly, the "correct" focused test has more power than any other test. The only exception is Moran's *I* having slightly more power against the MA(1) process than the LM test against moving average errors. The results for the robust tests allow for a more accurate assessment than the conclusion in Anselin *et al.* (1996, p. 100): "[t]he robust tests ... seem more appropriate to test for lag dependence in the presence of error correlation than for the reverse case." The power of the test for autoregressive errors in the presence of a spatial lag is not signiˇ cantly different from the power of the LM error test, in both the AR(1) and the MA(1) model. So, the use of either type of tests is equivalent. The power of the LM test against a spatial lag in the presence of autocorrelated errors does have signiˇ cantly more power than the unidirectional LM test against a spatial lag.

[15] Caution is necessary because the effect is computed assuming all other variables are zero, and because the variables related to density and connectivity of the weights matrix are implicitly dependent on sample size as well.

[16] One of the reasons for this occurring so frequently is that the speciˇ cation of the meta-regression makes that the intermediate step using the ˇ xed effects estimator to attain an estimate for the residual variance cannot be applied, because the ˇ xed effects estimator is a within-estimator. It is, however, also likely that the extensive speciˇ cation of differences within and between studies in the meta-regression sufˇ ciently accounts for the heterogeneity (see also Table 2.4).

Table 2.5. Weighted least squares results for focused unidirectional spatial dependence tests under known data generating processes[a]

Variable DGP	AR(1)	MA(1)	Lag	SEC	All
Constant	−2.542*	−3.260*	−2.425*	−2.686*	−0.811*
	(0.428)	(0.534)	(0.458)	(0.824)	(0.239)
Tests					
I	−0.067	0.391*	−0.479*	−1.532*	
	(0.054)	(0.119)	(0.156)	(0.112)	
KR	−0.450*	−0.664*	−0.451◇	−0.988*	−0.282*
	(0.144)	(0.151)	(0.216)	(0.098)	(0.072)
LM_λ			−0.823*	−1.633*	−0.299*
			(0.159)	(0.113)	(0.055)
LM_ξ	−1.460*	−1.981*			−1.314*
	(0.128)	(0.114)			(0.074)
LM^*_λ	−0.084	−0.083	−3.659*		−0.710*
	(0.144)	(0.115)	(0.204)		(0.077)
LM^*_ξ	−3.262*	−2.864*	0.739*		−2.286*
	(0.148)	(0.144)	(0.200)		(0.087)
LM_θ					1.134*
					(0.195)
Distribution and sample size					
Lognormal	0.076			−0.0071	0.067
	(0.052)			(0.080)	(0.066)
Exponential	0.236				0.404
	(0.760)				(1.116)

continued on next page

Table 2.5. Continued[a]

Variable DGP	AR(1)	MA(1)	Lag	SEC	All
Mixed normal				−0.076	0.001
				(0.080)	(0.149)
Monte Carlo sample size	0.012*	0.023*	0.021$^{\diamond}$	$2.5E-6$	0.001$^{\diamond}$
	(0.001)	(0.006)	(0.008)	$(1.4E-4)$	$(2.5E-4)$
Weights					
Density	−0.045*	0.080	0.092	−0.185*	−0.119*
	(0.011)	(0.257)	(0.096)	(0.014)	(0.007)
Connectedness	−0.026	−0.259*	0.059	0.317*	0.158*
	(0.024)	(0.093)	(0.127)	(0.026)	(0.017)
Irregular lattice	0.188$^{\diamond}$	−0.240°	0.023	0.721*	0.340*
	(0.075)	(0.146)	(0.198)	(0.082)	(0.046)
Spatial parameters					
λ_1	5.961*				5.040*
	(0.135)				(0.116)
θ_1		5.044*			3.941*
		(0.174)			(0.109)
ξ			7.939*		6.243*
			(0.363)		(0.201)
θ				0.222*	0.206*
				(0.012)	(0.018)
Misspeciˇ cations					
Heteroskedasticity	0.363*			4.001*	0.514*
	(0.054)			(0.739)	(0.068)

continued on next page

Table 2.5. Continued[a]

Variable DGP	AR(1)	MA(1)	Lag	SEC	All
Spatially correlated x	0.366		1.109◊		
	(0.639)		(0.488)		
Systems endogeneity	−1.247*				−1.107*
	(0.091)				(0.121)
Inference					
Variance error distribution	−0.238		−1.808*	1.507°	−0.973*
	(0.418)	(0.659)		(0.805)	(0.223)
Nominal p-value 0.025	−0.531				−0.594
	(0.805)				(1.190)
One-sided test Moran's I	0.297				−0.481
	(1.061)				(1.525)
BLUS residuals	−0.662				−0.484
	(1.100)				(1.625)
RELUS residuals	−0.814				−0.634
	(1.097)				(1.621)
Bootstrap, BCPM	1.484				1.972°
	(1.538)				(1.081)
Bootstrap, PM	1.008				1.531
	(1.533)				(1.053)
Bootstrap, PPM	0.362				0.901
	(1.587)				(1.086)

continued on next page

Table 2.5. Continued[a]

Variable DGP	AR(1)	MA(1)	Lag	SEC	All
n	1453	288	358	612	2711
R^2-adjusted	0.63	0.81	0.61	0.61	0.50
F	107.85*	122.91*	47.32*	79.87*	99.69*
Log-likelihood	−1808.75	−329.64	−484.89	−714.81	−3856.93
LM(REM)	b	c	b	b	b

[a] Estimated standard errors are in parentheses. Signiˇ cance is indicated by *, ◇ and ° for the 0.01, 0.05 and 0.10 level, respectively.
[b] The test is not available because the random effects model cannot be estimated due to a negative residual variance (see Greene, 1997, pp. 333–338, for details).
[c] The random effects model is not applicable here because the results are taken from one study (Anselin and Florax, 1995c).

Only limited results are available with respect to different error distributions and other types of misspeciˇ cation. The available results suggest that different distributional assumptions regarding the error term do not cause the power to be signiˇ cantly different, and they do not invalidate the above conclusions. Conversely, heteroskedasticity does have a signiˇ cant positive effect on the power of the test statistics, and systems endogeneity has a negative effect. The presence of spatially correlated exogenous variables leads to a greater power when combined with a spatially lagged dependent variable, but not for the combination with a spatial AR(1) process. The above implies that the familiar speciˇ cation strategy to select the alternative model for which the corresponding unidirectional LM test is highest, is likely to be appropriate even in situations in which heteroskedasticity and/or autocorrelated exogenous variables are present, and in the case where the spatial error component model is the "true" model. It is, however, remarkable that when we assume the DGP unknown, the LM test against spatial error components has the highest power – even higher than Moran's I. This warrants further attention.

Table 2.6. Weighted least squares results for diffuse and focused multidirectional tests against spatial dependence and heteroskedasticity for corresponding data generating processes, and a comparison with Moran's I and the LM test against spatial autoregressive errors[a]

Variable	KR^{H}	$LM_{\eta\lambda}$	Spatial AR(1) – Hetero model
Constant	−1.507*	−2.255*	−1.379*
	(0.343)	(0.311)	(0.218)
Tests			
I			−1.364*
			(0.115)
KR^{H}			−0.746*
			(0.127)
LM_{λ}			−1.625*
			(0.115)
Distribution and sample size			
Lognormal	−1.057*	0.304*	−0.102
	(0.134)	(0.112)	(0.075)
Monte Carlo sample size	0.008*	0.009*	0.010*
	(0.003)	(0.003)	(0.001)
Weights			
Density	−0.124◇	−0.079*	−0.040 **
	(0.053)	(0.023)	(0.016)
Connectedness	0.070	0.191*	−0.004
	(0.122)	(0.057)	(0.038)
Spatial parameters			
λ_1	4.173*	3.490*	4.932*
	(0.283)	(0.255)	(0.188)
Misspeciˇ cations			
continued on next page			

Table 2.6. Continued[a]

Variable	KR^{H}	$LM_{\eta\lambda}$	Spatial AR(1) – Hetero model
Heteroskedasticity low	0.625*	0.702*	0.348*
	(0.175)	(0.136)	(0.101)
Heteroskedasticity medium	1.735*	2.228*	0.852*
	(0.193)	(0.171)	(0.109)
Heteroskedasticity high	2.235*	3.215*	0.965*
	(0.199)	(0.311)	(0.115)
n	180	225	765
R^2-adjusted	0.63	0.63	0.51
F	38.79*	48.31*	74.45*
Log-likelihood	−383.70	−266.30	−1091.91
LM(REM)	c	b	b

[a] Estimated standard errors are in parentheses. Signiˇ cance is indicated by *, ◇ and ° for the 0.01, 0.05 and 0.10 level, respectively.

[b] The test is not available because the random effects model cannot be estimated due to a negative residual variance (see Greene, 1997, pp. 333–338, for details).

[c] The random effects model is not applicable here because the results are taken from one study (Kelejian and Robinson, 1998).

The results for the characteristics of the weights matrices are less coherent and somewhat surprising. In particular, connectedness is signiˇ cantly different from zero for the MA(1) model, but with a negative sign, and neither of the weights matrix characteristics seems to have an impact in the case of the spatial lag model. Finally, the results with respect to the statistical inference procedures are in line with the conclusions drawn for the diffuse misspeciˇ cation tests.

2.6.3 Combined Tests for Heteroskedasticity and Dependence

The last meta-regressions are concerned with multidirectional tests for heteroskedasticity and spatial dependence. We compare the focused multidirectional LM test and the diffuse KR^{H} test in isolation, as well as against Moran's I, and the LM test against spatial autoregressive errors. The data generating process is in all cases the spatial AR(1) model, with heteroskedasticity added beyond the heteroskedasticity that is intrinsic to the spatial error speciˇ cation. These tests are investigated in Anselin and Grifˇ th (1988), and in Kelejian and Robinson (1998). Although they use slightly different speciˇ cations for the heteroskedasticity, we code them similarly as low, medium, and high heteroskedasticity. Kelejian and Robinson (1998) point out that the power of the tests should not be related to the error variance.

Table 2.6 shows that the power of both multidirectional tests is sensitive to departures from normality for the error distribution. For the KR^{H} test, it decreases power and for the LM test against heteroskedasticity and dependence, it increases the power of the test. The tests are very sensitive to the value of the spatial autoregressive parameter, as well as to the extent of heteroskedasticity.

In the last column, we compare the performance of the multidirectional tests among each other and to Moran's I and the LM test for AR(1) errors. It demonstrates that the multidirectional LM test has the highest power, followed by the KR^{H} test. The power of the tests designed for this alternative is higher than for the diffuse Moran's I test and for the LM test against spatial autoregressive errors. Unfortunately, no simulation study is available in which concurrent results for the KR test are reported.

2.7 Conclusions

In this chapter, we analyze the experimental simulation literature regarding spatial dependence testing. We use a method that is akin to the response surface technique developed in mainstream econometrics. Response surface analyses are, however, usually confined to the analysis of experimental simulation results from one study. The meta-analysis technique used in this chapter extends to the analysis of quantitative results across studies. In order to account for heterogeneity in the experimental design across studies, we suggest the use of a random effects estimator. It becomes clear, however, that the addition of a random effect is not necessary, because the extensive representation of differences in research design through "fixed effects" sufficiently accounts for the heterogeneity.

The results of the meta-analysis are new in the sense that they compare results across studies. They are also new because they improve over current practice in response surface analyses by weighting the log of the odds ratio of the rejection probabilities with their associated estimated standard error. In addition, we account for the unbalanced nature of the "panel data" by using a weighting procedure ensuring each study is equally important in generating the meta-analysis results.

The extent to which a meta-analysis of experimental simulation studies concerning spatial dependence tests can gain new insights for practitioners, is still limited. This is caused by two factors. First, the output of simulation experiments is usually so abundant that only a fraction of the results is reported in, and can hence be extracted from, published sources. The sampling possibilities are hampered not only by space constraints in publication, but also by results being presented in graphs rather than in tabular form. Second, there are as of now still many combinations of tests under different DGPs and other simulation characteristics, for which no experimental results are available. For instance, experimental results of many tests under the spatial error component model are missing, the impact of heteroskedasticity and systems endogeneity is not yet complete, and we do not know how the KR test performs under heteroskedasticity.

The most notable results of the meta-analysis are as follows. First, among the diffuse tests, the Kelejian-Robinson test has lower power than Moran's I. Because the KR test also has power against heteroskedasticity, whereas Moran's I does not, we cannot conclude that Moran's I is uniformly more powerful than the KR test. In addition, the superiority of Moran's I is not uniform across DGPs. The conclusion holds for the AR(1) and the MA(1) model, but is reverse for the spatial lag model and the spatial error component model. These results are attained controlling for the fact that the *KR* test uses less information, because it is based on the comparison of uniquely deˇ ned pairs. Second, in almost all cases, density of the weights matrix has a negative effect on the power of the tests, whereas connectedness has a positive effect. This is an unexpected result, which needs further attention. Third, the *KR* test is much more sensitive to departures from the normally distributed errors assumption as compared to Moran's I, and LM tests. This is remarkable because the normality restriction is not applicable for the *KR* test. Fourth, the power of spatial dependence tests depends on sample size, and medium-sized samples are needed (n approaching 100) for an adequate performance of the test statistics with small magnitudes of spatial autocorrelation. Fifth, the classical speciˇ cation strategy based on unidirectional LM tests (i.e., choose the alternative corresponding to the LM test with the highest value) is likely to be adequate even when heteroskedasticity or autocorrelated exogenous variables are present, or the true model is the spatial error component model. More research into this issue is, however, warranted. Finally, for multidirectional test for spatial dependence and heteroskedasticity the corresponding LM test has more power than the multidirectional *KR* test, even when we account for the *KR* test using less information.

The results of the meta-analysis should be looked upon and used with caution, because we are only able to use the published tabulated results of a much larger sample of simulation results. A considerable improvement in the reliability and the warranty to generalize the results of a meta-analysis is feasible if the full simulation results can be obtained from the authors of the respective studies. But even under those circumstances, there are still considerable "holes" in the experiments that have to be ˇ lled.

The current meta-analysis pertains only to the power of the tests, and should be complemented with an analysis dealing with the size of the tests. Moreover, given that the meta-regression model is non-linear, it may also be useful if in a future meta-analysis a sense of the "elasticity" or sensitivity of the results is developed.

A future meta-analysis should also improve on the meta-regression speciˇ cation. We account for the difference in the amount of information used for the *KR* test versus tests employing the spatial weights matrix concept, but substantial improvements are still possible. One potential topic for further investigation is the operationalization of the characteristics of the weights matrix. In the current analysis, the density and the connectivity measure are related to sample size, which complicates the interpretation of the ˇ ndings. Moreover, in future research one may want to develop a ratio scale indicator (uniformly deˇ ned and applied over studies) of the extent of heteroskedasticity present in each experiment. Preferably, such an indicator should also be used to distinguish between heteroskedasticity intrinsic to spatially autocorrelated models, and additional heteroskedasticity introduced by the experimenter. Another potential extension is concerned with misspeciˇ cation of the weights matrix. An indicator signaling the extent to which sparseness and connectedness are over- or underestimated may be helpful. A ˇ nal example relates to Kelejian and Yuzefovich's (2001) observation that the R^2 across experiments should be kept constant. Instead of implementing their suggestion in the original Monte Carlo experiments, which puts serious restrictions on the parameters that can be compared, we can artiˇ cially control for these differences by including the R^2-value of each experiment in the meta-analysis.

Acknowledgments

This chapter is a considerably extended version of a paper presented at the North-American Regional Science Association International (RSAI) conference in Santa Fe, NM, U.S.A., in 1998, and the European RSAI conference in Dublin, Ireland, in 1999. The authors would like to thank John H.L. Dewhurst, Harry H. Kelejian, and an anonymous reviewer for comments on previous versions.

3 Moran-Flavored Tests with Nuisance Parameters: Examples

Joris Pinkse

The Pennsylvania State University

3.1 Introduction

Since Moran (1950b) originally proposed his test of correlation, many authors have investigated its properties under varying conditions. In this chapter I demonstrate how new technical results of Pinkse (1999) can be used to verify that the Moran test, or a *cross-correlation* variant thereof (see Box and Jenkins, 1976, for a detailed discussion of cross-correlation in time series models), indeed has a limiting normal distribution under the null hypothesis of independence.

Many tests for spatial dependence are based on the Moran test statistic, or can be written in the form of a Moran- avored test. A prime example of a test that often takes the form of a Moran- avored test is the Lagrange Multiplier (LM) or score test (Burridge, 1980, made this observation).[1] A general discussion and many useful references can be found in Anselin (1988, 1997). Other authors who have explored LM tests in the context of spatial regression models are Anselin and Rey (1991), Anselin and Florax (1995c) and Anselin *et al.* (1996). Pinkse and Slade (1998) propose a simulation-based test in probit models.

It is also possible to test for spatial independence nonparametrically. A nonparametric test of spatial independence rejects any alternative to the null hypothesis of spatial independence provided that the sample size is big enough. A nonparametric spatial independence test can be found in Brett and Pinkse (1997), which is based on a similar test for serial independence by Pinkse (1998).

The vast literature on testing for spatial dependence further includes Anselin and Kelejian (1997), Kelejian and Robinson (1995), and King (1981).

Cliff and Ord (1972, 1973, 1981) and Sen (1976) have studied the properties of the Moran test under fairly general conditions. Sen only studies the case where the variables whose correlation structure is being investigated are observed, although he deals with a minor nuisance parameter problem arising when the mean of these variables is unobserved. Cliff and Ord (1981) also consider the case in which the variables whose correlation is to be studied are errors in a linear regression model. They formally prove that the vector of nuisance parameters, in this case the vector of regression coef˜ cients, does not affect the limiting distribution.

The Moran test is used to detect the correlation between the same variable at different locations. Pinkse's (1999) test allows for the correlation to be tested between

---

[1] Although there is a conceptual distinction between the LM and score tests, they are in fact identical.

a variable at one location and a potentially different variable at another location, i.e. cross-correlation. To my knowledge Pinkse (1999) is the first to prove rigorously that the Moran test can be applied to most problems with a finite number of nuisance parameters in a spatial context. Pinkse (1999) details general yet weak conditions under which Moran-flavored tests have a limiting normal distribution under the null hypothesis.

The primary purpose of Pinkse (1999) is to formulate general conditions under which Moran-flavored tests have a limiting normal distribution. These conditions can then be used to verify that (new) Moran-flavored tests researchers encounter or formulate in models for which asymptotic normality has not yet been rigorously established indeed have a limiting normal distribution. Here, I illustrate Pinkse's conditions in six situations of interest to researchers involved in empirical work involving spatial data.

The outline of this chapter is as follows. In Sect. 3.2, I propose the test statistic. Section 3.3 through 3.5 discuss the conditions under which asymptotic normality obtains under the null hypothesis. Section 3.3 discusses conditions on the weights matrix. In Sect. 3.4, six example models are formulated and in each case the specific relationship of the model to the conditions on the nuisance parameter structure is explored. Section 3.5.1 discusses the required moment conditions and Sect. 3.5.2 further explores the most complicated of the six models. Section 3.6 concludes. A synopsis of Pinkse's (1999) conditions is provided in the Appendix.

3.2 Test Statistics

The test statistics considered have the form:

$$\hat{\tau} = \frac{\hat{A}' W \hat{U}}{\hat{\Gamma}_n}. \tag{3.1}$$

$\hat{U}_i$ and $\hat{A}_i$ are proxies for the zero mean identically distributed sequences U_i and A_i with variances σ^2_U, σ^2_A and covariance σ_{UA}. An example could have U_i the error in a regression model and $\hat{U}_i$ its corresponding residual. W is a weights matrix, discussed in detail below. Finally, $\hat{\Gamma}_n$ is a correction factor which ensures that $\hat{\tau}$ has a limiting $N(0,1)$ distribution, namely:

$$\hat{\Gamma}_n = \sqrt{\hat{\sigma}^2_{\hat{U}} \hat{\sigma}^2_{\hat{A}} \mathrm{tr}\,(\mathrm{WW}') + \hat{\sigma}^2_{\hat{\mathrm{U}}\hat{\mathrm{A}}} \mathrm{tr}\,(\mathrm{W}^2)}.$$

Here tr is the trace operator and $\hat{\sigma}^2_{\hat{U}}, \hat{\sigma}^2_{\hat{A}}, \hat{\sigma}_{\hat{U}\hat{A}}$ are sample variances and covariance.

My test statistic differs from the traditional Moran statistic in two respects. First, U and A are unobserved and second, U can be different from A. If the variables have nonzero means, they should be demeaned first, which generates a nuisance parameter (their population mean). Pinkse (1999) obtains similar results when one (but not both) of U and A has nonzero mean and is not demeaned. Nonzero means without demeaning lead to a more complicated form of the correction factor $\hat{\Gamma}_n$. Moreover, when nuisance parameters are present, nonzero means cause the approximation error (caused by the estimation of the vector of nuisance parameters) to affect the asymptotic distribution of the test statistic in a nontrivial manner. This requires a more structured set of conditions, which is beyond the scope of this chapter, but can be found in Pinkse (1999).

Under the null hypothesis U_i is independent of A_j for all $i \neq j$, and the alternative hypothesis is that of a given correlation structure implied by W. There are correlation structures which are captured neither by the null nor by the alternative hypothesis. Behavior of $\hat{\tau}$ under such correlation structures is undetermined. It would therefore be a mistake to think of the null hypothesis as being any correlation structure other than the correlation structure implied by W. The test statistic behavior, under spatial correlation which is different from that implied by W, is different from that under independence; most results only apply under independence. Similarly, tests do not necessarily have any power against alternatives different from the alternative for which they were constructed. Often they are *consistent*, i.e., will reject with certainty in a sample of inˇ nite size, against a wider class of alternatives than for which they were constructed but hardly ever against all such alternatives. A notable exception are some *nonparametric* tests (e.g., Brett and Pinkse, 1997).

I now proceed with a discussion of the conditions that are needed for asymptotic normality under the null hypothesis. A synopsis of the formal conditions can be found in the Appendix.

3.3 Weights Matrix

An important determinant as to whether the limiting distribution of $\hat{\tau}$ is indeed normal is the weights matrix (W) chosen. The weights matrix should be chosen to re ect the suspected spatial correlation structure of the data. There are some fairly weak conditions the weights matrix must satisfy in the limit, that is when the sample size increases to inˇ nity. The conditions in Pinkse (1999) are weaker than those in Sen (1976) and are:

$$\mathrm{tr}(\mathrm{W}^2 + \mathrm{W}\mathrm{W}') = O_e(n),$$

$$\max_{t \le n} \sum_{i=1}^{n} (|w_{it}| + |w_{ti}|) = o(\sqrt{n}).$$

where o means "order of", in the sense that the ratio of the left hand side to the argument of o tends to zero when the sample size n increases to inˇ nity. O_e means "exact order of," meaning that the ratio of the left hand side over the argument of O_e is bounded away from zero and inˇ nity in the limit. It is therefore different from the related common notation O. The w_{it}'s are the elements of the W matrix. The possible dependence of the weights on the sample size is here suppressed in the notation.

Virtually all weights matrices of practical interest satisfy Pinkse's (1999) conditions on the weights matrix, which allow for negative weights, asymmetric weights matrices and for the ratio of the maximum row sum to the average row sum to increase at a rate slightly slower than $\sqrt{n}$, instead of being bounded as in Sen (1976). Negative weights are of interest when correlation between one pair of observations is thought to be of the opposite sign of another pair of observations. Asymmetric weights matrices are only of interest when $A \neq U$; the correlation between A_1 and U_2 could well be different from that between A_2 and U_1. The weakening of the ratio of row sums condition could be relevant when one, perhaps centrally located, observation (say firm) is much more strongly affected by the addition of new observations (entry of competitors) than other observations (firms).

Weight matrix conditions are not informative about the kind of weights matrices for which the test statistic is approximately normal in small samples. It is generally best to select a weights matrix which is simple in structure but is nonetheless consistent against the spatial correlation structure of interest. In particular, the small sample distribution of the test statistic will be closest to normal when the number of nonzero elements in each row and column is roughly the same and small. In practice, this means that one should generally let the weights decline rapidly (perhaps a large power or exponentially) with distance or use a distance-based weights matrix with a cut-off.

Note that "misspecifying" W in a test statistic is nowhere near as serious as misspecifying the weights matrix in a spatial regression model. In a test statistic, (minor) misspecification can render the test statistic less powerful, in a regression model it usually causes the estimator to be inconsistent.

In a test statistic, misspecifying W by choosing a simpler structure may in fact increase the power of the test (see e.g., Florax and Rey, 1995). Stetzer (1982) finds in a Monte Carlo study that, although the choice of weights matrix has an effect on the performance of estimators in spatial regression models, other factors, including delineation of the geographical area studied, tend to be more important. Griffith (1995) addresses the boundary problem, i.e. the impact on regression results of spillover effects from locations outside the geographical area studied.

3.4 Nuisance Parameters

There are many reasons for testing for spatial correlation of the errors in a regression model. Spatial error correlation may be indicative of a failure to model the spatial data structure adequately. The structure of the spatial error correlation found may be informative about possibly omitted regressors. If the structure of the spatial error correlation is known, more efficient estimation procedures can be constructed. If the errors are spatially correlated and such spatial correlation is ignored it can lead to incorrect inferences.

For the test statistic to be applied to proxies rather than unobserved variables, the relationship between proxies and unobservables needs to be described. Here several single-equation examples are discussed. In each case, in order to fit in with Pinkse (1999), a Taylor series expansion is used. The Taylor expansion is based on the notion that $\hat{U}_i = u(\xi_i, \hat{\beta})$, $U_i = u(\xi_i, \beta)$, where generally $\xi_i = (Y_i, X_i)$. Thus:

$$\begin{aligned} \hat{U}_i - U_i &= u(\xi_i, \hat{\beta}) - u(\xi_i, \beta) \\ &= D_i'(\beta - \hat{\beta}) + (\beta - \hat{\beta})' Q_i(\tilde{\beta})(\beta - \hat{\beta}), \end{aligned} \tag{3.2}$$

for,

$$D_i = D_i(\beta) = \frac{\partial u}{\partial \beta}(\xi_i, \beta), \text{ and } Q_i(\beta) = \frac{\partial^2 u}{\partial \beta \partial \beta'}(\xi_i, \beta)/2,$$

with $\tilde{\beta}$ a vector between β and $\hat{\beta}$. A similar Taylor expansion gives D_{Ai} and Q_{Ai}.

Consider the following six models.

1. Linear regression model in which spatial error correlation is to be tested:

$$Y = X\beta + U. \tag{3.3}$$

The null hypothesis is independence of the errors. One often formulates spatial error correlation as $U = \psi W U + \varepsilon$ with ε an i.i.d. vector of errors. See Anselin and Rey (1991) for an elaborate discussion. For the linear regression model, $A_i = U_i$, and $\hat{U}_i - U_i = X_i'(\beta - \hat{\beta})$, such that $D_i = X_i$ and $Q_i = 0$.

2. Spatial regression model, estimated by Maximum Likelihood, in which $\psi = 0$ is to be tested:

$$Y = \psi \tilde{W} Y + X\beta + U. \tag{3.4}$$

To test $\psi = 0$, the above model only needs to be estimated under the null hypothesis provided the score test (see Anselin, 2001a) is used. Under the null hypothesis, the model reduces to $Y = X\beta + U$ and the Maximum Likelihood estimator under normality and homoskedasticity equates to the ordinary least squares estimator. Some tedious algebra shows that under the assumption of normality and homoskedasticity, the score is $2Y'\tilde{W}'\hat{U}$ with $\hat{U}$ the ordinary least squares residuals of a regression of Y on X. In this case, $W = \tilde{W}'$, $A_i = Y_i - \mu_Y$, $\hat{A}_i = Y_i - \hat{\mu}_Y$, and $\hat{U}_i - U_i = X_i'(\beta - \hat{\beta})$. An impressive survey, which includes a discussion of spatial lag dependence is Anselin and Bera (1998).

3. Nonlinear regression model to be estimated by nonlinear least squares and errors to be tested for spatial correlation (via the residuals):

$$Y_i = \beta_1 + X_{i2}\beta_2 + X_{i3}\beta_2\beta_3 + U_i. \tag{3.5}$$

Here $A_i = U_i$, $\hat{U}_i - U_i = D_i'(\beta - \hat{\beta}) + (\beta - \hat{\beta})' Q_i(\beta - \hat{\beta})$, with $D_i' = [1, X_{i2} + X_{i3}\beta_3, X_{i3}\beta_3]$ and Q_i is a 3 by 3 matrix with the $(2,3)$ and $(3,2)$ elements equal to $X_{i3}/2$ and all other elements zero. The model formulated here is somewhat simpliˇ ed in that all third derivatives of the regression function in the direction of the coefˇ cient vector are zero. In principle, virtually all nonlinear regression models can be dealt with but a stylized one facilitates the discussion. There has been relatively little work on nonlinear spatial regression models, but the issues involved are similar to linear regression models. See Davidson and MacKinnon (1993) for an excellent exposition on nonlinear regression models outside the spatial context.

4. A probit model:

$$Y_i = I(X_i'\beta + \varepsilon_i \geq 0), \tag{3.6}$$

with I the indicator function taking the value one if its argument is true and zero if it is false. Assume normality and homoskedasticity. Again, spatial error correlation is to be tested for and the score here is $2\hat{U}'W\hat{U}$, with:

$$\hat{U}_i = u_i(\hat{\beta}) = \phi(X_i'\hat{\beta})\frac{Y_i - \Phi(X_i'\hat{\beta})}{\Phi(X_i'\hat{\beta})[1-\Phi(X_i'\hat{\beta})]} \tag{3.7}$$

with $\hat{\beta}$ the probit Maximum Likelihood estimator and Φ and ϕ the distribution and density functions of a standard normal. Let $A_i = U_i = u_i(\beta)$. Then $\hat{U}_i - U_i = D_i'(\beta - \hat{\beta}) + (\beta - \hat{\beta})'Q_i(\beta - \hat{\beta})$ with $D_i = -u_i'(\beta)$ and $Q_i = u_i''(\tilde{\beta})/2$ with $\tilde{\beta}$ some vector between β and $\hat{\beta}$. It can be shown that:

$$u_i'(\beta) = X_i\left[\phi_i'\left(\frac{Y_i}{\Phi_i} - \frac{1-Y_i}{1-\Phi_i}\right) - \varphi_i^2\left(\frac{Y_i}{\Phi_i^2} - \frac{1-Y_i}{(1-\Phi_i)^2}\right)\right],$$

$$u_i''(\beta) = X_iX_i'\left[\phi_i''\left(\frac{Y_i}{\Phi_i} - \frac{1-Y_i}{1-\Phi_i}\right) - 3\varphi_i\varphi_i'\left(\frac{Y_i}{\Phi_i^2} - \frac{1-Y_i}{(1-\Phi_i)^2}\right)\right.$$
$$\left. +2\varphi_i^3\left(\frac{Y_i}{\Phi_i^3} - \frac{1-Y_i}{(1-\Phi_i)^3}\right)\right],$$

with $\varphi_i = \varphi(X_i'\beta)$, $\Phi_i = \Phi(X_i'\beta)$. The spatial probit model has been used extensively. The standard Maximum Likelihood estimator is inconsistent in the presence of spatial error correlation because of induced heteroskedasticity. Parametric approaches particular to the spatial probit problem include McMillen (1992) and Pinkse and Slade (1998). One generic semiparametric estimator which will likely work is the maximum score estimator of Manski (1975). Manski's estimator is cumbersome to compute (see Manski and Thompson, 1986; Pinkse, 1993).

5. Estimation by the generalized method of moments (GMM: Hansen, 1982) of the regression model:

$$Y_i = g(X_i, \beta) + U_i, \tag{3.8}$$

subject to the moment condition $\mathrm{E}\{Z_1[Y_1 - g(X_1,\beta)]\} = 0$, where g is a regression function which is known up to the parameter vector β and Z_1 is a vector of instruments whose dimension is equal to or greater than that of the vector of possibly endogenous regressors X_1. See Kelejian and Robinson (1993) for an interesting recent example of the use of GMM in spatial econometrics. Now, $\hat{U}_i = Y_i - g(X_i, \hat{\beta})$. Hence:

$$\hat{U}_i - U_i\frac{\partial g}{\partial \beta'}(\beta)(\beta - \hat{\beta}) + (\beta - \hat{\beta})'\frac{\partial^2 g}{\partial\beta\partial\beta'}(\tilde{\beta})(\beta - \hat{\beta})/2,$$

where $\tilde{\beta}$ is again a vector between $\hat{\beta}$ and β. Again $A_i = U_i$.

Table 3.1. Taylor expansion components for the six models

Model	D_i	Q_i
1	X_i	0
2	X_i	0
3	$\begin{bmatrix} 1 \\ X_{i2}+X_{i3}\beta_3 \\ X_{i3}\beta_3 \end{bmatrix}$	$\begin{bmatrix} 0 & 0 & 0 \\ 0 & 0 & X_{i3}/2 \\ 0 & X_{i3}/2 & 0 \end{bmatrix}$
4	$u_i'(\beta)$	$u_i''(\tilde{\beta})/2$
5	$\frac{\partial g}{\partial \beta}(\beta)$	$\frac{\partial^2 g}{\partial\beta\partial\beta'}(\tilde{\beta})/2$
6	$\begin{bmatrix} 1 \\ X_{i2}+X_{i3}\beta_3 \\ X_{i3}\beta_3 \end{bmatrix}$	$\begin{bmatrix} 0 & 0 & 0 \\ 0 & 0 & X_{i3}/2 \\ 0 & X_{i3}/2 & 0 \end{bmatrix}$

6. A spatially autoregressive probit model. This model is given by:

$$Y^* = \psi\tilde{W}Y^* + X\beta + \varepsilon,$$
$$Y_i = I(Y_i^* \geq 0),$$

where the errors ε_i are assumed independent $N(0,1)$ and $\psi = 0$ is to be tested. The vector Y^* is *latent*, i.e., unobserved. Here the score is $2(X\hat{\beta}+\hat{U})'W\hat{U}$, with $W = \tilde{W}'$. Note that unlike in the linear regression model, $X\hat{\beta}+\hat{U} \neq Y$. Here, $A_i = \beta'(X_i - \mu_X) + U_i$ and $\hat{A}_i = \hat{\beta}'(X_i - \hat{\mu}_X) + \hat{U}_i$.

The deˇ nitions of D_i and Q_i in equation (3.2) for the various models are represented in the Table 3.1.

In model 2, $A_i = Y_i - \mu_Y$ is not observed and hence replaced with $\hat{A}_i = Y_i - \hat{\mu}_Y$. A similar Taylor expansion can be applied to the approximation of A_i by $\hat{A}_i$, namely $\hat{A}_i - A_i = D_{Ai}'(\beta_A - \hat{\beta}_A) + (\beta_A - \hat{\beta}_A)' Q_{Ai}(\beta_A - \hat{\beta}_A)$, where in this case $D_{Ai} = 1$ and $\hat{\beta}_A = \hat{\mu}_Y, \beta_A = \mu_Y$. Model 2 thus contains an example in which one of the variables whose spatial correlation is to be investigated, Y_i, is observed but has nonzero mean. It is also possible that the variable of interest has nonzero mean *and* is unobserved and needs to be proxied. An example is a spatial autoregressive model in a probit model, i.e., model 6.

Model 6 has the additional problem that $\hat{A}_i - A_i$ is somewhat complicated. A detailed discussion of this case is found in Pinkse (1999), but for here it sufˇ ces to say that:

$$D_{Ai} = -\begin{bmatrix} u_i'(\beta) + X_i - \mu_X \\ -\beta \end{bmatrix}, \quad Q_{Ai} = \begin{bmatrix} u_i''(\tilde{\beta}) & -\iota \\ -\iota' & 0 \end{bmatrix},$$

where as before $\tilde{\beta}$ denotes some vector between β and $\hat{\beta}$.

Pinkse (1999) imposes some restrictions on D_i and Q_i (and hence D_{Ai} and Q_{Ai}). The conditions apply regardless of the model, but their meaning and implications depend on the form of the model. They are discussed below.

3.5 Conditions

3.5.1 Exogeneity

A condition which must hold under the *null hypothesis*, and which is all but unavoidable, is not much weaker than that of *strict exogeneity*. The concept of strict exogeneity was introduced by Engle *et al.* (1983) and essentially says that all regressors are independent of all errors. Contrary to strict exogeneity, dependence between regressors and errors at the same location is allowed for provided that the parameter vector can be estimated under such dependence.

For instance, a linear model with heteroskedastic errors is allowed. A model with endogenous right hand side variables does not pose a problem, unless these are spatially lagged endogenous variables. An example is Model 5 where X_i does not belong to neighboring observations but includes endogenous variables other than Y_i at the same location as Y_i.

The exogeneity condition which must hold under the null hypothesis excludes the possibility of spatially lagged dependent variables. Assuming independence between errors at one location and D_i, D_{Ai} at another location cannot be avoided. In particular, it cannot be replaced by a *weak dependence* condition such as *strong mixing* (Rosenblatt, 1956) on the process $\{D_i, D_{Ai}, U_i\}$, for instance. The reason is that if there is weak dependence, the asymptotic distribution will also depend on $E(D_iU_j)$ and $E(D_{Ai}U_j)$ for all values of i, j and may be nonstandard.

3.5.2 Moment and other Conditions

Some General Issues. The discussions here will focus on D_i, Q_i, where similar conditions apply to D_{Ai}, Q_{Ai}. The D_i's can have different distributions, but must have uniformly bounded second moments for the results to go through. In models 1–3 this simply means that the regressors have finite variances, in model 4 it is implied by finite regressor variances and in model 5 it depends on the functional form of g. If g is exponentially increasing or includes high powers of the X_i's, the condition can be problematic, depending on the exact structure of g.

The conditions on Q_i are necessarily much weaker. The Q_i's can depend on the sample size, but assuming $\sqrt{n}$-consistency of $\hat{\beta}$ for β it suffices that their maximum increases (in probability) at a rate slower than $n^{3/4}$. This divergence condition is trivially satisfied for models 1 and 2, extremely weak for models 3, 4 and 6, and weak for the most common specifications for g in model 5. Indeed, for model 3, the divergence condition is implied by the existence of moments of X_{i3} greater than the $4/3$-rd moment, which was already necessary to satisfy the conditions on D_i in that model. For model 5, it suffices to have g increase (decrease) at most quadratically in the right (left) tail. It is automatically satisfied for a twice continuously differentiable function on a compact support. As an illustration, I will demonstrate the most challenging case, that of the probit model (models 4 is used, model 6 can be done similarly) here. The illustration is somewhat technical and can be skipped without loss of continuity.

Technical Illustration for the Probit Model, Model 4. First consider for arbitrary j_1, j_2 :

$$\tilde{T}_i = T_i(\tilde{\beta}) = X_{ij_1}X_{ij_2}\tilde{\phi}_i''\left(\frac{Y_i}{\tilde{\Phi}_i} - \frac{1-Y_i}{1-\tilde{\Phi}_i}\right),$$

with $\tilde{\phi}_i = \varphi(X_i'\tilde{\beta})$, $\tilde{\Phi}_i = \Phi(X_i'\tilde{\beta})$. The other terms in the deˇ nition of u_i'' can be dealt with similarly. Write $\tilde{T}_i = \tilde{T}_{i1}Y_i + \tilde{T}_{i0}(1-Y_i)$. I ˇ rst determine when:

$$P\Big(\max_{i\le n}|\tilde{T}_{i1}|Y_i \ge \alpha_n\Big) \to 0, \text{ as } n \to \infty,$$

where the conditions depend on the properties of the sequence $\{\alpha_n\}$. Here, α_n should increase at a rate slower than $n^{3/4}$, as established in the previous paragraph. First, note that $\varphi''(t) = (t^2-1)\phi(t)$. Second, note that $\Phi(t)$ is well-approximated by $-\phi(t)/t$ when t is moderate to large negative. In particular, there are three ˇ xed ˇ nite numbers $C > 0, t^* < 0$ such that $\phi(t)/\Phi(t) < Ct$ for all $t < t^*$. Thus:

$$\begin{aligned}|\tilde{T}_{i1}| &\le |X_{ij_1}X_{ij_2}|\Big\{\Big|CX_i'\tilde{\beta}[(X_i'\tilde{\beta})^2-1]\Big|I(X_i'\tilde{\beta}\le t^*) + I(X_i'\tilde{\beta} > t^*)/\Phi(t^*)\Big\}\\ &\le |X_{ij_1}X_{ij_2}|\Big\{\Big|CX_i'\tilde{\beta}[(X_i'\tilde{\beta})^2-1]\Big|I(X_i'\tilde{\beta}\le t^*) + 1\Big\}\\ &= |\tilde{T}_{i1}^*|,\end{aligned}$$

for,

$$\tilde{T}_{i1}^* = X_{ij_1}X_{ij_2}\Big\{CX_i'\tilde{\beta}[(X_i'\tilde{\beta})^2-1]I(X_i'\tilde{\beta}\le t^*) + 1\Big\},$$

where I used the fact that $\phi''(t)$ has a maximum of e^{-1} at $t = \pm\sqrt{2}$ and a minimum of 1 at $t = 0$. Let $T_{i1}^* = X_{ij_1}X_{ij_2}\{CX_i'\beta[(X_i'\beta)^2-1]I(X_i'\beta \le t^*)+1\}$. Then:

$$\begin{aligned}P\Big(\max_{i\le n}|\tilde{T}_{i1}|Y_i \ge \alpha_n\Big) &\le P\Big(\max_{i\le n}|\tilde{T}_{i1}^*|Y_i \ge \alpha_n\Big)\\ &\le P\Big(\max_{i\le n}|T_{i1}^*|Y_i \ge \alpha_n\Big) + P\Big(\max_{i\le n}|\tilde{T}_{i1}^* - T_{i1}^*|Y_i \ge \alpha_n\Big).\end{aligned}$$

Now:

$$P\Big(\max_{i\le n}|T_{i1}^*|Y_i \ge \alpha_n\Big) \le \sum_{i=1}^{n} P(|T_{i1}^*|Y_i \ge \alpha_n) = \sum_{i=1}^{n} P(|T_{i1}^*|\Phi_i \ge \alpha_n) \to 0,$$

exponentially because $\{|T_{i1}^*|\,\Phi_i\}$ is uniformly bounded.

Now $P\left(\max_{i\le n}\left|\tilde{T}_{i1}^* - T_{i1}^*\right|Y_i \ge \alpha_n\right)$. The difference $\tilde{T}_{i1}^* - T_{i1}^*$ depends on a sum over products of functions of β and the difference between functions evaluated at $\tilde{\beta}$ and the same functions evaluated at β. A typical example of such a term is:

$$X_{ij_1}X_{ij_2}CX_i'\beta[(X_i'\beta)^2-1][I(X_i'\tilde{\beta}\le t^*) - I(X_i'\beta \le t^*)].$$

Now:

$$A \le B + C + D + E, \tag{3.9}$$

where,

$$A = \alpha_n^{-1} \max_{i \le n} \left| X_{ij_1} X_{ij_2} C X_i' \beta [(X_i' \beta)^2 - 1][I(X_i' \tilde{\beta} \le t^*) - I(X_i' \beta \le t^*)] \right|$$

$$B = \alpha_n^{-1} \max_{i \le n} \left| X_{ij_1} X_{ij_2} C X_i' \beta [(X_i' \beta)^2 - 1] \right| [I(X_i' \tilde{\beta} \le t^*) I(X_i' \beta > 0)]$$

$$C = \alpha_n^{-1} \max_{i \le n} \left| X_{ij_1} X_{ij_2} C X_i' \beta [(X_i' \beta)^2 - 1] \right| [I(X_i' \tilde{\beta} \le t^*) I(t^* < X_i' \beta \le 0)]$$

$$D = \alpha_n^{-1} \max_{i \le n} \left| X_{ij_1} X_{ij_2} C X_i' \beta [(X_i' \beta)^2 - 1] \right| [I(X_i' \tilde{\beta} > t^*) I(2t^* \le X_i' \beta < t^*)]$$

$$E = \alpha_n^{-1} \max_{i \le n} \left| X_{ij_1} X_{ij_2} C X_i' \beta [(X_i' \beta)^2 - 1] \right| [I(X_i' \tilde{\beta} > t^*) I(X_i' \beta < 2t^*)].$$

Clearly, $X_i' \beta [(X_i' \beta)^2 - 1]$ is bounded in any ˇ nite neighborhood of $X_i' \beta = t^*$. So, the second and third right hand terms are bounded by:

$$\Theta \alpha_n^{-1} \max_{i \le n} \left| X_{ij_1} X_{ij_2} \right|,$$

for some ˇ xed $\Theta > 0$. For $\Theta \alpha_n^{-1} \max_{i \le n} \left| X_{ij_1} X_{ij_2} \right|$ to converge to zero in probability, a fairly weak moment condition on the regressors sufˇ ces. For the ˇ rst and fourth terms in the last displayed equation $|X_i' (\tilde{\beta} - \beta)| > t^*$. But:

$$P \left\{ \alpha_n^{-1} n^{-1/2} \max_{i \le n} \left| X_{ij_1} X_{ij_2} C X_i' \beta [(X_i' \beta)^2 - 1] \right| \|X_i\| \right\} \to 0,$$

as $n \to \infty$ can also be satisˇ ed by a fairly weak moment condition.

3.6 Conclusions

In this chapter, I have discussed the conditions derived in Pinkse (1999) under which the Moran test, or cross-correlation variations thereof, have a limiting normal distribution under the null hypothesis, both on raw data and in the presence of nuisance parameters. Their impact is illustrated using six models frequently encountered in empirical work involving spatial data.

Because of the level of generality of the Pinkse (1999) results, the conditions are sometimes easy to verify and sometimes they do take some work. In the end, most conditions are moment conditions on model variables, conditions on the convergence rate of the parameter estimators, but usually a combination of both. Even when the conditions are relatively cumbersome to verify, it is far easier than proving asymptotic validity of the test from scratch, which can equate to formulating the Pinkse (1999) proofs for a speciˇ c case.

Acknowledgments

This research was ˇ nancially supported by the Social Sciences and Humanities Research Council of Canada. I thank the editors and one anonymous referee for useful comments. I thank Jennifer Innes for editorial suggestions.

Appendix: Synopsis of Conditions

All conditions listed here only apply under the null hypothesis.

A1.1 For Asymptotic Normality of Raw Data Statistic

In the absence of nuisance parameters, the following conditions are sufˇ cient for asymptotic normality of the test statistic under the null hypothesis. Below, tr denotes the trace operator (sum of eigenvalues or equivalently, sum of diagonal elements), and $|W|$ denotes the matrix whose elements are the absolute values of the elements of W.

1. A_i and U_i have moments greater than two.
2. W has diagonal elements equal to zero, $n^{-1}\mathrm{tr}(W^2 + WW')$ converges to a non-zero constant:

$$n^{-2}\mathrm{tr}(|\mathrm{W}||\mathrm{W}||\mathrm{W}'||\mathrm{W}'|)\mathrm{arrow}0,$$

   and,

$$n^{-1/2}\max_{t\leq n}\sum_{i=1}^{n}(|w_{it}|+|w_{ti}|)arrow0,\ \text{as narrow}\infty.$$

   In the special case in which A_i has mean different from zero, in addition:

$$n^{-1}\|W'\iota\|^2, \text{ and } n^{-1}\||W'|\iota\|$$

   converge to positive constants.

A1.2 Nuisance Parameters when A_i has Zero Mean

These additional conditions are needed for asymptotic normality in the presence of nuisance parameters provided A_i has zero mean.

3. D_i and D_{Ai} as deˇ ned in the Taylor expansions in equation (3.2) have ˇ nite second moments and are independent of (U_j, A_j) for all $j \neq i$.
4. The maximum over the largest elements (in absolute value) of $Q_i(\tilde{\beta})$ and $Q_{Ai}(\tilde{\beta})$, also deˇ ned in the Taylor expansions, increase with the sample size at a rate no faster than q_n which satisˇ es $n^{1/4}z_n^2 q_n \to 0$,as $n \to \infty$, where z_n is the convergence rate of $\hat{\beta}$ (most commonly $n^{-1/2}$), where z_n must satisfy $n^{1/4}z_n \to 0$ as $n \to \infty$.

A1.3 Nuisance Parameters when A_i has Non-Zero Mean

When A_i has mean different from zero, the correction factor is different and conditions are more difˇ cult to express. None of the examples in this chapter have A_i have mean different from zero. Even if A_i did have mean different from zero, one can often demean A_i ˇ rst. Please refer to Pinkse (1999) for an in-depth discussion of the issues.

4 The In uence of Spatially Correlated Heteroskedasticity on Tests for Spatial Correlation

Harry H. Kelejian[1] and Dennis P. Robinson[2]

[1] University of Maryland
[2] University of Arkansas at Little Rock

4.1 Introduction

In cross sectional regression models the possibility of spill-overs between neighboring units is increasingly being recognized in both the theoretical and applied literature.[1] Within a regression framework, typically recognized forms of such spill-overs relate to the model's dependent and independent variables, as well as to the error terms. General issues relating to spill-overs suggest that the model's error terms may be spatially correlated. Because the statistical properties of the regression parameter estimators depend upon whether or not the error terms are indeed spatially correlated, tests for such correlation are frequently considered.[2]

By far the most frequently considered test for spatial correlation is the test based on Moran's I statistic which is formulated in terms of regression residuals (see Cliff and Ord, 1972; Moran, 1950a). Under standard conditions, this test is locally best invariant (King, 1981). In addition, if the error terms are normally distributed the exact small sample distribution of Moran's I can, somewhat tediously, be determined (see e.g., Tiefelsdorf and Boots, 1995). Therefore, an exact small sample test can be considered. However, in practice an approximate computationally simple test is typically considered which is based on the asymptotic distribution of Moran's I under the null hypothesis of error independence (see e.g., Cliff and Ord, 1973; Sen, 1976; Terui and Kikuchi, 1994), and in a framework involving endogenous regressors (Anselin and Kelejian, 1997). Monte Carlo studies suggest that in many cases, these large sample tests have considerable power, and typically more so than other tests which are considered (see e.g., Bartels and Hordijk, 1977; Anselin and Rey, 1991; Anselin and Florax, 1995c; Kelejian and Robinson, 1995).

[1] Some recent theoretically oriented studies are Kelejian and Prucha (1999), Anselin *et al.* (1996), Anselin and Kelejian (1997), Kelejian and Robinson (1997), Brett and Pinkse (1997), and LeSage (1997a). Some recent studies which are primarily applied in nature are Case (1991), Case *et al.* (1993), Holtz-Eakin (1994), Shroder (1995), and Kelejian and Robinson (1998). Classic references are Cliff and Ord (1973, 1981) , Anselin (1988b), and Cressie (1993).

[2] An example relating to this is given in DeLong and Summers (1991). See also Dubin (1988), and Anselin and Kelejian (1997).

Tests for spatial correlation based on Moran's I assume the absence of heteroskedasticity.[3] In a Monte Carlo framework, Anselin and Griffith (1988) gave results which suggest that such tests may have some power (but weak) against heteroskedasticity; in another study Kelejian and Robinson (1995) gave Monte Carlo results which suggest the opposite in that they detected a slight loss of power. To date, there are no theoretical results which describe the influence of heteroskedasticity on tests for spatial correlation.

The purpose of this chapter is to provide theoretical results which describe the influence of heteroskedasticity on the asymptotic version of the test for spatial correlation which is based on Moran's I statistic (henceforth, MI). Because, under typical assumptions,[4] MI is identical to the Lagrangian multiplier test for spatial correlation (henceforth, LM) our results relate to LM as well. Interestingly, it turns out that the effect of heteroskedasticity on MI and LM depends upon whether or not that heteroskedasticity itself is spatially correlated, and, furthermore, whether that correlation is, in a manner to be defined, positive or negative. For instance, suppose a model's error term is heteroskedastic because its variance, conditional on the regressors of the model, is related to a certain variable. Suppose also that, unconditionally, the variable in question is spatially correlated. As one example, suppose the variable in question is income per capita. Then, one might not expect income per capita to be independently distributed over the cross sectional units. In such a case, the extent of heteroskedasticity would be spatially correlated. If, as an illustration, income per capita is positively spatially correlated in the sense that neighboring areas tend have similar incomes, then the extent of heteroskedasticity between neighboring units would be positively spatially correlated. Alternatively, if heteroskedasticity relates to a productivity index for a particular set of goods, that heteroskedasticity could be negatively spatially correlated if neighboring areas specialize in the production of different sets of goods and the productivity index in question is positively related to the degree of specialization. [5]

Our theoretical results suggest that MI and LM remain valid even if the error terms are heteroskedastic, as long as that heteroskedasticity is not itself spatially correlated. If it is, its effect on MI and LM depends upon whether that correlation is positive or negative. If it is positive, our results imply that a researcher is more likely to conclude that the error terms are spatially correlated, when they are not; the reverse is true if it is negative.

These results are important for at least two reasons. First, heteroskedasticity is often overlooked when testing for spatial correlation via MI or LM. If there is

[3] There are, of course, tests for "error term problems" that consider the possibility that the error terms may be both spatially correlated and heteroskedastic (see e.g., Anselin *et al.*, 1996; Kelejian and Robinson, 1997).

[4] The typical assumptions considered are model linearity, normality of the error term, the absence of spatial lags, and the absence of endogenous variables (see e.g., Burridge, 1980). Anselin and Kelejian (1997) show that the equivalence holds even if the model contains endogenous variables, as long as it does not contain spatially lagged dependent variables.

[5] We define positive and negative spatially correlated heteroskedasticity in a more formal way in Sect. 4.2.

heteroskedasticity it is reasonable to assume that it may be spatially correlated. If so, researchers may be lead to false conclusions concerning whether or not their error terms are spatial correlated. Therefore, inferences would be in error.

Secondly, our results suggest that if the error terms of a model are heteroskedastic a complete description of that model should entail the possible spatial correlation of that heteroskedasticity. Such an analysis would reveal the interactive nature of the model's uncertainty over neighboring units, and hence should be of interest in and of itself!

Finally we suggest a modiˇ cation of MI which, under reasonable conditions, should be valid whether or not the error terms are heteroskedastic, and whether or not that heteroskedasticity is spatially correlated.[6]

The model is speciˇ ed in Sect. 4.2; this section also contains a discussion of each assumption made. Our main results are given in Sect. 4.3. A summary and suggestions for further research are given in Sect. 4.4. Technical details are relegated to the Appendix.

4.2 The Model

For simplicity of presentation, in this section we specify a linear regression model which does not contain a spatial lag of the dependent variable, or other variables which must be viewed as endogenous (see e.g., Anselin and Kelejian, 1997; Kelejian and Robinson, 1997). Our central results are given in terms of this model. We then describe why those results generalize to models which contain endogenous variables which are not spatially lagged dependent variables. The analysis involving spatially lagged dependent variables is more complex and beyond the scope of this chapter.

Consider the model:

$$y = X\beta + \varepsilon, \tag{4.1}$$

$$\varepsilon = \rho W_n \varepsilon + D_\sigma^{1/2} u, \tag{4.2}$$

where y is an n by 1 vector of observations on the dependent variable, X is an n by k matrix of observations on k exogenous variables, β is a corresponding k by 1 vector of parameters, ρ is a scalar autoregressive parameter, W_n is a weights matrix, D_σ is an n by n diagonal matrix whose ith diagonal element is σ_i^2, and $u = (u_1, ..., u_n)'$ is a stochastic n by 1 vector. The subscript n on the weights matrix is meant to indicate the size of the matrix.

Our formal assumptions are given below. At this point we note that, essentially, the researcher wishes to test $H_0 : \rho = 0$ against $H_1 : \rho \neq 0$. In doing this the researcher assumes that u_i is *i.i.d.* $(0,1)$ and $D_\sigma = \sigma^2 I_n$ – i.e., he assumes that the elements of $D_\sigma^{1/2} u$ in equation 4.2 are homoskedastic, when they are not unless $\sigma_i^2 = \sigma^2, i =$

[6] Instead of considering a robust test for spatial correlation with respect to heteroskedasticity, one could also consider joint tests for both of these problems. For a very nice description of many joint tests for error term problems see Anselin *et al.* (1996), Anselin and Kelejian (1997), and Kelejian and Robinson (1997).

$1,...,n$. Our results relate to the effect that heteroskedasticity has on the test of H_0 against H_1. In doing this we consider the possibility that σ_i^2 itself may be spatially correlated. As an example, σ_i^2 may depend upon a variable which, as described further below, is spatially correlated. Finally, our list of assumptions, except for four, are a subset of the assumptions made in Anselin and Kelejian (1997) in their model which involved endogenous regressors. The four "new" assumptions relate to the nature of the heteroskedasticity, which was not considered in Anselin and Kelejian (1997). For the reader's convenience, we give a brief discussion of each assumption. A more complete discussion of the assumptions which were made in Anselin and Kelejian (1997) can be found in their study.

4.2.1 Statistical Assumptions and Interpretations

Except for the presence of heteroskedasticity, our model is a special case of the one considered in Anselin and Kelejian (1997). Because we are, as were Anselin and Kelejian (1997), interested in the asymptotic distribution of Moran's I statistic, the list of our assumptions which do not relate to heteroskedasticity, namely Assumptions 1–6 below, is a subset of the list of assumptions made by Anselin and Kelejian (1997).[7] A detailed discussion of these assumptions is given in Anselin and Kelejian (1997). Thus, in order to avoid repetition, our discussion relating to Assumptions 1–6 below is "brief."

The following notation will be used throughout this chapter. In general, let A_r be a matrix. Then, we will denote its i,jth element as $a_{r,ij}$, its ith row as $a_{r,i.}$, and its jth column as $a_{r,.j}$. Similarly, if v_r is a vector, we denote its ith element as $v_{r,i}$. As an illustration of this notation, some of our assumptions below relate to the matrix $M_n = D_\sigma^{1/2} W_n D_\sigma^{1/2}$, where D_σ is specified in (2); thus, $m_{n,ij} = w_{n,ij}\sigma_i\sigma_j$.

Let B be an n by n matrix, and let the above notation extend to B in an obvious way – i.e., its i,jth element is b_{ij}. Then, we will say that B is "absolutely uniformly summable" if:

$$\max_{1\le i\le n} \sum_{j=1}^{n} |b_{ij}| \le c_B \text{ and } \max_{1\le j\le n} \sum_{i=1}^{n} |b_{ij}| \le c_B \text{ for all } n, \tag{4.3}$$

where c_B is a finite constant which does not depend upon n.[8] For future reference we note that if B_1 and B_2 are n by n matrices which are "absolutely summable", then so is $B_3 = B_1B_2$. We also note that if L' is a g by n matrix whose elements are bounded for all n, and B is defined as above, then the elements of $n^{-1}L'BL$ are also

[7] Our assumptions are a subset of the assumptions made by Anselin and Kelejian (1997) because, unlike our model, theirs contained endogenous variables as well as spatially lagged dependent varaibles.

[8] For simplicity of presentation, we have presented our discussion in terms of square matrices. A more general presentation is given in Kelejian and Prucha (1999). On a somewhat intuitive level, we define a matrix to be absolutely uniformly summable if all of the sums of the absolute values of the elements in each row can be bounded by the same finite constant which does not depend upon n, and similarly for the columns of the matrix.

bounded for all n (see e.g., Kelejian and Prucha, 1999). Given these preliminaries, our list of assumptions are speciˇ ed below.

Assumption 1 *$w_{n,ij}$ does not depend upon n and so $w_{n,ij} = w_{ij}$ for all $n > 1$. Furthermore, $|w_{ij}| \leq c_w < \infty$ for $i, j = 1, ..., n$ and $n > 1$, where c_w is a ˇnite constant.*

This assumption implies that the elements W_n do not depend upon the sample size, and are bounded in absolute value by c_w. Therefore our large sample analysis is conditional upon a given sequence of weights matrices. One scenario which is consistent with this is the one in which the sample increases by augmentation – e.g., all the cross sectional units in a sample of size $n+1$, except for one, are represented in the sample of size n. A violation would be the case in which the sample of size $n+1$ corresponds to $n+1$ units randomly drawn, without replacement, from the population of all possible units. In this case all (or even none) of the units represented in the sample of size n need be represented in the sample of size $n+1$.

Assumption 2 *Let r_n be the number of rows in W_n that consist entirely of zero elements. Then, $0 \leq r_n \leq \lambda_1$ for all n, where λ_1 is a ˇnite constant.*

Essentially, this assumption rules out the case in which a researcher assumes that spatial correlation may be a problem but then speciˇ es a weights matrix that implies, in large samples, an unbounded number of error terms are independent of all others.

Assumption 3 *$w_{ij} \neq 0$ if and only if $w_{ji} \neq 0$. However, w_{ij} and w_{ji} need not be equal.*

This assumption implies that if the jth unit is viewed as a neighbor of the ith, then the ith unit is viewed as a neighbor of the jth. Hence, a violation of this assumption would be the case in which spill-overs are "causal" in that they are one directional.

Assumption 4 *The sequence of weights matrices W_n satisfy the following constraints:*

(a) $w_{i,i+j} = 0$ and $w_{i.}w'_{(i+j).} = 0$, for all i and $j > \lambda_2$, where $1 < i+j \leq n$, $n > 1$, and where λ_2 is a ˇnite constant.

(b) $w_{ii} = 0$, for all $i = 1, ..., n$ and $n > 1$.

(c) $\lim_{n\to\infty} n^{-1} \sum_{i=1}^{n} \sum_{j=1}^{n} w_{ij} = s_{1w}$, where s_{1w} is a ˇnite constant.

(d) $\lim_{n\to\infty} n^{-1} \text{tr}[(W_n + W'_n)(W_n + W'_n)] = s_{2w}$, where s_{2w} is a ˇnite constant.

Part (a) of Assumption 4 implies that, regardless of the sample size, a given error term is directly related to at most λ_2 "neighboring" error terms, none of which are further from it than λ_2 units in the sample. It also implies that two error terms will not have any "neighbors" in common if they are sufficiently far apart. Part (b) is a normalization of the model that implies that no unit is its own neighbor. Parts (c) and (d) are standard conditions in large sample analysis of spatial models, (see e.g., Cliff and Ord 1981, p. 19; Anselin and Kelejian, 1997) which limit the size of the elements of W_n.

Assumption 5 *The innovations u_i are independently and identically distributed, (i.i.d.), with mean* $\mathrm{E}(u_i) = 0$*, unit variance* $\mathrm{E}(u_i^2) = 1.0$*, and finite fourth moment* $\mathrm{E}(u_i^4) = \mu_4$.

Our analysis will focus on the large sample distribution of Moran's I statistic under the null hypothesis $H_0 : \rho = 0$. In this case $\varepsilon = D^{1/2}u$. In the absence of heteroskedasticity $\sigma_i^2 = \sigma^2, i = 1, ..., n$ and so under H_0 and Assumption 5 the elements of ε will be exactly as specified in Anselin and Kelejian (1997). The variance of u_i is taken to be unity without loss of generality. For example, if u_i were $(i.i.d)$ with mean and variance $(0, \sigma_u^2)$, then given $\rho = 0$, ε_i would be independently distributed with mean and variance $(0, \sigma_u^2\sigma_i^2) \equiv (0, \pi_i^2)$, where $\pi_i^2 = \sigma_u^2\sigma_i^2$ – i.e., σ_u^2 would be an unidentified scale factor.

Assumption 6 *X is nonstochastic, and rank$(X) = k$. Also, $|x_{ij}| \leq c_x$ where c_x is a finite constant, and $\lim_{n\to\infty} n^{-1}X'X = Q_{xx}$, where Q_{xx} is a finite nonsingular matrix, $i = 1, ..., n$ and $j = 1, ...k$.*

This assumption implies that the analysis is conditional on the realized values of the exogenous regressors. Furthermore, perfect multicollinearity is excluded by the rank condition. Finally, the bound of the elements of X and the limit condition are typical in large sample analysis (see e.g., Schmidt 1976, chapter 2; Kelejian and Prucha 1999).

As indicated above, Assumptions 1–6, or their equivalent, were also made by Anselin and Kelejian (1997) (among others). Assumptions 7–10 below are the additional assumptions we make in order to account for heteroskedasticity in determining the asymptotic distribution involved.

Assumption 7 *The diagonal elements of the matrix D_σ in (2) are such that*

(a) $0 < b_1 < \sigma_i^2 < b_2 < \infty$, $i = 1, 2...,$ *where b_1 and b_2 are constants.*

(b) $\lim_{n\to\infty} n^{-1} \sum \sigma_i^2 = \bar{\sigma}^2$, *where* $\bar{\sigma}^2 \neq 0$.

Part (a) of this assumption essentially specifies the variances as bounded constants, which are bounded away from zero. These are reasonable specifications because variances are typically assumed to be finite and bounded;[9] furthermore, variances that are zero effectively imply the absence of the corresponding error term. Part (b) seems reasonable in that, unless the sequence of variances is "peculiar", its average should converge in the limit. One such peculiar sequence would be: $(a,b,b,c,c,c,d,d,d,d,...)$.

Assumption 8 *$\lim_{n\to\infty} n^{-1}X'D_\sigma X = Q_{XDX}$, where Q_{XDX} is a finite nonsingular matrix.*

This is a standard condition in large sample theory involving regression models whose error terms are either heteroskedastic, autocorrelated, or both (see e.g., Schmidt, 1976, chapter 2; Judge *et al.*, 1985, chapter 5).

Assumption 9 *Let $v_i = \sigma_i^2 - \bar{\sigma}^2$, $i = 1,...,n$, and $D_v = \mathrm{diag}_{i=1}^n(v_i)$. Then, we assume*

(a) $\lim_{n\to\infty} n^{-1}\mathrm{tr}(W_nD_vW_n) = 0$; $\lim_{n\to\infty} n^{-1}\mathrm{tr}(W_nD_vW_n') = 0$

(b) $\lim_{n\to\infty} n^{-1}\mathrm{tr}(D_vW_nD_vW_n) = h_1$, *where h_1 is a finite constant which is not necessarily zero;*

(c) $\lim_{n\to\infty} n^{-1}\mathrm{tr}(D_vW_nD_vW_n') = h_2$, *where h_2 is a finite constant which is not necessarily zero.*

The three conditions in Assumption 9 are reasonable. To see this first note that Part (b) of Assumption 7 implies:

$$\lim_{n\to\infty} n^{-1}\sum_{i=1}^{n} v_i = 0. \tag{4.4}$$

Therefore, in a sense, v_i can be viewed as a "variance residual". Now note that:

$$\begin{aligned}\lim_{n\to\infty} n^{-1}\mathrm{tr}(W_nD_vW_n) &= \lim_{n\to\infty} n^{-1}\textstyle\sum_{i=1}^n v_i(w_{i.}w_{.i})\\ &= \lim_{n\to\infty} n^{-1}\textstyle\sum_{i=1}^n(\sigma_i^2-\bar{\sigma}^2)(w_{i.}w_{.i}-\bar{\delta}),\end{aligned} \tag{4.5}$$

where $\bar{\delta} = n^{-1}\sum_{i=1}^n(w_{i.}w_{.i})$. It follows that $n^{-1}\mathrm{tr}(W_nD_vW_n)$ can be viewed as the sample correlation between v_i and $(w_{i.}w_{.i})$. Similarly, the second assumption of Part (a) relates to the sample correlation between v_i and $(w_{i.}w_{i.}')$. Thus, the limiting conditions in Part (a) of Assumption 9 are reasonable unless the variances are somehow

[9] As an example of a violation, suppose $\sigma_i^2 = i$, $i = 1,...,n$. In this case each variance would be finite but they would not be bounded since $\sigma_n^2 \to \infty$ as $n \to \infty$.

correlated with the corresponding row/column and row/row products $(w_{i.}w_{.i})$ and $(w_{i.}w'_{i.})$.[10]

Now consider Part (b). The interpretation of this limiting condition is more complex because it involves quadratic terms in the variance residuals. Fortunately, a rather straightforward interpretation is available in a random parameter framework, which we now describe. It will become clear that the reasonableness of Part (b) of Assumption 9 does not depend upon the random parameter speciˇ cation.

Suppose that $\sigma_i^2, 1 = 1, ..., n$ is randomly determined and its mean is $\bar{\sigma}^2$:

$$E(\sigma_i^2) = \bar{\sigma}^2, i = 1, ..., n. \tag{4.6}$$

As above, let $v_i = \sigma_i^2 - \bar{\sigma}^2$ and note, in this setting, that $E(v_i) = 0$, $i = 1, ..., n$. Let the covariance between σ_i^2 and σ_j^2 be $c_{vij} = E(v_i v_j) = E(\sigma_i^2 - \bar{\sigma}^2)(\sigma_j^2 - \bar{\sigma}^2)$. Finally, let C_{vi} be the diagonal matrix whose diagonal elements are $c_{vi1}, c_{vi2}, ..., c_{vin}$:

$$C_{vi} = \begin{Bmatrix} c_{vi1} & 0 & . & . & . & . & 0 \\ 0 & c_{vi2} & 0 & . & . & . & 0 \\ . & 0 & . & 0 & . & . & 0 \\ . & . & . & . & & & . \\ . & . & . & . & & & . \\ . & . & . & & & . & 0 \\ 0 & 0 & 0 & . & . & . & c_{vin} \end{Bmatrix}. \tag{4.7}$$

Given these speciˇ cations and notation, consider the sum in Part (b) of Assumption 9 and note that:

$$n^{-1}\mathrm{tr}(D_v W_n D_v W_n) = n^{-1}\sum_{i=1}^{n} w_{i.}(v_i D_v) w_{.i}. \tag{4.8}$$

In light of (8) it follows, in this setting, that:

$$\begin{aligned} \mathrm{E}[n^{-1}\mathrm{tr}(D_v W_n D_v W_n)] &= n^{-1}\sum_{i=1}^{n} w_{i.}\mathrm{E}(v_i D_v) w_{.i} \\ &= n^{-1}\sum_{i=1}^{n} w_{i.} C_{vi}\, w_{.i}. \end{aligned} \tag{4.9}$$

Note from (4.9) that C_{vi} is diagonal and its jth diagonal element is the covariance between $v_i = \sigma_i^2 - \bar{\sigma}^2$ and $v_j = \sigma_j^2 - \bar{\sigma}^2$. If the heteroskedasticity is spatially correlated the elements of C_{vi} need not be zero. Thus, for example, if the heteroskedasticity is predominately positively (negatively) spatially correlated, the sum in equation (4.9), which corresponds to h_1, would (for large n) be positive (negative) if the elements of the weighting matrix are (as typically speciˇ ed) nonnegative. In the absence of spatial correlation of the heteroskedasticity, the only nonzero element of C_{vi} would be

[10] We account for a more general version of Part (a) of Assumption 9 in Sect. 4.3.2.

its ith diagonal element, $i = 1,...,n$. However the ith diagonal element of C_{vi} would be of no consequence in the sum because the ith element of both $w_{i.}$ and $w_{.i}$ are zero – see Assumption 4. Therefore, in this case we would expect $h_1 = 0$.

The same analysis can be applied to the expression in Part (c) of Assumption 9 by noting that:

$$n^{-1}\mathrm{tr}(D_v W_n D_v W_n') = n^{-1}\sum_{i=1}^{n} w_{i.}(v_i D_v)w_{i.}'. \tag{4.10}$$

Therefore, our expectations concerning h_2 are the same as those corresponding to h_1. Our expectations concerning both h_1 and h_2 are summarized in (4.11):

$$\begin{aligned} h_1 > 0 \text{ and } h_2 > 0 &\text{ if covariances are predominately positive,} \\ h_1 = h_2 = 0 &\text{ if covariances are predominately zero,} \\ h_1 < 0 \text{ and } h_2 < 0 &\text{ if covariances are predominately negative.} \end{aligned} \tag{4.11}$$

Recall that $M_n = D_\sigma^{1/2} W_n D_\sigma^{1/2}$. Our ˇ nal assumptions relate to M_n.

Assumption 10 *(a)* $\lim_{n\to\infty} n^{-1}\sum_{i=1}^{n}\sum_{j=1}^{n} m_{ij} = s_{1m}$, *where* s_{1m} *is a ˇ nite constant.*

(b) $\lim_{n\to\infty} n^{-1}\mathrm{tr}[(M_n + M_n')(M_n + M_n')] = s_{2m}$, *where* s_{2m} *is a ˇ nite constant.*

Clearly this assumption corresponds to Parts (c) and (d) of Assumption 4 and should hold because each element of M_n is just a scaled version of the corresponding element of $W_n : m_{ij} = w_{ij}\sigma_i\sigma_j$.

4.3 Basic Results

4.3.1 Standard Cases

Consider Moran's I statistic which is formulated in terms of least squares residuals:

$$I = \frac{\hat{\varepsilon}' W_n \hat{\varepsilon}}{\hat{s}_{1w}(\hat{\varepsilon}'\hat{\varepsilon}/n)}, \tag{4.12}$$

where,

$$\hat{\varepsilon} = y - X\hat{\beta}, \hat{\beta} = (X'X)^{-1}X'y,$$

$$\hat{s}_{1w} = n^{-1}\sum_{i=1}^{n}\sum_{j=1}^{n} w_{ij}.$$

Then the proof of the following theorem is given in the Appendix.

Theorem 1. *Assume that y is generated by the model speciˇed in Sect. 4.2, and Assumptions 1–10 hold. Then, under* $H_0 : \rho = 0$*:*

$$n^{-1/2} I \xrightarrow{D} N(0, \sigma_I^2), \tag{4.13}$$

where,

$$\sigma_I^2 = \frac{\bar{\sigma}^4 s_{2w} + 2h_1 + 2h_2}{2\, s_{1w}^2 \bar{\sigma}^4},$$

and where s_{1w}, s_{2w}, h_1 *and* h_2 *are speciˇed in Assumptions 4 and 9.*

Remark 1. Theorem 1 indicates that Moran's I statistic is, under H_0, asymptotically normally distributed even if the disturbance terms are heteroskedastic. Furthermore, if the heteroskedasticity is not spatially correlated, $h_1 = h_2 = 0$ (see equation 4.11), and hence the variance of that distribution, σ_I^2, reduces to $s_{2w}/2s_{1w}$. This variance is exactly the same as the one given in Anselin and Kelejian (1997, p. 163)[11] for the case in which the disturbance terms are homoskedastic. It follows that the asymptotic distribution of Moran's I is the same whether or not the disturbance terms are heteroskedastic, as long as that heteroskedasticity is not spatially correlated. This implies that the standard tests for spatial correlation based on Moran's I, or the LM statistic, are valid even if there is heteroskedasticity as long as it is not spatially correlated. For later reference, we note that the standard test based on Moran's I assuming homoskedasticity would be:

$$\text{Reject } H_0 : \rho = 0 \text{ if} : \left| \frac{n^{-1/2} I}{(\hat{s}_{2w}/2\hat{s}_{1w})^{1/2}} \right| > 1.96, \tag{4.14}$$

where,

$$\hat{s}_{2w} = n^{-1} \mathrm{tr}[(W_n + W_n')(W_n + W_n')].$$

Remark 2. Assume now that heteroskedasticity is present, and it is predominately positively spatially correlated so that $h_1 > 0$ and $h_2 > 0$. Suppose also that the standard test in (4.14) is considered which is based on the assumption of homoskedasticity. In this case one would expect the empirical type one error to exceed the theoretical type one error. The reason for this is that the standard deviation which is being considered, say $sd = [s_{2w}/2s_{1w}]^{1/2}$, is less than the one which should be considered, namely σ_I, which is deˇned by (4.13). For example, let $\alpha = \sigma_I/sd$ and note that $\alpha > 1$. Then, in the large sample it follows from (4.13) that:

$$\begin{aligned} \text{Prob}\left(\left| \frac{n^{-1/2} I}{sd} \right| > 1.96 \right) &= \text{Prob}\left(\left| \frac{I\alpha}{\sigma_I} \right| > 1.96 \right) \\ &= \text{Prob}\left(\left| \frac{I}{\sigma_I} \right| > \frac{1.96}{\alpha} \right) \\ &> 0.05. \end{aligned} \tag{4.15}$$

[11] To see this note that Anselin and Kelejian (1997) demonstrate that the term A in their equation (4.11) is zero if the model does does not contain a spatial lag.

Thus, if a researcher ignores heteroskedasticity which is predominately positively correlated, that researcher is more likely to conclude that his error terms are spatially correlated even though they are not.

Remark 3. Clearly in the above framework, if the heteroskedasticity is predominately negatively spatially correlated, the reverse will be true – i.e., the empirical type one error should be less than the theoretical type one error.

Remark 4. Consider now the case in which the regression in (4.1) is expanded to include endogenous regressors, but no spatially lagged dependent variables. Assume also that the equations determining these endogenous regressors do not contain spatially lagged dependent variables, or spatially correlated error terms. Finally, assume that a set of instruments is available which can be used to estimate (4.1), and that set of instruments satisˇ es the conditions speciˇ ed in Anselin and Kelejian (1997). Then, in the Appendix we demonstrate that the result in (4.13) still holds – i.e., our results are not effected by the presence of endogenous variables!

4.3.2 A Heteroskedastic Robust Version of MI

Although Part (a) of Assumption 9 is very reasonable it may not hold for some models. Therefore, in giving a heteroskedastic robust version of the spatial correlation test based on Moran's I statistic we do not maintain Part (a) of Assumption 9. Instead, we only assume

Assumption 11 $\lim_{n\to\infty} n^{-1}\mathrm{tr}(W_n D_v W_n) = h_3$; $\lim_{n\to\infty} n^{-1}\mathrm{tr}(W_n D_v W_n') = h_4$ *where* h_3 *and* h_4 *are ˇ nite constants, which may or may not be zero.*

It should be clear from Preliminary 4 and the proof of Theorem 1 in the Appendix that under Assumption 11:

$$n^{-1/2} I \xrightarrow{D} \mathrm{N}(0, \sigma_{II}^2), \tag{4.16}$$

where,

$$\sigma_{II}^2 = \frac{s_{2m}}{2 s_{iw}^2 \bar{\sigma}^4},$$
$$s_{2m} = 2 \lim_{n\to\infty} n^{-1} [\mathrm{tr}(D_\sigma W_n D_\sigma W_n) + \mathrm{tr}(D_\sigma W_n D_\sigma W_n')].$$

The results in (A17) and (A18) of the Appendix also make it clear that:

$$s_{2m} = \bar{\sigma}^4 s_{2w} + 2h_1 + 2h_2 + 4\bar{\sigma}^2 h_3 + 4\bar{\sigma}^2 h_4. \tag{4.17}$$

Now consider the case in which the variances, $\sigma_i^2, i = 1, ..., n$ are modeled in such a way that they can be consistently estimated as, say $\hat{\sigma}_i^2$. Suppose also that the consistency is uniform in the sense that:

$$|\hat{\sigma}_i^2 - \sigma_i^2| \leq K||H_n||, \; i \geq 1, \tag{4.18}$$

where K is a ˇ nite constant and H_n is a ˇ nite dimensional vector such that $H_n \xrightarrow{P} 0$.[12] Let:

$$\hat{\sigma}^2 = n^{-1}\sum_{i=1}^{n}\hat{\sigma}_i^2,\ \hat{v}_i = \hat{\sigma}_i^2 - \hat{\sigma}^2,\ i = 1,...,n,$$
$$\hat{D}_v = \text{diag}_{i=1}^{n}(\ \hat{v}_i),$$
$$\hat{h}_1 = n^{-1}\text{tr}(\hat{D}_v W_n \hat{D}_v W_n),\ \hat{h}_2 = n^{-1}\text{tr}(\hat{D}_v W_n \hat{D}_v W_n'),$$
$$\hat{h}_3 = n^{-1}\text{tr}(W_n \hat{D}_v W_n),\ \hat{h}_4 = n^{-1}\text{tr}(W_n \hat{D}_v W_n'). \tag{4.19}$$

Let:

$$\hat{\sigma}_{II}^2 = \frac{\hat{\sigma}^4 \hat{s}_{2w} + 2\hat{h}_1 + 2\hat{h}_2 + 4\hat{\sigma}^2\hat{h}_3 + 4\hat{\sigma}^2\hat{h}_4}{2\ \hat{s}_{1w}^2\hat{\sigma}^4}, \tag{4.20}$$

where $\hat{s}_{1w}$ and $\hat{s}_{2w}$ are deˇ ned by (4.12) and (4.14). In the Appendix we demonstrate that:

$$\hat{\sigma}_{II}^2 \xrightarrow{P} \sigma_{II}^2. \tag{4.21}$$

Then, given (4.16) the obvious test for spatial correlation, sssuming the possibility of heteroskedasticity is:

$$\text{Reject H}_0 : \rho = 0 \text{ if } \left|\frac{n^{-1}I}{\hat{\sigma}_{II}}\right| > 1.96. \tag{4.22}$$

Because the test in (4.22) is based on the general result in (4.16), it should be robust, in large samples, with respect to heteroskedasticity. To be more speciˇ c, the empirical and theoretical type one errors should be the same whether or not the error terms are heteroskedastic, and if heteroskedastic, whether or not that heteroskedasticity is spatially correlated.

4.4 Conclusions

Researchers have often considered the possibility that the error terms of a regression model are heteroskedastic. We have argued that in many of these cases, the extent of this heteroskedasticity may be spatially correlated. If so, its description should be

[12] As an illustration, one such formulation would be $\sigma_i^2 = f(z_i\phi)$, where ϕ is a vector of parameters (typically regression parameters), z_i is an exogenous vector of observable variables, and f is a function whose ˇ rst derivative is bounded. Then, if ϕ can be consistently estimated, as say $\hat{\phi}$ by, e.g., the Maximum Likelihood procedure, the condition in (4.18) will hold. For example, the mean value theorem implies that, taking $\hat{\sigma}_i^2 = f(z_i\hat{\phi})$:

$$\hat{\sigma}_i^2 = f(z_i\phi) + f'(z_i\tilde{\phi})(\hat{\phi} - \phi),$$

where $\tilde{\phi}$ is between, element by element, $\hat{\phi}$ and ϕ. In this case $|f'(z_i\tilde{\phi})| < K$ and $||\hat{\phi} - \phi|| = ||H_n||$ as in (4.18). Concerning the norm in (4.18), let A be a matrix or vector. Then, $||A|| = \{\text{tr}(A'A)\}^{1/2}$. We note that this norm is sub multiplicative in that the sense that $||A_1A_2|| \leq ||A_1\ |\ ||A_2||$ (see e.g., Kelejian and Prucha, 1999).

part of the model; among other things, this may help to explain interrelationships between the extent of uncertainty over the regions considered.

We have also given results which describe how heteroskedasticity effects the type one error of the large sample test for spatial correlation which is based on the Moran I statistic. Because of the equivalence of this test to the one based on the LM statistic, our results apply to that test as well. These results suggest that a researcher is more likely to accept the hypothesis of spatial correlation if heteroskedasticity is positively correlated over the cross sectional units, and less likely to do so if that correlation is negative. We also show that in the absence of spatially correlated heteroskedasticity the empirical and theoretical type one errors of the standard test for spatial correlation based on Moran' I statistic are the same. Finally, because researchers may not know the exact nature of heteroskedasticity we give a robust version of the test based on Moran's I.

Our results are in a large sample framework; therefore, they may or may not hold in small or even moderate samples. Furthermore, it is not clear what the "cost" of large sample robustness is in terms of small sample power. An obvious suggestion for further work therefore is to study the small sample properties of the standard test based on Moran's I statistic, as well as those of the robust version we suggest under various scenarios involving heteroskedasticity.

Another, and perhaps more innovative area of future research relates to our suggestion that heteroskedasticity may itself be spatially correlated. As an example, on a theoretical level if heteroskedasticity relates to a set of variables which may be spatially correlated, models which describe that spatial correlation should be developed, along with corresponding tests for its existence. Finally empirical work, perhaps based on descriptive methods, suggesting the absence or presence such heteroskedasticity would also be of interest.

Acknowledgments

We would like to thank, without implicating, a referee and the editors of this volume for helpful comments on an earlier version of this chapter. We would also like to thank, without implicating, Robert Pietrowsky, Navigation Division Chief of the U.S. Army Engineers Institute for Water Resources (IWR), for support in the preparation of this manuscript. Finally, the views expressed in this chapter are those of the authors and not necessarily those of the US Army Corps of Engineers.

Appendix

In this Appendix we prove Theorem 1. We do this in terms of a series of preliminary results.

A1.1 Preliminaries

Preliminary 1: $n^{1/2}\Delta_n = O_P(1)$, where $\Delta_n = \hat{\beta} - \beta$.

Proof: Since $\hat{\beta} = (X'X)^{-1}X'y = \beta + (X'X)^{-1}X'\varepsilon$, and $\varepsilon = D_\sigma^{1/2}u$, it follows that:

$$\begin{aligned} n^{1/2}\Delta_n &= n(X'X)^{-1}X'\varepsilon \\ & n(X'X)^{-1}n^{-1/2}(X'D_\sigma^{1/2})u. \end{aligned} \tag{A1}$$

By Assumptions 6–8:

$$\begin{aligned} n(X'X)^{-1} &\to Q_{xx}, \text{ where } Q_{xx}^{-1} \text{ exists}, \\ n^{-1}X'D_\sigma X &\to Q_{XDX}, \text{ where } Q_{XDX}{}^{-1} \text{ exists}, \\ X'D_\sigma^{1/2} &: \text{ has bounded elements}. \end{aligned} \tag{A2}$$

By Assumption 5, the elements of u are *i.i.d.* $(0,1)$ and have ˇ nite third absolute moments. It follows from the Lindeberg-Feller central limit theorem that[13] $n^{-1/2}X'D_\sigma^{1/2}u \to \mathrm{N}(0, Q_{XDX})$ and so:

$$n^{1/2}\Delta_n \xrightarrow{D} \mathrm{N}(0, Q_{xx}^{-1}Q_{XDX}Q_{xx}^{-1}). \tag{A3}$$

Preliminary 1 follows from (A3).

Preliminary 2: $n^{-1}\hat{\varepsilon}'\hat{\varepsilon} \xrightarrow{P} \bar{\sigma}^2$.

Proof: From (4.12) in the text:

$$\begin{aligned} \hat{\varepsilon} &= y - X\hat{\beta} = y - X(\hat{\beta} - \beta + \beta) \\ &= y - X(\Delta_n + \beta) \\ &= \varepsilon - X\Delta_n. \end{aligned} \tag{A4}$$

Therefore:

$$\begin{aligned} n^{-1}\hat{\varepsilon}'\hat{\varepsilon} &= n^{-1}(\varepsilon - X\Delta_n)'(\varepsilon - X\Delta_n) \\ &= n^{-1}\varepsilon'\varepsilon + n^{-1}\Delta_n'X'X\Delta_n - 2\Delta_n'(n^{-1}X'\varepsilon). \end{aligned} \tag{A5}$$

The probability limit of the last two terms in (A5) is zero. To see this, note ˇ rst that Preliminary 1 implies that:

$$\begin{aligned} n^{-1}\Delta_n'X'X\Delta_n &= n^{-1}(n^{1/2}\Delta_n')(n^{-1}X'X)(n^{1/2}\Delta_n) \\ & n^{-1}O_P(1)(n^{-1}X'X)(O_P(1)). \end{aligned} \tag{A6}$$

In light of Assumption 6, $n^{-1}X'X \to Q_{xx}$ and so it follows from (A6):

$$n^{-1}\Delta_n'X'X\Delta_n \xrightarrow{P} 0. \tag{A7}$$

---

[13] A simple presentation of this theorem is given in Judge *et al.* (1985, pp. 156–157) For more detail, see Davidson (1994, chapter 23).

Now consider the last term in (A5). Let $\delta_1 = (n^{-1}X'\varepsilon)$. Then it should be clear that $\mathrm{E}(\delta_1) = 0$ and $\mathrm{E}(\delta_1\delta_1') = n^{-1}(n^{-1}X'D_\sigma X)$. By Assumption 8, $n^{-1}X'D_\sigma X \to Q_{XDX}$. It follows that $\mathrm{E}(\delta_1\delta_1') \to 0$, and so via Tchebyshev's inequality $n^{-1}X'\varepsilon \xrightarrow{P} 0$. Since via Preliminary 1 $\Delta_n = O_P(n^{-1/2})$, we have $\Delta_n \xrightarrow{P} 0$ and so our claim concerning the last term holds.

Finally denote the ˇ rst term in (A5) as δ_2 :

$$\delta_2 = n^{-1}\varepsilon'\varepsilon$$
$$= n^{-1}\sum_{i=1}^{n}\varepsilon_i^2. \tag{A8}$$

Then, by (4.2) in the text $\varepsilon_i = \sigma_i u_i$ and so ε_i has mean zero, $\mathrm{E}(\varepsilon_i) = 0$, variance $\mathrm{E}(\varepsilon_i^2) = \sigma_i^2$, ˇ nite fourth moment $\mathrm{E}(\varepsilon_i^4) = \sigma_i^4\,\mu_4$, and is independently distributed over $i = 1,\ldots,n$. Thus:

$$\mathrm{E}(\delta_2) = n^{-1}\sum_{i=1}^{n}\sigma_i^2,$$
$$\mathrm{Var}(\delta_2) = n^{-2}\sum_{i=1}^{n}\mathrm{Var}(\varepsilon_i^2),$$
$$= n^{-2}\sum_{i=1}^{n}(\sigma_i^4\mu_4 - \sigma_i^4). \tag{A9}$$

Assumptions 5 and 7 imply that $[\sigma_i^4\mu_4 - \sigma_i^4]$ is bounded. It follows from (A9) that $\mathrm{Var}(\delta_2) \to 0$ and hence by Tchebyshev's inequality: $\delta_2 = n^{-1}\varepsilon'\varepsilon \xrightarrow{P} \bar{\sigma}^2$. Preliminary 2 therefore follows.

Preliminary 3: $n^{-1/2}\hat{\varepsilon}'W_n\hat{\varepsilon} - n^{-1/2}\varepsilon'W_n\varepsilon \xrightarrow{P} 0$.

Proof: Using (A4) we have:

$$n^{-1/2}\hat{\varepsilon}'W_n\hat{\varepsilon} = n^{-1/2}(\varepsilon'W_n\varepsilon) + n^{-1/2}(\Delta_n'X'W_nX\Delta_n) - 2n^{-1/2}\Delta_n'X'W_n\varepsilon. \tag{A10}$$

It should be clear from (A10) that the proof of Preliminary 3 requires:

$$[n^{-1/2}(\Delta_n'X'W_nX\Delta_n) - 2n^{-1/2}\Delta_n'X'W_n\varepsilon] \xrightarrow{P} 0. \tag{A11}$$

Let δ_3 denote the ˇ rst term in (A11), and express it as:

$$\delta_3 = n^{-1/2}(n^{1/2}\Delta_n)(n^{-1}X'W_nX)(n^{1/2}\Delta_n). \tag{A12}$$

Assumptions 1, 3, 4a, and 4b imply that W_n has only a bounded number of bounded elements in each row and column and hence is an absolutely summable matrix. Therefore, given Assumption 6 and the discussion concerning (4.3), the elements of $n^{-1}X'W_nX$ remain bounded for all n. It then follows from Preliminary 1 and (A12) that $\delta_3 \xrightarrow{P} 0$.

Let δ_4 denote the second term in (A11) and express it as:

$$\delta_4 = 2(n^{1/2}\Delta_n')(n^{-1}X'W_n\varepsilon).$$

Let $\delta_5 = (n^{-1}X'W_n\varepsilon)$. Then, $E(\delta_5) = 0$, and $E(\delta_5\delta_5') = n^{-1}(n^{-1}X'W_nD_\sigma W_nX)$. Because D_σ is a diagonal matrix with bounded elements, it is absolutely summable. Since W_n is also absolutely summable, the results relating to (4.3) imply that $W_nD_\sigma W_n$ is absolutely summable, and hence the elements of $n^{-1}X'W_nD_\sigma W_nX$ are bounded. It follows that $E(\delta_5\delta_5') \to 0$ and hence, by Preliminary 1, $\delta_4 \xrightarrow{P} 0$, which in turn implies Preliminary 3.

Preliminary 4: Recalling the expression for Moran's I in (4.12):

$$n^{-1/2}I \to N(0,\ s_{2m}/2s_{1w}^2\bar{\sigma}^4),$$

where,

$$s_{2m} = \lim_{n\to\infty} n^{-1}\mathrm{tr}[(M_n+M_n')(M_n+M_n'),\ M_n = D_\sigma^{1/2}W_nD_\sigma^{1/2},$$

and where s_{1w} is deˇ ned in Assumption 4.

Proof: Preliminaries 2 and 3, and Assumption 4c imply:

$$\left(n^{-1/2}I - n^{-1/2}\frac{\varepsilon'W_n\varepsilon}{s_{1w}\bar{\sigma}^2}\right) \xrightarrow{P} 0. \tag{A13}$$

Therefore, if $n^{-1/2}(\varepsilon'W_n\varepsilon)\ /\ s_{1w}\bar{\sigma}^2)$ has a limiting distribution, $n^{-1/2}I$ converges in distribution to the same distribution. To obtain this distribution, ˇ rst note that:

$$\begin{aligned} n^{-1/2}\varepsilon'W_n\varepsilon &= n^{-1/2}u'D_\sigma^{1/2}W_nD_\sigma^{1/2}u \\ &= n^{-1/2}u'M_nu. \end{aligned} \tag{A14}$$

Assumptions 1, 2, 3, 4, 7, and 10 imply that M_n satisˇ es all of the assumptions Anselin and Kelejian (1997) made concerning their weights matrix, W_n. In addition, the elements of u satisfy all of the assumptions Anselin and Kelejian (1997) made concerning their disturbance vector, ε. Therefore, it follows from the results Anselin and Kelejian (1997, p. 180) give that:

$$n^{-1/2}\varepsilon'W_n\varepsilon \xrightarrow{D} N\left(0, \frac{s_{2m}}{2}\right), \tag{A15}$$

where,

$$s_{2m} = \lim_{n\to\infty} n^{-1}\mathrm{tr}[(M_n+M_n')(M_n+M_n')].$$

Preliminary 4 trivially follows from (A13) and (A15).

Proof of Theorem 1: Recall that $\hat{s}_{2m} = n^{-1}\mathrm{tr}[(M_n+M_n')(M_n+M_n')]$, and note that:

$$\begin{aligned} \hat{s}_{2m} &= 2n^{-1}\mathrm{tr}[(M_nM_n+M_nM_n')] \\ &= 2n^{-1}\mathrm{tr}(M_nM_n)+2n^{-1}\mathrm{tr}(M_nM_n'). \end{aligned} \tag{A16}$$

Since $M_n = D_\sigma^{1/2} W_n D_\sigma^{1/2}$, $\hat{s}_{2m}$ can be expressed in terms of W_n as:

$$\begin{aligned}\hat{s}_{2m} &= 2n^{-1}\mathrm{tr}(D_\sigma^{1/2} W_n D_\sigma^{1/2} D_\sigma^{1/2} W_n D_\sigma^{1/2}) \\ &\quad +2n^{-1}\mathrm{tr}(D_\sigma^{1/2} W_n D_\sigma^{1/2} D_\sigma^{1/2} W_n' D_\sigma^{1/2}) \\ &= 2n^{-1}\mathrm{tr}(D_\sigma W_n D_\sigma W_n) + 2n^{-1}\mathrm{tr}(D_\sigma W_n D_\sigma W_n') \\ &\equiv \hat{s}_{21m} + \hat{s}_{22m}, \end{aligned} \tag{A17}$$

where $\hat{s}_{21m}$ and $\hat{s}_{22m}$ are deˇ ned, respectively, as the ˇ rst and second terms in the second line of (A17). Assumption 9 implies that $D_\sigma = \bar{\sigma}^2 I + D_v$. Using this expression for D_σ, $\hat{s}_{21m}$ can be expressed as:

$$\begin{aligned}\hat{s}_{21m} &= 2n^{-1}\mathrm{tr}[(\bar{\sigma}^2 I + D_v) W_n (\bar{\sigma}^2 I + D_v) W_n) \\ &\quad 2n^{-1}\bar{\sigma}^4 \mathrm{tr}(W_n W_n) + 4n^{-1}\bar{\sigma}^2 \mathrm{tr}(W_n D_v W_n) \\ &\quad +2n^{-1}\mathrm{tr}(D_v W_n D_v W_n). \end{aligned} \tag{A18}$$

Given Assumption 9 we have:

$$[\hat{s}_{21m} - 2n^{-1}\bar{\sigma}^4 \mathrm{tr}(W_n W_n) - 2h_1] \to 0. \tag{A19}$$

A similar argument will demonstrate that:

$$[\hat{s}_{22m} - 2n^{-1}\bar{\sigma}^4 \mathrm{tr}(W_n W_n') - 2h_2] \to 0. \tag{A20}$$

It then follows from (A16–A20), and Assumption 4 that:

$$\begin{aligned} &[\hat{s}_{2m} - 2(n^{-1}\bar{\sigma}^4 \mathrm{tr}(W_n W_n) + n^{-1}\bar{\sigma}^4 \mathrm{tr}(W_n W_n')) - 2h_1 - 2h_2] \\ &= [\hat{s}_{2m} - n^{-1}\bar{\sigma}^4 \mathrm{tr}[(W_n + W_n')(W_n + W_n')] - 2h_1 - 2h_2] \\ &= [\hat{s}_{2m} - n^{-1}\bar{\sigma}^4 \mathrm{tr}[(W_n + W_n')(W_n + W_n')] - 2h_1 - 2h_2] \\ &= [\hat{s}_{2m} - \bar{\sigma}^4 s_{2w} - 2h_1 - 2h_2] \to 0. \end{aligned} \tag{A21}$$

Theorem 1 follows from (A21), Preliminary 4.

Demonstration Relating to Remark 4: Consider now the case in which the model in (4.1) contains endogenous variables and appropriate instruments are available for consistent estimation based on all of the assumptions in Anselin and Kelejian (1997) (also see Kelejian and Prucha, 1997). For ease of presentation again let β denote the parameter vector, and let $\hat{\beta}$ be its consistent instrumental variable estimator. In this case an analysis which is quite similar to that in Kelejian and Prucha (1997) will demonstrate that $n^{1/2}(\hat{\beta} - \beta)$ will typically be $O_P(1)$, and hence Preliminary 1 would still hold. An argument which is virtually identical to that given above would then demonstrate that Preliminary 2 holds. The results given in Anselin and Kelejian (1997) then imply that Preliminary 3 holds since, in the absence of spatial lags the term A in Anselin and Kelejian is zero. Preliminary 4 and the proof of the claim in Remark 4 then follow from the above analysis.

Proof of (4.21): Consider the components of $\hat{\sigma}^2_{II}$ in (4.20). Assumption 4 implies that $\hat{s}_{1w} \to s_{1w}$ and $\hat{s}_{2w} \to s_{2w}$. Now consider $\hat{\sigma}^2$ and express it as:

$$\hat{\sigma}^2 = n^{-1}\sum_{i=1}^{n}[(\hat{\sigma}_i^2 - \sigma_i^2) + \sigma_i^2]$$
$$n^{-1}\sum_{i=1}^{n}\sigma_i^2 + n^{-1}\sum_{i=1}^{n}(\hat{\sigma}_i^2 - \sigma_i^2). \tag{A22}$$

Assumption 7 implies that $n^{-1}\sum_{i=1}^{n}\sigma_i^2 \to \bar{\sigma}^2$. The condition in (4.18) implies that:

$$\begin{aligned}\underset{n\to\infty}{\text{plim}}\ |n^{-1}\sum_{i=1}^{n}(\hat{\sigma}_i^2 - \sigma_i^2)| &\le \underset{n\to\infty}{\text{plim}}\ n^{-1}\sum_{i=1}^{n}|\hat{\sigma}_i^2 - \sigma_i^2| \\ &\le \underset{n\to\infty}{\text{plim}}\ K||H_n|| = 0.\end{aligned} \tag{A23}$$

It follows that $\hat{\sigma}^2 \xrightarrow{P} \bar{\sigma}^2$, and so:

$$\underset{n\to\infty}{\text{plim}}\ \hat{\sigma}^4 = \bar{\sigma}^4. \tag{A24}$$

Thus, our proof is complete if the remaining terms in the numerator of (4.20) converge in probability to their respective counterparts.

Consider $\hat{h}_1$. It is evident from (4.8) that:

$$\hat{h}_1 = n^{-1}\sum_{i=1}^{n} w_{i.}(\hat{v}_i\hat{D}_v)w_{.i}. \tag{A25}$$

Since $\hat{v}_i = \hat{\sigma}_i^2 - \hat{\sigma}^2$ we have:

$$\begin{aligned}\hat{v}_i &= \sigma_i^2 + (\hat{\sigma}_i^2 - \sigma_i^2) - \bar{\sigma}^2 - (\hat{\sigma}^2 - \bar{\sigma}^2) \\ &= v_i + (\hat{\sigma}_i^2 - \sigma_i^2) - (\hat{\sigma}^2 - \bar{\sigma}^2) \\ &= v_i + \delta_i - \Delta_n,\end{aligned} \tag{A26}$$

where $\delta_i = (\hat{\sigma}_i^2 - \sigma_i^2)$ and $\Delta_n = (\hat{\sigma}^2 - \bar{\sigma}^2) \xrightarrow{P} 0$. Since $\hat{D}_v = \text{diag}_{i=1}^{n}(\hat{v}_i)$, it follows that:

$$\hat{D}_v = D_v + D_n - \Delta_n I;\ D_n = \text{diag}_{i=1}^{n}(\delta_i). \tag{A27}$$

It follows from (A25–A27) that:

$$\begin{aligned}\hat{h}_1 &= n^{-1}\sum_{i=1}^{n} w_{i.}(v_i + \delta_i - \Delta_n)(D_v + D_n - \Delta_n I)w_{.i} \\ &= n^{-1}\sum_{i=1}^{n} w_{i.}(v_i D_v)w_{.i} + P_n;\ P_n = \hat{h}_1 - n^{-1}\sum_{i=1}^{n} w_{i.}(v_i D_v)w_{.i}.\end{aligned} \tag{A28}$$

It follows that $\hat{h}_1 \xrightarrow{P} h_1$ if $P_n \xrightarrow{P} 0$.

To see that this is indeed the case consider one of the components of P_n namely:

$$\begin{aligned} q_n &= n^{-1} \sum_{i=1}^{n} w_{i.} \delta_i D_v w_{.i} \\ &= n^{-1} \sum_{i=1}^{n} \sum_{t=1}^{n} w_{it} \delta_i v_t w_{ti}. \end{aligned} \tag{A29}$$

Assumption 7 implies that v_t is bounded and so $|v_t| < c_v, t = 1, ...,$ where c_v is a ˇ nite constant. Assumptions 1,3, and 4 imply:

$$\sum_{t=1}^{n} |w_{it} w_{ti}| \leq \lambda_2 c_w^2; \; n > 1. \tag{A30}$$

Given the bound on v_t, and (A30) it follows that from (4.16):

$$\begin{aligned} \operatorname*{plim}_{n\to\infty} |q_n| &\leq \operatorname*{plim}_{n\to\infty} n^{-1} \sum_{i=1}^{n} \sum_{t=1}^{n} |w_{it}| \, |\delta_i| \, |v_t| \, |w_{ti}| \\ &\leq c_v \operatorname*{plim}_{n\to\infty} n^{-1} \sum_{i=1}^{n} \sum_{t=1}^{n} |w_{it}| \, |w_{ti}| \, |\delta_i| \\ &\leq c_v K \operatorname*{plim}_{n\to\infty} n^{-1} \sum_{i=1}^{n} \sum_{t=1}^{n} |w_{it}| \, |w_{ti}| \, ||H_n|| \\ &\leq c_v K \lambda_2 c_w^2 \operatorname*{plim}_{n\to\infty} n^{-1} \sum_{i=1}^{n} ||H_n|| \\ &\leq c_v K \lambda_2 c_w^2 \operatorname*{plim}_{n\to\infty} ||H_n|| = 0. \end{aligned} \tag{A31}$$

A similar analysis will demonstrate that the remaining terms deˇ ning P_n have zero probability limits, and so $\hat{h}_1 \xrightarrow{P} h_1$ since $P_n \xrightarrow{P} 0$. Given this, it should be evident that $\hat{h}_i \xrightarrow{P} h_i, i = 2, 3, 4$.

5 A Taxonomy of Spatial Econometric Models for Simultaneous Equations Systems

Sergio J. Rey[1] and Marlon G. Boarnet[2]

[1] San Diego State University
[2] University of California, Irvine

5.1 Introduction

The spatial econometric literature has developed a large number of approaches that can handle spatial dependence and heterogeneity, yet almost all of these approaches are single equation techniques. For many regional economic problems there are both multiple endogenous variables and data on observations that interact across space. To date, researchers have often been in the undesirable position of having to choose between modeling spatial interactions in a single equation framework, or using multiple equations but losing the advantages of a spatial econometric approach. This chapter establishes a framework for applying spatial econometrics within the context of multi-equation systems. Speci˘ cally, we discuss the need for multi-equation spatial econometric models and we develop a general model that can subsume many interesting special cases. We also examine the small sample properties of common estimators for speci˘ c cases of the general model.

This chapter is organized as follows. In Sect. 5.2 we overview recent research that has relied on spatial econometric methods applied to multi-equation systems. We then present the general taxonomy of spatial econometric models in simultaneous equations systems and outline a number of the key distinctions between some of the more interesting models within the taxonomy. Section 5.4 highlights a number of estimation issues associated with their implementation. This is followed by an empirical evaluation of alternative estimators in a series of Monte Carlo simulations, the design of which is laid out in Sect. 5.5 and the results discussed in Sect. 5.6. In the ˘ nal section we summarize the key ˘ ndings and suggest an agenda for future research on the taxonomy.

5.2 Recent Applications of Spatial Econometrics in a Multi-Equation Framework

There have been a small number of applications of spatial econometrics in multi-equation frameworks. While the estimators are sometimes ad-hoc and have not been examined in detail, those applications provide insight into the motivation for combining spatial econometrics and simultaneous systems.

One of the earliest combinations of a spatial (but not explicitly spatial econometric) approach with simultaneous systems techniques was the intra-urban population and employment model of Steinnes and Fisher (1974). Steinnes and Fisher developed a model of population and employment levels, which they estimated with data from 100 Chicago community areas and suburbs for 1960.[1] Both population and employment were endogenous variables, and since Steinnes and Fisher's work it has been commonly accepted that population and employment are both endogenous in urban models (e.g., Boarnet, 1994a,b; Deitz, 1993; Steinnes, 1977).

Steinnes and Fisher (1974) also innovated by developing potential variables that aggregated community area population and employment into larger units. This was done to provide some degree of spatial interaction. In their model, community area population depended on a weighted average of employment in all community areas in the data set, and community area employment was similarly a function of a weighted average of population in the community areas. Steinnes and Fisher did not use spatial econometrics to estimate their system, instead they assumed the potential variables were predetermined in line with the usual treatment of lagged variables in time series analysis. In a footnote, they did, however, acknowledge the questionable validity of this assumption and argued that a fuller consideration of this assumption would lead to "the relatively new ˇ eld of stochastic processes over space" (p. 71). Ironically, the importance of the potential variables and the associated issue of spatial simultaneity in their speciˇ cation were largely overlooked in later work.[2]

Twenty years later, Boarnet (1994b) proposed an adaptation of a model developed by Carlino and Mills (1987) which integrated the use of potential variables and spatial econometrics in a two equation model of population and employment growth in New Jersey municipalities. Speciˇ cally, Boarnet estimated two equations relating the population and employment change between two time periods (1988 and 1980):

$$\begin{aligned} P\Delta_{1988} &= \alpha_0 + \alpha_1 T_{P_{1980}} + \alpha_2 Z_{P_{1980}} + \alpha_3 (I+W) E_{1980} \\ &\quad + \frac{\alpha_3}{\lambda_e} (I+W) E\Delta_{1988} - \lambda_p P_{1980} + \mu, \end{aligned} \tag{5.1}$$

$$\begin{aligned} E\Delta_{1988} &= \beta_0 + \beta_1 T_{E_{1980}} + \beta_2 Z_{E_{1980}} + \beta_3 (I+W) P_{1980} \\ &\quad + \frac{\beta_3}{\lambda_p} (I+W) P\Delta_{1988} - \lambda_e E_{1980} + \upsilon, \end{aligned} \tag{5.2}$$

where $P\Delta_{1988}$ is an n by 1 vector of observations on the change in population (P) for each municipality i, $P\Delta_{i,1988} = P_{i,1988} - P_{i,1980}$, $E\Delta_{1988}$ is an n by 1 vector of observations on the change in employment (E), $E\Delta_{i,1988} = E_{i,1988} - E_{i,1980}$, $T_{P_{1980}}$ a vector of transportation access variables which in uence municipal population growth, measured in 1980, $Z_{P_{1980}}$ a vector of other amenity variables which in uence municipal population growth, $T_{E_{1980}}$ a vector of transportation access variables which in uence municipal employment growth, $Z_{E_{1980}}$ a vector of other amenity

[1] While the model in Steinnes and Fisher (1974) included equations for both population and employment levels, they only reported the results for the population regression.

[2] An exception is Carlino and Mills (1986), who use potential variables for employment to examine the effect of agglomeration economies on county population and employment growth.

variables which influence municipal employment growth, I is an n by n identity matrix, W is an n by n matrix of spatial weights (where each element is $1/d_{i,j}^{\theta}$), P_{1980} is a vector of observations of municipal populations in 1980, E_{1980} is similarly defined for employment, μ and υ are stochastic error terms, and n equals the number of municipalities.

The above equations are an example of a spatial cross-regressive model. That is, the right-hand side of each equation contains spatial lags of the endogenous variable from the other equation, creating spatial links across equations. Boarnet's motivation for using the spatial cross-regressive structure in equations (5.1) and (5.2) was that New Jersey municipalities are too small to be their own labor market.[3] Population growth in a municipality depends, in part, on the growth in job opportunities in a surrounding commuter-shed. Similarly, employment growth in a municipality depends on changes in the available labor pool in a surrounding commuter-shed or labor market area. Those labor market, or commuter-shed, relationships are mediated by commuting patterns which demonstrate how residents (or employers) in a municipality depend on job opportunities (or labor supply) in surrounding municipalities. Specifically, the spatially cross-regressive variables in the model of Boarnet (1994b) include a dampening parameter which reflects the decline in strength of the labor market relationship with distance.

In the case of urban econometric models, there is thus a strong theoretical justification for both spatial interaction and multiple endogenous variables. Alternatively, there are other forms of spatial dependence that could be combined with multi-equation systems to examine urban problems. For example, the classic problems of spread and backwash could be studied by including a spatial crossregressive lag term[4] in each of the two equations in a population and employment model:

$$P\Delta_t = \alpha_0 + \alpha_1 T_{P_{t-1}} + \alpha_2 Z_{P_{t-1}} + \alpha_3 W(E\Delta_t) - \lambda_p P_{t-1} + \mu_t, \qquad (5.3)$$

$$E\Delta_t = \beta_0 + \beta_1 T_{E_{t-1}} + \beta_2 Z_{E_{t-1}} + \beta_3 W(P\Delta_t) - \lambda_e E_{t-1} + \upsilon_t, \qquad (5.4)$$

where the time period subscripts have been changed to t and $t-1$ for generality.

The coefficients on the spatial crossregressive lag terms could test the extent to which municipalities capture excess growth from nearby areas (spread) or the extent to which localities lose growth to nearby locations (backwash). Alternatively, Henry *et al.* (1997) adapted the model in (5.1) and (5.2) to examine spread and backwash without including spatial crossregressive lag terms. Instead, they included interaction terms, shown in the model below, to test how differential population levels across core, fringe, and hinterland regions within functional economic areas affect the growth of rural hinterland census tracts in three southern U.S. states, using

[3] For a more detailed justification, see Boarnet (1992, chap. 3 and 6).

[4] The spatial crossregressive lag term pertains to a spatial lag of the endogenous variable from a different equation. This is distinct from the spatial autoregressive lag which is the spatial lag of the dependent variable from the *same* equation. It is also distinct from the spatial crossregressive variable which reflects a spatial lag of an *exogenous* variable.

data from 1980 to 1990:

$$P\Delta_{1988} = \alpha_0 + \alpha_1 T_{P_{1980}} + \alpha_2 Z_{P_{1980}} + [\alpha_3 + \alpha_4 g_1 + \alpha_5 g_2](I+W)E_{1980} + \left[\frac{\alpha_3}{\lambda_e} + \alpha_6 g_1 + \alpha_7 g_2\right](I+W)E\Delta_{1988} - \lambda_p P_{1980} + \mu, \quad (5.5)$$

$$E\Delta_{1988} = \beta_0 + \beta_1 T_{E_{1980}} + \beta_2 Z_{E_{1980}} + [\beta_3 + \beta_4 g_1 + \beta_5 g_2](I+W)P_{1980} + \left[\frac{\beta_3}{\lambda_p} + \beta_6 g_1 + \beta_7 g_2\right](I+W)P\Delta_{1988} - \lambda_e E_{1980} + \upsilon, \quad (5.6)$$

where g_1 is the ratio of 1990 to 1980 population for the urban core of the functional economic area that contains the census tract, and g_2 is the ratio of 1990 to 1980 population for the urban fringe of the functional economic area.

While the need to combine spatial econometrics and simultaneous systems has been most closely examined in the context of urban systems, it is also evident in other problems. For example, the large literature on production function studies of public capital was originally speciˇ ed in a single equation context without any spatial modeling (Aschauer, 1989; Munnell, 1990a). Yet some authors have recently begun to examine both the spatial nature of infrastructure investments (Holtz-Eakin and Schwartz, 1995; Boarnet, 1998) and the need to examine multiple endogenous variables (Duffy-Deno and Eberts, 1991; de Frutos and Pereira, 1993). This activity acknowledges both that public capital investments create spillovers across geographic areas and that the right hand side variables in a production function (typically labor, private capital, and public capital) are best modeled as variables endogenous in a larger system. Yet so far, no author has combined spatial econometrics with a system of simultaneous equations to study public infrastructure. More generally, spatial econometric techniques have recently been applied to a host of applied economic problems, including (but not limited to) regional economic convergence (Rey and Montouri, 1999), analyses of state and local public expenditures (Case *et al.*, 1993; Murdoch *et al.*, 1993), strategic behavior relating to international environmental issues (Murdoch *et al.*, 1997) and the adoption of technology in developing countries (Case, 1992). While the overwhelming majority of the applications have been in single equation models, there is certainly the possibility that many applications can beneˇ t from a combination of spatial econometrics and simultaneous equations. The rest of this chapter lays the groundwork for combining those two approaches.

5.3 Taxonomy

It will be useful to view the existing approaches reviewed above as speciˇ c cases of a more general framework. To motivate this framework, consider the classic regression model:

$$y_1 = X\beta_1 + \varepsilon_1. \quad (5.7)$$

The notation is as follows: y_1 is the n by 1 vector of observations on the first dependent variable; X is an n by k matrix of observations on k exogenous variables with associated parameters in the k by 1 vector β_1, and ε_1 is an n by 1 disturbance vector. We examine the more general situation that arises from the consideration of both *feedback simultaneity* and *spatial simultaneity*. Considering spatial simultaneity first, equation (5.7) is extended through the addition of a spatial autoregressive lag term:

$$y_1 = X\beta_1 + \rho_{11} W y_1 + \varepsilon_1, \tag{5.8}$$

where ρ_{11} is the scalar spatial autoregressive coefficient and W is an n by n spatial weights matrix. Feedback simultaneity arises when (5.7) is specified as part of a multi-equation system and is expanded to include a second endogenous variable, y_2, also measured in each of the n locations:

$$y_1 = X\beta_1 + \gamma_{21} y_2 + \varepsilon_1, \tag{5.9}$$

where this second variable is also determined within the system,

$$y_2 = X\beta_2 + \gamma_{12} y_1 + \varepsilon_2. \tag{5.10}$$

We allow for two additional sources of spatial simultaneity. The first is represented by the inclusion of a spatial autoregressive lag term in (5.10):

$$y_2 = X\beta_2 + \gamma_{12} y_1 + \rho_{22} W y_2 + \varepsilon_2, \tag{5.11}$$

while the second arises from the addition of a spatial cross-regressive term in each of the system equations. The resulting system would be specified as:

$$y_1 = X\beta_1 + \gamma_{21} y_1 + \rho_{21} W y_2 + \rho_{11} W y_1 + \varepsilon_2, \tag{5.12}$$

$$y_2 = X\beta_2 + \gamma_{12} y_1 + \rho_{12} W y_1 + \rho_{22} W y_2 + \varepsilon_2. \tag{5.13}$$

Collecting both equations, we express the system using matrix notation as follows:

$$Y\Gamma = WYP + XB + E, \tag{5.14}$$

where $Y = [y_1, y_2]$, $B = (\beta_1, \beta_2)$, $E = (\varepsilon_1, \varepsilon_2)$, and

$$\Gamma = \begin{pmatrix} 1 & -\gamma_{12} \\ -\gamma_{21} & 1 \end{pmatrix}, \; P = \begin{pmatrix} \rho_{11} & \rho_{12} \\ \rho_{21} & \rho_{22} \end{pmatrix}.$$

The error terms have the following properties:

$$\text{Cov}\,[\varepsilon_{1,i},\ \varepsilon_{2,i}] = 0, \qquad \forall i, \tag{5.15}$$

$$\text{Cov}\,[\varepsilon_{l,i},\ \varepsilon_{l,j}] = 0, \qquad \forall i \neq j, \text{ and } l = 1,2, \tag{5.16}$$

$$\text{Cov}\,[\varepsilon_{1,i},\ \varepsilon_{2,j}] = 0, \qquad \forall i \neq j. \tag{5.17}$$

These properties imply that the errors do not display any cross-equation covariance, are not spatially autocorrelated within a given equation and are not spatially autocorrelated across equations. The error terms and the exogenous variables are also independent.[5]

As it stands, the system in (5.14) has a "two sided reduced form" in the sense that the matrix of endogenous variables would be both pre- and post-multiplied by two distinct coefficient matrices. Thus, the system does not lend itself to the application of traditional order and rank rules for checking for identification. We return to this issue below. We can, however, isolate one of the two equations to provide a more detailed view of its reduced form. Starting with (5.14), we define the matrices $A = (I - \rho_{11}W)$ and $B = (I - \rho_{22}W)$.

We then have:

$$y_1 = A^{-1}(\rho_{21}W + \gamma_{21})y_2 + A^{-1}(X\beta_1 + \varepsilon_1), \tag{5.18}$$
$$y_2 = B^{-1}(\rho_{12}W + \gamma_{12})y_1 + B^{-1}(X\beta_2 + \varepsilon_2). \tag{5.19}$$

The system can be expressed more succinctly as:

$$(G - H)y = Dx + \varepsilon, \tag{5.20}$$

where, $G = (\Gamma' \otimes I)$, $H = (P' \otimes W)$, $D = (B' \otimes I)$, $x = \mathrm{Vec}(X)$, $\varepsilon = \mathrm{Vec}(E)$, and $y = \mathrm{Vec}(Y)$. A "single-sided" reduced form for the system is then:

$$y = \Theta^{-1}Dx + \Theta^{-1}\varepsilon, \tag{5.21}$$

where $\Theta = (G - H)$.

This general form nests the 35 different specifications, listed in Table 5.1, as specific cases. These arise by imposing zero restrictions on various model parameters. To structure the taxonomy it is useful to note that there are essentially three dimensions to consider: *feedback* simultaneity; *spatial autoregressive lag* simultaneity; and *spatial crossregressive lag* simultaneity. With respect to the two equation system, each dimension can be expressed in one of three ways irrespective of how the other two dimensions are specified. For example, feedback simultaneity can be totally absent from the system, take a recursive form, or take on a full feedback structure. Similarly the spatial lags can be totally absent, present in only one equation or present in both equations. Similar specifications hold for the spatial cross-regressive terms as well. In addition to these specifications, a number of additional possibilities arise when two of the dimensions are present in the intermediate form (i.e., recursive, lag in one equation, cross-regressive in one equation).

Ten of the equations include the traditional notion of feedback simultaneity, eight equations omit any form of traditional simultaneity (either in a feedback or recursive form) but do include either spatial lag or cross-lag simultaneity. Sixteen models are recursive in the a-spatial endogenous variables. Finally, a classic two-equation regression model without spatial or traditional simultaneity rounds out the

[5] These are also necessary conditions for identification, as is outlined below. We also limit the system to two equations in this initial presentation as well as omit the possibility of including spatial lags of the exogenous variables as dimensions of our taxonomy. Future work will extend the taxonomy to consider these issues.

taxonomy. Models 2–5 include only one form of simultaneity, either through feedback endogeneity or through a particular form of spatial dependence. In contrast, the remaining 30 models include at least two distinct forms of simultaneity.

The presence of the two types of spatial simultaneity raises a number of issues. At first glance, the spatial cross-regressive term appears to play a similar role to the spatial lag, since it provides for a form of spatial spillover to enter the system. Given that the spatial lag term is sometimes viewed as simply another form of an endogenous variable in a simultaneous equations system (Murdoch *et al.*, 1993), it would appear that the cross-regressive term could be viewed in a similar way. However, the form of endogeneity introduced by the spatial lag is fundamentally distinct from that due to the appearance of "non-spatial" endogenous variables on the right hand side of an equation. This is because, in the model with a spatial lag, each observation of the dependent variable is related to all values (associated with each of the observations) of all the error terms (one for each equation). In the model with only feedback endogeneity, each observation of the dependent variable is related not only to its own error term but also to the error terms of other endogenous variables. This is only for the same observational unit, however. In other words, the feedback simultaneity is expressed through variable to variable interaction for the same observation, while spatial lag simultaneity is expressed through observation to observation interactions for the same variable. Moreover, the cross-regressive term actually embodies both spatial and feedback simultaneity within the same variable. An interesting issue is the extent to which this cross regressive term captures the systematic relations of the spatial autoregressive lag and feedback variables.

5.4 Estimation Issues

The combination of spatial and traditional simultaneity found in many of the models in the taxonomy creates a number of complications that require further attention. Chief among these are the questions of whether or not each equation is identified in a specific model, the choice of estimator and the treatment of instruments. Because of the two-sided nature of the reduced form, the traditional rank and order rules for checking the identification of the models cannot be applied. Identification can be checked, however, if the models are viewed as special cases of simultaneous equations that are nonlinear in the endogenous variables. Following Kelejian and Oates (1989), we distinguish between basic endogenous variables that appear as left hand side variables, and additional endogenous variables, that appear on the right hand side of an equation and are functions of the basic endogenous variables. In the taxonomy presented above, the basic endogenous variables would include y_1 and y_2, while the additional endogenous variables would include their spatial lags, both in the own and cross forms.

Table 5.1. Model taxonomy

Model	Equations	Simultaneity
1	$y_1 = X\beta_1 + \varepsilon_1$ $y_2 = X\beta_2 + \varepsilon_2$	None
2	$y_1 = X\beta_1 + \varepsilon_1$ $y_2 = X\beta_2 + \gamma_{12}y_1 + \varepsilon_2$	Recursive
3	$y_1 = X\beta_1 + \rho_{11}Wy_1 + \varepsilon_1$ $y_2 = X\beta_2 + \rho_{22}Wy_2 + \varepsilon_2$	Two Lags
4	$y_1 = X\beta_1 + \rho_{11}Wy_1 + \varepsilon_1$ $y_2 = X\beta_2 + \varepsilon_2$	One Lag
5	$y_1 = X\beta_1 + \rho_{21}Wy_2 + \varepsilon_1$ $y_2 = X\beta_2 + \rho_{12}Wy_1 + \varepsilon_2$	Two Cross-Lags
6	$y_1 = X\beta_1 + \rho_{11}Wy_1 + \varepsilon_1$ $y_2 = X\beta_2 + \rho_{22}Wy_2 + \rho_{12}Wy_1 + \varepsilon_2$	Two Lags & One Cross-Lag
7	$y_1 = X\beta_1 + \rho_{11}Wy_1 + \rho_{21}Wy_2 + \varepsilon_1$ $y_2 = X\beta_2 + \varepsilon_2$	One Lag & One Cross-Lag Same Eq.
8	$y_1 = X\beta_1 + \rho_{11}Wy_1 + \varepsilon_1$ $y_2 = X\beta_2 + \rho_{12}Wy_1 + \varepsilon_2$	One Lag & One Cross-Lag Different Eq.
9	$y_1 = X\beta_1 + \rho_{11}Wy_1 + \rho_{21}Wy_2 + \varepsilon_1$ $y_2 = X\beta_2 + \rho_{12}Wy_1 + \varepsilon_2$	One Lag & Two Cross-Lags
10	$y_1 = X\beta_1 + \rho_{11}Wy_1 + \rho_{21}Wy_2 + \varepsilon_1$ $y_2 = X\beta_2 + \rho_{12}Wy_1 + \rho_{22}Wy_2 + \varepsilon_2$	Two Lags & Two Cross-Lags
11	$y_1 = X\beta_1 + \gamma_{21}y_2 + \varepsilon_1$ $y_2 = X\beta_2 + \gamma_{12}y_1 + \varepsilon_2$	Feedback
12	$y_1 = X\beta_1 + \gamma_{21}y_2 + \rho_{11}Wy_1 + \varepsilon_1$ $y_2 = X\beta_2 + \gamma_{12}y_1 + \varepsilon_2$	Feedback & One Lag
13	$y_1 = X\beta_1 + \gamma_{21}y_2 + \rho_{11}Wy_1 + \varepsilon_1$ $y_2 = X\beta_2 + \gamma_{12}y_1 + \rho_{22}Wy_2 + \varepsilon_2$	Feedback & Two Lags
14	$y_1 = X\beta_1 + \gamma_{21}y_2 + \rho_{11}Wy_1 + \rho_{21}Wy_2 + \varepsilon_1$ $y_2 = X\beta_2 + \gamma_{12}y_1 + \rho_{22}Wy_2 + \varepsilon_2$	Feedback & Two Lags & One Cross-Lag
15	$y_1 = X\beta_1 + \gamma_{21}y_2 + \rho_{11}Wy_1 + \rho_{21}Wy_2 + \varepsilon_1$ $y_2 = X\beta_2 + \gamma_{12}y_1 + \rho_{12}Wy_1 + \rho_{22}Wy_2 + \varepsilon_2$	Feedback & Two Lags & Two Cross-Lags
16	$y_1 = X\beta_1 + \gamma_{21}y_2 + \rho_{11}Wy_1 + \rho_{21}Wy_2 + \varepsilon_1$ $y_2 = X\beta_2 + \gamma_{12}y_1 + \rho_{12}Wy_1 + \varepsilon_2$	Feedback & One Lag & Two Cross-Lags
17	$y_1 = X\beta_1 + \gamma_{21}y_2 + \rho_{11}Wy_1 + \rho_{21}Wy_2 + \varepsilon_1$ $y_2 = X\beta_2 + \gamma_{12}y_1 + \varepsilon_2$	Feedback & One Lag & One Cross-Lag (Same eq.)
18	$y_1 = X\beta_1 + \gamma_{21}y_2 + \rho_{11}Wy_1 + \varepsilon_1$ $y_2 = X\beta_2 + \gamma_{12}y_1 + \rho_{12}Wy_1 + \varepsilon_2$	Feedback & One Lag & One Cross-Lag (Different eq.)
19	$y_1 = X\beta_1 + \gamma_{21}y_2 + \rho_{21}Wy_2 + \varepsilon_1$ $y_2 = X\beta_2 + \gamma_{12}y_1 + \varepsilon_2$	Feedback & One Cross-Lag
20	$y_1 = X\beta_1 + \gamma_{21}y_2 + \rho_{21}Wy_2 + \varepsilon_1$	Feedback & Two Cross-Lags

continued on next page

Table 5.1. Continued

Model	Equations	Simultaneity
	$y_2 = X\beta_2 + \gamma_{12}y_1 + \rho_{12}Wy_1 + \varepsilon_2$	
21	$y_1 = X\beta_1 + \gamma_{21}y_2 + \rho_{11}Wy_1 + \varepsilon_1$ $y_2 = X\beta_2 + \varepsilon_2$	Recursive & One Lag (Same eq.)
22	$y_1 = X\beta_1 + \gamma_{21}y_2 + \varepsilon_1$ $y_2 = X\beta_2 + \rho_{22}Wy_2 + \varepsilon_2$	Recursive & One Lag (Different eq.)
23	$y_1 = X\beta_1 + \gamma_{21}y_2 + \rho_{11}Wy_1 + \varepsilon_1$ $y_2 = X\beta_2 + \rho_{22}Wy_2 + \varepsilon_2$	Recursive & Two Lags
24	$y_1 = X\beta_1 + \gamma_{21}y_2 + \rho_{21}Wy_2 + \varepsilon_1$ $y_2 = X\beta_2 + \varepsilon_2$	Recursive & One Cross-Lag (Same eq.)
25	$y_1 = X\beta_1 + \gamma_{21}y_2 + \varepsilon_1$ $y_2 = X\beta_2 + \rho_{12}Wy_1 + \varepsilon_2$	Recursive & One Cross-Lag (Different eq.)
26	$y_1 = X\beta_1 + \gamma_{21}y_2 + \rho_{21}Wy_2 + \varepsilon_1$ $y_2 = X\beta_2 + \rho_{12}Wy_1 + \varepsilon_2$	Recursive & Two Cross-Lags
27	$y_1 = X\beta_1 + \gamma_{21}y_2 + \rho_{11}Wy_1 + \rho_{21}Wy_2 + \varepsilon_1$ $y_2 = X\beta_2 + \varepsilon_2$	Recursive & One Lag & One Cross-Lag (Same eq.)
28	$y_1 = X\beta_1 + \gamma_{21}y_2 + \rho_{11}Wy_1 + \varepsilon_1$ $y_2 = X\beta_2 + \rho_{12}Wy_1 + \varepsilon_2$	Recursive & One Lag & One Cross-Lag (Different eq.)
29	$y_1 = X\beta_1 + \gamma_{21}y_2 + \rho_{21}Wy_2 + \varepsilon_1$ $y_2 = X\beta_2 + \rho_{22}Wy_2 + \varepsilon_2$	Recursive & One Cross-Lag & One Lag (Different eq.)
30	$y_1 = X\beta_1 + \gamma_{21}y_2 + \varepsilon_1$ $y_2 = X\beta_2 + \rho 21Wy_1 + \rho_{22}Wy_2 + \varepsilon_2$	Recursive (Different eq.) & One Lag & One Cross-Lag
31	$y_1 = X\beta_1 + \gamma_{21}y_2 + \rho_{11}Wy_1 + \rho_{21}Wy_2 + \varepsilon_1$ $y_2 = X\beta_2 + \rho 22Wy_2 + \varepsilon_2$	Recursive & Two Lags & One Cross-Lag (Same eq.)
32	$y_1 = X\beta_1 + \gamma_{21}y_2 + \rho_{11}Wy_1 + \varepsilon_1$ $y_2 = X\beta_2 + \rho_{12}Wy_1 + \rho 22Wy_2 + \varepsilon_2$	Recursive & Two Lags & One Cross-Lag (Different eq.)
33	$y_1 = X\beta_1 + \gamma_{21}y_2 + \rho_{11}Wy_1 + \rho_{21}Wy_2 + \varepsilon_1$ $y_2 = X\beta_2 + \rho_{12}Wy_1 + \varepsilon_2$	Recursive & Two Cross-Lags & One Lag (Same eq.)
34	$y_1 = X\beta_1 + \gamma_{21}y_2 + \rho_{21}Wy_2 + \varepsilon_1$ $y_2 = X\beta_2 + \rho_{12}Wy_1 + \rho_{22}Wy_2 + \varepsilon_2$	Recursive & Two Cross-Lags & One Lag (Different eq.)
35	$y_1 = X\beta_1 + \gamma_{21}y_2 + \rho_{11}Wy_1 + \rho_{21}Wy_2 + \varepsilon_1$ $y_2 = X\beta_2 + \rho_{12}Wy_1 + \rho_{22}Wy_2 + \varepsilon_2$	Recursive & Two Cross-Lags & Two Lags

The necessary conditions for identiˇ cation of an equation in such a model are:

1. The disturbance terms of each equation have zero means and are not spatially autocorrelated.
2. All the basic endogenous variables in the model can be expressed in terms of the disturbance terms, the exogenous variables and the additional endogenous variables.
3. The solution of the model for the basic endogenous variables in terms of the exogenous variables and the disturbance terms is unique.

4. The number of basic endogenous variables appearing on the right hand side of an equation must be less than or equal to the number of exogenous and additional endogenous variables appearing in the model but not in that equation.

As is well known, the presence of endogenous variables on the right hand side (RHS) of an equation in the system results in a nonzero covariance between the regressors and the disturbance term. This leads to the inconsistency of ordinary least squares (OLS). At the same time, there are a wide number of estimators that are consistent in such settings. We subsequently refer to these as Simultaneous Equations Systems Estimators (SESE). However, from an applied perspective knowing that an estimator is consistent may only be cold comfort in situations in which sample sizes are moderate or small, as is the case for many regional economic studies. While the trade off between the inconsistency of OLS relative to the consistency but larger (or non-existent) variance of system methods in small samples has attracted much attention in the mainstream econometrics literature, there is still the question of whether the results of these studies carry over to the models in this taxonomy. Of particular interest is the question of how large the sample size must be before the asymptotic properties of the systems approaches are reflected. We examine these issues in the next section.

There is also the issue of implementation of the SESEs in systems that contain not only the traditional feedback endogeneity but also the simultaneity introduced through the spatial lag and/or cross lag. Kelejian and Robinson (1993), in the context of a single equation model with a spatially lagged dependent variable and spatially autocorrelated error term, suggest a Generalized Methods of Moments Estimator in which the instrument matrix is composed of a subset of the linearly independent columns of (X, WX). This two stage estimator would proceed with the following sequence of steps:[6]

1. Obtain the calculated values of each basic endogenous variable that appears on the RHS of the equation by regressing that variable on the predetermined variables, and their spatial lags.
2. Obtain the calculated values of the additional endogenous variables in the same manner as step 1.
3. Replace the basic and the additional endogenous variables in the ith equation with their calculated values, and then estimate the parameters of the equation using OLS.

Kelejian and Robinson (1993) also suggest that the instrument matrix could be expanded to include higher order terms such as W^2X and W^3X which may improve on the efficiency of the first stage estimator. However, in practice finite sample sizes may limit the number of higher order terms that can be considered.[7] This is because

[6] Extensions to this estimator have been presented in Kelejian and Prucha (1998, 1999) and Kelejian and Robinson (1997).

[7] Use of a subset of the principal components of the matrix of instruments with the higher order terms may be a way to mitigate the small sample problem.

the two stage estimator becomes more like OLS, which is inconsistent in these settings, as the number of instruments used in the ˇ rst stage approaches the sample size. In a more general context, Kelejian and Oates (1989) have noted that the optimal ratio between the sample size and the number of variables used in the ˇ rst stage remains an open question.

In implementing the two stage estimator for any model involving either the own spatial lag or cross spatial lag, there are two possible instruments that can be used for the lags in the ˇ rst stage. The ˇ rst, suggested by Anselin (1980), is the spatial lag of the predicted values of the dependent variable:

$$W\hat{y} = WX(X'X)^{-1}X'y = WX\hat{\beta}. \tag{5.22}$$

The second is to use the predicted value of the spatial lag as its instrument:

$$\widehat{Wy} = X(X'X)^{-1}X'Wy = X\widehat{\beta_w}. \tag{5.23}$$

This second approach is in the same spirit as the traditional treatment in simultaneous equation settings, where each endogenous variable (including any spatial lag) is regressed on the complete set of exogenous variables to form its instrument. In the ˇ rst approach, the initial regression uses only the original endogenous variables (excluding the lags) and then the lag of the predicted variables are used to form the instrument for the spatial lag.

The two approaches will not be equivalent which can be shown as follows. The difference between the two instruments for the spatial lag is:

$$W\hat{y} - \widehat{Wy} = \left[WX(X'X)^{-1}X' - X(X'X)^{-1}X'W\right]y. \tag{5.24}$$

It is obvious that the term in the brackets would have to be 0 in order for the two instruments to be equal. This will not be the case for either row standardized or unstandardized weight matrices. For an unstandardized symmetric weights matrix, the two terms in the brackets become each others' transpose.[8] However, this property does not hold for a row standardized weights matrix.

5.5 Monte Carlo Experiments

To provide some empirical insight to the issues raised above, we carried out a series of Monte Carlo simulations. A consideration of all the models in the taxonomy is clearly outside the scope of the current chapter and, instead, we focus on the model with both traditional simultaneity as well as spatially lagged dependent variables which correspond to model 13 from Table 5.1:

$$\begin{aligned} y_1 &= X\beta_1 + \gamma_{21}y_2 + \rho_{11}Wy_1 + \varepsilon_1, \\ y_2 &= X\beta_2 + \gamma_{12}y_1 + \rho_{22}Wy_2 + \varepsilon_2. \end{aligned} \tag{5.25}$$

[8] A referee pointed out that not all unstandardized matrices need to be symmetric.

Table 5.2. Parameter values for experiments

DGP	1	2	3	4	5	6	7	8	9	10
ρ_{11}	0.0	0.0	0.0	0.0	0.3	0.3	0.3	0.6	0.6	0.8
ρ_{22}	0.0	0.3	0.6	0.8	0.3	0.6	0.8	0.6	0.8	0.8

The system is specified so that the identification conditions are satisfied by means of zero restrictions on the β parameters. In particular, the matrix of exogenous variables is n by k, where $k = 5$, while β_1 and β_2 are 5 by 1 parameter vectors each with nonzero elements for the constant term and two other elements: the second and third for β_1, and the fourth and fifth for β_2. The observations on the exogenous variables are taken from a uniform distribution on the interval 1–10 and are kept constant for all replications. Following Anselin and Kelejian (1997), we set $\gamma_{21} = 1.0$ and $\gamma_{12} = 0.10$, with all nonzero elements of β_1 and β_2 equal to 1, and ε_1 and ε_2 are uncorrelated standard normal error terms.

The layout of the observations in the experiments is based on regular lattices of varying sizes that have been used extensively in previous spatial econometric research (Anselin and Rey, 1991; Anselin and Florax, 1995b; Florax and Rey, 1995). The weight matrices employed are based on the rook criterion of contiguity, with each matrix being row standardized. Because the system consists of two equations, each with n observations, there is an additional computational burden associated with the experiments relative to the single equation case. Therefore, we analyze a limited number of sample sizes: $n = 25, 81, 225$ for the 10 different combinations of the spatial parameters that are summarized in Table 5.2. For each data generating process (DGP) we generate 5,000 realizations. We examine the small sample properties of five different estimators: Ordinary Least Squares (OLS); Non-Spatial (standard) Two Stage Least Squares (2SLS); Spatial Two Stage Least Squares (S2SLS); and two versions of the Kelejian-Robinson-Prucha estimator. The first includes the exogenous variables and their lags to construct the instrument matrix (KRP_1) and the second includes the second power of these variables as well (KRP_2). More explicitly, focusing on the first equation in the system with the parameter vector $\theta_1' = [\beta_1', \gamma_{12}, \rho_{11}]$, the estimators examined are as follows:

$$\hat{\theta}_{OLS} = (Z_1'Z_1)^{-1}Z_1'y_1, \tag{5.26}$$

where $Z_1 = [x_1, y_2, Wy_1]$,

$$\hat{\theta}_{2SLS} = (Z_1'Z_1)^{-1}Z_1'y_1, \tag{5.27}$$

where $Z_1 = [x_1, \hat{y}_2, \widehat{Wy_1}]$, with $\hat{y}_2 = Qy_2$, $Q = X(X'X)^{-1}X'$, $X = [x_1, x_2]$, and $\widehat{Wy_1} = QWy_1$,

$$\hat{\theta}_{S2SLS} = (Z_1'Z_1)^{-1}Z_1'y_1, \tag{5.28}$$

with $Z_1 = [x_1, \hat{y}_2, W\hat{y}_1]$, and $W\hat{y}_1 = WQy_1$,

$$\hat{\theta}_{KRP1} = (Z_1'Z_1)^{-1}Z_1'y_1, \tag{5.29}$$

where $Z_1 = [x_1, \hat{y}_2, \widehat{Wy_1}]$, $\hat{y}_2 = Qy_2$, $Q = X(X'X)^{-1}X'$, $X = [x_1, x_2, WX]$, $\widehat{Wy_1} = QWy_1$, and X includes all the columns of $[x_1, x_2]$ with the exception of the constant terms,

$$\hat{\theta}_{KRP2} = (Z_1'Z_1)^{-1}Z_1'y_1, \tag{5.30}$$

with $Z_1 = [x_1, \hat{y}_2, \widehat{Wy_1}]$, $\hat{y}_2 = Qy_2$, $Q = X(X'X)^{-1}X'$, $X = [x_1, x_2, WX, WWX]$, $\widehat{Wy_1} = QWy_1$, and X is as in (5.29).

5.6 Results

Tables 5.3–5.8 summarize the results of our experiments for several different characteristics of the distributions of the five estimators. Following Kelejian and Prucha (1999) our measure of Bias is defined as the absolute difference between the median and the true parameter value under the DGP. Our second measure is closely related to the Root Mean Squared Error (RMSE) and is defined as:

$$RMSE = \left[Bias^2 + \left(\frac{IQ}{1.35} \right)^2 \right]^{(1/2)}, \tag{5.31}$$

where IQ is the inter-quartile range. As Kelejian and Prucha (1999) note, if the distribution is normal then $IQ/1.35$ is equal to the standard deviation, however, unlike the traditional measures of RMSE and Bias, the measures used here are assured to exist.

Tables 5.3 and 5.4 provide a comparison of the alternative estimators for slope coefficients from the two separate equations. Several findings emerge. The SESE approaches dominate OLS with respect to bias, with the exceptions of S2SLS in the first equation (Table 5.3) and 2SLS in the second equation (Table 5.4). At the same time, however, with respect to RMSE, OLS dominates all the SESEs in the first equation, but only the 2SLS estimator in the second. The consistency property of the KRP estimators is reflected in all sample sizes and for both equations. This is not the case for the other two SESEs approaches, 2SLS and S2SLS, for which the relative performance with respect to bias changes across the two equations. With respect to the KRP estimators, KRP_1 does better on average in both equations relative to KRP_2 with respect to bias. The impact of including higher order terms in the instrument matrix appears to have mixed results, as the KRP_2 estimator has a slightly lower RMSE on average in the first equation, but higher in the second relative to KRP_1 which only includes the exogenous variables and their spatial lags in the instrument matrix. Tables 5.5 and 5.6 contain a similar comparison for the estimators on the coefficients on the feedback variables y_2 and y_1, respectively. The results are in general agreement with those found in Tables 5.3 and 5.4. Again, the SESE approaches dominate OLS with respect to bias, with the exceptions of S2SLS in the first equation and 2SLS in the second equation, while OLS has a lower RMSE than each of the SESEs in the first equation, but only dominates 2SLS in the second

Table 5.3. Bias and RMSE $\beta_{2,1}$, OLS=1.

n	ρ_{11}	ρ_{22}	Bias				RMSE			
			2SLS	S2SLS	KRP_1	KRP_2	2SLS	S2SLS	KRP_1	KRP_2
25	0.0	0.0	0.166	0.190	0.428	0.568	1.278	1.026	1.023	1.009
25	0.0	0.3	0.254	0.671	0.562	0.591	1.235	1.037	1.018	1.004
25	0.0	0.6	0.019	1.397	0.699	0.702	1.207	1.071	1.028	1.007
25	0.0	0.8	0.278	5.571	0.429	0.570	1.253	1.384	1.037	1.008
25	0.3	0.3	0.148	4.560	0.586	0.761	1.261	1.266	1.033	1.019
25	0.3	0.6	0.308	8.824	0.439	0.635	1.339	1.645	1.027	1.028
25	0.3	0.8	0.210	20.891	0.575	0.746	1.394	2.629	1.026	1.004
25	0.6	0.6	0.010	35.915	0.312	0.507	1.515	4.090	1.032	1.015
25	0.6	0.8	0.664	105.205	0.396	0.623	1.630	8.432	1.024	1.010
25	0.8	0.8	0.074	515.734	0.556	0.641	2.079	43.081	1.025	1.025
81	0.0	0.0	0.205	0.584	0.405	0.362	1.481	1.025	1.010	1.007
81	0.0	0.3	0.088	1.875	0.372	0.375	1.394	1.032	0.996	0.991
81	0.0	0.6	0.293	4.450	0.482	0.504	1.380	1.079	0.997	1.003
81	0.0	0.8	0.538	12.570	0.131	0.083	1.368	1.195	1.018	1.014
81	0.3	0.3	0.478	0.720	0.226	0.229	1.579	1.050	1.001	0.998
81	0.3	0.6	0.094	7.716	0.256	0.206	1.520	1.196	1.012	1.009
81	0.3	0.8	1.923	38.803	0.026	0.103	1.319	1.482	1.018	1.012
81	0.6	0.6	1.110	2.613	0.405	0.268	1.475	1.554	0.999	1.010
81	0.6	0.8	0.286	15.354	0.450	0.584	1.233	15.724	1.011	0.999
81	0.8	0.8	0.542	0.786	0.409	0.156	1.348	14.759	1.006	0.992
225	0.0	0.0	0.183	0.368	0.549	0.549	1.610	0.999	0.999	1.002
225	0.0	0.3	0.223	0.470	0.190	0.190	1.628	0.997	0.999	1.001
225	0.0	0.6	0.014	0.908	0.350	0.270	1.666	1.013	1.007	1.007
225	0.0	0.8	0.585	4.131	0.030	0.077	1.785	1.023	1.003	1.004
225	0.3	0.3	0.839	6.254	0.136	0.152	2.539	1.234	1.013	1.013
225	0.3	0.6	0.893	11.794	0.170	0.059	2.384	1.412	1.017	1.011
225	0.3	0.8	0.882	45.602	0.083	0.089	2.791	2.543	1.017	1.001
225	0.6	0.6	1.222	35.766	0.310	0.261	3.606	5.548	1.002	0.999
225	0.6	0.8	0.336	122.637	0.082	0.411	1.359	11.156	1.026	1.004
225	0.8	0.8	0.953	57.755	0.576	0.629	1.089	11.150	1.007	1.009
Col. Median			0.290	5.913	0.400	0.393	1.435	1.325	1.015	1.007

equation. Also repeated is the relatively lower bias of the KRP estimators in both equations, with KRP_1 doing slightly better than KRP_2 on this criterion. Here the consistency property appears more strongly as the bias now tends to decline with increasing sample size, in contrast to the case for the slope parameters for which there was no discernible trend.

Finally, Tables 5.7 and 5.8 compare the performance of the estimators for the spatial lag parameters in each of the equations. The patterns found in comparing

Table 5.4. Bias and RMSE $\beta_{4,2}$, OLS=1.

n	ρ_{11}	ρ_{22}	Bias				RMSE			
			2SLS	S2SLS	KRP_1	KRP_2	2SLS	S2SLS	KRP_1	KRP_2
25	0.0	0.0	0.084	0.640	0.422	0.639	1.330	0.845	0.807	0.843
25	0.0	0.3	0.253	0.640	0.403	0.642	1.464	0.839	0.810	0.838
25	0.0	0.6	1.647	0.656	0.394	0.633	2.384	0.841	0.801	0.835
25	0.0	0.8	3.394	0.674	0.419	0.643	3.306	0.837	0.800	0.833
25	0.3	0.3	0.463	0.624	0.414	0.642	1.455	0.826	0.821	0.844
25	0.3	0.6	1.807	0.581	0.361	0.594	2.138	0.798	0.779	0.821
25	0.3	0.8	2.704	0.581	0.384	0.627	2.453	0.779	0.792	0.827
25	0.6	0.6	1.881	0.485	0.381	0.584	1.783	0.803	0.862	0.870
25	0.6	0.8	2.233	0.331	0.352	0.574	1.800	0.702	0.842	0.859
25	0.8	0.8	1.944	0.313	0.337	0.558	1.315	0.788	0.901	0.910
81	0.0	0.0	2.266	0.200	0.119	0.201	3.361	0.507	0.500	0.508
81	0.0	0.3	2.069	0.203	0.120	0.203	3.148	0.538	0.532	0.540
81	0.0	0.6	1.715	0.222	0.115	0.199	2.580	0.540	0.529	0.532
81	0.0	0.8	1.802	0.290	0.104	0.185	2.956	0.577	0.522	0.523
81	0.3	0.3	2.119	0.215	0.135	0.216	2.597	0.509	0.506	0.519
81	0.3	0.6	1.677	0.215	0.113	0.197	2.160	0.538	0.526	0.533
81	0.3	0.8	1.743	0.334	0.140	0.224	2.663	0.597	0.507	0.515
81	0.6	0.6	1.381	0.223	0.139	0.230	1.643	0.522	0.531	0.545
81	0.6	0.8	2.018	1.832	0.162	0.229	3.620	5.475	0.627	0.609
81	0.8	0.8	1.828	0.808	0.142	0.207	1.939	2.211	0.837	0.821
225	0.0	0.0	0.227	0.088	0.057	0.088	0.708	0.311	0.305	0.312
225	0.0	0.3	0.161	0.076	0.042	0.076	0.698	0.302	0.298	0.301
225	0.0	0.6	1.758	0.086	0.040	0.073	3.414	0.300	0.296	0.297
225	0.0	0.8	2.848	0.129	0.047	0.078	4.009	0.319	0.299	0.301
225	0.3	0.3	0.079	0.068	0.034	0.066	0.722	0.309	0.307	0.308
225	0.3	0.6	1.584	0.080	0.034	0.067	2.497	0.307	0.301	0.303
225	0.3	0.8	1.397	0.061	0.037	0.074	2.001	0.316	0.305	0.306
225	0.6	0.6	0.948	0.038	0.038	0.074	1.420	0.370	0.377	0.378
225	0.6	0.8	3.119	1.431	0.067	0.106	3.806	2.666	0.472	0.455
225	0.8	0.8	1.105	1.381	0.119	0.132	2.823	1.254	0.838	0.812
Col. median			1.750	0.301	0.127	0.205	2.272	0.587	0.530	0.536

Table 5.3 versus 5.4 and Table 5.5 versus 5.6 are repeated in the comparison of the estimates for the lag parameters.

Taking the results in Tables 5.3–5.8 together, several general conclusions can be reached. On average the KRP estimators dominate the other SESE approaches for all of the parameter values based on a RMSE criterion. It is also the case that the switch in the relative performance of the 2SLS and S2SLS estimators is uniform in that, with respect to bias, the former estimator is superior to the latter for the

ˇ rst equation but the situation is reversed in the second equation. This is robust to the particular parameter under consideration. A similar result holds for the RMSE values for KRP_2 versus KRP_1, with the former dominating the latter in the ˇ rst equation yet not in the second equation. There is also a pattern to the dominance of the RMSE of OLS over all other estimators for parameters in the ˇ rst equation, while in the second equation OLS dominates only the 2SLS estimator.

The bias of the S2SLS estimator for the coefˇ cients for the ˇ rst equation is very sensitive to the value of the spatial autoregressive parameters under the DGP. In particular, when one of the autoregressive parameters reaches a value of 0.8, while the other parameter is non-zero, the bias of the S2SLS estimator increases dramatically. This is true for all of the coefˇ cient estimates (Tables 5.3, 5.5 and 5.7). The bias is also markedly larger in the ˇ rst equation compared to the second. We think that this alternating pattern may be related to the difference in the parameters on the basic endogenous variables, which are set to unity in the ˇ rst equation and 0.10 in the second. It may be that the linear combination of this coefˇ cient from the ˇ rst equation with the larger values of the spatial autoregressive lag coefˇ cients approaches a critical value that affects the S2SLS estimator, while in the second equation the smaller value of the coefˇ cient on the basic endogenous variable keeps this linear combination below the critical value. This may also provide an explanation for why the KRP estimators clearly dominate OLS in the second equation but not in the ˇ rst equation, although further research into the causes of these patterns is needed.

5.7 Conclusions

This chapter has explored some of the issues that arise in the application of spatial econometric methods in the context of simultaneous equation systems. We suggest a taxonomy of 35 models that combine three sources of simultaneity: feedback, spatial autoregressive and spatial cross-regressive. These models have the potential to open up new avenues of applied spatial econometric research in urban and regional economics.

The results of our experiments suggest that care must be taken in distinguishing between the simultaneity due to the presence of spatial variables and that due to the traditional endogenous variables. Estimators which take that distinction in mind utilize spatially explicit instruments which leads to clear gains in lower bias and generally lower RMSE than estimators which omit any spatial instruments. Additionally, we ˇ nd that the way in which the instruments for the spatial lag variable are constructed matters, in that predicting the spatial lag of the dependent variable is to be preferred to constructing the lag of the predicted dependent variable.

Our chapter is an initial foray into what appears to be a potentially rich area for further investigation. We have only touched on one of the models in the taxonomy and we are interested to see to what extent the ˇ ndings from our experiments carry over to these other models. We also hope to expand the taxonomy in a number of dimensions such as incorporating spatial lags of the exogenous variables, consider-

Table 5.5. Bias and RMSE $\gamma_{2,1}$, OLS=1.

n	ρ_{11}	ρ_{22}	Bias				RMSE			
			2SLS	S2SLS	KRP$_1$	KRP$_2$	2SLS	S2SLS	KRP$_1$	KRP$_2$
25	0.0	0.0	0.161	0.081	0.007	0.335	1.833	1.252	1.186	1.123
25	0.0	0.3	0.374	1.009	0.163	0.434	1.530	1.262	1.152	1.073
25	0.0	0.6	0.243	1.314	0.182	0.450	1.391	1.313	1.133	1.081
25	0.0	0.8	0.062	0.681	0.325	0.627	1.431	1.207	1.087	1.066
25	0.3	0.3	0.231	2.720	0.204	0.488	1.418	1.493	1.139	1.107
25	0.3	0.6	0.080	4.013	0.259	0.523	1.412	1.652	1.145	1.112
25	0.3	0.8	0.012	5.322	0.213	0.490	1.666	1.733	1.145	1.106
25	0.6	0.6	0.045	18.258	0.169	0.363	1.563	3.316	1.141	1.088
25	0.6	0.8	0.428	27.060	0.075	0.357	1.864	4.546	1.129	1.089
25	0.8	0.8	0.268	82.676	0.058	0.189	2.207	15.394	1.181	1.111
81	0.0	0.0	0.255	0.062	0.052	0.019	2.020	1.153	1.137	1.116
81	0.0	0.3	0.002	0.609	0.070	0.133	1.571	1.237	1.123	1.094
81	0.0	0.6	0.300	1.389	0.083	0.170	1.966	1.372	1.062	1.045
81	0.0	0.8	0.138	2.280	0.079	0.204	2.517	1.593	1.062	1.047
81	0.3	0.3	0.096	2.958	0.036	0.090	1.683	1.712	1.148	1.156
81	0.3	0.6	0.292	8.141	0.062	0.054	2.555	2.579	1.177	1.153
81	0.3	0.8	0.213	17.758	0.130	0.068	2.866	4.249	1.224	1.203
81	0.6	0.6	1.743	74.998	0.848	0.484	2.500	6.174	1.235	1.203
81	0.6	0.8	0.006	71.405	0.154	0.192	1.843	27.205	1.283	1.234
81	0.8	0.8	0.041	95.754	0.200	0.185	1.546	28.224	1.237	1.191
225	0.0	0.0	0.035	0.008	0.010	0.033	1.560	0.958	0.950	0.947
225	0.0	0.3	0.196	0.046	0.035	0.069	3.640	0.906	0.938	0.936
225	0.0	0.6	0.216	0.243	0.002	0.042	6.356	0.836	0.922	0.916
225	0.0	0.8	0.075	0.624	0.032	0.085	7.420	0.791	0.928	0.924
225	0.3	0.3	0.385	2.644	0.046	0.026	6.622	1.635	1.043	1.032
225	0.3	0.6	0.079	6.083	0.004	0.064	8.566	2.593	1.127	1.101
225	0.3	0.8	0.175	10.259	0.062	0.070	7.435	4.054	1.155	1.111
225	0.6	0.6	1.042	47.472	0.155	0.174	7.936	10.426	1.222	1.151
225	0.6	0.8	0.060	57.937	0.245	0.188	2.067	20.924	1.212	1.182
225	0.8	0.8	0.010	60.658	0.005	0.036	1.248	25.952	1.125	1.077
Col. median			0.168	3.486	0.077	0.180	1.915	1.682	1.140	1.104

Table 5.6. Bias and RMSE $\gamma_{1,2}$, OLS=1.

n	ρ_{11}	ρ_{22}	Bias				RMSE			
			2SLS	S2SLS	KRP_1	KRP_2	2SLS	S2SLS	KRP_1	KRP_2
25	0.0	0.0	0.930	0.616	0.359	0.612	1.059	0.721	0.596	0.717
25	0.0	0.3	1.086	0.634	0.378	0.624	1.195	0.725	0.598	0.717
25	0.0	0.6	1.861	0.664	0.387	0.624	1.902	0.742	0.599	0.713
25	0.0	0.8	2.849	0.678	0.372	0.624	2.734	0.756	0.603	0.728
25	0.3	0.3	1.039	0.604	0.356	0.605	1.082	0.705	0.592	0.713
25	0.3	0.6	1.701	0.594	0.331	0.585	1.653	0.690	0.589	0.697
25	0.3	0.8	2.242	0.585	0.346	0.593	2.107	0.659	0.604	0.704
25	0.6	0.6	1.501	0.464	0.291	0.535	1.367	0.592	0.623	0.702
25	0.6	0.8	1.934	0.330	0.312	0.533	1.723	0.412	0.632	0.703
25	0.8	0.8	2.126	0.077	0.297	0.519	1.669	0.219	0.736	0.772
81	0.0	0.0	0.717	0.177	0.089	0.179	1.048	0.342	0.318	0.342
81	0.0	0.3	0.939	0.192	0.108	0.187	1.352	0.342	0.327	0.338
81	0.0	0.6	1.916	0.205	0.092	0.181	2.302	0.354	0.321	0.344
81	0.0	0.8	3.367	0.253	0.103	0.184	3.479	0.382	0.321	0.345
81	0.3	0.3	1.059	0.191	0.095	0.186	1.269	0.350	0.328	0.349
81	0.3	0.6	1.875	0.208	0.102	0.186	1.988	0.345	0.326	0.349
81	0.3	0.8	2.791	0.242	0.116	0.206	2.771	0.361	0.327	0.356
81	0.6	0.6	1.675	0.192	0.116	0.210	1.663	0.326	0.352	0.367
81	0.6	0.8	2.191	0.020	0.136	0.206	2.096	0.317	0.424	0.423
81	0.8	0.8	3.410	0.170	0.079	0.132	2.864	0.404	0.611	0.568
225	0.0	0.0	0.199	0.069	0.034	0.069	0.427	0.198	0.192	0.198
225	0.0	0.3	0.575	0.069	0.032	0.067	1.095	0.200	0.196	0.199
225	0.0	0.6	2.461	0.064	0.025	0.061	2.854	0.197	0.193	0.196
225	0.0	0.8	3.815	0.035	0.032	0.068	3.821	0.210	0.203	0.205
225	0.3	0.3	0.924	0.069	0.032	0.066	1.135	0.196	0.193	0.196
225	0.3	0.6	2.198	0.056	0.031	0.066	2.227	0.192	0.188	0.195
225	0.3	0.8	2.802	0.279	0.025	0.061	2.783	0.354	0.199	0.204
225	0.6	0.6	1.747	0.092	0.025	0.058	1.728	0.224	0.226	0.227
225	0.6	0.8	2.319	0.539	0.048	0.080	2.284	0.763	0.257	0.251
225	0.8	0.8	3.580	0.332	0.039	0.063	3.322	0.515	0.415	0.401
Col. median			1.896	0.206	0.103	0.186	1.815	0.354	0.327	0.353

Table 5.7. Bias and RMSE $\rho_{1,1}$, OLS=1.

n	ρ_{11}	ρ_{22}	Bias				RMSE			
			2SLS	S2SLS	KRP$_1$	KRP$_2$	2SLS	S2SLS	KRP$_1$	KRP$_2$
25	0.0	0.0	3.311	0.760	0.167	0.538	5.738	1.638	1.056	1.016
25	0.0	0.3	1.696	11.555	0.128	0.568	5.111	2.186	1.057	1.014
25	0.0	0.6	0.356	18.988	0.160	0.516	4.667	4.126	1.042	1.007
25	0.0	0.8	0.069	23.411	0.316	0.647	3.954	6.864	1.045	1.035
25	0.3	0.3	3.865	43.222	1.432	1.094	5.228	3.044	1.079	1.032
25	0.3	0.6	0.721	61.203	0.208	0.375	4.771	5.679	1.094	1.055
25	0.3	0.8	0.297	65.222	0.064	0.435	3.681	9.660	1.104	1.047
25	0.6	0.6	10.436	1131.073	3.673	2.436	4.212	10.077	1.142	1.062
25	0.6	0.8	0.630	139.193	0.198	0.293	3.243	15.605	1.157	1.104
25	0.8	0.8	0.364	134.187	0.080	0.234	1.886	24.232	1.169	1.114
81	0.0	0.0	130.100	3.400	23.000	17.800	10.279	1.742	1.125	1.085
81	0.0	0.3	4.863	19.075	0.569	0.458	11.268	2.686	1.116	1.105
81	0.0	0.6	1.061	25.407	0.336	0.511	11.683	5.065	1.193	1.100
81	0.0	0.8	0.279	25.064	0.228	0.381	9.414	8.872	1.262	1.140
81	0.3	0.3	1.056	6.234	0.142	0.226	10.670	3.228	1.157	1.088
81	0.3	0.6	2.715	25.950	0.200	0.241	11.481	6.952	1.303	1.186
81	0.3	0.8	10.888	539.481	1.350	1.913	8.123	15.386	1.459	1.264
81	0.6	0.6	1.298	28.571	0.391	0.463	8.534	14.848	1.426	1.151
81	0.6	0.8	0.074	272.100	0.112	0.173	1.780	73.757	1.287	1.192
81	0.8	0.8	0.262	225.364	0.331	0.308	2.236	71.008	1.367	1.255
225	0.0	0.0	5.818	1.153	0.182	0.005	34.894	1.809	1.112	1.126
225	0.0	0.3	3.681	17.805	0.194	0.301	34.017	4.159	1.156	1.146
225	0.0	0.6	1.493	29.938	0.143	0.178	27.564	10.388	1.139	1.085
225	0.0	0.8	0.164	45.229	0.045	0.126	18.842	22.092	1.108	1.022
225	0.3	0.3	1.542	10.847	0.061	0.075	29.263	5.923	1.099	1.024
225	0.3	0.6	0.215	64.366	0.013	0.011	24.972	16.909	1.275	1.140
225	0.3	0.8	0.598	251.231	0.074	0.047	16.173	47.642	1.362	1.209
225	0.6	0.6	2.289	133.728	0.084	0.139	20.365	49.694	1.414	1.159
225	0.6	0.8	0.251	448.774	0.350	0.297	3.386	144.289	1.448	1.311
225	0.8	0.8	0.167	365.777	0.011	0.002	1.941	134.991	1.281	1.190
Col. median			1.059	36.580	0.188	0.305	8.329	9.266	1.156	1.104

Table 5.8. Bias and RMSE $\rho_{2,2}$, OLS=1.

n	ρ_{11}	ρ_{22}	Bias				RMSE			
			2SLS	S2SLS	KRP$_1$	KRP$_2$	2SLS	S2SLS	KRP$_1$	KRP$_2$
25	0.0	0.0	57.525	0.439	0.663	0.710	7.703	1.167	1.278	1.142
25	0.0	0.3	12.605	0.328	0.263	0.472	9.147	1.112	1.229	1.108
25	0.0	0.6	2.982	0.498	0.368	0.610	10.751	0.888	0.985	0.920
25	0.0	0.8	7.009	0.530	0.369	0.619	9.886	0.735	0.768	0.793
25	0.3	0.3	1.044	0.586	0.429	0.656	5.144	0.900	0.943	0.916
25	0.3	0.6	3.191	0.520	0.363	0.608	6.012	0.766	0.762	0.822
25	0.3	0.8	4.091	0.471	0.348	0.593	4.126	0.614	0.672	0.737
25	0.6	0.6	2.550	0.396	0.327	0.563	2.691	0.636	0.722	0.770
25	0.6	0.8	2.606	0.229	0.336	0.547	2.343	0.396	0.667	0.725
25	0.8	0.8	2.021	0.033	0.287	0.522	1.585	0.230	0.735	0.773
81	0.0	0.0	44.558	0.384	0.287	0.381	25.531	1.171	1.178	1.150
81	0.0	0.3	69.784	0.106	0.040	0.069	33.346	1.214	1.285	1.213
81	0.0	0.6	59.497	0.784	0.288	0.280	61.229	1.389	1.432	1.247
81	0.0	0.8	6.113	0.832	0.129	0.019	54.196	1.199	0.890	0.728
81	0.3	0.3	21.705	0.069	0.090	0.134	21.157	0.927	1.004	0.929
81	0.3	0.6	21.863	0.204	0.033	0.036	29.479	0.837	0.898	0.788
81	0.3	0.8	2.744	0.560	0.000	0.097	19.997	0.793	0.537	0.468
81	0.6	0.6	8.426	0.160	0.063	0.155	10.042	0.581	0.597	0.546
81	0.6	0.8	2.676	1.207	0.138	0.215	2.859	1.177	0.437	0.429
81	0.8	0.8	3.228	1.329	0.112	0.169	2.835	1.316	0.591	0.561
225	0.0	0.0	35.728	0.193	0.007	0.153	45.930	1.311	1.322	1.314
225	0.0	0.3	78.971	0.164	0.143	0.093	89.724	1.281	1.327	1.269
225	0.0	0.6	76.798	0.203	0.016	0.019	88.859	0.777	0.764	0.701
225	0.0	0.8	26.974	0.627	0.002	0.042	28.458	0.765	0.344	0.303
225	0.3	0.3	28.182	0.027	0.021	0.050	34.924	0.633	0.644	0.631
225	0.3	0.6	25.376	0.090	0.024	0.058	26.814	0.424	0.416	0.380
225	0.3	0.8	8.915	0.821	0.021	0.056	9.086	0.900	0.251	0.241
225	0.6	0.6	7.144	0.349	0.018	0.054	7.246	0.487	0.309	0.304
225	0.6	0.8	2.747	1.330	0.037	0.070	2.981	1.585	0.251	0.249
225	0.8	0.8	4.488	0.504	0.020	0.056	4.388	0.777	0.442	0.418
Col. median			7.785	0.417	0.121	0.154	9.964	0.863	0.749	0.754

ing more complicated error processes and relaxing the assumptions that the weight matrices are identical for all spatial lags (both cross and own).

In addition to extensions of the taxonomy there are other interesting research directions we plan to explore. There is a wider set of estimators beyond the ones we utilized here that need to be evaluated within the taxonomy. From a substantive perspective, the spatial spillover and multiplier properties of the different models should be investigated. Finally, there are a host of issues related to developing new diagnostic tests for spatial effects as well as exogeneity for models in the taxonomy. Thus far only Anselin and Kelejian (1997) have examined the properties of tests for spatial error dependence in the presence of a-spatial endogenous regressors. Their focus was on the tests applied to a single equation that had spatially dependent error terms and either a spatial lag or another (traditional) endogenous variable. Spatial diagnostics that had previously been developed for single equation settings were found to perform poorly in the presence of endogeneity. The questions related to the generalization of these single-equation results to settings outlined above remain for future research.

Acknowledgments

A previous version of this chapter was presented at a special session on spatial econometrics during the 45th North American Meetings of the Regional Science Association International, Santa Fe, New Mexico. The authors wish to thank participants of that session as well as the editors and anonymous reviewers for helpful comments on this work. The usual caveats apply. Rey received funding from the College of Arts and Letters Faculty Development Program at the San Diego State University. Boarnet received funding from the U.S. and California Departments of Transportation, through a grant administered by the University of California Transportation Center.

6 Exploring Spatial Data Analysis Techniques Using R: The Case of Observations with No Neighbors

Roger S. Bivand[1] and Boris A. Portnov[2]

[1] Norwegian School of Economics and Business Administration
[2] Ben-Gurion University of the Negev, Israel

6.1 Introduction

It is widely acknowledged that one of the impediments to a broader acceptance of techniques for spatial data analysis is that handling spatial data involves more insight and possibly the use of additional applications than other forms of data (Anselin, 2000, p. 217). We are perhaps more familiar with the potential difficulties caused by the inadequate mapping of data into temporal reference frameworks, such as the predicted complications attributed to the year 2000 problem, when a circular measure ($99+1=0$) was treated as linear. Spatial data come with many assumptions about their reference frameworks, including projection metadata, and are often derived from geographical information systems or other archives of spatial position data. Some of these are also time-specific, where boundary segments are introduced to or removed from maps of polygon representations of spatial objects.

While it is possible to abstract spatial data from their reference framework, in practice the analyst may very well wish to be able to supplement attribute data for spatial objects at a later stage, to move results to another application retaining positional data, or to interpolate to some other set of spatial objects. In this sense, the objects are very different from more characteristic statistical observations, or database records, in that they are linked to positional information which is not arbitrary, and which also indicates the relative positions of the objects in relationship to each other. While it is clear that time series and spatial series are not analogous (Ripley, 1988, pp. 1–8), it is not difficult to grasp that $01-99=2$ and $99+2=01$ in a two-digit representation of years. Getting spatial data into representations that are both "well-known" and adequately documented seems more challenging. Added reasons for treating spatial reference frameworks seriously are the ever-increasing volume of attribute data with geography, and the realization that spatial indices may permit otherwise incompatible data to be "fused."

While interactive data-analysis environments such as S-PLUS and R, both implementations of the underlying S language (Becker *et al.*, 1998; Chambers and Hastie, 1992; Chambers, 1998), require that the user be willing to work within a programming language, the fact that the functions they offer may be extended by the user also allows us to explore some of the consequences of choices of ways of representing relationships between spatial objects. Following on from an earlier review (Bivand and Gebhardt, 2000), we will here be concerned to show how the flexibility offered by working within a well-supported programming environment can permit the exploration of a chosen issue within spatial data analysis. Our choice has been motivated by the need to examine the potential usefulness of spatial econometric techniques in relation to studying urban clustering in sparsely populated regions.

In this contribution, we will introduce brie y the facilities now available in R for creating and manipulating spatial weights objects, and show how they permit exploration of varying approaches, including differing weighting schemes. Following this, we will describe one of the consequences of some deˇ nitions of spatial neighborhood, that some spatial objects have no "neighbors" under the scheme chosen. This in turn generates artifacts to which our attention may be drawn, for instance in Moran scatterplots (Anselin, 1996), with potential consequences for inferences. Finally, we turn to our urban clustering case, to see how these technical questions may affect the analytical choices we would prefer to make based on domain knowledge.

6.2 Implementing Spatial Weights Objects in R

Basic information on the R programming environment itself may be found in the initial source (Ihaka and Gentleman, 1996), and the project site[1]. The functions to be discussed here are now available in a single contributed package named **spdep** on CRAN. Some of them utilize dynamically loaded compiled C code, but most are coded in R directly. In addition to the package documentation, available online on CRAN, aspects of some of the functions have been discussed in Bivand (2001, 2002a).

Cliff and Ord (1973, pp. 11–13) provide the initial formalization of the relationships as a generalized weighting matrix, most usually termed W. In a more recent study reviewing the use of different forms of weighting matrices, Grifˇ th (1996) has demonstrated that a parsimonious speciˇ cation of the relationships between observations is to be preferred to one making assumptions about say distance decay. Brett and Pinkse (1997) also note differences in inference which can occur in using distance bands and contiguities, which they call "Hotelling neighbors" for obvious reasons.

It is usual in the literature to deˇ ne the contiguity relation in terms of sets $N_{(i)}$ of neighbors of zone or site i, where $c_{ij} = 1$ if i is linked to j and $c_{ij} = 0$ otherwise. This implies no use of other information than that of neighborhood set membership. Set membership may be deˇ ned on the basis of shared boundaries, of centroids lying within distance bands, or other a priori grounds. The functions in **spdep** provide for most of these membership deˇ nitions: boundary contiguities of polygons using `poly2nb()`, and distance bands by `dnearneigh()`, but include others of potential interest. These are neighbors by Delaunay triangulation (`tri2nb`), derived from the **tripack** package maintained by Albrecht Gebhardt, `soi.graph()` based on an initial triangulation, and `gabrielneigh()` and `relativeneigh()` contributed by Nicholas Lewin-Koh, and `knn2nb()` providing k'th nearest neighbors.

The internal representation of the $N_{(i)}$ set of neighbor indices is kept very simple, as a list of length n with list elements integer vectors of spatial object indices. These

[1] http://www.r-project.org

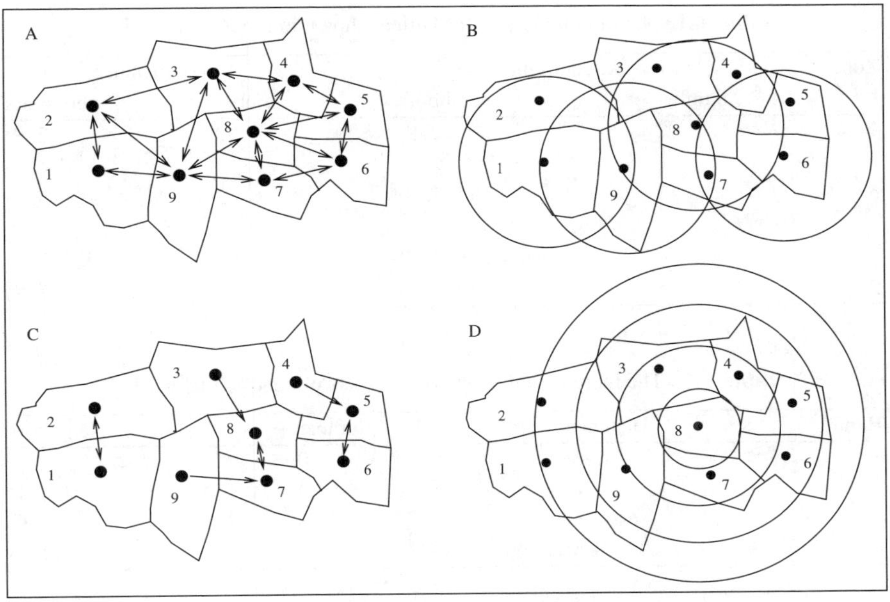

Fig. 6.1. Selected neighborhood schemes for polygon and point spatial objects – A: contiguous neighbors, B: distance neighbors, C: nearest neighbors, D: distance band neighbors.

neighbor lists have a region ID attribute, through which the indices may be manipulated if necessary. Functions are also provided for ˇ nding higher order neighbors (`nblag()`), for editing the neighbor relationships interactively (`edit.nb()`), for carrying out set operations on neighbors lists (due to Nicholas Lewin-Koh), for subsetting neighbors lists (`subset.nb()`), and dropping neighbor links non-interactively (`droplinks()`). Finally, utility functions are provided for displaying summaries of neighbors lists, and if spatial object coordinates are available, for plotting a map of the neighbors links.

Figure 6.1 A shows the way in which the sets of contiguous neighbors of each zone are constructed; in Fig. 6.1 B, neighbors are deˇ ned within a ˇ xed distance from the zone in question. In table form, the sets of neighbors for selected zones are shown in Table 6.1.

As Getis and Ord (1992, p. 190) point out, there are good reasons for examining patterns of spatial dependence at a more local scale. If we do not have good reason to suppose that the process in question is spatially stationary, it seems natural to apply distance-based tests to the observed spatial series. For use with distance statistics, one deˇ nes a symmetric one/zero spatial weighting matrix using the distance between the coordinates of a point associated with the observations. The choice of point for non-site series is not arbitrary, nor is the choice of the distance metric. Here the administrative centres of the observation units have been taken as adequately representing the location of the observation. Distance has been assumed to

Table 6.1. Neighborhood sets for lattices shown in Fig. 6.1 A and B.

Zone	A: contiguity		B: distance	
	number	neighbors	number	neighbors
1	2	(2, 9)	2	(2, 9)
...				
6	3	(5, 7, 8)	2	(5, 7)
...				
8	6	(3, 4, 5, 6, 7, 9)	4	(3, 4, 7, 9)
9	5	(1, 2, 3, 7, 8)	3	(1, 7, 8)

Table 6.2. The incremental neighborhood sets of zone 8 (Fig. 6.1 D).

Band	Distance	Number	Neighbors
1	< 30	0	
2	30 – 60	3	(3, 4, 7)
3	60 – 90	3	(5, 6, 9)
4	90 – 120	2	(1, 2)

be the simple Euclidean distance between points, ignoring barriers and other factors. Distance has further been banded on the basis of the frequencies of interpoint distances, and the furthest nearest neighbor distance as shown in Fig. 6.1. A typical element of the non-standardized spatial weight matrix $C(d)$ for distance d is deˇ ned as:

$$c_{ij}(d) = \begin{cases} 1 \text{ if hypot}(i,j) \leq d, i \neq j \\ 0 \text{ otherwise} \end{cases},$$

and,

$$\text{hypot}(i,j) = \sqrt{(x_i - x_j)^2 + (y_i - y_j)^2}.$$

The extent to which results are affected by the choice of points representing zones, and the choice of a simple representation of distance is unknown. Distance banded spatial weight matrices may be stored in the same fashion as contiguity matrices, and may also be represented as sliced increments, again reducing storage requirements.

In Fig. 6.1 C, the nearest neighbors of each zone are shown. It is zone 9 that has the furthest nearest neighbor distance, at 50 km from zone 7, while zone 3 is 39 km from zone 8. Figure 6.1 D illustrates the use of distance bands, at 30, 60, 90, and 120 km. Table 6.2 shows the incremental neighborhood sets for zone 8 for these bands. If zones were permitted to be their own neighbors, then zone 8 would belong to the set of neighbors for band 1.

These are coded in the form of a weights matrix W, most often with a zero diagonal, and the off-diagonal non-zero elements often scaled to sum to unity in each row (also known as standardized weights matrices), with typical elements:

$$w_{ij} = \frac{c_{ij}}{\sum_{j=1}^{n} c_{ij}}.$$

Alternative coding styles are described by Tiefelsdorf *et al.* (1999) and Tiefelsdorf (2000, pp. 29–31). This is done in function `nb2listw()`, which permits the speciˇ cation of the required weighting style and, if desired, the introduction of general rather than binary weights. It is at this point, and in the case of other helper functions calling `nb2listw()`, such as `nb2mat()` to create a full weights matrix, that we meet the question of what to do with spatial objects with no neighbors. In the present implementation, neighbors list elements for such objects are coded with an integer vector of length 1 with a value of $\{0\}$ – an out-of-bounds index, and are retrieved as having no neighbors by `card()`.

The `nb2listw()` function returns a list with three elements: the neighbors list used to generate it, a corresponding weights list, and the style employed. It is then used in turn in the `lag.listw()` function for calculating spatial lags of numeric vectors, and in `joincount()` for counting same-color neighbors, as well as for calculating constants for tests for spatial autocorrelation.

```
> x <- c(10, 12, 15, 17, 19, 18, 17, 16, 14)
> neigh8 <- c(3, 4, 5, 6, 7, 9)
> x[8]
[1] 16
> mean(x[neigh8])
[1] 16.66667
```

We can exemplify the spatial lag using the neighborhood set for zone 8 from Fig. 6.1 and Table 6.1. Here we are just using standard R to illustrate the lag operation; `x` is the vector of numeric attribute vlues of the spatial objects, and `neigh8` is an integer vector of the indices of the neighbors of zone 8 in the chosen scheme. In R, square brackets are used to retrieve values from vectors, so that `x[neigh8]` retrieves the values of `x` for the neighbors of zone 8. We take the mean here to give each neighbor an equal weight, with the row sum of weights equal to 1, and ˇ nd the spatially lagged values for this weighting scheme to be 16.67, which corresponds closely to the observed value of 16.0.

6.3 Spatial Lags: Consequences of Observations with No Neighbors

Say that we have generated a weighting scheme that is logical for the researcher with domain knowledge, but which produces a list of neighbors in which at least one of the spatial objects has no neighbors. In the very simple case we have just examined, we set the neighbors of zone 8 to the empty set:

```
> x <- c(10, 12, 15, 17, 19, 18, 17, 16, 14)
> neigh8 <- NULL
> mean(x[neigh8])
[1] NaN
```

Since the length of `x[neigh8]` is zero, and its sum is zero, the standard function `mean.default()` quite sensibly returns 0/0 as `NaN` – not a number. But if we recast the operation in terms of the row of a full weights matrix corresponding to zone 8, with all elements set to zero, here `ids`:

```
> ids <- rep(0, length(x))
> t(ids) %*% x
     [,1]
[1,]    0
```

we see that the lagged value is set to numeric zero, which may have meaning, or may be a marked outlier among lagged values for other zones with non-empty sets of neighbors. For this reason, many of the functions in **spdep** have been furnished with an argument: `zero.policy`, which is set to `FALSE` by default. The analyst will thus be obliged to set it to `TRUE` if functions terminate with the error message, and if the lack of neighbors is both known and accepted:

```
> data(columbus)
> col.listw <- nb2listw(col.gal.nb)
> card(col.gal.nb)[21]
[1] 3
> col.21 <- droplinks(col.nb.gal, 21)
> card(col.21)[21]
[1] 0
> col.21.listw <- nb2listw(col.21)
Error in nb2listw(col.21) : Empty neighbor sets found
> col.21.listw <- nb2listw(col.21, zero.policy=TRUE)
>
```

The `droplinks()` function serves to remove all links to and from the speci- ˇ ed zone (only links from a zone, corresponding to row entries, if argument `sym` is `FALSE`), creating a new neighbors list, in which zone 21 has no neighbors. The function itself was added to replicate results due to Fingleton (1999c, pp. 5–6) on methods for generating a spatial unit root – his Table 1 is reproduced by `example(droplinks)`, for which links from the central cell on a square grid to its neighbors are dropped to remove circularity.

The presence of spatial objects with no neighbors requires care in the calculation of the weights, and the implementation for the S and C style coding schemes now replaces the number of observations in total n by the number of observations with non-empty sets of neighbors (Tiefelsdorf, 2000, equations 3.6 and 3.10). With this substitution, the spatial weights constants S_0, S_1, and S_2 used in tests for spatial autocorrelation (Cliff and Ord, 1981, p.19), are the same for these coding schemes

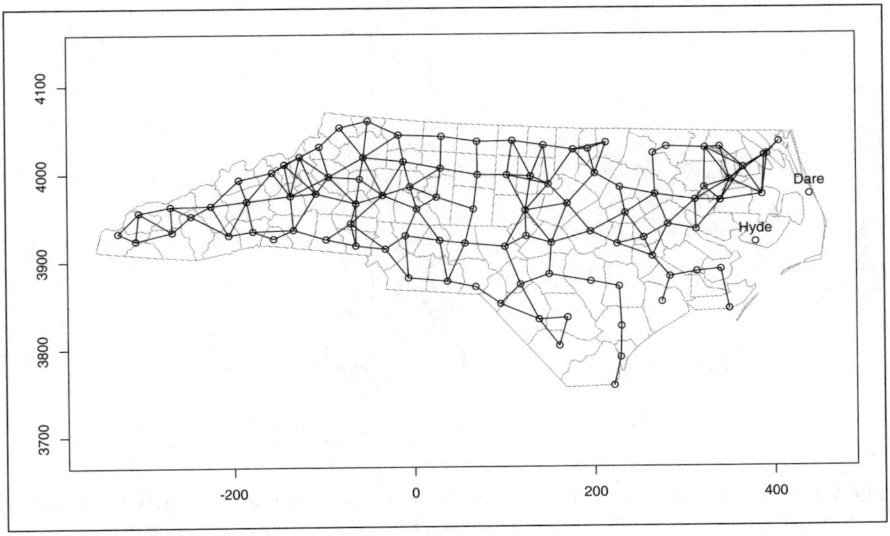

Fig. 6.2. North Carolina: neighbors links between county seats, maximum distance 30 miles (Cressie, 1993, pp. 386–389).

for complete neighbors lists and neighbors lists subsetted to exclude spatial objects with empty sets of neighbors. Since the weights coding schemes for binary (or uncoded general) weights, and row-standardised W style weights do not involve n, no changes are needed in these cases. In all cases, n differs between the complete lists and those that have been subsetted to remove spatial objects with empty neighbors sets, potentially affecting the calculation of estimates of parameters and test values.

With these modiˇ cations, differences in tests for spatial autocorrelation between subsetted data sets dropping no-neighbor spatial objects and full data sets retaining them will be in n, and in other calculations such as the mean of the variable being tested, its sum of squares of deviations from the mean, and kurtosis. For tests of spatial dependence in regression residuals, the difference between means for the full and subsetted data sets becomes the differences in estimated coefˇ cients and cross-product matrices. Subsetting the data just to test for residual autocorrelation when using a list of neighbors with no-neighbor objects seems unnecessarily intrusive, but as in the case of tests for autocorrelation on a single variable, zero is a value with substantive meaning. In the single variable case, a lagged value of zero implies that the imputed neighbors of an object which actually has no neighbors are given the global mean value of no deviations.

In the classic North Carolina sudden infant death syndrome data set discussed in Cressie (1993), a criterion for neighborhood of a distance between county seats of less than 30 miles. As has been noted by others, this leaves two counties (28 Dare, 48 Hyde, both on the Atlantic coast, sharing the Cape Natteras National Seashore) with no neighbors, since as can be seen, their nearest neighbors lie a little over 30

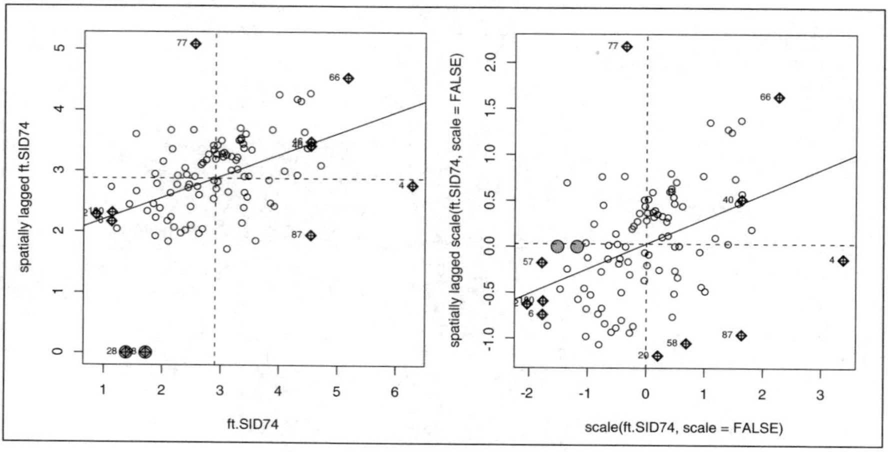

Fig. 6.3. Moran scatterplots for the Freeman-Tukey square root transformed SIDS by county in North Carolina, 1974–78, non-centered variable (left), centered variable (right); no-neighbor objects marked by grey disks.

miles away. In Cressie and Read (1985), county boundary contiguities are given as the neighborhood criterion.

```
> data(nc.sids)
> plotpolys(nc.utm.polys, nc.utmbbs, border\index{border}="grey")
> plot(sidsorig.nb, utm18.countyseats, add=TRUE)
> text(utm18.countyseats[card(sidsorig.nb) == 0,],
+   rownames(nc.sids)[card(sidsorig.nb) == 0], pos=3)
> milecoords <- cbind(nc.sids$east, nc.sids$north)
> nndists <- unlist(nbdists(knn2nb(knearneigh(milecoords)),
+   milecoords))
> nndists[card(sidsorig.nb) == 0]
[1] 32.01562 30.47950
```

Using Moran scatterplots (Anselin, 1996) of observed variable values – here for the Freeman-Tukey square root transformed SIDS incidence rates, we can see that the two spatial objects appear with their lags set to zero. This may be compared, in the context of Moran's I, with the difference in the range of summation of the numerator and the denominator in the Durbin-Watson test of time series regression residuals. In the left-hand plot in Fig. 6.3, the values are shown as observed, in the right-hand plot as deviations from the mean.

```
> ft.SID74 <- sqrt(1000)*(sqrt(nc.sids$SID74/nc.sids$BIR74) +
+   sqrt((nc.sids$SID74+1)/nc.sids$BIR74))
> moran.plot(ft.SID74,
+   nb2listw(sidsorig.nb, zero.policy = TRUE),
```

```
+   zero.policy = TRUE)
> moran.plot(scale(ft.SID74, scale=FALSE),
+   nb2listw(sidsorig.nb, zero.policy = TRUE),
+   zero.policy = TRUE)
```

A further artifact of the inclusion of spatial objects with no neighbors in the adopted weighting scheme is that the mean of the local Moran's I_i no longer equals the global Moran's I, unless n is reduced to the effective number of observations, that is those with neighbors. This is because of the change in the order of summations, with local Moran's I_i set to zero for spatial objects with no neighbors. Alternatively, the mean of the local Moran's I_i could be taken over spatial objects with neighbors. This does not however alter the conclusion that the lack of neighbors for one or more zones does affect the calculation of statistics of spatial dependence, and at least potentially inference from them. In the time series case, it is argued that with increasing series length, the impact of differing ranges of summarion for the numerator and denominator reduces in the Durbin-Watson test. In spatial data this may also be assumed, so that a few such observations among many may not affect conclusions. It may however be appropriate to make the analyst aware that permitting spatial objects to have no neighbors does lead to a number of choices in the implementation of functions for testing dependence.

The relationship between this practical data analysis issue and the use of the R data analysis environment is that exploring what happens in different settings is made relatively easy. This applies both with regard to the writing of new functions, to modifying functions for local use (using `fix()`), and having access to a complete toolbox of other non-spatial functions. These include list, vector and matrix functions, and can be used to prototype alternative implementations such that the impact of previously unarticulated assumptions becomes clearer. In this case, the assumption is that the weighted sum of an empty set of neighbors should be set to zero rather than set missing, if we simply move from a list to a matrix representation of spatial weights.

6.4 Case Study: Clusters of Towns in an Urban System with Sparsely Populated Regions

An "urban cluster" (UC) is a group of urban settlements located in close proximity to each other and connected by strong socio-economic and functional links (Portnov and Erell, 2001). Theoretically, any urban contiguity can be considered a cluster of towns in which inter-town distances are fairly small. Let us assume, however, that these inter-town distances increase to 20, 40, or 200 km. Do urban localities in such a cluster still perform as a single functional unit, or do they split into functionally independent urban formations? To what extent are the development levels exhibited by individual towns in such diffuse UCs still interlinked?

However, a number of questions, pertinent to the phenomenon of urban clustering remain largely unanswered. They include:

- How large is a geographic area within which the effect of aerial proximity of urban places on the development of individual towns is distinctively felt?
- Is there any difference in the spatial extent and performance of UCs in centrally located and peripheral regions?

This case starts with a brief overview of previous studies of the phenomenon of urban clustering. The general patterns of urban development in Israel are then discussed in brief. This discussion is followed by an analysis of spatial links that neighboring urban localities in Israel tend to exhibit in their development.

6.4.1 Studies of Urban Clustering

Somewhat surprisingly, following the publication of Christaller's and Lösch's landmark studies in the 1930s, there have been only isolated attempts to examine further the nature of urban clustering and the effect of this phenomenon on the development of individual towns. In one of such studies, Golany (1982) emphasises the role of urban clusters as a means of reducing the perception of isolation in peripheral regions. He suggests that in addition to psychological effects, the clustering of towns in sparsely populated areas may result in additional economic beneˇ ts, normally associated with the initial phase of urban agglomeration, such as lowering the per capita costs of infrastructure and transportation.

In a case study of two metropolitan regions of the U.S. the North Carolina Piedmont cluster of dispersed towns and the Philadelphia cluster, which has a more centralised pattern of settlements, Krakover (1987) went somewhat further, focusing his analysis on both comparative advantages and disadvantages of urban clustering. As he argues, UCs undergo two distinctive phases of growth. When towns in such clusters are relatively small, their prevailing economic, technological, and spatial conditions are conducive to economies of agglomeration. However, at the later phase, when cities pass a certain population threshold, diseconomies of excessive concentration may establish themselves earlier in the larger city than in a cluster of smaller towns, since an increasing number of entrepreneurs might realise advantages of moving their enterprises to suburban locations.

Fujita and Mori (1997) developed a theoretical model of the dynamic formation of urban places. This model is based on the assumption that new cities are created periodically as a result of what they termed the "catastrophic bifurcation" of existing settlements. According to this model, as the number of cities increases, the urban system may approach a highly regular central place system. However, the model in question has no clear spatial dimension: it neither indicates the physical dimensions of cities and clusters at which the catastrophic bifurcation occurs, nor does it explain the interdependency of development processes observed in individual towns in such clusters.

Portnov and Erell (2001) focused their analysis on the performance of UCs in core and peripheral areas of selected countries: Israel, Norway and New South Wales, Australia. As the authors of this analysis suggest, the effect of urban clustering on the patterns of urban growth is twofold:

- In sparsely populated peripheral areas, the presence of small neighboring towns may mutually increase their chances to attract potential investors and migrants due to socio-economic interaction and inter-urban exchanges;
- In core areas, where a major population centre dominates social and economic life of adjacent towns, dense clusters of small urban localities may reduce the attractiveness of individual towns to both investors and migrants due to inter-town competition and overcrowding.

The goods, people and information may spread in space through both interaction and diffusion. As a result, events and circumstances at one place can affect conditions at other places if the places interact. In UCs, such an interaction, which presumably results in the development interdependency of individual towns, may be attributed to two different factors hierarchical choices of migrants and location preferences of firms and entrepreneurs:

1. Hierarchical Choices of Migrants
   - Migrants often choose their destinations hierarchically: first, among clusters of localities, and then among individual towns in such clusters. As Fotheringham (1991) argues, the reason is that migrants do not have all the information necessary to analyse every possible destination prior to making a decision on where to move, specifically when the overall number of possible destinations is large. Therefore, migrants tend to process spatial information hierarchically, first evaluating clusters or groups of alternatives and then evaluating only alternatives within a preferred cluster.
2. Location Preferences of Firms and Entrepreneurs
   - In the process of location decision-making, both firms and individual entrepreneurs may prefer clusters of towns, rather than individual settlements. Within a cluster of small but closely located towns, they may expect to find a larger pool of skilled labor and consumers, compared with that available in a single-town. The establishment of a new industrial enterprise in a given urban cluster may, in turn, trigger a chain reaction leading to further concentration of firms, the effect which Myrdal (1958) termed the process of "cumulative causation". More recent studies (see inter alia Shilton and Craig, 1999; Walcott, 1999; Swann *et al.*, 1998) also suggest that in the case of industries, the positive effect of clustering is attributed to information sharing, joint research, better opportunities for networking and international trade.

Since both migrants and entrepreneurs may consider a cluster of neighboring towns as an integrated functional unit, a strong interdependency of development processes in individual towns located in such a cluster can thus be expected. However, if such hypothetical interdependency does occur, it should have certain spatial limits. For instance, migrants are unlikely to perceive a town as a part of a particular UC, if distances, which separate this town from the rest of the cluster, are fairly large. In the case of firms and individual entrepreneurs, the possibilities of hiring skilled employees from adjacent localities may also be restricted, if inter-town distances surpass are greater than those considered practicable for daily commuting.

These assumptions (viz. development interdependency of individual towns in UCs, and commuting distances as spatial limits of UCs) can be tested using the techniques of spatial analysis.

6.4.2 Patterns of Urban Development in Israel

Israel's urban system, which is selected for the present analysis, is formed by publically designated urban localities, of which we will be using 157. They have populations varying between the largest cities of Jerusalem (645,800), Tel Aviv-Yafo (350,530) and Haifa (268,130), and many small localities, of which 69 have less than 10,000 residents. The population ˇ gures used here are three-year averages for 1994–1996 and 1998–2000. Most of the country's urban settlements are concentrated along the Mediterranean coast, in close proximity to Tel Aviv and Haifa. The set of urban localities changes over time, with new entities being created, but all are deˇ ned as urban rather than rural for the purposes of ofˇ cial statistics. They are a data set that is not as adequate for our present purposes as would be gridded population data, because of the very great differences in character between the largest cities and the smallest localities.

The overall population of these population centres along with their immediate hinterland (the Tel Aviv, Central, Haifa districts) amounts to some 3.2 million residents, or nearly 60 percent of the country's population. Urban settlement in this part of the country is extremely dense. For example, in the Tel Aviv district, the overall density of population exceeds 6,700 residents per km^2. In contrast, in peripheral areas of the country, urban settlement is sparse, speciˇ cally in the south, where average population density does not exceed 35 residents per km^2 (ICBS, 1999). This spatial inequality of urban development is considered an advantage for the present analysis, for which diverse patterns of urban settlement are desirable.

As Fig. 6.4 shows, the data set varies considerably in density, with many locations in the central coastal belt very near one another, while in southern half of the country settlement is very sparse. As Portnov and Erell (2001) demonstrate, these varied settlement pattern densities are frequently in areas where climatic pressure impacts land use, be it cold or heat. And in these conditions extra care is needed with respect to giving advice on sustainable urban development, so that simply abandoning areas posing practical difˇ culties for data analysis is not feasible. The left hand map expresses the unevenness of the positioning of the locations in rug plots on the eastings and northings axes. On the eastings axis, we can see that all are within a 100 km span, denser toward the centre, by with no outliers. On the northings axis, however, one location is somewhat isolated to the north, and the southern half of the country is characterised by a completely different density.

The right hand map in Fig. 6.4 presents the basic data set of percentage population changes, extending from a few cases of decline in population, through to increases by over 1000 percent (only two locations grew by more than 100 percent in the 1994–1996 to 1998–2000 period). There are two reasons for smoothing using three year periods: the smallest locations do have missing data, but should be

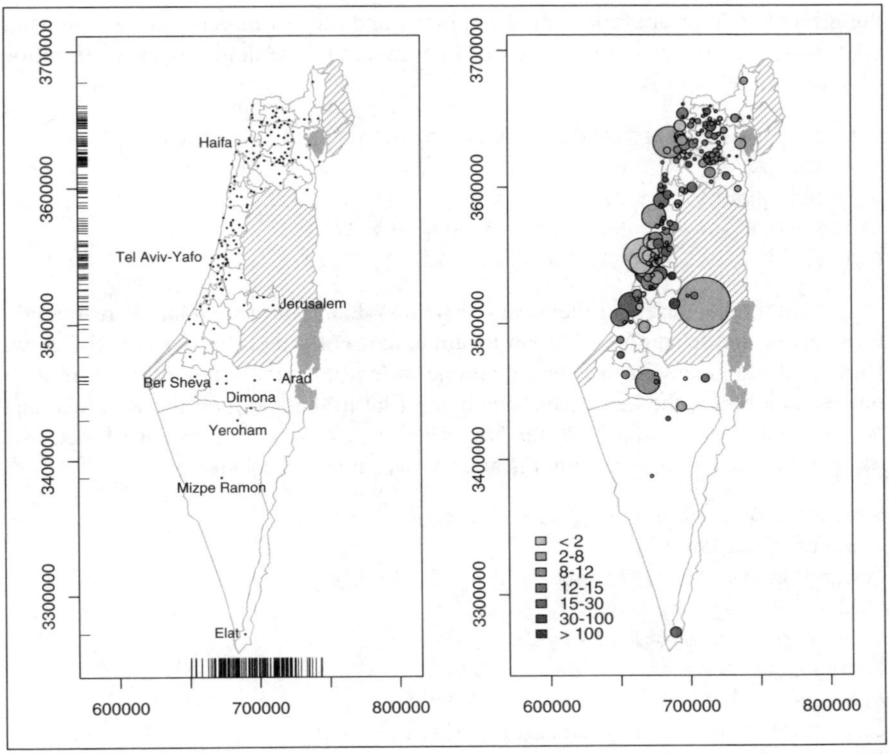

Fig. 6.4. Urban locations in Israel, UTM zone 36 (background regions represent varying natural conditions); left map: positions and axes rug plots; right map: locations marked by circles proportional to their population size in 1998–2000 and shaded by percentage population change 1994–96 to 1998–2000.

retained in the analysis, and in more general terms Israel has experienced very substantial immigration, leading to substantial influx in some locations, especially those to which migrants are initially directed, and thus spikes in population levels not representative of longer term trends.

From the map we can see that localities close to central Tel Aviv-Yafo experienced least growth, with suburban localities growing more strongly. A second area of stronger growth in smaller, more rural, localities may be seen to the south-east of Haifa. But in both these cases, the rapidly growing smaller urban localities are in the north and centre of the country, and appear to be close to one another.

6.4.3 Use of R Functions

We will first turn to the construction of lists of neighbors for the set of urban localities. Two types of approaches will be used, distance based, and graph based, since the urban localities are represented as points, and are not in general contiguous as administrative districts, often separated by rural entities. Examining the distribution of nearest neighbor distances:

```
> nndists <- unlist(nbdists(knn2nb(knearneigh(ul.coords)),
+   ul.coords))/1000
> round(quantile(nndists, seq(0,1,0.1)), digits=1)
0% 10% 20% 30% 40% 50% 60% 70% 80% 90% 100%
0.8 1.7 2.2 2.5 3.0 3.3 3.7 4.6 5.7 9.5 117.1
```

About three quarters of the locations lie less than 5km from their nearest neighbors, given the deˇ nition of urban localities currently used by the Israeli Central Bureau of Statistics. Further, less than one in ten lie further than 10km from their nearest neighbors, the key exceptions being Elat in the south on the Red Sea, and Mizpe Ramon in the middle of the Negev desert. Constructing distance-based lists of neighbors for 5km maximum distance between neighbors yields:

```
> ul5km.nb <- dnearneigh(ul.coords, 0, 5000)
> summary(ul5km.nb)
Connectivity of ul5km.nb:
Number of regions: 157
Number of nonzero links: 318
Percentage nonzero weights: 1.290113
Average number of links: 2.025478
Link number distribution\index{distribution}:
 0  1  2  3  4  5  6  7  8
37 42 26 18 18  5  6  3  2
> t5 <- table(n.comp.nb(ul5km.nb)$comp.id)
> t5[t5 > 1]
1 2 6  7 9 11 12 13 14 18 21 22 23 24 26 28 29 32 37 45 46 47 49
2 4 2 21 3 24  2  8 15  3  2  6  2  3  5  2  2  2  2  2  2  4  2

> ul10km.nb <- union.nb(ul5km.nb, ul5.10km.nb)
> summary(ul10km.nb)
Connectivity of ul10km.nb:
Number of regions: 157
Number of nonzero links: 1178
Percentage nonzero weights: 4.779099
Average number of links: 7.503185
Link number distribution\index{distribution}:
 0  1  2  3  4  5  6  7  8  9 10 11 12 13 14 15 16 17 18 19
14  6 14  4  9 10  8 11 13 13 12  8  8 11  4  3  4  3  1  1
> t10 <- table(n.comp.nb(ul10km.nb)$comp.id)
> t10[t10 > 1]
  1   2   4   5  16
  3 131   3   3   3
```

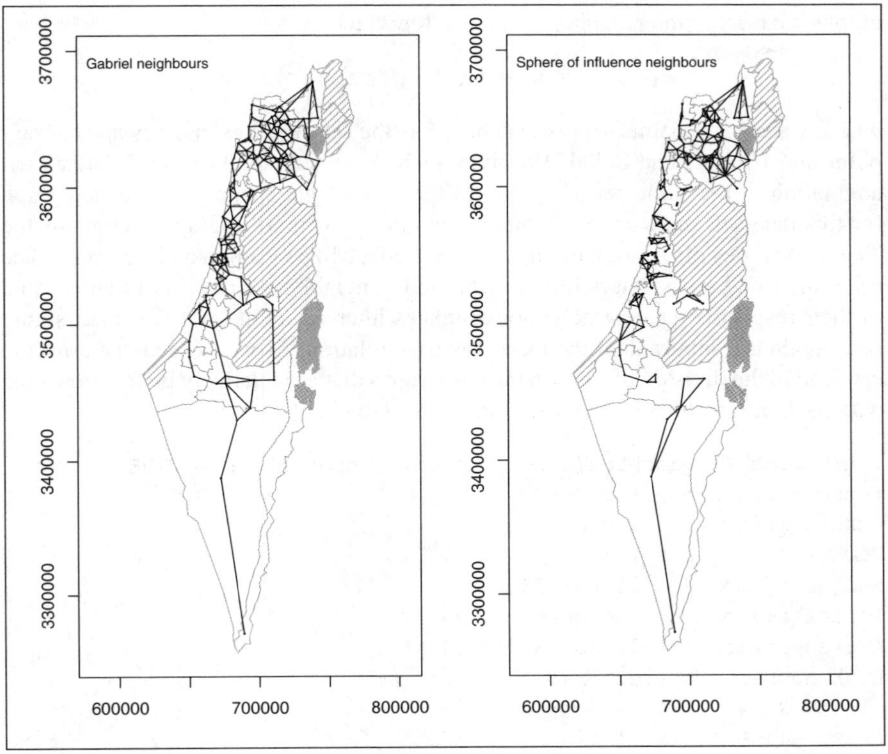

Fig. 6.5. Graph based neighborhood criteria: Gabriel graph (left), sphere of in uence graph (right).

Here 37 of 157 urban localities are without neighbors, and 42 have only one neighbor, but both Ganne Tiqwa and Or Yehuda each have as many as 8 neighbors within 5km. It has as many as 60 disjoint connected subgraphs, and after removing the 37 isolated localities, 23 remain of which only 3 have 15 or more localities belonging to them. Adding a further 5km, that is using a distance of between 5km and 10km as the criterion for being a neighbor, reduces the number of isolated localities to 16, and the union of these sets to 14. Both the 5–10km band and the union 0–10km have one dominant connected subgraph with 131 localities, a set which we will use below. However, some places are now heavily connected, with Bet Dagan having 19 links.

Two alternative graph based neighborhood criteria[2] are shown in Fig. 6.5. Both of these by deˇ nition include all spatial objects, and the Gabriel graph in addition ensures that all objects are included in a single graph – there are no disjoint sub-

[2] Code and documentation for graph based neighborhood relationships was contributed to **spdep** by Nicholas Lewin-Koh.

graphs. Gabriel graph neighbors are those for which:

$$d(x,y) \leq \min((d(x,z)^2 + d(y,z)^2)^{1/2})|z \in S,$$

where x and y are points, $d()$ is distance, S is the set of points and z is an arbitrary point in S (Matula and Sokal, 1980); as such it is a subgraph of the Delaunay triangulation of the same set of points. In the case of the sphere of in uence graph for this data set, there are 8 disjoint subgraphs, of which subgraph 3 contains the Negev localities of: Arad, Dimona, Elat, Kuseife, Mizpe Ramon and Yeroham. The criterion used here is that points are admitted as neighbors if circles of radius equal to their respective nearest neighbor distances intersect in at least two places, and once again is a subgraph of the Delaunay triangulation. As we can see, the criterion can lead to the division of a graph into subgraphs that are relatively better connected with each other than with the rest of the set of points.

```
> ulGab.nb <- graph2nb(gabrielneigh(ul.coords), sym=TRUE)
> summary(ulGab.nb)
Connectivity of ulGab.nb:
Number of regions: 157
Number of nonzero links: 552
Percentage nonzero weights: 2.239442
Average number of links: 3.515924
Link number distribution:
 1  2  3  4  5  6  7
 2 22 54 55 21  2  1

> ulSoI.nb <- graph2nb(soi.graph(tri2nb(ul.coords), ul.coords),
+      sym=TRUE)
Loading required package: tripack
> summary(ulSoI.nb)
Connectivity of ulSoI.nb:
Number of regions: 157
Number of nonzero links: 516
Percentage nonzero weights: 2.093391
Average number of links: 3.286624
Link number distribution:
 1  2  3  4  5  6  7  9
11 35 50 34 17  8  1  1
> table(n.comp.nb(ulSoI.nb)$comp.id)
 1  2  3  4  5  6  7  8
 4 93  6  3 15 25  2  9
```

The next empirical issue to address is that the variable of interest, percentage population change in the second half of the 1990s in Israeli urban localities, is awkwardly distributed:

```
> round(quantile(ul.pop$ppopch, seq(0, 1, 0.1)), digits=1)
  0%   10%   20%   30%   40%   50%   60%   70%   80%   90%    100%
-3.9   1.1   6.5   8.3  10.1  11.6  13.2  14.1  16.3  28.1  1561.5
> stem(ul.pop$ppopch)

  The decimal point is 1 digit(s) to the right of the |

  -0 | 4322222210000
   0 | 00112333334444
   0 | 5666677777888888888899999999
   1 | 00000000111111111122222222222333333333333334444444444444
   1 | 555556666777889
   2 | 00023
   2 | 56689
   3 | 11234
   3 |
   4 | 34
   4 | 578
   5 | 04
   5 | 6

outliers: 466, 1561

> pch.f <- as.ordered(cut(ul.pop$ppopch,
+     breaks=c(-4, 2, 8, 12, 15, 30, 100, Inf),
+     include.lowest=TRUE))
> table(pch.f)
pch.f
[-4,2]   (2,8]   (8,12]   (12,15]   (15,30]   (30,100]  (100,Inf]
    17      25       42        35        23         13          2
```

Using the factor constructed above – also used for the class intervals of the shaded proportional circle map shown in Fig. 6.4 – we can use join counts to make an initial assessment of spatial dependence. Here we drop the highest class, which only has two members, and which are not neighbors under any of the neighbor criteria presented above. By counting same-color joins for each of the percentage population change classes, and testing under non-free sampling for the estimated standard deviate of the statistic to be greater than its expectation for each of the four neighbor criteria and for the binary (B) and row-standardised (W) weighting schemes, we obtain the results shown in Table 6.3.

Using the `joincount.test()` function with selected neighbors lists:

```
> joincount.test(pch.f, nb2listw(ul5km.nb, style="B",
+     zero.policy=TRUE), zero.policy=TRUE)
```

Table 6.3. Same-color join count statistics for percentage population change classes by neighborhood criterion and weighting scheme: standard deviates and probability values under non-free sampling.

			Percent population change					
Neighbors	Weights		< 2	2–8	8–12	12–15	15–30	30–100
< 5 km	W	Std. dev.	1.733	−1.114	−0.158	0.797	1.102	2.443
		p. value	0.042	0.867	0.563	0.213	0.135	0.007
	B	Std. dev.	1.848	−0.857	0.150	0.601	0.535	1.953
		p. value	0.032	0.804	0.440	0.274	0.296	0.025
5–10 km	W	Std. dev.	1.845	−1.228	−0.488	0.904	−0.570	−0.277
		p. value	0.033	0.890	0.687	0.183	0.716	0.609
	B	Std. dev.	3.583	−1.260	−0.285	1.891	−0.619	−0.511
		p. value	0.000	0.896	0.612	0.029	0.732	0.695
Gab	W	Std. dev.	3.696	−0.689	1.548	3.168	1.560	3.979
		p. value	0.000	0.755	0.061	0.001	0.059	0.000
	B	Std. dev.	3.060	−0.766	1.576	4.100	1.545	3.407
		p. value	0.001	0.778	0.058	0.000	0.061	0.000
SoI	W	Std. dev.	4.058	−0.163	0.477	3.649	2.226	3.545
		p. value	0.000	0.565	0.317	0.000	0.013	0.000
	B	Std. dev.	3.825	−0.138	0.773	3.945	1.244	2.731
		p. value	0.000	0.555	0.220	0.000	0.107	0.003

Using the distance neighbor criteria and either of the weighting schemes leads to the conclusion that spatial dependence is most evident for the urban localities with lowest percentage rates of population change. Since many of these, like Tel Aviv-Yafo, Bat Yam, Holon, or Ramat Gan, are large cities in the most densely populated parts of the country, where further growth is well-nigh impossible because density is already very high, this is in line with our hypotheses. But it is disappointing that the distance-based criteria fail to distinguish some of the features that seem to be present in Fig. 6.4, especially the apparently clear clustering of more rapid growth east of Haifa or inland of Tel Aviv-Yafo. Maybe one can attach some meaning to the 12–15 percent class in the 5–10 km band for the binary weighting scheme, or to the 30–100 percent class in the 0–5 km band for both schemes (all 12 localities are small, for example Binyamina and Zikhron Yaaqov north of Hadera), but this is perhaps trying to force our perception onto the test results for the distance neighbors criteria.

Our infererences for the class with lowest rates are similar for the two graph based neighbor criteria – urban localities with declining or stable populations are very likely to neighbor each other. It also still seems that the 2–8 percent growth class displays no signiˇ cant spatial dependence, and that traces of dependence for the Gabriel graph criterion for the 8–12 percent and 15–30 percent classes are at best marginal, especially considering that we are applying multiple tests (for exploratory

Table 6.4. Moran's I statistic for ranks of percentage population change

Neighbors	Weights	Moran's I	Std. deviance	Prob. value
< 5 km	W	0.212	2.346	0.009
	B	0.253	3.356	0.000
5–10 km	W	0.060	1.181	0.119
	B	0.110	2.514	0.006
< 10 km	W	0.112	2.362	0.009
	B	0.163	4.318	0.000
Gabriel	W	0.235	3.895	0.000
	B	0.231	3.982	0.000
Sphere of in uence	W	0.215	3.338	0.000
	B	0.177	2.985	0.001

purposes) to the same data. For the remaining classes, 12–15 percent and 30–100 percent, we conclude that dependence is present, perhaps lessening the doubts expressed above for the distance based criteria. For the important class of 12–15 percent change, we can note that both larger coastal cities such as Ashqelon, Hadera and Nahariyya are present, as is the smaller north Negev town of Arad.

An alternative approach is to use the adaptation of Moran's I for ranks suggested by Cliff and Ord (1981, p. 46), with an appropriate replacement for the sample kurtosis coef˜ cient in the variance expression. The R code used, typically:

```
> moran.test(rank(ul.pop$ppopch), nb2listw(ul5km.nb,
+      zero.policy=TRUE), zero.policy=TRUE, rank=TRUE)
```

yields the results shown in Table 6.4 for the same neighbors and weights alternatives, supplemented with the distance criterion of up to 10 km. Once again, for the distance criteria it is necessary to take account of urban locations without neighbors, effectively dropping these places from the results.

Table 6.4 shows very clearly for both types of neighborhood criterion that we can, on balance, expect neighboring urban localities to have similar rank percentage population change for the latter half of the 1990s. The only neighborhood criterion that does not bear out this conclusion is for the row-standardised 5–10 km distance criterion weights, but here the difference between the binary and row-standardised schemes would suggest that where localities have many neighbors in the 5–10km band, they are more likely to have similar ranks than when they have few such neighbors – the "W" scheme weights up objects with few neighbors.

Finally, we return to the interesting subgraph in the 10 km distance neighbors object noted above. The 131 localities form a belt running north up the coast from Ashdod, and reaching east of Haifa into Galilee. Outside the belt are all localities south of a line drawn between Ashqelon and Jerusalem, and the six small north-eastern localities of Bet Shean, Qiryat Shemona, Rosh Pinna, Tamra, Tuba Zangariyye and Zefat. In many ways, it splits out the core/periphery structure of the urban system, and will now let us subset the data to permit us to use the rank variant of Moran's *I* to test localities within and outside the set derived from the subgraph separately, here just using the Gabriel graph neighborhood criterion and row-standardised weights.

```
> comp.10km <- n.comp.nb(ul10km.nb)
> t10 <- table(comp.10km$comp.id)
> t10[t10 > 1]
  1   2   4   5  16
  3 131   3   3   3
> clump <- comp.10km$comp.id == 2
> summary(subset(ulGab.nb, clump))
Link number distribution:
 1  2  3  4  5  6  7
 4 15 49 44 16  2  1
> moran.test(rank(subset(ul.pop$ppopch, clump)),
+     nb2listw(subset(ulGab.nb, clump)), rank=TRUE)
> summary(subset(ulGab.nb, !clump))
Link number distribution:
 1  2  3  4  5
 5 12  6  2  1
> moran.test(rank(subset(ul.pop$ppopch, !clump)),
+     nb2listw(subset(ulGab.nb, !clump)), rank=TRUE,
+     alternative="less")
```

For the core, the subset of the Gabriel graph neighbors gives a value of Moran's *I* statistic of 0.274, with a standard deviate of 4.128, and a probability value of 0.00002 for a null hypothesis that the observed statistic is equal to its expectation, and an alternative that it is greater. In the core, it seems using this approach that there is strong spatial dependence in rank percentage population change – we know from the fact that the localities were less than 10 km from their nearest neighbors in the underlying 10 km distance representation of neighborhood that they are also close to each other. The values of the statistic and its standard deviate are both higher than for the whole unsubsetted data set as reported in Table 6.4. For the periphery, however, the value of the statistic is -0.300, with a standard deviate of -1.355, and a probability value of 0.088 for the alternative that the observed value of the statistic is less than the expected. The peripheral subset of the Gabriel graph has relatively fewer links than the core subset, but conclusions from the binary weighting scheme are similar. Neighboring peripheral urban locations, relatively distant from one another, do not show similar rank percentage population change, but rather the reverse: they seem to differ weakly from one another, as though they were perhaps competing for the available growth.

6.5 Conclusions

It would be rash to claim that analyses such as those exempliˇed in this discussion could not be undertaken in other programming environments, naturally much the same could have been done in many other systems, especially in S-PLUS. It is however possible that few systems would have been sufficiently open – both in terms of access to the source code of interpreted and compiled functions, and in terms of richness of underlying system capabilities – for such analyses to have been accomplished in this way. It has to be admitted that some experience both of the R command line user interface, as well as the ability to write at least script-style programs, is needed to do some of the things attempted here. It should also be remarked that it is specifically the example of the greatly varying density of the Israeli urban localities system that has driven the relatively comprehensive incorporation of arguments and procedures for handling spatial objects with no neighbors under the chosen weighting scheme.

It is also worth noting that the basic presumptions of free software for R in general and the **spdep** package in particular (both are licensed under the terms of the GNU General Public License Version 2) have also been realised. Shortly after an early release, Nicholas Lewin-Koh contributed the very useful graph based neighborhood criteria functions, as an improvement on the initial simple Delaunay triangulation function, and more complete set operations on neighbors lists to extend an initial function to report differences between lists. As can be seen in the above examples, these contributions have broadened the applicability of the package, and together with interactive editing using `edit.nb()`, now provide an extendable workbench for creating and exploring neighborhood relationships. Others have also contributed through suggestions and bug reports, so that the package is becoming a community project. Since all are in any case invited to read and share, and to write if so motivated, there is no obvious disadvantage even if it turns out that these R prototypes can be better implemented in alternative environments.

With regard to the chosen case – with empirically realistic but challenging distributions both of the urban locations themselves, and of the variable of interest, it has been possible to explore the possible spatial dependence of percentage changes in population, and point to some tentative conclusions. At this stage it is too early to address the key policy question of whether sustainable clusters of smaller towns are more likely to lead to endogenous growth in a sparsely populated region with a harsh climate than say a single large city, not least because the Negev at present has so few urban localities. We have however established beyond doubt that population change does display spatial dependence for the chosen data set and criteria for neighborhood, and as a by-product, we have been able to make a relatively robust core-periphery classification based on proximity.

Whether the absence of neighbors for a number of spatial objects in a data set under examination will impact our conclusions remains an open question. The number of such objects is important, as is their relative placing. While the distance neighborhood criterion is clearly the main reason for no-neighbor objects appearing, they can also be created by subsetting neighbors lists and other such operations. It is thus advisable to be able to access summary measures of the structure of neighbors lists, and to use this information to set appropriate argument flags where relevant or feasible. That this has now been demonstrated in R provides an opportunity for other platforms for the analysis of potentially dependent spatial data to revisit this practical issue.

Part II

Discrete Choice and Bayesian Approaches

7 Techniques for Estimating Spatially Dependent Discrete Choice Models

Mark M. Fleming

Fannie Mae Foundation

7.1 Introduction

Much has been written on the techniques for dealing with spatial dependence, spatial lag and spatial error, in continuous econometric models (e.g., Anselin, 1980, 1990; Anselin and Bera, 1998; Griffith, 1987; Kelejian and Prucha, 1998, 1999). The study of spatial dependence in discrete choice models, particularly in the context of the spatial probit model (e.g., Case, 1992; McMillen, 1992, 1995a; Bolduc *et al.*, 1997; Pinkse and Slade, 1998, and Chapter 8 in this volume), has received less attention in the literature. This may be in part due to the added complexity that spatial dependence introduces into discrete choice models and the resulting need for more complex estimators.

Many techniques have been proposed to deal with discrete choice estimation when spatial dependence is present. The inconsistency of the standard probit model, if the spatial dependence causes heteroskedasticity, and the efficiency implications of not using all the information in the non-spherical variance-covariance structure have both been considered.

Authors who have addressed the heteroskedasticity caused by spatial dependence in discrete choice models include Case (1992), and Pinkse and Slade (1998).[1] The heteroskedasticity is dealt with through innovative specification of the spatial dependence (Case, 1992), or a Generalized Method of Moments (GMM) technique that uses the spatial structure to determine the heteroskedastic variance terms (Pinkse and Slade, 1998). Concentrating on the heteroskedasticity induced by the spatial dependence results in estimates of the parameters of the likelihood function that remain consistent, assuming independence of the error terms. However, the likelihood is no longer efficient because it does not use the information in the off-diagonal terms of the variance-covariance matrix. In return, the need to estimate an n-dimensional integral is reduced to the simpler product of independent density functions.

If one wants to address the heteroskedasticity induced by spatial dependence and utilize the additional information in the off-diagonal elements of the variance-covariance matrix the problem of multidimensional integration must be solved in the

---

[1] McMillen (1992) considers discrete choice models with heteroskedastic error structures, but they are not specifically derived from the spatial autocorellated error structure described here. A functional form for the heteroskedasticity is specified and the model is estimated as one of the class of Non-Linear Weighted Least Squares Estimators.

estimation technique. The EM algorithm, simulation methods, and Bayesian methods all offer solutions to this problem. The EM algorithm (e.g., Dempster *et al.*, 1977) and Bayesian techniques, particularly Gibbs sampling (e.g., Bolduc *et al.*, 1997; LeSage, 2000; Albert and Chib, 1993; Geman and Geman, 1984), [2] indirectly solve the multidimensional likelihood function based on the underlying principle that there is a way to determine a possible outcome of the unobserved latent variable. Simulation methods (Beron and Vijverberg, 2003; Geweke, 1989; Keane, 1994; McFadden, 1989; Hajivassiliou, 1990) compute the multidimensional likelihood function and its derivatives by developing parameter probability distributions. Parameter estimates are derived from these distributions rather than from the multidimensional likelihood function directly. All of these spatially correlated techniques utilize the complete variance-covariance matrix, but at the cost of computational and conceptual complexity.

An alternative to the heteroskedastic estimators and the spatially correlated techniques is to describe the spatially dependent discrete choice problem as a weighted non-linear version of the linear probability model (e.g., Greene, 1997; Maddala, 1983; Amemiya, 1985; Judge *et al.*, 1985) with a general variance-covariance matrix. Amemiya (1985) discusses Non-Linear Weighted Least Squares estimators that are based on the ˇ rst order conditions of the basic probit Maximum Likelihood function. The approach discussed here describes the same group of non-linear weighted least squares models as a GMM estimator (Hansen, 1982) and extends them to discrete choice models with spatial dependence. In so doing, the higher order integration problem that arises in a spatially dependent likelihood function is avoided. This approach also avoids calculation of the n by n determinants (a computation intensive procedure for large samples) that are found in the Maximum Likelihood function of the underlying latent models used in the EM algorithm and Gibbs sampler, or in the heteroskedastic approach of Pinkse and Slade (1998).

In addition to the expanding literature on methods of estimation, there are also an increasing number of techniques designed to test for the presence of spatial dependence in discrete choice models (Pinkse and Slade, 1998; Pinkse, 1999; Kelejian and Prucha, 2001). While a discussion of these techniques is not in the scope of this chapter, testing discrete choice models for spatial dependence is clearly essential to determining the necessity of the estimation techniques discussed here.

The goal of this chapter is to bring together the literature on spatial discrete choice estimation methods, provide a cohesive description with critical insights, and compare the different techniques. There are a variety of problems in economics that could beneˇ t from these spatial discrete choice econometric techniques, such as land use change, deforestation, migration, local government interaction, and technology adoption. It is hoped that this chapter will spur increased use and testing of these methods, particularly Monte Carlo studies of estimator properties.

[2] Gibbs sampling has already found acceptance and application in other disciplines such as epidemiology (e.g., Clayton, 1991; Gilks *et al.*, 1996).

7.1.1 The Problem of Spatial Dependence

Following the basic framework in any econometrics text (see e.g., Greene, 1997; Maddala, 1983; Amemiya, 1985; Judge *et al.*, 1985), the binary discrete choice probit model begins with a model speciˇ ed in latent form, as:

$$y_i^* = x_i\beta + \varepsilon_i \quad \varepsilon \sim (0,\Omega), \tag{7.1}$$

where y_i^* is an unobserved latent variable, X is an n by k matrix of regressors with individual rows x_i, β is the corresponding k by 1 parameter vector, ε_i is a normally distributed stochastic error with zero mean and is the ith element in a vector, ε, with variance-covariance matrix $\mathrm{E}[\varepsilon\varepsilon'] = \Omega$.

The basic Maximum Likelihood function for this model assumes that the variance-covariance structure is uncorrelated and homoskedastic, e.g., $\varepsilon \sim \mathrm{N}(0,\Omega)$, where $\Omega = \sigma^2 I$. The latent dependent variable is not observed directly, but an indicator of the latent variable is observed as:

$$\begin{aligned} y_i &= 1 \text{ if } y_i^* \geq 0, \\ y_i &= 0 \text{ otherwise}, \end{aligned} \tag{7.2}$$

where y_i is the observed counterpart to the continuous dependent variable. The probability that the latent variable is greater than zero is expressed as $P(y^* \geq 0) = P(\varepsilon < X\beta) = \Phi(X\beta)$, where $\Phi(\cdot)$ is a cumulative normal distribution function. Dropping the subscript i implies the vector notation for the stacked model, $i=1,\ldots n$. The Maximum Likelihood function is derived from the underlying assumption that each observation is drawn from a Bernoulli distribution with success probability, $F(\cdot)$. Assuming independence of the ε's, as stated above, and therefore independence of the y's, yields the likelihood:

$$P(Y_1 = y_1, Y_2 = y_2, \ldots, Y_n = y_n) = \prod_{i=1}^{n} \int_{-\infty}^{a_i} \phi(\varepsilon_i)d\varepsilon_i, \tag{7.3}$$

where $a_i = [2y_i - 1]\frac{x_i\beta}{\sigma_{\varepsilon_i}}$, and $\phi(\cdot)$ is the normal density function associated with $\Phi(\cdot)$, a standard probit formulation.

If instead, the errors are correlated and distributed normally (e.g., Ω is non-diagonal) then independence of the y's cannot be assumed and the likelihood function becomes:

$$P(Y_1 = y_1, Y_2 = y_2, \ldots, Y_n = y_n) = \int_{-\infty}^{a_1} \cdots \int_{-\infty}^{a_n} f(\varepsilon)d\varepsilon, \tag{7.4}$$

where

$$f(\varepsilon) = (2\pi)^{-n/2} |\Omega|^{-1} \exp^{(-1/2)(\varepsilon'\Omega^{-1}\varepsilon)}.$$

Evaluation of this likelihood function requires multidimensional integration because of the error correlation.

7.1.2 A Spatial Discrete Choice Speciˇ cation

The spatial models under consideration in this chapter are a class of spatial lag and spatial error models that express spatial dependence in an autoregressive form.[3] In both spatial models, the autoregressive nature of the dependence is the spatial equivalence of time series autoregressive models. The spatial autoregressive lagged dependent variable model (SAL) includes spatially lagged dependent variables. The spatial autoregressive error model (SAE) includes spatially correlated errors and is a special case of regression models with non-spherical variance-covariance matrices. Mathematically, the underlying latent model speciˇ cation with spatial dependence becomes:

$$y_i^* = \rho \sum_{j=1}^{n} w_{ij} y_j^* + x_i\beta + \mu_i, \text{ for the SAL model,}$$

$$y_i^* = x_i\beta + \varepsilon_i, \text{ where, } \varepsilon_i = \lambda \sum_{j=1}^{n} w_{ij}\varepsilon_j + \mu_i, \text{ for the SAE model,} \tag{7.5}$$

with,

$$\begin{aligned} y_i &= 1 \text{ if } y_i^* \geq 0, \\ y_i &= 0 \text{ otherwise,} \end{aligned} \tag{7.6}$$

where y_i^* is the unobserved latent version of the observed dependent variable, y_i, w_{ij} is an element in the postulated weights matrix W, the spatial autoregressive lag coef˜ cient is ρ, or the spatial autoregressive error coef˜ cient is λ, and μ is an *iid* normal random variable with mean zero and variance σ_μ^2.

These two spatial models can be rearranged and written in matrix form as:

$$\begin{aligned} y^* &= (I - \rho W)^{-1}(X\beta + \mu) \text{ for the SAL model,} \\ y^* &= X\beta + (I - \lambda W)^{-1}\mu \text{ for the SAE model.} \end{aligned} \tag{7.7}$$

The variance-covariance matrices for these two spatial models are:

$$\begin{aligned} \Omega &= (I - \rho W)^{-1}(I - \rho W)^{-1\prime}\sigma_\mu^2 \text{ for the SAL model,} \\ \Omega &= (I - \lambda W)^{-1}(I - \lambda W)^{-1\prime}\sigma_\mu^2 \text{ for the SAE model,} \end{aligned} \tag{7.8}$$

and the probit likelihood function given either variance-covariance structure is:

$$P(Y_1 = y_1, Y_2 = y_2, \ldots, Y_n = y_n) = \int_{-\infty}^{a_1} \cdots \int_{-\infty}^{a_n} \phi(\mu)d\mu, \tag{7.9}$$

where,

$$\phi(\mu) = (2\pi)^{-n/2}|\Omega|^{-1}\exp^{(-1/2)(\mu'\Omega^{-1}\mu)}.$$

---

[3] Excellent references for spatial econometrics in general and spatial econometric model speciˇ cation include Anselin (1988b), and Anselin and Bera (1998).

This model differs substantially from the non-spatial specification because the spatially correlated covariance structure does not allow the simplification of the multivariate distribution into the product of univariate distributions. These spatial covariance structures also imply heteroskedastic variances and therefore cause inconsistency of the standard estimator for a non-spatial discrete choice model in the presence of either form of spatial dependence (McMillen, 1992; Beron and Vijverberg, 2003).

To achieve consistency the method of estimation must account for heteroskedasticity and assume the off-diagonal terms of the variance-covariance matrix are zero. If full use of the spatial information is also required, then the estimation technique must be able to account for the off-diagonal variance-covariance terms and the resulting n-dimensional integration problem. The proposed techniques to deal with these spatial dependence structures can be divided into two groups: solutions that focus on the heteroskedasticity induced by the spatial model structures, and solutions that consider the full variance-covariance structure and the associated n-dimensional integration.

7.2 Heteroskedastic Estimators

Case (1992) addressed the heteroskedasticity in an SAE model by specifying a specialized form for the spatial weights such that W implies a heteroskedastic variance-covariance matrix. Estimation is performed by normalizing the model by the non-constant variances implied by the spatial correlation in a similar fashion to the standard heteroskedasticity correction methods described in basic econometrics texts (e.g., Greene, 1997; Judge *et al.*, 1985).

Pinkse and Slade (1998) propose the use of a Generalized Method of Moments (GMM) estimator based on the moment conditions implied by the likelihood function for a probit model that accounts for the heteroskedasticity caused by a spatially autoregressive error structure (SAE), as described in equation (7.8) above. The authors show that the score vector from the maximum likelihood function for a discrete choice model is a set of moment conditions that can be used in a GMM framework. The extension of this to account for spatial error autocorrelation results in the estimation of a GMM model with heteroskedastic variances.

The heteroskedastic Maximum Likelihood function for this model is:

$$\ln L = \sum_{i=1}^{n} \left\{ y_i \ln \Phi \left(\frac{x_i \beta}{\sigma_i} \right) + (1 - y_i) \ln \left[1 - \Phi \left(\frac{x_i \beta}{\sigma_i} \right) \right] \right\}, \tag{7.10}$$

where σ_i^2 is the variance based on Ω with the spatial parameter, λ. The moments used in the GMM model are derived by taking the first order conditions of the likelihood function with respect to β and setting them equal to zero.

The moments for the heteroskedastic probit model are written as:

$$\bar{m}(\beta, \lambda) = \frac{1}{n} \sum_{i=1}^{n} h_i' \left[\frac{(y_i - \Phi)\phi}{\Phi(1 - \Phi)} \right], \tag{7.11}$$

where,

$$\Phi = \Phi\left(\frac{x_i\beta}{\sigma_i}\right), \phi = \phi\left(\frac{x_i\beta}{\sigma_i}\right),$$

and h_i is the ith row of a matrix of instruments, H. The GMM estimator minimizes the criteria:

$$S(\beta,\lambda) = \bar{m}(\beta,\lambda)' M^{-1}\bar{m}(\beta,\lambda),$$

where M is any positive deˇ nite matrix. If the observation speciˇ c variances are known (e.g., λ is known) then each observation can be divided by its own standard deviation and a standard probit model estimated. If the variances are unknown, they are deˇ ned as a function of the spatial weights matrix and the unknown spatial parameter, λ. Therefore, the GMM model must estimate all the parameters together, which requires the evaluation of Ω for any candidate choice of λ as part of the non-linear optimization of the minimization criteria. Clearly, because of the complex form of Ω, that includes inverses of n by n matrices dependent on the spatial parameter, the optimization problem can become quite difˇ cult.

The authors do not report the covariance estimates because of concern about asymptotic properties not holding for the small sample used to demonstrate the method. Given the concern about the size of the sample for the covariance matrix properties, the parameter estimates themselves may also be questionable, because the model relies on the use of large sample asymptotic properties to describe the consistency of the estimates as well as the asymptotic normality of the GMM estimator.

For this model, the regularity conditions for consistency require the spatial correlation to be structured such that the variances are ˇ nitely bounded. This bounding condition is based on the asymptotic domain increasing such that observations are added at the edges, or increasing domain asymptotics (Cressie, 1993). Whether this is a reasonable assumption will depend on the particular empirical application and the chosen spatial dependence structure. For lattice based data (census tracts, states, counties, etc.) this approach seems plausible because it is not possible to "inˇ ll" these geographic units. For micro level data (economic agents, environmental sampling locations, etc.) the data may be bounded by a particular geography and the more appropriate asymptotic approach is to "inˇ ll" the domain with more and more observations, or inˇ ll asymptotics, rather than increase the boundary of the domain (Cressie, 1993). Obviously, this has very different effects on the spatial structure, as more observations become potential "neighbors" when the density of the data increases. It is unclear that consistency still holds for inˇ ll asymptotics.[4] The asymptotic normality of the GMM estimator further relies on the condition that the dependence relationship dies as distance increases. This regularity condition is more restrictive than the similar conditions in the autoregressive time-series models, because the speed with which the relationship dies must account for the two-dimensional nature of the data.

[4] Lahiri (1996) discusses regularity conditions and consistency with inˇ ll asymptotics for spatial data.

Because of these asymptotic conditions the practitioner of this estimation technique must pay careful attention to the choice of spatial weights matrix because not all speciˇ cations will necessarily satisfy these conditions. Furthermore, the complexity of the optimization of the moment conditions makes practical application more difˇ cult.

7.3 Full Spatial Information Estimators

7.3.1 The EM Algorithm

The EM algorithm was ˇ rst described by Dempster *et al.* (1977) for models in time series. Ruud (1991) provides a survey of the general method and shows the wide variety of models to which the EM algorithm can be applied. For the binary discrete choice probit speciˇ cation a model is speciˇ ed with an unobserved latent variable that is observed according to an observation rule. The EM algorithm uses the likelihood function corresponding to the latent model as the basis for estimation. The two step process includes an E or expectation step and an M or maximization step. The E-step takes the expectation of the likelihood function for the latent variable conditional on the observed variable and a starting value for the parameter vector. The M step maximizes the resulting expected likelihood function for the parameter vector. The E and M steps are then repeated until the parameter vector converges. The estimated parameter vector converges to the Maximum Likelihood estimator of the original multidimensional likelihood function.

The process can be simpliˇ ed by using the EM algorithm to estimate the simple discrete choice model. The E-step simply becomes the expected value of the latent variable given the observed variable. Therefore, the EM algorithm reduces to a straightforward expectation calculation and maximization of the likelihood function corresponding to the linear latent model. For the non-spatial discrete choice probit model described in equations (7.1) and (7.2), the expected value of the latent variable is given by:

$$\begin{aligned} \mathrm{E}\left[y_i^*|y_i=1\right] &= x_i\beta+\mathrm{E}\left[\varepsilon_i|\varepsilon_i>-x_i\beta\right]=x_i\beta+\sigma\frac{\phi\left(x_i\beta/\sigma\right)}{\Phi\left(x_i\beta/\sigma\right)},\\ \mathrm{E}\left[y_i^*|y_i=0\right] &= x_i\beta+\mathrm{E}\left[\varepsilon_i|\varepsilon_i<-x_i\beta\right]=x_i\beta-\sigma\frac{\phi\left(x_i\beta/\sigma\right)}{1-\Phi\left(x_i\beta/\sigma\right)}, \end{aligned} \tag{7.12}$$

where σ is set equal to one because it cannot be identiˇ ed in a regular probit model. Replacing the unobserved latent variable with its expected value makes the latent equation a simple linear regression model that can be estimated by OLS. Therefore, the EM algorithm consists of constructing the expectations in equation (7.12) with initial parameter values, regressing the calculated $\hat{y}_i^*$ on x_i for a new parameter vector, β, and iterating this procedure until convergence occurs. The resulting estimates are asymptotically Maximum Likelihood probit estimates.

Generalizing the EM algorithm to discrete choice models with spatially lagged dependent variables and spatial error autocorrelation, as in equations (7.5) and (7.6), requires reformulating the E-step and using the appropriate continuous Maximum Likelihood model with the estimated latent variable in the M-step. McMillen (1992) generalizes the EM algorithm to these spatial cases and notes increased complexity in both the E-step and M-step. To keep the notation clear, the following simpliˇ cation is used:

$$\text{let } \delta_{1ij} \text{ be a typical element of } (I-\rho W)^{-1},$$

$$\text{let } \delta_{2ij} \text{ be a typical element of } (I-\lambda W)^{-1},$$

$$\begin{aligned} x_i^* &= \sum_{j=1}^{n} \delta_{1ij} x_j \beta, \\ \sigma_i^2 &= \sigma_\mu^2 \sum_{j=1}^{n} \delta_{1ij} \text{ for the SAL model,} \\ \sigma_i^2 &= \sigma_\mu^2 \sum_{j=1}^{n} \delta_{2ij} \text{ for the SAE model.} \end{aligned} \tag{7.13}$$

The expected values for the SAL model are:

$$\begin{aligned} \mathrm{E}\left[y_i^* | y_i = 1\right] &= x_i^* \beta + \mathrm{E}\left[\varepsilon_i | \varepsilon_i > -x_i^* \beta\right] = x_i^* \beta + \sigma_i \frac{\phi\left(x_i^* \beta / \sigma_i\right)}{\Phi\left(x_i^* \beta / \sigma_i\right)}, \\ \mathrm{E}\left[y_i^* | y_i = 0\right] &= x_i^* \beta + \mathrm{E}\left[\varepsilon_i | \varepsilon_i < -x_i^* \beta\right] = x_i^* \beta - \sigma_i \frac{\phi\left(x_i^* \beta / \sigma_i\right)}{1 - \Phi\left(x_i^* \beta / \sigma_i\right)}, \end{aligned} \tag{7.14}$$

and for the SAE model,

$$\begin{aligned} \mathrm{E}\left[y_i^* | y_i = 1\right] &= x_i \beta + \mathrm{E}\left[\varepsilon_i | \varepsilon_i > -x_i \beta\right] = x_i \beta + \sigma_i \frac{\phi\left(x_i \beta / \sigma_i\right)}{\Phi\left(x_i \beta / \sigma_i\right)}, \\ \mathrm{E}\left[y_i^* | y_i = 0\right] &= x_i \beta + \mathrm{E}\left[\varepsilon_i | \varepsilon_i < -x_i \beta\right] = x_i \beta - \sigma_i \frac{\phi\left(x_i \beta / \sigma_i\right)}{\Phi\left(x_i \beta / \sigma_i\right)}. \end{aligned} \tag{7.15}$$

Rather than using OLS in the M-step, the underlying spatial model is estimated via Maximum Likelihood with the likelihood function:

$$\ln L = \left(\frac{-n}{2}\right) \ln \pi - \left(\frac{1}{2}\right) \ln |\Omega| - \left(\frac{1}{2}\right) \mu' \mu, \tag{7.16}$$

where $\mu = (I - \rho W)\hat{y}^* - X\beta$ for the SAL model and $\mu = (I - \lambda W)\left[\hat{y}^* - X\beta\right]$ for the SAE model, Ω is described in equation (7.8) for each model, and $\hat{y}^*$is the set of predicted latent values from the E-step (equation (7.14) or (7.15) depending on the model).

There remains a problem in obtaining estimates of parameter dispersion from the covariance matrix. The EM algorithm avoids n-dimensional integration in parameter estimation, but the Maximum Likelihood model in equation (7.9) is the true likelihood function for which the EM algorithm estimated parameters have converged. Therefore, the relevant covariance matrix needs to be estimated from the

n-dimensional dependence structure. Clearly, this is intractable. McMillen (1992) offers a covariance matrix based on interpreting the probit model as a non-linear weighted least squares model, conditional on the spatial parameter. This approach yields estimates of the parameter standard errors that are biased because the covariance matrix is determined based on the assumption of a ˇ xed spatial parameter in the conditional non-linear weighted least squares formulation. Therefore, any covariance between the spatial parameter and other parameters in the model is not accounted for as would be the case if one could estimate a covariance matrix for the n-dimensional Maximum Likelihood function directly.

Another drawback to the EM algorithm is that there is substantial computational burden in the repetitions of the algorithm. Each iteration through the M-step requires estimation of a spatial Maximum Likelihood model, requiring calculation of the determinant of an n by n matrix as many times as is necessary, to achieve convergence in the likelihood function for each M-step. For large n, calculation of the determinant is time consuming, but in the EM algorithm the likelihood function is maximized for every pass through the M step. Therefore, the EM algorithm requires many evaluations of n by n determinants.

7.3.2 The RIS Simulator

Beron and Vijverberg (2003) propose a recursive importance sampling (RIS) estimator to evaluate n-dimensional normal distributions. The RIS is the more general form of the better known (Geweke, 1989; Hajivassiliou, 1990; Keane, 1994) (GHK) smooth recursive estimator for multivariate normal probabilities that is one of the more successful simulation methods.[5] These simulation methods are based on McFadden's (1989) argument that evaluation of a multidimensional likelihood function is not necessarily the problem at hand. Instead, computing the likelihood function and its derivatives is actually an exercise in estimating a mean. In other words, individual terms in the likelihood may differ positively and negatively from the mean. If it is possible to build a probability distribution that re ects the positive and negative errors around the mean, it should be possible to obtain estimates of the likelihood function that are close to the actual likelihood value. This realization forms the basis for many simulation-based multivariate probability approximation techniques.[6]

Beron and Vijverberg (2003) describe a simulation procedure that can be applied to both the SAL and SAE models as described in equations (7.5) through (7.9) above. This RIS simulator is the only approach reviewed here that directly deals with the spatial, n-dimensional integration problem in the discrete choice likelihood function.

For μ distributed normally, deˇ ne $v_i = (1-2y_i)\mu_i$, to be a normally distributed error term. Because v_i is a linear transformation of a normally distributed error term

[5] Bolduc *et al.* (1997) implement the GHK estimator to study in a multinomial probit study of location choice with a spatial autoregressive error structure. They also compare this estimator to the Gibbs Sampler approach, which is described in more detail below.

[6] For an excellent review of simulation methods consult the symposium on simulation methods published in *Review of Economics and Statistics*, Vol. 76, November 1994.

it is also normally distributed. Based on the discrete choice censoring rule described in equation (7.6), v_i can be rewritten in vector notation as:

$$v < -C(I-\rho W)^{-1}X\beta \text{ for the SAL model,}$$
$$v < -CX\beta \text{ for the SAE model,} \qquad (7.17)$$

where C is an n by n matrix with diagonal elements, $C_{ii} = (1-2y_i)$. The variance-covariance matrix for v is $C\Omega C'$, where Ω is from equation (7.8) for either the SAL or SAE model. The right hand side of the inequalities in (7.17) can be described as a vector of upper bounds, V, such that the goal of the RIS simulator is to evaluate the probability $[v_i < V_i]$ for all i. For example, in the case of the SAE model this is equivalent to evaluating $[v_i < -x_i\beta]$ for $y_i = 0$ and $[v_i < x_i\beta]$ for $y_i = 1$.

Given this characterization of the SAL and SAE models, the RIS simulator[7] can be used to evaluate the n-dimensional integral for the spatially correlated probit model in equation (7.9). Because both of the spatial models are transformed as in equation (7.7) the error structure is spatially dependent with the variance-covariance matrix as in equation (7.8). One can deˇ ne a decomposition of Ω as $A'A = \Omega^{-1}$ so that $\eta = Av$ implies η is *iid* normally distributed. Deˇ ne $B = A^{-1}$, an upper triangular matrix with all positive diagonal elements, then $B\eta = v$. Because the model has a variance-covariance structure, the upper bound for one v_i is dependent on the other v_j based on the covariance structure. Therefore, the upper bounds in matrix form and for the jth η are:

$$\eta < B^{-1}V,$$
$$\eta_j < \sum_{i=j}^{n} b_{ji}V_i = \eta_{j0}, \qquad (7.18)$$

where the summation comes from the upper triangular form of B caused by the correlated error structure. Let $\phi(\eta_j)$ be a normal density function with an associated cumulative distribution function (*cdf*), Φ. Given:

$$g^c(\eta_j) = \frac{\phi(\eta_j)}{\Phi(\eta_{j0})} \text{ for } \eta_j \le \eta_{j0},$$

then the n-dimensional probit probability to be evaluated is,

$$\int_{-\infty}^{\eta_{n0}} \frac{\phi(\eta_n)}{g^c(\eta_n)} \left[\int_{-\infty}^{\eta_{n-1,0}} \frac{\phi(\eta_{n-1})}{g^c(\eta_{n-1})} \cdots \left(\int_{-\infty}^{\eta_{2,0}} \frac{\phi(\eta_2)}{g^c(\eta_2)} \Phi(\eta_{10}) g^c(\eta_2)\, d\eta_2 \right) \cdots \right] g^c(\eta_n)\, d\eta_n. \qquad (7.19)$$

The RIS simulator is implemented by drawing a large number, R, of random vectors of η from the chosen distribution, $\phi(\eta_j)$, satisfying the condition that the

---

[7] For more details on the RIS simulator see Vijverberg (1997), and Beron and Vijverberg (2003). The RIS simulator based on normal distributions is also known as the GHK simulator that is described in Hajivassiliou (1993).

drawn value be within the upper bound, $\eta_j \leq \eta_{j0}$. The recursive nature of this simulator is made apparent by the fact that the bounds in equation (7.18) are backwards determined. For every drawing of the random vector, r, given η_{n0}, a value $\tilde{\eta}_{nr}$ is drawn. Then $\tilde{\eta}_{n-1,0,r}$ is calculated using $\tilde{\eta}_{nr}$. This process is repeated until $\tilde{\eta}_{1,0,r}$ is calculated. The simulated probability is:

$$\hat{p} = \frac{1}{R}\sum_{r=1}^{R}\left(\Phi\left[\tilde{\eta}_{1,0,r}\right]\prod_{j=2}^{n}\frac{\phi(\tilde{\eta}_{k,r})}{g^c(\tilde{\eta}_{k,r})}\right). \qquad (7.20)$$

This approach to estimating spatially dependent discrete choice models is attractive because it directly considers the problem as one of evaluating the n-dimensional probit likelihood function. No other method described in this chapter deals directly with this likelihood function. Furthermore, this approach allows for the use of Likelihood Ratio tests on the model specifications, an advantage that is only available in this method because of the fact that the actual dependent probit likelihood function is evaluated. The RIS simulator generates standard errors based on the distribution of the R random draws, but because the simulator is recursive and taking into account the full dependence of the spatial model there is no need to condition standard errors on fixed values of spatial parameters, as in the EM algorithm. Based on the Monte Carlo study that Beron and Vijverberg (2003) performed, the primary concern with this technique is the overall computational burden of the method. While shown to be quite accurate in the Monte Carlo study, every doubling in size of the sample leads to an increase in computation time by a factor of approximately 3.5. Beron and Vijverberg (2003) studied the RIS simulator with R equal to 1000 on a 300 Mhz Pentium II machine, where a sample of 50 observations took 2.5 minutes and a sample of 200 observations took thirty minutes to finish. Based on these timing numbers a moderately sized sample of 1600 observations, for example, would require over twenty-one hours to compute a simulated probability. The authors note that computational time can be reduced by lowering R, but this comes at the cost of increasing the dispersion of the parameter estimates. Therefore, the RIS simulator can provide an accurate way in which to deal with the n-dimensional integration in the spatial discrete choice likelihood function, but the computational costs associated with accurate estimates is high.

7.3.3 The Gibbs Sampler

The Gibbs sampler technique has been applied in a variety of contexts including epidemiology (e.g., Albert and Chib, 1993; Clayton, 1991; Gilks *et al.*, 1996) and image analysis (Geman and Geman, 1984). More generally, Gibbs sampling is a Markov Chain Monte Carlo (MCMC) technique that relies on the concept that a large sample of values for the parameters in the posterior distribution can be used to approximate a probability density for the parameters. MCMC techniques have been applied in a variety of applications.[8] Bolduc *et al.* (1997) compare the Gibbs

[8] See, e.g., Besag *et al.* (1995), and Waller *et al.* (1997).

Sampler for a multinomial probit model with an SAE structure to the previously described RIS simulator and conclude that both approaches yield similar results, but note the relative computational and conceptual simplicity of the Gibbs sampler in comparion to the RIS simulator.

Bayesian spatial discrete choice methods (Bolduc *et al.*, 1997; LeSage, 2000) are similar to the EM approach in that they formulate a likelihood function as if the dependent variable were continuous and use estimates of the latent unobserved variable to estimate the parameters. The Bayesian approach is different, however, in the way it formulates the likelihood function and the estimates of the unobserved latent variable. In addition, this method overcomes the problems encountered in estimating standard errors in the EM algorithm because parameter standard errors are derived from the posterior parameter distributions directly. The Bayesian Gibbs sampler approach to estimating spatial discrete choice models (both SAL and SAE) is proposed in detail in LeSage (2000), and is an extension of the Gibbs sampling methods of Geman and Geman (1984) and a Bayesian Gibbs sampler for non-spatial discrete choice models by Albert and Chib (1993).

LeSage (2000), based on Geweke (1993), extends the SAL and SAE models even further by incorporating heteroskedastic error terms independent of spatial error dependence. This is important because, as stated before, heteroskedasticity causes inconsistency in discrete choice models (e.g., Greene, 1997). In the above discussion the heteroskedastic consistent methods assumed that after controlling for the spatial dependencies the error structure would no longer exhibit heteroskedasticity. In this framework, after controlling for spatial dependencies the error is still allowed to be heteroskedastic, ensuring that parameter inconsistency is not driven by heteroskedastic in uences.

Geman and Geman (1984) introduced Gibbs sampling as a technique for characterizing posterior distributions. The Gibbs sampler uses conditional posterior distributions to achieve estimates of the parameters in the unconditional posterior distribution. They show that a Markov chain that unfolds via the Gibbs sampler accurately characterizes the joint posterior distribution. More speciˇ cally, given a k by 1 parameter vector, θ, and a joint posterior distribution, $p\,[\theta \mid D]$, where D is data, and conditional distributions, $p\,[\theta_k \mid D, (\forall\theta_l, l \neq k)]$, then Gibbs sampling proceeds as follows:

$$\text{Initialize sampling with } \theta^0,$$

$$\text{For } t = 0 \text{ to } T,$$

$$\begin{aligned}
&\text{Sample } \theta_1^{t+1} \sim p\left[\theta_1 \mid D, \left(\forall\theta_l^t, l \neq 1\right)\right],\\
&\text{Sample } \theta_2^{t+1} \sim p\left[\theta_2 \mid D, \left(\forall\theta_l^t, l \neq 2\right)\right],\\
&\cdots\cdots\cdots\cdots\\
&\text{Sample } \theta_k^{t+1} \sim p\left[\theta_k \mid D, \left(\forall\theta_l^t, l \neq k\right)\right],\\
&\quad t = t + 1.
\end{aligned} \tag{7.21}$$

Gelfand and Smith (1990) outline the proof that Gibbs sampling, with the complete set of conditional distributions for all the parameters in a model, produces a sample set that converges in the limit to the true joint posterior distribution of the parameters. Measures of parameter dispersion are easily calculated from the sample conditional distributions.

Based on the SAL and SAE models described in equations (7.5) through (7.8) with the independent error speciˇ ed as heteroskedastic:

$$\mu \sim, \mathrm{N}(0, \sigma_\mu^2 V),\ V = \mathrm{diag}\,(v_1, v_2, \ldots, v_n),$$

LeSage (2000) describes a Bayesian model with diffuse priors that leads to the set of conditional distributions necessary to implement the Gibbs sampler. Because the Gibbs sampler is ˇ nding increasing application in the literature and is computationally and conceptually preferred to the RIS simulator (Bolduc *et al.*, 1997) and the EM algorithm (LeSage, 2000) it is worth describing the conditional distributions for the SAL and SAE models in detail. The following discussion is based on LeSage (2000), except for an alternative approach to sampling the underlying latent dependent variable that uses a variance decomposition as suggested in Bolduc *et al.* (1997).

The conditional distribution for σ is:

$$p\,[\sigma_\mu \mid \rho, \beta \text{ or } \lambda, V] \propto \sigma_\mu^{-(n+1)} \exp\left[-\left(\frac{1}{2\sigma_\mu^2}\right) e'V^{-1}e\right], \tag{7.22}$$

for the SAL or SAE model, where e is$(I - \lambda W)\,y^* - X\beta$ for the SAL model and $(I - \lambda W)\,[y^* - X\beta]$ for the SAE model. This posterior is a conditional χ^2 distribution with n degrees of freedom. The conditional distribution for β is a standard multivariate normal:

$$p\,[\beta \mid \rho, \sigma_\mu^2, V] \sim \mathrm{N}\left[(X'V^{-1}X)^{-1}X'V^{-1}\bar{y}, \sigma^2(X'V^{-1}X)^{-1}\right] \text{ for the SAL model,}$$
$$p\,[\beta \mid \lambda, \sigma_\mu^2, V] \sim \mathrm{N}\left[(\bar{X}'V^{-1}\bar{X})^{-1}\bar{X}'V^{-1}\bar{y}, \sigma^2(\bar{X}'V^{-1}\bar{X})^{-1}\right] \text{ for the SAE model,}$$

with,

$$\bar{X} = (I - \lambda W)\,X \text{ for the SAE model,}$$
$$\bar{y} = (I - \rho W)\,y^*$$

or,

$$\bar{y} = (I - \lambda W)\,y^*, \tag{7.23}$$

for the SAL and SAE models respectively.

Based on Geweke (1993), independent priors are assumed for the unknown heteroskedastic terms, $\pi\,(v_i)$. The prior distribution is assumed to be:

$$\pi\,(v_i^{-1} \mid q) \sim \mathrm{ID}\,\frac{\chi^2\,(q)}{q}\ \forall i,$$

where q is a hyperparameter that controls the distribution of v_i. As the value for q changes, the resulting distribution for v_i changes. When q is large, the distributions of v_i are homoskedastic and when q is small the distributions are heteroskedastic. This approach to the heteroskedastic disturbances also reduces the number of parameters that need to be estimated in the model. Rather than estimate all v_i, the parameter of the χ^2 distribution, q, is set and the v_i terms are determined based on the variability of the distribution. The conditional posterior distribution for the heteroskedastic variances is:

$$p\left[v_i \mid \rho \text{ or } \lambda, \beta, \sigma_\mu^2\right] \propto \frac{\left(\sigma_\mu^{-2} e_i^2 + q\right)}{v_i} \sim \chi^2\left(q+1\right). \tag{7.24}$$

The conditional posterior distributions for the spatial parameters are conditioned on σ_μ, β, and all v_i so that everything in the joint posterior can be placed in the constant of proportionality. The conditional posterior for ρ is:

$$p\left[\rho \mid \beta, \sigma_\mu^2, V\right] \propto |I - \rho W| \exp\left[-\left(\frac{1}{2\sigma_\mu^2}\right) e'V^{-1}e\right], \tag{7.25}$$

and the conditional posterior for λ is,

$$p\left[\lambda \mid \beta, \sigma_\mu^2, V\right] \propto |I - \lambda W| \exp\left[-\left(\frac{1}{2\sigma_\mu^2}\right) e'V^{-1}e\right]. \tag{7.26}$$

These two conditional distributions have an unknown form making the prospect of Gibbs sampling dif˜ cult. To overcome this problem Metropolis sampling is used, a technique that is useful when a conditional distribution is mathematically expressible, but of unknown form.[9] Metropolis *et al.* (1953) showed that a Markov chain stochastic process for a parameter, where the chain of sampled values is indexed by $t\,(\theta_t, t > 0)$ with the same set of possible values as the true parameter value, can be drawn from the posterior distribution for the parameter (e.g., Casella and George, 1992; Gilks *et al.*, 1996). This approach to analyzing posterior distributions was further generalized and popularized by Hastings (1970), who was able to show that any Markov chain process that was in state θ^t can be characterized by a conditional distribution in period $t+1$. Hastings' iterative procedure is also known as Metropolis sampling. Repeating this process a suf˜ cient number of times allows one to build a distribution for each of the spatial parameters.

The ˇ nal conditional distribution to be analyzed is the one associated with the unobserved latent variable. This conditional posterior distribution is the key to the Gibbs sampling estimation algorithm for discrete choice models, because all of the other conditional posterior distributions are derived from the underlying continuous likelihood model. This data augmentation step provides the linkage between the discrete dependent variable and its latent continuous counterpart. This is also the step that re ects the conceptual approach of the EM algorithm where the E-step

9 Both LeSage (2000) and Bolduc *et al.* (1997) use this technique to simulate spatial autoregressive parameters.

is providing the same discrete to continuous linkage in the EM algorithm as the conditional distribution for the unobserved latent variable in the Gibbs sampler.

Chib (1992) and Albert and Chib (1993) show that the missing information on the dependent variable in non-spatial tobit and probit models respectively, can be characterized by truncated normal distributions of the form $N(x_i\beta, 1)$. The tobit model requires truncation in accordance with the type of tobit (e.g., left, right, or double truncation depending on the cause). The probit model requires normal distributions truncated at the left by 0 if $y = 1$ and truncated at the right by 0 if $y = 0$. To extend this to the SAL and SAE models note that the underlying latent models in equation (7.7) with LeSage's heteroskedasticity included imply the following distributions for the dependent latent variable:

$$\begin{aligned} y^* &\sim N\left(\tilde{X}\beta, \sigma_\mu^2 AVA'\right) \text{ for the SAL model,} \\ y^* &\sim N\left(X\beta, \sigma_\mu^2 BVB'\right) \text{ for the SAE model,} \\ A = (I-\rho W)^{-1}, & \; B = (I-\lambda W)^{-1}, \tilde{X} = AX. \end{aligned} \tag{7.27}$$

LeSage (2000) proposes the use of univariate truncated normal distributions based on equation (7.27) where the individual variance terms of the variance-covariance matrices are used. This approach loses the information found in the covariance terms of the multivariate normal distribution of y^*. Bolduc *et al.* (1997) suggest instead that the underlying latent models be transformed using the Cholesky root of the inverted error covariance matrices. This takes advantage of the conditional nature of the Gibbs sampler, because when the conditional posterior for y^* is evaluated it uses Gibbs sampler estimates of the other parameters. In particular, estimates of ρ or λ, σ_μ^2, and V can be used to construct an estimate of Ω and a Cholesky root of $\Omega^{-1} = D$. This allows the latent independent variable to be transformed such that it is distributed independently. Therefore, letting $\widehat{y}_i^*$, $\widehat{\tilde{x}}_i$ for the SAL model, and $\widehat{x}_i$ for the SAE model be the Cholesky transformed dependent and independent variables, the truncated distributions to be sampled are:

$$f\left(\widehat{y}_i^* \mid \rho, \beta, \sigma_\mu^2, V\right) = \left\{ \begin{array}{l} N(\widehat{\tilde{x}}_i\beta, 1) \text{ truncated at the left by 0 if } y_i = 1 \\ N(\widehat{\tilde{x}}_i\beta, 1) \text{ truncated at the right by 0 if } y_i = 0 \end{array} \right\}, \tag{7.28}$$

for the SAL model, and,

$$f\left(\widehat{y}_i^* \mid \lambda, \beta, \sigma_\mu^2, V\right) = \left\{ \begin{array}{l} N(\widehat{x}_i\beta, 1) \text{ truncated at the left by 0 if } y_i = 1 \\ N(\widehat{x}_i\beta, 1) \text{ truncated at the right by 0 if } y_i = 0 \end{array} \right\}, \tag{7.29}$$

for the SAE model. These conditional distributions are used to "predict" the continuous value of the underlying latent variable conditional on the parameters of the model.

The Gibbs sampler procedure based on this set of conditional distributions is started with an arbitrary set of initial parameters, (ρ^0 or $\lambda^0, \beta^0, \sigma_\mu^0, v_i^0$). The conditional distribution in equation (7.22) is calculated based on these starting values. This result, as well as the remaining starting parameter values, are then used in the conditional distribution in equation (7.23). The parameter estimates derived in equations (7.22) and (7.23) and any remaining starting values are used in equation (7.24) to calculate estimates of the heteroskedastic terms. A Metropolis sampling technique is then applied to the conditional distribution using $\left(\beta^1, \rho^0 \text{ or } \lambda^0, \sigma^1, v_i^1\right)$ for equations (7.25) or (7.26). Finally, the conditional distribution for the latent variable is sampled based on equations (7.28) or (7.29). Having completed one pass of the Gibbs sampler this process is repeated a large number of times to derive conditional distributions for all of the parameters. The mean of the conditional distribution is the final parameter estimate and the standard deviation of the distribution is used for inference.

Apart from Bolduc *et al.* (1997) and LeSage (2000), spatial Bayesian Gibbs samplers have not been extensively tested in empirical applications or Monte Carlo studies. Because the technique is a sampling method it is important to understand its behavior in varying sample size settings. LeSage (2000) compares his Gibbs sampler to the EM algorithm on the relatively small Anselin (1988b) neighborhood crime data in Columbus, Ohio, and finds that while the β coefficients are similar across techniques the spatial coefficients can vary more substantially. Given these results, a Monte Carlo study of the EM algorithm, RIS simulator, and Gibbs sampler may be able to shed some light on the strengths and weaknesses of the different techniques. All three methods are computationally burdensome as they deal with the complex spatial dependence structures. Again, Monte Carlo simulations may shed some light on the true computational costs of these different methods. From a purely informative perspective, the RIS simulator and Gibbs sampler are preferable to the EM algorithm as they both are capable of providing standard errors for all the parameters instead of conditionally on the spatial parameters.

7.4 Weighted Non-Linear Least Squares Estimators

The above discussion of heteroskedastic and spatially correlated techniques for estimating spatial discrete choice models are all based on the formulation of a Maximum Likelihood function. Case (1992) uses a heteroskedasticity consistent Maximum Likelihood function. Pinkse and Slade (1998) do not estimate a Maximum Likelihood function, but derive the necessary GMM moment equations from the likelihood function. Both approaches rely on a spatial autoregressive error structure to define a variance-covariance matrix from which heteroskedastic variances can be derived. The EM algorithm and Gibbs sampler use the Maximum Likelihood function associated with the related latent model and the RIS simulator forms the multidimensional likelihood function, but uses simulation techniques to derive parameter estimates.

This section describes a spatially dependent discrete choice methodology that considers the problem as a weighted non-linear version of the linear probability model (e.g., Greene, 1997; Maddala, 1983; Amemiya, 1985; Judge *et al.*, 1985) with a general variance-covariance matrix that can be estimated with a Generalized Method of Moments (GMM) estimator (Hansen, 1982). The estimators are described using a GMM methodology, but turn out to be weighted non-linear forms of the more familiar two stage least squares (2SLS) and feasible generalized least squares estimators.

This approach eliminates the higher order integration problem that arises in a spatially dependent likelihood function and the need to calculate n by n determinants found in the Maximum Likelihood function of the underlying latent models used in the EM algorithm and Gibbs sampler. For the SAL model this approach allows speciˇ cation of the discrete choice model in the form of an instrumental variable or 2SLS procedure. For the SAE model this approach extends the literature on multi-period probit models with dependence over time (e.g., Avery *et al.*, 1983; Poirier and Ruud, 1988) and speciˇ es the discrete choice model as a weighted non-linear feasible generalized least squares procedure.

7.4.1 Spatial Lag Dependence – A 2SLS Estimator

The endogenous spatially lagged dependent variable in the SAL model in this GMM framework is treated as any non-spatial endogenous variable would be in a GMM model. Standard instrumental variables or 2SLS estimation techniques are GMM models and have been discussed in the context of spatially lagged dependent variables by a number of authors (Anselin, 1980, 1988b, 1990; Kelejian and Prucha, 1998). As Kelejian and Prucha (1998) show, the ideal set of instruments for the spatially dependent lag are the increasing in order linear combinations of the exogenous variables and the spatial weights matrix $[X, WX, W^2X, ...]$. Therefore, for the SAL model under consideration here, the GMM estimator described below is a weighted non-linear version of the 2SLS (or instrumental variables) estimator described by Kelejian and Prucha (1998).

7.4.2 Spatial Error Dependence – A Feasible Generalized Least Squares Estimator

Avery *et al.* (1983) consider a multi-period probit model with serial correlation. Therefore, the Maximum Likelihood approach requires higher order integration dependent upon the persistence of the correlation. This alternative is a less ef˜ cient, but consistent, approach to estimation using a generalized method of moments estimator based on the weighted non-linear least squares speciˇ cation of a discrete choice model. The advantage of this formulation is that the estimates remain consistent with the incorrect assumption of no correlation. Furthermore, the weights are chosen so that the moment conditions are of the same form as the normal equations from the ordinary probit model. Under the ordinary probit assumptions the same estimated values are achieved via GMM, albeit with a differing variance-covariance matrix. This consistent special case is coined pseudo Maximum Likelihood.

Conley (1999) extends the GMM estimators of Hansen (1982) to the case of spatially correlated error structures. In this model parameters are estimated using the GMM minimization of sample moment conditions and the spatially correlated variance-covariance structures are estimated with non-parametric techniques, a spatial analog to Newey and West (1987). This spatial "Newey and West" approach is not suited to all types of spatial processes. In fact, the spatial autoregressive processes considered here do not satisfy the covariance stationarity requirements necessary for the non-parametric estimators.

Kelejian and Prucha (1999) suggest a moments estimator (ME) for estimating the spatial parameter in spatial autoregressive error processes with continuous dependent variables.[10] This approach requires consistent residuals estimated in a first stage model and spatial weights matrices that are bounded and finite. The row and column sums of the weights matrix must asymptotically approach finite numbers. Most spatial structures will meet this requirement, including the spatial autoregressive processes being considered here, as long as the spatial weights matrix is specified as a process with fading dependence. Therefore, for the SAE model under consideration here, the GMM estimator described below is a weighted non-linear feasible generalized least squares estimator. While the significance of the spatial parameter estimate cannot be assessed, it is considered to be a nuisance parameter that must be accounted for to improve the efficiency of regression coefficients and consistency of standard errors.

7.4.3 Spatial Discrete Choice GMM Estimators

The motivation for these models is not in the formulation of likelihood functions formulated as draws from a Bernoulli distribution, but a modification of the linear probability model. The model is estimated by determining the probability that the value of the indicator variable is either one or zero. In other words:

$$\Pr(y_i = 1) = F(x_i\beta) \text{ and } \Pr(y_i = 0) = 1 - F(x_i\beta). \tag{7.30}$$

The *cdf* can be thought of as a transformation of the latent process, $x_i\beta$, which is not bounded by zero and one, to the probabilistic range of zero and one. Therefore, if $x_i\beta$ goes to infinity, the probability that the indicator variable is one goes to one. If $x_i\beta$ goes to negative infinity the probability that the indicator variable is one goes to zero. This transformation deals with the chief complaint about the linear probability model that predictions are not restricted to the unit interval, causing the possibility of negative variances. In the spirit of regression, where the dependent variable is described by its conditional mean and an error term (Greene, 1997), the implied non-linear model is:

$$y = \mathrm{E}[y \mid X] + (y - \mathrm{E}[y \mid X]) = F(X\beta) + \varepsilon. \tag{7.31}$$

The expectation is the dependent variable conditional on the regressors. Because of the binary nature of the dependent variable, the error term is conditionally heteroskedastic (Greene, 1997). Using non-linear least squares with heteroskedastic robust standard errors, an exactly identified GMM estimator, is one way in which this

[10] An example of this approach is applied in Bell and Bockstael (2000).

model can be estimated. As Judge *et al.* (1985) notes, the ˇ tted relationship is very sensitive to the values of the exogenous variables. This sometimes causes difˇ culty in convergence of the non-linear minimization algorithm. A weighted non-linear least squares approach, following the spirit of Avery *et al.* (1983) in choosing the weights, helps to scale the exogenous variables and reduce problems with convergence.

Including spatial dependence in this general speciˇ cationof the model is straightforward. Both the spatially lagged dependent and variable model and the spatial error model can be speciˇ ed as:

$$\begin{aligned} y &= F(Z\delta) + \mu, \text{ for the SAL model,} \\ y &= F(X\beta) + \varepsilon, \\ \varepsilon &= \lambda W\varepsilon + \mu, \text{ for the SAE model,} \end{aligned} \tag{7.32}$$

where,

$$\mu \sim \left(0, \sigma_\mu^2 I\right),$$

for both models.

For the SAL model, Z is an n by k matrix of regressors with individual rows z_i, δ is the corresponding k by 1 parameter vector, μ is an *iid* stochastic error term with zero mean, and $F(\cdot)$ is the transformation *cdf*, assumed to be the normal *cdf* for a probit speciˇ cation. Z contains the spatial lag as an endogenous variable (e.g., $Z = [Wy^*, X]$ and $\delta = (\rho, \beta')'$). For the SAE model the transformation function includes only the exogenous variables and associated parameters, $X\beta$, but the variance-covariance matrix is spatial because of the spatial autoregressive error structure (e.g., for the SAE model $\delta = (\lambda, \beta')'$).

Using a GMM approach to this problem the speciˇ c form for the moments based on the models described in equation (7.32) is:

$$\begin{aligned} \mathrm{E}\{h_i A_i [y_i - F(z_i\delta)]\} &= 0, l = 1....L \text{ for the SAL model} \\ \mathrm{E}\{x_i A_i [y_i - F(x_i\beta)]\} &= 0, l = 1...., k \text{ for the SAE model} \end{aligned} \tag{7.33}$$

where A is an n by n diagonal matrix with individual speciˇ c weights, A_i, of the form,

$$\frac{f(\cdot)}{F(\cdot)[1 - F(\cdot)]},$$

where $f(\cdot)$ is a normal *pdf* and $F(\cdot)$ is a normal *cdf*, both with arguments $z_i\delta$ or $x_i\beta$ depending on the spatial model. For the SAL model H is an n by L matrix of instruments for the matrix of regressors, Z, where h_i is the ith row of $H = [X, WX, W^2X, ...]$.[11] The sample analogs to these moment conditions are,

$$\begin{aligned} \bar{m}(\delta) &= \frac{1}{n} H' A [y - F(Z\delta)] = 0 \text{ for the SAL model,} \\ \bar{m}(\delta) &= \frac{1}{n} X' A [y - F(X\beta)] = 0 \text{ for the SAE model.} \end{aligned} \tag{7.34}$$

[11] In practice, the higher order combinations are not included in H.

The GMM approach minimizes a weighted least squares criteria:

$$S(\delta) = \bar{m}(\delta)' M^{-1} \bar{m}(\delta),$$

where M is any positive deˇ nite matrix. The efˇ cient positive deˇ nte choice for M is the asymptotic variance of the moment conditions (Hansen, 1982):

$$\begin{aligned} M_{GMM} &= \text{Asy.Var}[\bar{m}(\delta)] = \text{E}[\bar{m}(\delta)\bar{m}(\delta)^{'}] \\ &= \frac{1}{n^2} H^{'} A \Omega A^{'} H \text{ for the SAL model,} \\ M_{GMM} &= \text{Asy.Var}[\bar{m}(\delta)] = \text{E}[\bar{m}(\delta)\bar{m}(\delta)^{'}] \\ &= \frac{1}{n^2} X^{'} A \Omega A^{'} X \text{ for the SAE model.} \end{aligned} \tag{7.35}$$

In practice, the non-linear speciˇ cation of the discrete choice model is heteroskedastic. Therefore, Ω in equation (7.35) for the SAL model incorporates White's heteroskedastic consistent variance-covariance matrix, $\Omega = \hat{\Psi}$. For the SAE model $\Omega = (I - \lambda W)^{-1} \hat{\Psi} (I - \lambda W)'^{-1}$, which takes into account the heteroskedasticity as well as the spatial error structure.

For both spatial models the weighting matrix is not available at the outset of estimation because it depends on parameters in the model. Any positive deˇ nite M, such as an identity matrix, $H^{'}H$, or $X^{'}X$, can be used to achieve consistent estimates in a ˇ rst iteration of the procedure, a more efˇ cient choice of M constructed, and the process further iterated until convergence of the parameter estimates.

For the SAE model the optimal weighting matrix additionally depends on the spatial error autoregressive parameter, λ. Kelejian and Prucha (1999) have derived a Moments Estimator (ME) for estimating the spatial parameter in an SAE model with continuous dependent variables. This approach requires ˇ rst stage estimation of consistent residuals and spatial weighting matrices that are bounded and ˇ nite (the row and column sums of the weighting matrix must asymptotically approach a ˇ nite number). Most spatial structures will meet this requirement.

The proposed discrete choice GMM model detailed here differs from the continuous model described by Kelejian and Prucha in that the linear model is replaced by a non-linear model. Because the GMM methodology provides consistent residuals with any choice of positive deˇ nite weighting matrix, the ˇ rst stage GMM residual estimates can be applied to solve for a spatial error autoregressive parameter, λ, for use in a second stage weighting matrix, M.

The three moment conditions derived in Kelejian and Prucha (1999) are used to construct a non-linear least squares estimator based on a three-equation system:

$$\begin{bmatrix} \frac{2}{n}\tilde{\varepsilon}'\tilde{\tilde{\varepsilon}} & \frac{-1}{n}\tilde{\tilde{\varepsilon}}'\tilde{\tilde{\varepsilon}} & 1 \\ \frac{2}{n}\tilde{\tilde{\tilde{\varepsilon}}}'\tilde{\tilde{\varepsilon}} & \frac{-1}{n}\tilde{\tilde{\tilde{\varepsilon}}}'\tilde{\tilde{\tilde{\varepsilon}}} & \frac{1}{n}\text{Tr}(W'W) \\ \frac{1}{n}(\tilde{\varepsilon}'\tilde{\tilde{\tilde{\varepsilon}}} + \tilde{\tilde{\varepsilon}}'\tilde{\tilde{\varepsilon}}) & \frac{-1}{n}\tilde{\tilde{\varepsilon}}'\tilde{\tilde{\tilde{\varepsilon}}} & 0 \end{bmatrix} - \begin{bmatrix} \frac{1}{n}\tilde{\varepsilon}'\tilde{\varepsilon} \\ \frac{1}{n}\tilde{\tilde{\varepsilon}}'\tilde{\tilde{\varepsilon}} \\ \frac{1}{n}\tilde{\varepsilon}'\tilde{\tilde{\varepsilon}} \end{bmatrix} = \kappa(\lambda, \sigma^2), \tag{7.36}$$

where $\tilde{\varepsilon}$ is a vector of consistent model residuals, $\tilde{\tilde{\varepsilon}} = W\tilde{\varepsilon}$, and $\tilde{\tilde{\tilde{\varepsilon}}} = WW\tilde{\varepsilon}$. The ME follows from the minimization of $[\kappa(\lambda, \sigma^2)'\kappa(\lambda, \sigma^2)]$.

A consistent estimate of the spatial parameter, λ, estimates of the A_i weights, and $\hat{\Psi}$ based on the same set of residuals used to estimate λ can be used to construct Ω and M for the SAE model. One may iteratively improve the ef˜ ciency of the parameters used to construct the spatial parameter, λ, the A_i weights, and $\hat{\Psi}$ until convergence of the parameters, β, occurs in the minimization described below.

Combining the moments in equation (7.34) with the weighting matrix in equation (7.35) leads to the minimization criteria:

$$S(\delta) = \left[\frac{1}{n}H'A(y - F(Z\delta))\right]' \left[\frac{1}{n^2}H'A\Omega A'H\right]^{-1} \left[\frac{1}{n}H'A(y - F(Z\delta))\right], \quad (7.37)$$

for the SAL model, and,

$$S(\delta) = \left[\frac{1}{n}X'A(y - F(X'\beta))\right]' \left[\frac{1}{n^2}X'A\Omega A'X\right]^{-1} \left[\frac{1}{n}X'A(y - F(X'\beta))\right], \quad (7.38)$$

for the SAE model.

The asymptotic variance-covariance matrix used in practice is:

$$VC_{GMM} = [G'M^{-1}G],$$

where G is a matrix of derivatives with jth row,

$$G^j = \frac{\partial \bar{m}(\delta)}{\partial \delta'}.$$

Therefore, for the SAL model described here the variance-covariance matrix is:

$$VC_{GMM}(\delta) = \left[G'\left(\frac{1}{n^2}H'A\Omega A'H\right)^{-1} G\right],$$

$$G^j = \frac{\partial \bar{m}(\delta)}{\partial \delta'} = \frac{1}{n}H'\frac{\partial (A\varepsilon)}{\partial \delta'}, \quad (7.39)$$

and for the SAE model the variance-covariance matrix is,

$$VC_{GMM}(\delta) = \left[G'\left(\frac{1}{n^2}X'A\Omega A'X\right)^{-1} G\right],$$

$$G^j = \frac{\partial \bar{m}(\delta)}{\partial \delta'} = \frac{1}{n}X'\frac{\partial (A\varepsilon)}{\partial \delta'}. \quad (7.40)$$

The term $A\varepsilon$ is often referred to as the generalized residual, $\hat{\mu}$, due to Cox and Snell (1968).[12]

[12] Using the generalized residual notation greatly simpli˘ es the expression of the GMM model. For example, in the SAL model denoting the generalized residual as $\hat{\mu} = A\varepsilon$, the moment conditions become $\mathrm{E}[h_i\hat{\mu}_i] = 0$ with sample analog $\bar{m}(\delta) = 1/n \cdot (H'\hat{\mu})$, minimization criterion $S(\delta) = (H'\hat{\mu})'(H'A\Omega A'H)^{-1}(H'\hat{\mu})$, and variance-covariance matrix

$$VC_{GMM}(\delta) = \left[\left(H'\frac{\partial \hat{\mu}}{\partial \delta'}\right)' (H'A\Omega A'H)^{-1} \left(H'\frac{\partial \hat{\mu}}{\partial \delta'}\right)\right]$$

where $\partial\hat{\mu}/\partial\delta'$ is the matrix of derivatives of the generalized residual.

The two GMM estimators described here are weighted non-linear 2SLS and feasible generalized least squares estimators. For the SAL model the regularity conditions for consistency and asymptotic normality are the same as for non-linear 2SLS with the addition of ˇ nite row and column sums in the limit. This condition is met by most spatial dependence processes that fade with distance. For the SAE model the conditions are the same as for non-linear feasible generalized least squares with the same row and column sum conditions on the spatial process.

These estimators minimize moments equivalent to the probit log likelihood score vector when the error is *iid* and no spatially lagged dependent variables exist, and is consistent in the presence of spatial autoregressive error dependence. Therefore, one can compare consistent "probit" estimates to the SAL or SAE GMM estimators. Furthermore, these estimators do not require the calculation of n by n determinants and avoid the need for a large number of simulation passes through the model. One drawback to the GMM SAE estimator is that it treats the spatial error autoregressive parameter as a nuisance parameter and therefore standard error estimates are not provided.

7.5 Conclusions

The study of spatial dependence in discrete choice models, particularly in the context of the spatial probit model, has received less attention in the literature relative to spatial continuous models. Possible reasons for the lack of attention include the added complexity that spatial dependence introduces into discrete choice models and the need for more complex estimators. Many techniques have been proposed that focus on either the inconsistency of the standard probit model, if the spatial dependence causes heteroskedasticity, or the use of the information in the non-spherical variance-covariance structures.

The methods that deal with heteroskedasticity and ignore off-diagonal dependence (Case, 1992; Pinkse and Slade, 1998) are consistent and less computationally intensive. Pinkse and Slade (1998) still require the calculation of n by n determinants, but doesn't require the large number of simulation passes. The GMM estimators described here do not require n by n determinant calculations or many simulation passes, but the gains in computational ease come at the expense of an estimate of the spatial error parameter standard error for the SAE model. The EM algorithm (McMillen, 1992), the RIS simulator (Beron and Vijverberg, 2003), and the Gibbs Sampler (Bolduc *et al.*, 1997; LeSage, 2000) all rely on simulation techniques for estimating the parameters of the n-dimensional integral in the spatially dependent Maximum Likelihood function. Therefore, all three methods are computationally intensive and can be time consuming for moderate to large sample sizes. Both the RIS simulator and the Gibbs sampler provide unbiased estimates of the standard errors for all the model parameters, as opposed to the biased estimates from the EM algorithm. The Gibbs sampler is the most exible of the spatially dependent models because it can incorporate spatial lag dependence and spatial error dependence in addition to general heteroskedasticity of unknown form. Table 7.1 summarizes the different estimator costs and bene˘ ts.

The purpose of this chapter was to bring together the literature on spatial discrete choice estimation methods and provide a cohesive description and comparison of the different techniques. Because of the wide variety of potential economic applications of these econometric techniques, it is hoped that there will be increased use and testing of these methods, particularly Monte Carlo studies of the different estimator properties.

Acknowledgments

The author wishes to thank two anonymous referees and Luc Anselin for invaluable comments on an earlier version of this work. The views expressed in this chapter are not necessarily those of Fannie Mae. No Fannie Mae data sources were used in this chapter.

Table 7.1. Summary of Estimator Differences

	Computational Burden	Requires Calculation of n by n Determinant	Provides Spatial Parameter Standard Errors	Solves Problem of Spatially induced Heteroskedasticity	Solution for n-dimenional Integration
Pinkse & Slade (SAE)	high	yes[1]	yes[2]	yes	no
Non-Linear Least Squares (SAL)	low	no	yes	yes	*
Non-Linear Least Squares (SAE)	moderate	no	no[3]	yes	*
EM Algorithm (SAL)	higher	yes[4]	no[5]	yes	yes
EM Algorithm (SAE)	higher	yes[4]	no[5]	yes	yes
RIS Simulator (SAL)	highest	yes[4]	yes[6]	yes	yes
RIS Simulator (SAE)	highest	yes[4]	yes[6]	yes	yes
Gibbs Sampler (SAL)	higher	yes[4]	yes[6]	yes	yes
Gibbs Sampler (SAE)	higher	yes[4]	yes[6]	yes	yes

[1] As many times as needed for convergence.
[2] More accurate in large samples.
[3] Non-spatial parameter standard errors are unbiased.
[4] For every iteration.
[5] Non-spatial parameter standard errors are biased.
[6] Accuracy improving with number of iterations.
* Not necessary for least squares speciˇ cations.

8 Probit in a Spatial Context: A Monte Carlo Analysis

Kurt J. Beron and Wim P.M. Vijverberg

University of Texas at Dallas

8.1 Introduction

Data are often observed in a binary form: vote for or vote against; buy or don't buy; build or don't build; move or don't move, etc. In classical econometrics this situation has been extensively studied and appropriate procedures developed to handle the nature of the data. The standard model however does not allow for spatial processes to drive the choices made by decision makers. For example, whether one city increases its sales tax may depend the actions of neighboring cities. Whether one jurisdiction subsidizes the construction of a new sports arena depends on the options that are offered to the sports enterprise by other jurisdictions – which has been occurring with increasing frequency in the United States, at the threat of the team moving elsewhere. In both of these cases, the conventional probit model fails to account for interdependencies.

There is, of course, no reason that the data generating process could not involve a spatial component such as a spatial lag or spatial error. The spatial linear model that deals with continuous, as opposed to binary, situations has been analyzed and refined (for an overview, see Anselin, 1988b; Anselin and Bera, 1998), but the counterpart of a spatial probit has only been discussed in specific cases. The objective of this chapter is to provide a general discussion of the spatial probit model and to demonstrate a spatial probit model that allows for spatial lag or spatial error. We construct an estimation strategy based on Monte Carlo simulation that demonstrates the ability of the spatial probit to capture the true underlying model and we comment on the findings. Finally, we compare the spatial probit to the conventional linear spatial estimator that does not account for the binary dependent variable. In the course of this comparison we provide some benchmarks that may help the researcher decide how the lower cost linear model may be suggestive of what a spatial probit analysis would find.

8.2 Probit Models

8.2.1 Standard Probit

The standard probit model is familiar to any applied econometrician. One assumes that the data (y_i, X_i) for $i = 1, \ldots, n$ are generated by the following process:

$$
\begin{aligned}
y_i^* &= X_i'\beta + u_i, && (8.1)\\
y_i &= 1 \ \text{ if } \ y_i^* \geq 0,\\
&= 0 \ \text{ if } \ y_i^* < 0, && (8.2)
\end{aligned}
$$

where u_i is independently and identically distributed N(0,1). The variable y_i^* is only partially observed: one knows whether it is positive or negative. Deˇ ne the indicator function $N(y^*)$ as $N = 1$ whenever $y^* \geq 0$ and $N = 0$ when $y^* < 0$. Equation (8.2) can be restated as $y = N(y^*)$, and is also equivalent to:

$$
\begin{aligned}
y_i &= 1 \ \text{ if } \ u_i \geq -X_i'\beta,\\
&= 0 \ \text{ if } \ u_i < -X_i'\beta. && (8.3)
\end{aligned}
$$

For the purpose of similarity with the spatial probit model and the exposition of the simulator that permits one to estimate the spatial probit model, we restate equation (8.3) with upper bounds only. Deˇ ne $v_i = (1 - 2y_i)u_i$. Thus, for $y_i = 0$, we have $v_i = u_i$ and $v_i < -X_i'\beta$; and for $y_i = 1$, we have $v_i = -u_i$ and $v_i \leq X_i'\beta$. It also follows that v_i is distributed N(0,1). Thus, since the equality $u_i = -X_i'\beta$ happens with probability 0, the inequality in equation (8.3) can be restated more concisely as:

$$
v_i < -(1 - 2y_i)X_i'\beta \ \text{ for } \ i = 1, \ldots, n. \qquad (8.4)
$$

Deˇ ne Z as a $n \times n$ matrix with $Z_{ii} = (1 - 2y_i)$ and $Z_{ij} = 0$. Note that Z is a diagonal matrix, with the property of $ZZ' = I_n$, the $n \times n$ identity matrix. Thus, the condition on v_i can be stated in matrix form as $v < -ZX\beta$, and the log-likelihood function is written as:

$$
\ln L = \ln \Phi_n[-ZX\beta; 0, I_n], \qquad (8.5)
$$

where $\Phi_n[U; \mu, \Sigma]$ is, in general terms, an n-dimensional normal cumulative distribution function with upper bound vector U, mean vector μ and variance matrix Σ.

8.2.2 Spatial Probit

The spatial probit model comes in two forms. The ˇ rst permits spatial error autocorrelation among the disturbances. This model is written in matrix form as:

$$
y^* = X\beta + u, \ \text{ where } \ u = \rho W u + \varepsilon. \qquad (8.6)
$$

The matrix W contains the information that causes spatial error autocorrelation, such as contiguity or distance. The parameter ρ measures the importance of the spatial dependence: $\rho = 0$ returns the model to standard probit. The observed variable y relates to y^* in the same way as above: $y = N(y^*)$ where the indicator function now operates on an n-dimensional vector.

The disturbance u can be expressed as:

$$u = (I_n - \rho W)^{-1}\varepsilon. \tag{8.7}$$

Let us denote the expression $(I_n - \rho W)^{-1}$ by Γ_ρ. Therefore $u = \Gamma_\rho \varepsilon$. Assuming that ε is distributed N(0,I_n), the mean of u is 0, and the variance is $\mathrm{Var}(u) = \Gamma_\rho \Gamma'_\rho$. As above, de˘ ne $v = Zu$, so that the observation of y as a vector of zeroes and ones implies that $v < -ZX\beta$. Moreover, $\mathrm{Var}(v) = Z\mathrm{Var}(u)Z' \equiv \Omega_\rho$. The log-likelihood function becomes:

$$\ln L = \ln \Phi_n[-ZX\beta; 0, \Omega_\rho]. \tag{8.8}$$

When there is a spatial lag, y^* is assumed to depend on y^*-values of spatially-related observations (e.g., neighbors).[1] Thus:

$$y^* = \alpha W y^* + X\beta + \varepsilon, \tag{8.9}$$

or, rewritten,

$$y^* = (I_n - \alpha W)^{-1} X\beta + (I_n - \alpha W)^{-1}\varepsilon. \tag{8.10}$$

De˘ ne $\Gamma_\alpha = (I_n - \alpha W)^{-1}$, and $u = \Gamma_\alpha \varepsilon$. Then with ε distributed N(0,I_n), we have $\mathrm{Var}(u) = \Gamma_\alpha \Gamma'_\alpha$. Once again, de˘ ne $v = Zu$: $\mathrm{Var}(v) = Z\mathrm{Var}(u)Z' \equiv \Omega_\alpha$, and, as before, the observation of $y = N(y^*)$ leads to an upper limit on v: $v < -Z\Gamma_\alpha X\beta$. With all this, the log-likelihood function is written as:

$$\ln L = \ln \Phi_n[-Z\Gamma_\alpha X\beta; 0, \Omega_\alpha]. \tag{8.11}$$

To estimate the parameters, one must have some way to evaluate an n-dimensional normal probability. There is no analytical solution for even a univariate normal cumulative distribution function (cdf), let alone for a multivariate one. Section 8.3 will brie y describe a simulator that can approximate an n-dimensional normal probability with remarkable precision.

8.2.3 Previous Literature

The extensions to the standard probit model described above are not entirely novel. There are several links with existing literature. A number of studies have recognized the inadequacy of the standard probit model when the data are generated by a

[1] One might wish to model y_i^* as a spatially lagged function of y_j for $j \neq i$. This model is infeasible. Indirectly, y_i^* would be a function of y_i, but y_i is determined by y_i^* through $y_i = N(y_i^*)$.

process that contains spatial effects. McMillen (1992) notes that both the spatially dependent error model and the spatial lag model imply heteroskedastic disturbances, which cause the parameter estimates to be inconsistent. A subsequent study illustrates other consequences by means of a Monte Carlo analysis (McMillen, 1995b): with smaller sample sizes it is difˇ cult to reject a homoskedastic probit model; yet, the marginal effect of X on the probability that y equals 1 is better estimated with the heteroskedastic probit model. Of course, heteroskedastic probit is not the same as spatial probit as in equations (8.8) or (8.11) above. In essence, consider a spatial error autocorrelation model: the variance of u_i is Ω_{ii}, a diagonal element of Ω_ρ in equation (8.8). With heteroskedastic probit, the likelihood function to be maximized is given by:

$$\ln L = \ln \Phi_n[-ZX\beta; 0, \tilde{\Omega}], \tag{8.12}$$

where $\tilde{\Omega}_{ii} = \Omega_{ii}$ for $i = 1, \ldots, n$ and $\tilde{\Omega}_{ij} = 0$ for $i \neq j$. This model does yield consistent estimates of β, even while the correlation among u is ignored, but the standard error of $\hat{\beta}$ is biased (Poirier and Ruud, 1988; Avery *et al.*, 1983). Conceptually, since $\tilde{\Omega}_{ii}$ depends on ρ, one could even attempt to estimate ρ. McMillen (1992, 1995b) speciˇ es a functional relationship for $\tilde{\Omega}_{ii}$ in terms of observable variables that are actually unrelated to the spatial matrix W. When the equation for y^* contains a spatial lag, y_i^* depends not only on $X_i'\beta$ but, as seen in equation (8.10), also on many if not all other $X_j'\beta$ for $j \neq i$. Maximizing the log-likelihood function in equation (8.12) can no longer yield consistent estimates of β: α is not a mere nuisance parameter. Even so, McMillen's solution is helpful when data from a large sample contain spatial error autocorrelation. An application of this technique is found in Case (1992), where the adoption of new technology among farmers depended on the actions taken by neighbors. She actually uses a contiguity matrix W of a particular form[2] that allows a signiˇ cant simplication in the way the spatial lag model is expressed and estimated.

The spatial probit model examines choices of n individuals under the assumption of spatial interaction. A spatial probit model is analytically closely related to the multinomial probit model. In a multinomial probit model, the behavior of individuals in the sample is assumed to be uncorrelated, and each individual selects one of J alternative actions. The attractiveness of alternative j could be modeled as $U_{ji} = X_{ji}\beta + u_{ji}$. The alternative-speciˇ c disturbances u_{ji} may well be correlated across alternatives; indeed this is the motivation behind the nested multinomial logit model that one could use to estimate β. But while a multinomial logit model yields such correlation patterns implicitly, a multinomial probit model permits one to specify them explicitly. Thus, in one application, Bolduc *et al.* (1996, 1997) examine the locational choice of general physicians across $J = 18$ provinces in Canada and specify a spatial dependence error structure based on distance between provinces.

[2] Case's weights matrix is block diagonal, measuring residence of farmers within districts. Each block consists of ones except for zeroes along the main diagonal. This allows for an algebraic expression for the inverse of $(I - \alpha W)$, but at the cost of excluding correlation across districts.

The likelihood function is similar to equations (8.8) and (8.11), in that it involves the evaluation of a multidimensional normal probability for each individual in the sample. The ˇ rst of these two studies estimates the model with a multinomial logit model mixed with a spatially correlated normal disturbance; the second study uses the GHK simulator which is a special case of the RIS simulator that will be discussed below.

The spatial probit model is also akin to the probit model applied to panel data of individuals who make a 0–1 choice in each of the J periods of the panel. The likelihood function contains an expression like equation (8.8), replacing n with J and summing this expression across sample individuals. Obviously, the correlation among disturbances across the panel for an individual is not spatially motivated. Rather, standard time-related serial correlation patterns are more appropriate. Several studies have examined this type of model. Avery *et al.* (1983) developed an orthogonality condition estimator that avoided the evaluation of multivariate probabilities. Keane (1994) used the GHK simulator discussed below in a Monte Carlo study of the *Methods of Simulated Moments* estimator and the *Simulated Maximum Likelihood* estimator. Lee (1998) also used the GHK simulator and the Simulated Maximum Likelihood technique in a Monte Carlo study of a number of dynamic models applicable to panel data.[3]

To our knowledge, there is only one study that has implemented a spatial probit model accounting for the full structure of the spatial dependence. Beron *et al.* (2003) analyzed the ratiˇ cation decision of the Montreal Protocol on ozone by 89 countries. They speciˇ ed a weights matrix that measured countries' economic interaction by means of international trade ows and estimated this model with the help of the GHK simulator.

8.2.4 Interpretation of the Parameters

In the standard probit model, the parameter β_j represents the impact of a one-unit change in X_{ji} on y_i^*. This information is difˇ cult to digest since y_i^* is not observed. Thus, it is common to express the impact of a change in X_{ji} on the probability that y_i equals 1. That is, the object of interest that can be more easily interpreted is:

$$\frac{d\Pr[y_i = 1|X_i]}{dX_i} = \phi(X_i\beta)\beta, \tag{8.13}$$

where ϕ is the standard normal univariate probability density function. This measure of marginal impact depends on X_i and is different for each observation in the sample. For this reason, one often substitutes the average of X into the argument of ϕ. That this is not always a satisfactory shortcut is obvious when X is highly variable but yields an average of $X\beta$ near 0: the marginal impact seems to be large but is in fact

[3] In a general formulation of these models, y^*_{it} is allowed to depend on X_{it}, $y^*_{i,t-j}$, and $y_{i,t-j}$ for $j = 1,2,\ldots$ That is, in a time series context, past choices ($y_{i,t-j}$) are permitted to have an impact on the current partially observable y^*_{it}, since there is no feedback effect from the present to the past. This feedback is the unique feature of spatial lag models.

much smaller for some observations. One might therefore compute the marginal impact for each observation and average over this set of values.[4]

With spatial dependence, the observations are no longer independent. In the case of spatial error autocorrelation, this does not make much of a difference. As mentioned in Sect. 8.2.2, y_i equals 1 iff $y_i^* > 0$ or iff $v_i < X_i\beta$. Since v_i has a $N(0, \Omega_{\rho,ii})$ distribution, the impact of X_i on the probability that y_i equals 1 is:

$$\frac{d\Pr[y_i = 1|X_i, W]}{dX_i} = \phi\left(\Omega_{\rho,ii}^{-1/2} X_i\beta\right)\Omega_{\rho,ii}^{-1/2}\beta. \tag{8.14}$$

In the case of a spatial lag model, the situation is more complicated. Let us first consider the impact of X_i on y_i^*. Let $D^{(i)}$ indicate the change in the vector $X\beta$ occasioned by a variation in X_i: all elements of $D^{(i)}$ equal 0 except for element i which equals $d(X_i\beta)$. The impact on the index variable y^* is $dy^* = \Gamma_{\alpha W} D^{(i)}$, and if y_i^* crosses the threshold of 0, y_i changes. This implies:

$$\frac{d\Pr[y_i = 1|X, W]}{dX_i} = \phi\left(\Omega_{\alpha,ii}^{-1/2}\left[\Gamma_\alpha X\beta\right]_i\right)\Omega_{\alpha,ii}^{-1/2}\Gamma_{\alpha,ii}\beta, \tag{8.15}$$

where $[\Gamma_\alpha X\beta]_i$ denotes the ith element of the vector that results from the expression inside the brackets.

Figure 8.1 illustrates the marginal impact of X_i on $\Pr[y_i = 1|X, W]$ for one of the weights matrix structures that we will use later on in the simulations, namely one that underlies the data structure of the T set with $n = 100$ observations. There is a single explanatory variable, ranging from 0 to 1 (implying a range for $X\beta$ from -1.5 to 1.5). The figure uses a value of $\rho = 0.50$ to compute the expression in equation (8.14) and $\alpha = 0.50$ to evaluate equation (8.15). The standard probit marginal effect is smooth, as equation (8.13) suggests. The variations evident in the marginal impact computed from the spatial error autocorrelation and spatial lag probit models derive from the variations in contiguity in the weights matrix that enters into the Ω and Γ matrices.

One may push the analysis of marginal impacts one step further. The weights matrix W has zeroes on the diagonal. On the basis of equation (8.10), one may distinguish a direct impact and an indirect impact of $X_i\beta$ on y_i^* for each i. The direct impact is $d(X_i\beta)$; the indirect effect is found as element (i,i) of the matrix $(I_n - \alpha W)^{-1} - I_n$ multiplied with $d(X_i\beta)$. This indirect effect is caused by the spatial interdependence among the observations: "How I feel (y_i^*) about an action determines how you feel (y_j^*) about yours, which in turn changes how I feel (y_i^*), which affects you (y_j^*), which ..." The indirect effect shows how i's action is, in the aggregate, influenced by others. Notice that this is of course a feature of all spatial lag models. The spatial lag probit model requires one to compute how y is impacted, and the magnitude of this

[4] For ease of interpretation, one may want to multiply equation (8.13) with the standard deviation of X. The result would indicate by how many percentage points the probability rises when X increases by one standard deviation. This is akin to developing an elasticity to measure the impact of X.

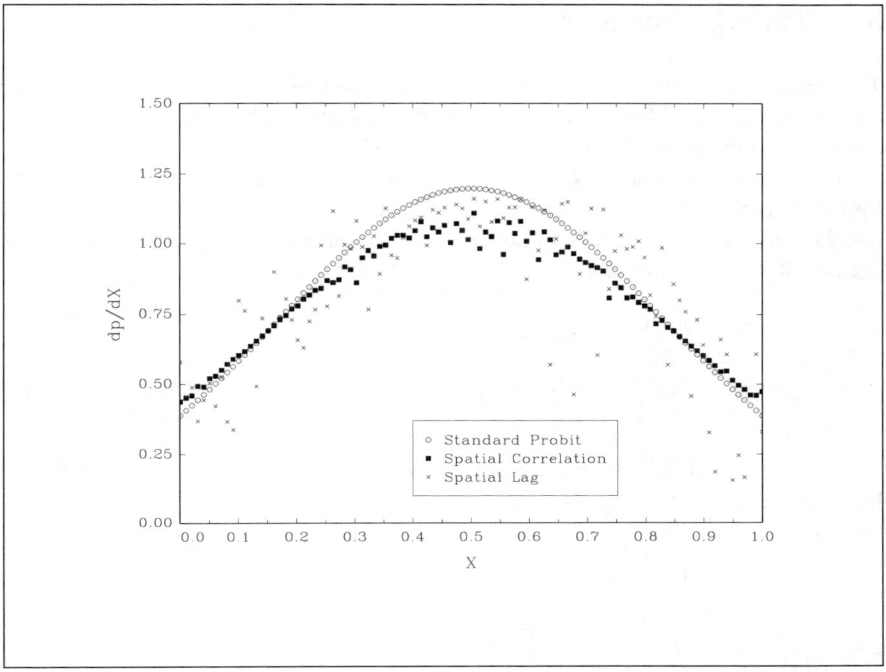

Fig. 8.1. Marginal effect of X on the probability that $y = 1$

impact is shown in equation (8.15). The point is that a share of $1/\left[(I_n - \alpha W)^{-1}\right]_{ii}$ of this impact is a direct impact and the remainder is due to spatial lag interactions with other observations.

As a ˇ nal note, equation (8.15) indirectly illustrates as well that a variation in X_j for any $j \neq i$ also causes a change probability that y_i equals 1. It is thus unrealistic to substitute the average of X into (8.15). Rather, for the given sample values, one should compute the marginal impact for each observation and summarize this by averaging. Furthermore, one may raise the question which observation j has the greatest impact on the outcome for i. There is much interesting detail to be gained from this, but note that it requires the evaluation of the expression:

$$\frac{d\Pr[y_i = 1|X, W]}{dX_j}, \tag{8.16}$$

for each pair (i, j) with $i \neq j$. It might not be immediately clear from equation (8.15) why one should not condition on other y_j for $j \neq i$. Equation (8.16) indicates that the actions of other observations are endogenously responding to a change in X_i. Thus, it would not be proper to condition on y_j for $j \neq i$.

8.3 The RIS Simulator

This section describes a simulator that can be used to evaluate an n-dimensional normal probability. This so-called recursive importance sampling or RIS simulator is developed in greater detail in Vijverberg (1997).

Let v be distributed as N(0,Ω).[5] We desire to evaluate $\Pr[v < V]$. Let A be an upper triangular matrix such that $A'A = \Omega^{-1}$, and let $\eta = Av$. Then η is *iid* standard normal. Deˇ ne $B = A^{-1}$; B is an upper triangular matrix with $b_{jj} > 0$ for all j. The bounds of the inequality $B\eta = v < V$ can be written as:

$$\begin{aligned} \eta_n &< b_{nn}^{-1} V_n \equiv \eta_{n0}, \\ \eta_j &< b_{jj}^{-1} \left[V_j - \sum_{i=j+1}^{n} b_{ji}\eta_i \right] \equiv \eta_{j0}(V_j, \eta_{j+1}, \ldots, \eta_n) \equiv \eta_{j0}. \end{aligned} \tag{8.17}$$

Let $g(\eta_j)$ be a suitably chosen density function that allows $-\infty < \eta_j < \infty$, and let G be the associated cdf. Deˇ ne $g^c(\eta_j) = g(\eta_j)/G(\eta_{j0})$ for $\eta_j \leq \eta_{j0}$. Then:

$$\begin{aligned} p = \Pr[v < V] &= \int_{-\infty}^{V} \phi_n(v; 0, \Omega) dv, \\ &= \int_{-\infty}^{\eta_{n0}} \cdots \int_{-\infty}^{\eta_{1,0}} \prod_{j=1}^{n} \phi_1(\eta_j) d\eta_1 \ldots d\eta_n, \\ &= \int_{-\infty}^{\eta_{n0}} \frac{\phi(\eta_n)}{g^c(\eta_n)} \left(\int_{-\infty}^{\eta_{n-1,0}} \frac{\phi_1(\eta_{n-1})}{g^c(\eta_{n-1})} \cdots \right. \\ &\quad \left(\int_{-\infty}^{\eta_{2,0}} \frac{\phi_1(\eta_2)}{g^c(\eta_2)} \Phi(\eta_{10}) g^c(\eta_2) d\eta_2 \right) \ldots \Bigg) g^c(\eta_n) d\eta_n. \end{aligned} \tag{8.18}$$

The RIS simulator consists of drawing R random vectors of η (excepting η_1) satisfying the condition $\eta_j \leq \eta_{j0}$ from the distribution deˇ ned by g. Thus, for $r = 1, \ldots, R$, given η_{n0}, draw $\tilde{\eta}_{n,r}$; determine $\tilde{\eta}_{n-1,0,r}$ from equation (8.17) by using $\tilde{\eta}_{n,r}$ in the place of η_n; given $\tilde{\eta}_{n-1,0,r}$, draw $\tilde{\eta}_{n-1,r}; \ldots$; given $\tilde{\eta}_{2,0,r}$, draw $\tilde{\eta}_{2,r}$; and determine $\tilde{\eta}_{1,0,r}$ from equation (8.17). Then the simulated value for p is:

$$\hat{p} = \frac{1}{R} \sum_{r=1}^{R} \left(\Phi\left[\tilde{\eta}_{1,0,r}\right] \prod_{j=2}^{n} \frac{\phi(\tilde{\eta}_{k,r})}{g^c(\tilde{\eta}_{k,r})} \right). \tag{8.19}$$

Suitable density functions that can be used for g are the logit, normal, t, and a transform of the Beta(2,2) (Vijverberg, 1997). Generating random variables is done fastest when the logit distribution is used, and relatively slow when the normal or

[5] The simulator applies whether Ω is standardized or not. If Ω is not standardized, let Ω_{ii} be the square root of the ith diagonal element of Ω. Let $\tilde{\Omega}$ be the standardized form of Ω; let $\tilde{A}'\tilde{A} = \tilde{\Omega}$, and let $\tilde{B} = \tilde{A}^{-1}$. The ith column of $\tilde{A}$ is equal to the ith column of A multiplied by Ω_{ii}, and the ith row of $\tilde{B}$ is the same as the ith row of B divided by Ω_{ii}. It is easily seen that η is still *iid* standard normal with the same bounds as in equation (8.17).

t distribution is used. However, one should be more interested in the variability of $\hat{p}$ or, since the spatial probit models utilize p in logarithmic form, in the variability of $\ln \hat{p}$. While it certainly is not a given that the normal density generates the lowest variability, tests on the basis of a variety of upper bounds and correlation patterns did suggest that the RIS-normal simulator is often preferred.[6] Since a Monte Carlo study consumes great quantities of computer time, we only employ the RIS-normal simulator. In an actual application of this technique, where the likelihood function is maximized only a few times, one should try different RIS simulators. Note that since the numerator and denominator in (8.19) cancel when $g^c(\tilde{\eta}_{k,r}) = \phi(\tilde{\eta}_{k,r})/\Phi[\tilde{\eta}_{k,0,r}]$ the RIS-normal simulator simpliˇ es to the following expression:

$$\hat{p} = \frac{1}{R} \sum_{r=1}^{R} \left(\prod_{j=1}^{n} \Phi[\tilde{\eta}_{j,0,r}] \right). \tag{8.20}$$

The RIS-normal simulator is identical to what is sometimes called the GHK simulator which is described in, among others, Börsch-Supan and Hajivassiliou (1993), Hajivassiliou (1993), Keane (1993), Hajivassiliou *et al.* (1996), and Stern (1997).

For our Monte Carlo study, we use either $R = 1000$ or $R = 2000$ draws and incorporate a simple antithetical sampling strategy (Vijverberg, 1997). For illustrative purposes, we took the ˇ rst of our Monte Carlo samples that was generated without spatial error autocorrelation or spatial lag and evaluated the log-likelihood function of the spatial error autocorrelation model (equation (8.8)) for different values of ρ and that for the spatial lag model (equation (8.11)) for various values of α, using the true values of β and the weights matrix underlying the S samples (Sect. 8.4). We simulated $\ln \hat{p}$ 100 times (rather than just once as one does when estimating the model). Figure 8.2 shows the standard deviation of these 100 simulated values; the inset illustrates their average. Figure 8.2 also points out that for this particular Monte Carlo sample, the estimated value of ρ and α is likely to be positive, even if the sample was generated with $\rho = \alpha = 0$. Estimation requires iterative search over values of β and either ρ or α, and thus the Maximized Likelihood function will reach a higher maximum than is shown in Fig. 8.2. It is shown that for values of ρ or α in the range $[-0.6, 0.6]$, the standard deviation is less than 0.02, which is tiny compared to average values around -30. Moreover, comparing models by means of Likelihood Ratio tests will be quite reliable.

[6] Vijverberg (1999) reports substantial increases in efˇ ciency when the observations are sorted such that the upper bounds decrease from $i = 1$ to $i = n$, of course sorting the weights matrix W in a similar way. Moreover, the general superiority of the normal kernel erodes by this sorting, and other RIS-simulators become relatively more efˇ cient.

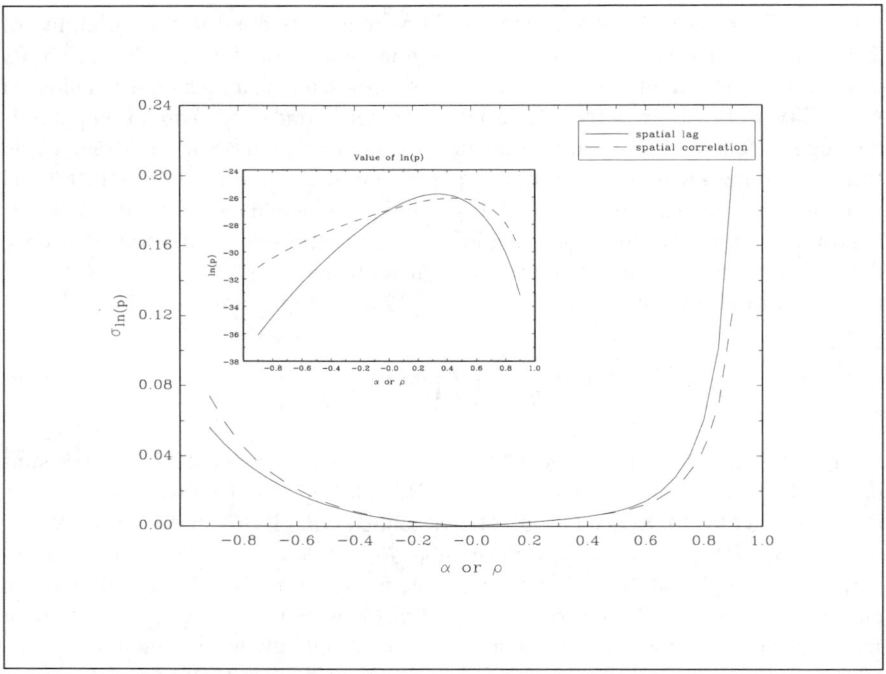

Fig. 8.2. Measuring accuracy in the simulation of ln$p$

8.4 Monte Carlo Data

In our Monte Carlo analysis, we examine the following model, stated in its most general form:

$$y^* = \alpha W y^* + X\beta + u, \text{ where } u = \rho W u + \varepsilon, \tag{8.21}$$

$$y = N(y^*), \tag{8.22}$$

where the indicator function N has been deˇ ned in Sect. 8.2.1. We study situations where either α or ρ is nonzero but not both at the same time.

There is a single X variable, constructed in the following way. Deˇ ne $\tilde{X}$ as an $n \times 1$ vector with elements increasing from $\tilde{X}_1 = 0$ to $\tilde{X}_n = 1$ in equal steps of $1/(n-1)$. X is a randomly scrambled version of $\tilde{X}$; the purpose of scrambling is to avoid any systematic correlation between X and the weights matrix W. Every Monte Carlo sample of size n uses the same X vector.

Parameter values are selected as follows. Throughout, we set $\beta_0 = -1.5$ and $\beta_1 = 3$. This implies that the deterministic part of y^* (i.e., $X\beta$) ranges from -1.5 to 1.5. In the context of a standard probit model, this means that the probability that y_i equals 1 varies from 0.0668 to 0.9332. Furthermore, by assumption, ε is distributed N(0,I_n).

Two types of Monte Carlo samples are constructed. The first uses a weights matrix that is the row-standardized contiguity matrix of the 50 states of the U.S.A., where Alaska and Hawaii are coded as non-contiguous to any other state.[7] There are five sets of parameter values for (α, ρ): (0,0) representing the standard probit conditions, (0.25,0) and (0.50,0) representing increasing spatial lag conditions, and (0,0.25) and (0,0.50) representing increasing degrees of spatial error autocorrelation.[8] For each of these parameter sets, 100 Monte Carlo samples are created, based on the same 100 random N(0,1) vectors of ε. We shall refer to these sets of samples as $S_{\alpha,\rho}$ with the values of α and ρ specified, e.g., as $S_{0.50,0}$. Thus, there are a total of 500 Monte Carlo samples of the first type. For each sample, we estimate the standard probit, the spatial error autocorrelation probit, and the spatial lag probit models, based on the RIS-normal simulator with $R = 2000$.

Using the U.S. state contiguity structure as the weights matrix has the advantage that the Monte Carlo simulations are informative for applied research that examines a dichotomous choice across states. Examples of such research would be the implementation of a state income tax, the election of a Republican for the governor's office, the pursuit of a particular regulatory initiative. The disadvantage is that one is limited to a simulation with $n = 50$: evidence on large sample properties eludes. For that reason, we construct a second type of Monte Carlo samples by means of a random contiguity matrix and samples sizes $n = 50, 100, 200$.[9] Let (z_{1i}, z_{2i}) be an uncorrelated random pair of coordinates, each selected from the uniform $[0,1]$ distribution, with $i = 1, \ldots, n$. Let d_{ij} be the distance between observations i and j. Define the elements of W prior to row-standardization as $W_{ij} = 1$ if $d_{ij} < d^{(n)}$ and $= 0$ otherwise. By varying the upper bound $d^{(n)}$ with n, we control the pervasiveness of contiguity. We use $d^{(50)} = 0.21$, $d^{(100)} = 0.15$, and $d^{(200)} = 0.10$. With these values, it turns out that, in our Monte Carlo samples, an observation is contiguous to an average of five other observations, with a minimum of 1 and a maximum of between 10 and 14. Thus, increasing n leads to more observations of a similar kind, not to simultaneously greater contiguity interactions. One may note that this random contiguity matrix has no structure, unlike the state contiguity matrix or the typical weights matrix that might be used in empirical applications. Indeed, this is one reason why we choose to present and compare the results of both types: from

7 The inclusion or exclusion of “islands” (Alaska and Hawaii) should have no bearing on the main conclusions of this chapter. In some applications, the substantive issue may dictate that Alaska and Hawaii be omitted. The model estimated in this chapter assumes that every state makes a discrete choice which may depend on a spatial factor ($\alpha W y^*$ or $\rho W u$) which drops to 0 when no neighbors are present and the particular row of W contains only 0. Including islands is akin to estimating parameters on two pooled subsamples: pooling increases the efficiency of the estimator of the nonspatial parameters. Therefore, obviously, the inclusion of Alaska and Hawaii has some effect on the estimates of the nonspatial parameters.

8 We focus on positive spatial dependence parameters as these are more often found in the literature and have a more “intuitive” interpretation (Anselin and Bera, 1998).

9 Selecting $n = 50$ allows us to examine whether the use of the U.S. state contiguity matrix forces any particular conclusion.

Table 8.1. Characteristics of the weights matrices: number of connections among observations (in percents)

Number of links	State Contiguity Matrix		Randomized Contiguity Matrix	
0	4	0	0	0
1	2	2	3	2
2	8	12	9	6
3	18	12	8	10.5
4	22	18	13	16
5	18	22	15	15.5
6	20	10	21	17.5
7	6	6	12	12
8	2	6	9	7
9	0	0	5	4.5
10	0	10	4	4.5
11	0	2	1	3
12	0	0	0	0.5
13	0	0	0	0.5
14	0	0	0	0.5
Average number of links	4.28	5.16	5.50	5.70
Dimension of matrix	50	50	100	200

the random contiguity matrix we gain insight into the theoretical properties of the spatial probit models, while from the state contiguity matrix we learn about the in uence of structure.

As to parameter values, we restrict ourselves only to the two (α, ρ) combinations of (0.50,0) and (0,0.50). As before, these Monte Carlo samples are created with $\beta_0 = -1.5$ and $\beta_1 = 3$, and ε has a $N(0,I_n)$ distribution. Sets of Monte Carlo samples of this type will be denoted as $T_{\alpha,\rho}(n)$, and there are obviously six of these sets, each with 100 samples. Because of the higher value of n, spatial probit models for the T sets are estimated with $R = 1000$.

To help understand the difference in the Monte Carlo outcomes, Table 8.1 summarizes the information contained in the weights matrices by means of the number of connections (contiguities) among the observations. For example, the W matrix that represents contiguity among U.S. states contains an average of 4.28 links per state, or, prior to row standardization, an average of 4.28 ones per row and per column. The number of connections among the simulated weights matrix is slightly larger, and the frequency distribution shows a few more observations with a large number of contiguities.

A major concern with simulation processes is the amount of processing time. On a 300MHz Pentium II computer, the spatial probit models with the state contiguity matrix take about 6 minutes. When the number of random draws in the simulator

(R) is halved, the standard deviation of $\ln \hat{p}$ increases by a factor of $\sqrt{2}$ and computation time is also halved. (This shows that the major computational burden is the simulation itself and not the triangularization of Ω to get B; see Sect. 8.3.) When the dimension (n) of the sample rises, the computation time increases dramatically: one $T_{\alpha,\rho}(n)$ sample with $R = 1000$ takes about 2.5 minutes for $n = 50$, 8.8 minutes for $n = 100$ and 30.5 minutes for $n = 200$. Doubling the sample size increases computation time by a factor of about 3.5.

8.5 Monte Carlo Results

The first question to ask is whether one is able to detect spatial dependence in probit models. Table 8.2 summarizes Likelihood Ratio tests for spatial error autocorrelation, denoted by LR_ρ, and spatial lag, denoted by LR_α based on the spatial probit models estimated by means of the RIS procedure. Since these tests are about a single parameter, the critical value at the 5 percent significance level is $\chi^2_{0.05}(1) = 3.84$.[10] The first row focuses on Monte Carlo sample set $S_{0,0}$, with data that contain no spatial lag or correlation. Indeed, for 90 out of 100 samples, we fail to reject the null hypothesis of no spatial error autocorrelation as well as the null hypothesis of no spatial lag. A more detailed check of the 100 LR_α and LR_ρ values reveals that spatial error autocorrelation per se is suspected in only 6 cases, and spatial lag in 7 cases, both of which are roughly consistent with a test at a 5 percent significance level.

The second row indicates that it is very difficult to detect mild cases of spatial lag with probit models. When spatial lag structure becomes more pronounced, as in the third and fourth rows, one is more likely to reject the standard probit model. The power of the test improves when the number of observations increases (rows 5 and 6 in Table 8.2). The same overall conclusions apply when the data are generated with a spatial error autocorrelation structure (rows 7 through 11 in Table 8.2).

If standard probit is rejected, which spatial dependence model should be focused on? Test statistics are not at all clear, as the right hand portion of Table 8.2 illustrates. Figures 8.3 and 8.4 show scatterplots of LR_α and LR_ρ in the two cases with serious spatial error autocorrelation ($\rho = 0.50$) and spatial lag ($\alpha = 0.50$), respectively, with the U.S. contiguity matrix and $n = 50$. Rejection of the hypothesis of no spatial error autocorrelation is indicated by the vertical line at $LR_\rho = 3.84$ (which may be extended further than drawn); rejection of the hypothesis of no spatial lag is shown by the horizontal line at $LR_\alpha = 3.84$. The diagonal line splits the remainder of the quadrant into areas where the spatial error autocorrelation model (below the diagonal) and spatial lag model (above the diagonal) is favored.

[10] Strictly taken, use of the $\chi^2(1)$ distribution to find the critical value is merely an assumption, as both the small sample properties and the asymptotic properties of this spatial model are unknown. On basis of the set $S_{0,0}$, a goodness-of-fit test showed that the Monte Carlo distribution of LR_ρ was well approximated by a $\chi^2(1)$ distribution (p-value=0.93), but that the approximation for LR_α was only fair (p-value=0.075). Further, our results are approximate in that we treat LR_ρ and LR_α as independent and do not test them jointly.

Table 8.2. Likelihood Ratio tests for spatial error autocorrelation and spatial lag, probit estimators

	LR_ρ		LR_α			Decision	
	Mean	St.Dev.	Mean	St.Dev.	Neither	Error	Lag
$S_{0,0}$	1.00	1.37	1.26	1.64	90	4	6
$S_{0.25,0}$	1.15	1.32	1.33	1.43	90	4	6
$S_{0.50,0}$	3.11	3.22	4.03	3.88	58	9	33
$T_{0.50,0}(50)$	3.79	3.69	5.48	4.91	41	11	48
$T_{0.50,0}(100)$	5.57	4.36	7.72	4.99	21	18	61
$T_{0.50,0}(200)$	12.57	8.28	15.89	9.38	7	17	76
$S_{0,0.25}$	1.00	1.23	1.05	1.31	92	3	5
$S_{0,0.50}$	2.22	2.44	1.87	2.35	75	17	8
$T_{0,0.50}(50)$	2.71	2.92	2.33	2.72	68	22	10
$T_{0,0.50}(100)$	5.04	4.37	3.48	3.27	48	44	8
$T_{0,0.50}(200)$	10.81	6.71	8.04	5.32	13	65	22

As the thick scatter in the lower left corner indicates, Likelihood Ratio tests often conclude that there is no hint of spatial dependence. When the sample size increases, spatial dependence becomes more evident (Figs. 8.5 and 8.6). Yet, in sample where there is evidence of spatial dependence, the nature of it is often not all that clear: many dots cluster near the 45 degree line. A simple decision rule stating that spatial error autocorrelation (or lag) exists whenever $LR_\rho > (<) LR_\alpha$ is nevertheless the best one can do, in view of the location of the scatterplots in Figs. 8.3 through 8.6.

Why does the Likelihood Ratio test have such low power? The foremost reason is that the samples, with 50 observations, are small.[11] This is exactly the reason why we developed random contiguity matrices that allow Monte Carlo ($T_{\alpha,\rho}(n)$) samples of larger size. But apart from this, note that what is available at the time of estimation are two vectors of values, y and X, and the weights matrix. If one were to observe y^*, any variation in either α or ρ would be noticeable. In the case of a dichotomous dependent variable, only when the variation in α or ρ causes y^* to change sign does one observe a difference in y. Therefore, one can speculate that it is more dif˜ cult to observe spatial dependence in probit models. Furthermore, one may expect that it is harder to observe spatial error autocorrelation than spatial lag structures: in a spatial error autocorrelation model, spatial changes in y come about only through variations in the disturbance term in contiguous states, but in a spatial lag model they can also be caused by variations in neighboring X values-compare equations (8.6)–(8.7) with (8.10).

A comparison of the realized y in the Monte Carlo sample sets illustrates this: of the 100 samples in the S layout, 23 of the $S_{0,0.25}$ samples are identical to $S_{0,0}$ and so are 5 of $S_{0,0.50}$. The problem is less among spatial lag models: 13 of $S_{0.25,0}$, and 0 of $S_{0.50,0}$. Across the 400 samples in the four spatially dependent sets, about

[11] For instance, see Anselin and Florax (1995c), and Anselin and Bera (1998).

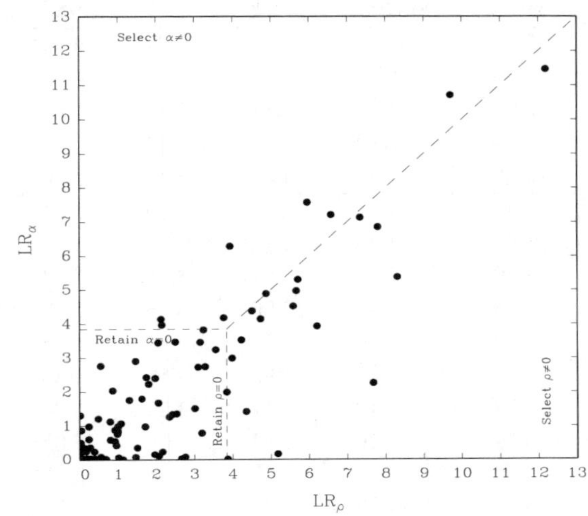

Fig. 8.3. Test results for spatial lag and spatial error autocorrelation, $S_{0,0.50}$

47 observations (out of 50) are on average the same as in the parallel sample in the $S_{0,0}$ set. That is, on average for 47 observations, it does not matter that spatial lag or spatial error autocorrelation is introduced; the outcome of y_i is still the same. Needless to say, that makes it difficult to detect spatial dependence. Only when the number of observations increases does it become easier.

Next, consider the estimates for the model parameters (β_1, and for the spatial models, α or ρ), summarized in Tables 8.3 and 8.4 for the S samples and in Tables 8.5 and 8.6 for the T samples. First, we focus on the β_1 parameter in the S samples. The estimates and descriptive statistics are reported in Table 8.3 for all combinations of estimators and spatial parameters. Specifically, for each of the three estimators (standard probit, spatial error probit and spatial lag probit), the results are given for models that are correctly specified as well as models that are misspecified with respect ot the spatial effect.[12] The estimates of β_1 vary around the true value of 3, with a standard deviation of roughly 1. Given that these statistics are summaries based on only 100 Monte Carlo samples, the standard error of the mean of

[12] For example, standard probit applied in $S_{0,0.50}$, i.e., with $\rho = 0.50$, represents a misspecified model with "ignored" spatial error autocorrelation; spatial error applied in $S_{0.50,0}$, i.e., with $\alpha = 0.50$, represents a misspecified model where the correct spatial effect is of the lag variety, not the error variety.

Table 8.3. Estimates for β_1, S samples

				Percentile			
Estimator	Sample	Mean	St.Dev.	5th	50th	95th	RMSE
classic	$S_{0,0}$	3.28	0.91	2.10	3.09	4.81	0.95
	$S_{0,0.25}$	3.25	0.97	2.13	3.04	4.92	1.00
	$S_{0,0.50}$	2.99	1.11	1.81	2.75	4.83	1.11
	$S_{0.25,0}$	3.10	0.97	1.98	2.86	4.92	0.98
	$S_{0.50,0}$	2.69	0.76	1.65	2.60	3.80	0.82
error	$S_{0,0}$	3.49	1.04	2.14	3.30	5.23	1.15
	$S_{0,0.25}$	3.53	1.43	2.12	3.18	5.36	1.52
	$S_{0,0.50}$	3.39	1.41	2.01	3.05	5.54	1.47
	$S_{0.25,0}$	3.35	1.27	2.00	3.09	5.52	1.32
	$S_{0.50,0}$	3.02	1.06	1.77	2.86	5.25	1.06
lag	$S_{0,0}$	3.30	1.02	1.98	3.18	4.83	1.06
	$S_{0,0.25}$	3.38	1.10	2.14	3.23	5.04	1.16
	$S_{0,0.50}$	3.29	1.20	1.93	3.06	5.05	1.23
	$S_{0.25,0}$	3.32	1.10	1.97	3.18	5.11	1.14
	$S_{0.50,0}$	3.29	1.08	1.76	3.20	5.39	1.12

Table 8.4. Estimates for α and ρ, S samples

				Percentile			
Estimator	Sample	Mean	St.Dev.	5th	50th	95th	RMSE
lag (α)	$S_{0,0}$	−0.14	0.35	−0.79	−0.09	0.36	0.38
	$S_{0,0.25}$	0.03	0.30	−0.54	0.08	0.45	0.31
	$S_{0,0.50}$	0.22	0.28	−0.29	0.23	0.62	0.35
	$S_{0.25,0}$	0.14	0.29	−0.43	0.21	0.48	0.31
	$S_{0.50,0}$	0.41	0.22	−0.00	0.44	0.71	0.24
error (ρ)	$S_{0,0}$	−0.17	0.41	−0.92	−0.12	0.40	0.45
	$S_{0,0.25}$	0.11	0.38	−0.67	0.16	0.59	0.41
	$S_{0,0.50}$	0.32	0.36	−0.28	0.40	0.75	0.41
	$S_{0.25,0}$	0.32	0.36	−0.28	0.40	0.75	0.49
	$S_{0.50,0}$	0.42	0.32	−0.20	0.50	0.80	0.53

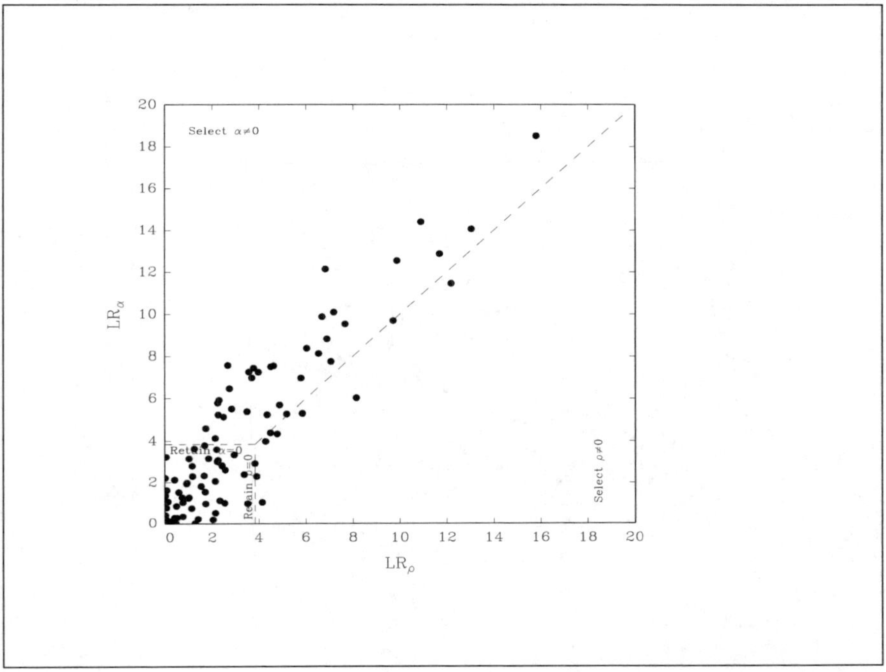

Fig. 8.4. Test results for spatial lag and spatial error autocorrelation, $S_{0.50,0}$

the estimates is one tenth of the standard deviation reported in the table. Consider the first row: this shows how even the standard probit model applied to a properly constructed but small sample yields biased estimates; the bias of 0.28 exceeds the standard error of the mean of 0.091 by a factor of almost 3. The bias of the standard probit model seems to vary with the nature of the data: the bias turns negative for $S_{0.50,0}$. The estimate $\hat{\beta}_1$ is usually more biased when the spatial error autocorrelation probit model is implemented, even when the data have a spatial error autocorrelation structure (i.e., even when the model is properly specified). The bias for esimates based on the spatial lag probit model is positive and fairly stable across data structures. Overall, the root mean squared error (computed as the sum of the variance and the squared bias) is largest for the spatial error autocorrelation probit estimates and smallest for the standard probit estimates, regardless of data structures. The major component of the root mean squared error is the variance of the estimator, not the bias.

Table 8.4 shows estimates of α and ρ obtained with the spatial probit RIS estimators. In sets $S_{0,0}$, $S_{0.25,0}$ and $S_{0.50,0}$, $\hat{\alpha}$ is somewhat biased downward. When the data have a spatial error autocorrelation structure, a spatial lag model is obviously a misspecification, but one may encounter statistically significant estimates of α anyway, as was already clear in Table 8.2. The downward bias in $\hat{\rho}$ is more serious.

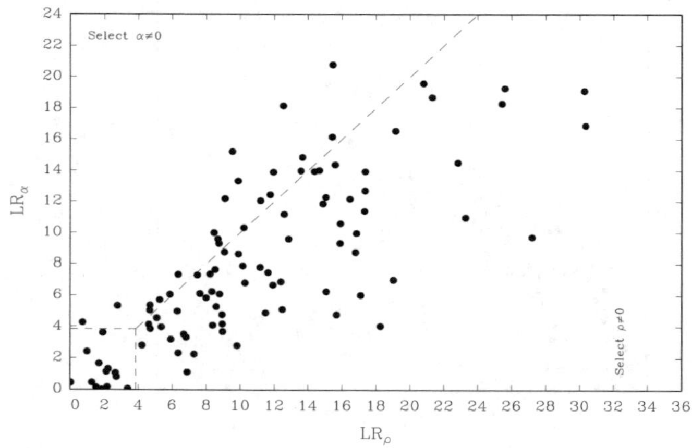

Fig. 8.5. Test results for spatial lag and spatial error autocorrelation, $T_{0,0.50}(200)$

Interestingly, even spatial lag data structures are likely to generate large estimates for a spatial error autocorrelation coefˇ cient. Note that the root mean squared errors of $\hat{\alpha}$ and $\hat{\rho}$ are smaller, respectively, when the spatial estimator is applied to the correctly speciˇ ed model.

In Table 8.5, the weights matrix re ects random contiguities and the number of observations n ranges from 50 to 200. The mean estimate of $\hat{\beta}_1$ declines as n increases. The large sample bias of the classic probit estimator in the misspeciˇ ed models becomes evident, as does the bias of the spatial lag when the data structure contains spatial error autocorrelation. Given the spatial error autocorrelation probit results for $T_{0.50,0}(n)$, it is likely that the bias turns negative when n increases further. When the spatial probit model is correctly speciˇ ed, the bias virtually disappears even for $n = 200$.

Table 8.6 shows how $\hat{\alpha}$ and $\hat{\rho}$ are impacted by sample size and model misspeciˇ cation. As differences in the root mean squared error indicate, bias becomes important now. When the spatial effects in the probit model are speciˇ ed correctly, the bias in $\hat{\alpha}$ and $\hat{\rho}$ disappears. However, model misspeciˇ cation leads to substantially positive values of $\hat{\alpha}$ and especially $\hat{\rho}$, suggesting once again that it is difˇ cult to detect the correct data structure. For example, a large and statistically signiˇ cant

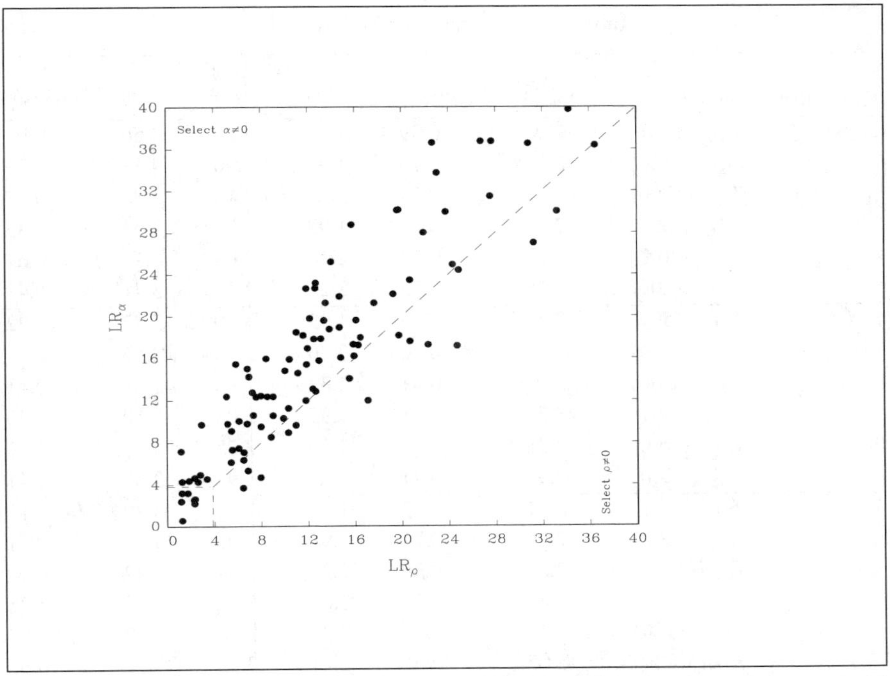

Fig. 8.6. Test results for spatial lag and spatial error autocorrelation, $T_{0.50,0}(200)$

estimate of ρ need not be an indication of spatial error autocorrelation, but can also be the result of a strong spatial lag.

8.6 Spatial Linear Probability Model

There is a relatively high cost in both computational time and effort to carry out a spatial probit estimation. It is therefore of some interest to compare the results of the spatial probit with the lower cost option of a standard linear spatial analysis. In this section we estimate and compare the linear counterparts of the probit models. Thus, we estimate a linear equation:

$$y = X\beta + \varepsilon, \tag{8.23}$$

to parallel the standard probit model of equations (8.1) and (8.2), a linear spatial error autocorrelation model,

$$y = X\beta + u, \quad \text{where } u = \rho W u + \varepsilon, \tag{8.24}$$

to parallel the spatial error autocorrelation probit of equation (8.6), and a linear spatial lag model,

$$y = \alpha W y + X\beta + \varepsilon, \tag{8.25}$$

Table 8.5. Estimates for β_1, T samples

				Percentile			
Estimator	Sample	Mean	St.Dev.	5th	50th	95th	RMSE
classic	$T_{0,0.50}(50)$	2.97	0.99	1.73	2.77	4.50	0.99
	$T_{0,0.50}(100)$	2.88	0.70	1.94	2.90	4.12	0.71
	$T_{0,0.50}(200)$	2.67	0.34	2.17	2.68	3.35	0.47
	$T_{0.50,0}(50)$	2.98	0.83	1.93	2.92	4.33	0.83
	$T_{0.50,0}(100)$	3.05	0.64	2.14	3.03	4.13	0.64
	$T_{0.50,0}(200)$	2.76	0.32	2.27	2.75	3.41	0.41
error	$T_{0,0.50}(50)$	3.40	1.15	1.92	3.10	5.37	1.22
	$T_{0,0.50}(100)$	3.25	0.70	2.30	3.18	4.56	0.74
	$T_{0,0.50}(200)$	3.02	0.41	2.37	2.98	3.69	0.41
	$T_{0.50,0}(50)$	3.34	1.21	1.92	3.08	5.60	1.26
	$T_{0.50,0}(100)$	3.28	0.72	2.20	3.22	4.50	0.77
	$T_{0.50,0}(200)$	2.99	0.41	2.33	2.99	3.65	0.41
lag	$T_{0,0.50}(50)$	3.11	1.12	1.76	2.90	4.79	1.12
	$T_{0,0.50}(100)$	2.96	0.71	1.99	2.96	4.20	0.71
	$T_{0,0.50}(200)$	2.81	0.35	2.28	2.84	3.42	0.40
	$T_{0.50,0}(50)$	3.40	1.18	2.03	3.17	5.18	1.25
	$T_{0.50,0}(100)$	3.30	0.78	2.28	3.28	4.62	0.84
	$T_{0.50,0}(200)$	3.05	0.38	2.41	3.03	3.77	0.39

Table 8.6. Estimates for α and ρ, T samples

				Percentile			
Estimator	Sample	Mean	St.Dev.	5th	50th	95th	RMSE
lag (α)	$T_{0,0.50}(50)$	0.25	0.27	−0.21	0.29	0.62	0.37
	$T_{0,0.50}(100)$	0.29	0.19	−0.09	0.30	0.56	0.35
	$T_{0,0.50}(200)$	0.36	0.17	0.05	0.38	0.55	0.39
	$T_{0.25,0}(50)$	0.46	0.19	0.16	0.49	0.75	0.19
	$T_{0.50,0}(100)$	0.45	0.13	0.24	0.48	0.64	0.14
	$T_{0.50,0}(200)$	0.48	0.12	0.25	0.50	0.66	0.12
error (ρ)	$T_{0,0.50}(50)$	0.37	0.32	−0.26	0.43	0.76	0.35
	$T_{0,0.50}(100)$	0.42	0.23	−0.00	0.46	0.71	0.25
	$T_{0,0.50}(200)$	0.48	0.13	0.24	0.49	0.67	0.13
	$T_{0.50,0}(50)$	0.48	0.13	0.24	0.49	0.67	0.49
	$T_{0.50,0}(100)$	0.47	0.20	0.15	0.51	0.71	0.51
	$T_{0.50,0}(200)$	0.50	0.13	0.24	0.51	0.70	0.52

as a parallel to the spatial lag probit of equation (8.10). This exercise is analogous to comparing the linear probability model to a non-spatial probit analysis. One similarity that carries over is the interpretation of the mean of y_i as the probability that $y_i = 1$, which results from the assumption that $E[\varepsilon] = 0$.

The problems associated with using a linear model (OLS) in place of the standard probit are well documented (Greene, 1997). The disturbance ε_i is assumed to be independently distributed. However, due to the dichotomous nature of the dependent variable, it cannot be identically distributed. In fact, it is binomial and is heteroskedastic. This presumably carries over to the spatial realm as well, but here we ˇ nd other peculiarities. Consider the spatial error autocorrelation linear model, rewritten as:

$$y = X\beta + (I_n - \rho W)^{-1}\varepsilon = X\beta + \Gamma_\rho \varepsilon. \tag{8.26}$$

If ε is indeed independently distributed, one must be able to conceive of a situation where only observation i receives a positive random shock $\Delta\varepsilon_i$. The impact of this shock on y_i equals $\Gamma_{\rho,ii}\Delta\varepsilon_i$; since y_i only takes on values of 0 and 1, this implies that $\Delta\varepsilon_i = 1/\Gamma_{\rho,ii}$ (if y_i equals 0 to start with). However, this same shock causes changes in other y's as well: $\Delta y_j = \Gamma_{\rho,ji}\Delta\varepsilon_i$. Of course, the magnitude of this change can only be $+1$ or -1. This implies that $\Gamma_{\rho,ij} = \pm\Gamma_{\rho,ii}$ or else equals 0. It also implies that a change in ε_i restricts the possible changes in other ε_j: they cannot both change and increase y_j at the same time. Thus, ε cannot be independently distributed, and the distribution that satisˇ es the requirements of a spatial error autocorrelation model, if it exists, would incorporate Γ_ρ in some form. Apart from the question whether one can indeed properly specify spatial linear probability models, standard spatial econometrics software like *SpaceStat* (Anselin, 1992) cannot be expected to account for these complications.

But let us ignore these issues: let us ask the question whether spatial linear probability models can be a time-efˇ cient informative substitute. Our focus here is in how well the linear spatial model does relative to the spatial probit estimator in three ways. First, how effective compared to the spatial probit is the linear model in picking up spatial dependence when it occurs either as a spatial lag or a spatial error for a binary dependent variable? Second, how similar are the predicted probability estimates of the linear model to the spatial probit model? Even though the linear model is conceptually inappropriate for binary data, if the results are similar to those of the appropriate model, then it may be an acceptable method to use, given the costs of estimation. Third, how close are the estimates of the spatial parameters α and ρ in the different speciˇ cations?

The data that were generated and analyzed in the preceding sections are now reanalyzed with the linear model. Recall that we have 500 samples of 50 observations, each based on the row-standardized contiguity matrix of the U.S. states, and 600 samples based on row-standardized simulated contiguity matrices with 50, 100, and 200 observations. The Monte Carlo simulations in this section involve OLS, linear spatial lag and linear spatial error estimations for each sample, ignoring the fact that the dependent variable is binary. We use the *SpaceStat* software to carry

Table 8.7. Likelihood Ratio tests for spatial error autocorrelation and spatial lag, linear model estimators

	LR_ρ		LR_α		Decision		
	Mean	St.Dev.	Mean	St.Dev.	Neither	Error	Lag
$S_{0,0}$	3.43	2.4	1.35	1.92	74	21	5
$S_{0.25,0}$	3.34	1.78	1.22	1.55	76	22	2
$S_{0.50,0}$	5.67	4.31	3.54	3.76	42	54	4
$T_{0.50,0}(50)$	4.04	4.25	5.45	5.07	45	8	47
$T_{0.50,0}(100)$	5.33	4.28	6.88	4.58	30	10	60
$T_{0.50,0}(200)$	12.47	7.96	15.12	8.92	10	11	79
$S_{0,0.25}$	3.23	1.63	1.09	1.50	78	19	3
$S_{0,0.50}$	4.61	3.21	1.96	2.45	58	41	1
$T_{0,0.50}(50)$	2.81	3.35	2.74	3.34	67	15	18
$T_{0,0.50}(100)$	4.86	4.40	4.00	3.67	52	34	14
$T_{0,0.50}(200)$	10.54	6.08	9.22	5.45	14	57	29

out the estimations (Anselin, 1992). The results presented are thus based on 3300 estimations. Note the considerable difference in time required between the two procedures. In the $n = 200$ case, the spatial probit RIS procedure took over 30 minutes. In contrast, the linear spatial model applied to the same case took less than a minute.

In the discussion that follows, we can distinguish two cases. The first focuses on the differences between the simulated weights matrices, the T data sets, and the state weights matrix, the S data sets. The second deals with differences between the results when the correct model is estimated (given the null hypothesis), versus the situations where misspecification occurs and an incorrect model is estimated.

We begin by comparing the results of the Likelihood Ratio tests for the spatial and linear models. The linear results are given in Table 8.7, which are to be compared to those listed in Table 8.2. Consider, for example, the first data set given, $S_{0,0}$, for the state weights matrix without any spatial component. The Likelihood Ratio test for the spatial probit (Table 8.2) is able to pick this up 90 percent of the time. The linear model, however, is only able to pick this up 74 percent of the time. The bulk of the misspecification of using a linear model to account for the binary dependent variable is attributed to the spatially correlated model, where we find 21 cases pointing to this (incorrect) result. This compares poorly with the spatial probit that only finds this in 4 cases.

Continuing to look at the state weights matrix data sets, we see that the linear model, in the presence of spatially generated data, favors a decision suggesting a spatially correlated error alternative over a lag model. This tends to occur both when this is correct (for $S_{0,0.25}$ and $S_{0,0.50}$) and when it is not (for $S_{0.25,0}$ and $S_{0.50,0}$). The number of correct decisions is higher for the linear model than the spatial probit when there is spatial error. However, the spatial probit, particularly for the higher α and ρ values, properly distinguishes between the lag and the error model alternatives, which is not the case for the estimates based on the linear model. The linear model with the state data does outperform the spatial probit in detecting that *something* is wrong, based on the higher correct rejections of the null hypothesis of no spatial component. However, its predisposition to favor the spatial error model makes its use at diagnosing the problem suspect.

When we turn to the simulated weights matrix analysis the conclusions change somewhat. A comparison of the results in Tables 8.2 and 8.7 for $T_{\alpha,\rho}$ suggests much more similarity between the spatial probit and the linear model. It is no longer the case that the linear model favors the spatial error model alternative over the spatial lag. It is now able to correctly separate the two models about as well as the spatial probit.

As discussed previously, the nature of only observing the 0/1 outcome will obscure some portion of the spatial structure of the model. When we use the simulated (pseudo-randomized) weights matrix we ˇ nd much closer results between the spatial probit and the linear model than when we use a speciˇ c weights matrix. To the extent that the randomization of the weights matrix ends up offsetting some of the otherwise possible extreme values that might occur in either Wy or Wu, this is not too surprising. Some of the power of the probit model compared to the linear model is in detecting changes that occur further from the mean. If these have been "averaged" away by randomization, then the two procedures become more similar.

A strategy suggested by the above results would be to examine carefully the weights matrix for an analysis. The more seemingly randomized the pattern, the greater the likelihood that a linear model can be used to give at least preliminary results to guide further analysis. However, the nature of most weights matrices is, by deˇ nition, *not* to be random. In these cases a spatial probit can be used to test for the presence of any spatial component. Failing to reject both types of spatial dependence in the data will provide some measure of comfort that there is likely to be no, or only a little, spatial dependence in the model. Otherwise the spatial probit provides some modest evidence as to which type of spatial dependence is likely to exist in the data.

We turn now to examining the predictions from the models. We saw previously how the β coefˇ cient estimates compared with the true underlying model. Now, in order to compare the linear model $\hat{\beta}$'s with the spatial probit $\check{\beta}$'s, we calculate the marginal impact of X on the predicted probabilities from the spatial probit by means of equations (8.13), (8.14), and (8.15). We do so for each observation in the sample, average across the sample, and then average across the samples of a given set (S or T). These predictions are shown as the last column of Table 8.8 for both types of spatial layouts (the S sets and the T sets). The other columns show the estimates obtained for the linear probability model, using OLS, the spatial error estimator and the spatial lag estimator (as indicated in the ˇ rst column of Table 8.8).

When we compare these estimated predictions to the mean coefˇ cient from the linear model simulations, given in the columns labeled "Mean," we see that the linear model consistently yields higher values. This is true for both the state and simulated weights matrices. In addition, in almost every case the probit prediction lies within one standard deviation of the linear model. If a researcher is primarily interested in the predicted probability from a model with a binary dependent variable and spatially generated data, a simple strategy suggests itself. Given the results from Table 8.8, the linear model seems to provide a reasonably accurate upper and lower bound for what the spatial probit would find.

The estimates for the spatial autoregressive parameters α and ρ, obtained using the linear probability model are reported in Table 8.9, for the two types of spatial layouts (S and T). In order to facilitate comparison with the spatial probit estimates, the last column of the table also repeats the mean estimates from Tables 8.4 and 8.6.

For both types of weights matrices and both parameters, we find that the means of the linear estimates are below those of the spatial probit estimates. This is useful information when we know the form of the spatial dependence *a priori*. For example, if we knew that the true model is of the spatially lagged variety, the results suggest that the estimates from a linear probability spatially lagged model will underestimate those from the spatial probit and so provide a lower bound.

Again we observe a difference between the results for the simulated weights matrices (T) and those for the state weights matrix (S). While the lower bound idea holds true in both cases, under the correct null hypothesis the estimates obtained for S are within one standard deviation of the spatial probit model. In the T cases, they are often within two standard deviations of the spatial probit. Since, in practice, weights matrices are more likely to be patterned, as with the state weights matrix, rather than "pseudo random," this allows a more precise bound to be obtained from the linear model.

If a researcher incorrectly estimates the wrong type of spatial dependence for a model then the results will be the opposite. Since the linear estimates are below the probit estimates for both the correct *and* the incorrect models, estimating an incorrectly specified model paradoxically leads to the linear estimates being closer to the truth. As always, this points up the importance of understanding the underlying process that is being modeled.

8.7 Conclusions

We have demonstrated the unique nature of binary data in a setting where spatial dependence is present and showed that a conventional probit analysis is inappropriate. We illustrate a method to estimate the parameters for both a spatial lag and a spatial error probit model. We explore the power of the Likelihood Ratio test for these forms of spatial dependence. The Likelihood Ratio test is not particularly powerful in small datasets. For example, our simulation suggests that a study where the units of analysis are the states of the U.S. is not likely to find evidence of spatial dependence. One needs a substantial number of observations to detect this.

Our simulations further point out that a weights matrix that contains more regularity facilitates detection of spatial dependence: this is borne out by both the spatial probit and spatial linear model analysis. The weights matrix based on contiguity among U.S. states has a more defined pattern and is less regular. This may

Table 8.8. Comparison of linear and probit estimates for β_1

		Linear		Probit
Estimator	Sample	Mean	St.Dev.	Marginal
OLS	$S_{0,0}$	0.99	0.15	0.85
	$S_{0,0.25}$	0.97	0.15	0.84
	$S_{0,0.50}$	0.90	0.17	0.80
	$S_{0.25,0}$	0.94	0.16	0.82
	$S_{0.50,0}$	0.86	0.16	0.77
error	$S_{0,0}$	0.98	0.16	0.85
	$S_{0,0.25}$	0.97	0.16	0.84
	$S_{0,0.50}$	0.90	0.18	0.79
	$S_{0.25,0}$	0.94	0.17	0.82
	$S_{0.50,0}$	0.81	0.18	0.74
lag	$S_{0,0}$	0.96	0.17	0.83
	$S_{0,0.25}$	0.98	0.16	0.85
	$S_{0,0.50}$	0.94	0.17	0.84
	$S_{0.25,0}$	0.95	0.17	0.85
	$S_{0.50,0}$	0.89	0.17	0.85
OLS	$T_{0,0.50}(50)$	0.90	0.18	0.80
	$T_{0,0.50}(100)$	0.92	0.15	0.81
	$T_{0,0.50}(200)$	0.89	0.08	0.81
	$T_{0.50,0}(50)$	0.92	0.16	0.81
	$T_{0.50,0}(100)$	0.96	0.13	0.84
	$T_{0.50,0}(200)$	0.91	0.07	0.82
error	$T_{0,0.50}(50)$	0.90	0.18	0.80
	$T_{0,0.50}(100)$	0.93	0.14	0.82
	$T_{0,0.50}(200)$	0.90	0.08	0.81
	$T_{0.50,0}(50)$	0.83	0.21	0.77
	$T_{0.50,0}(100)$	0.91	0.14	0.81
	$T_{0.50,0}(200)$	0.87	0.09	0.79
lag	$T_{0,0.50}(50)$	0.90	0.17	0.79
	$T_{0,0.50}(100)$	0.91	0.14	0.81
	$T_{0,0.50}(200)$	0.90	0.08	0.82
	$T_{0.50,0}(50)$	0.87	0.18	0.80
	$T_{0.50,0}(100)$	0.92	0.13	0.83
	$T_{0.50,0}(200)$	0.89	0.08	0.84

Table 8.9. Comparison of linear and probit estimates for α and ρ

		Linear		Percentile (Linear)			Probit
Estimator	Sample	Mean	St.Dev.	5th	50th	95th	Mean
lag (α)	$S_{0,0}$	−0.08	0.21	−0.46	−0.03	0.21	−0.14
	$S_{0,0.25}$	0.03	0.19	−0.33	0.05	0.30	0.03
	$S_{0,0.50}$	0.15	0.17	−0.13	0.16	0.41	0.22
	$S_{0.25,0}$	0.07	0.18	−0.27	0.10	0.32	0.14
	$S_{0.50,0}$	0.25	0.16	−0.03	0.27	0.52	0.41
	$T_{0,0.50}(50)$	0.20	0.17	−0.09	0.22	0.47	0.25
	$T_{0,0.50}(100)$	0.21	0.13	−0.01	0.21	0.40	0.29
	$T_{0,0.50}(200)$	0.25	0.09	0.04	0.26	0.36	0.36
	$T_{0.50,0}(50)$	0.32	0.15	0.06	0.33	0.59	0.46
	$T_{0.50,0}(100)$	0.28	0.10	0.11	0.30	0.44	0.45
	$T_{0.50,0}(200)$	0.31	0.09	0.14	0.32	0.45	0.48
error (ρ)	$S_{0,0}$	−0.08	0.23	−0.47	−0.08	0.26	−0.17
	$S_{0,0.25}$	0.06	0.21	−0.42	0.10	0.35	0.11
	$S_{0,0.50}$	0.20	0.20	−0.16	0.23	0.47	0.32
	$S_{0.25,0}$	0.07	0.21	−0.31	0.10	0.36	0.32
	$S_{0.50,0}$	0.26	0.19	−0.08	0.27	0.55	0.42
	$T_{0,0}(50)$	0.23	0.19	−0.12	0.24	0.49	0.37
	$T_{0,0.50}(100)$	0.25	0.14	−0.00	0.25	0.45	0.42
	$T_{0,0.50}(200)$	0.29	0.09	0.11	0.30	0.40	0.48
	$T_{0.50,0}(50)$	0.29	0.20	−0.07	0.31	0.62	0.48
	$T_{0.50,0}(100)$	0.28	0.12	0.05	0.29	0.43	0.47
	$T_{0.50,0}(200)$	0.31	0.10	0.13	0.32	0.47	0.50

be the norm rather than the exception among empirical applications. For example, distance-based weights matrices may exhibit even more pattern and less regularity (e.g., distance vs. contiguity among states in the U.S.A.). More research is necessary on this issue.

We compare our results to using a linear model that attempts to proxy for the more elaborate data generating process. A linear spatial model is much easier to estimate than a spatial probit model and therefore might be a substitute in the same way that the linear probability model was a substitute for the probit model when computational power was limited. However, we show the drawbacks of a linear spatial model. It fails to take into account the dichotomous nature of the dependent variable and, as well, cannot capture the spatial dependence in a theoretically adequately way. The classic probit model captures the dichotomous nature of the dependent variable but ignores spatial structure, and therefore yields biased and inconsistent parameter estimates. We ˇ nd support that the spatial probit model is superior to the linear model and the standard probit model, but there may be times where these simpler models are useful for exploratory purposes. No doubt, the linear spatial model will become obsolete as accessibility to spatial probit software becomes widespread.

9 Simultaneous Spatial and Functional Form Transformations

R. Kelley Pace[1], Ronald Barry[2], V. Carlos Slawson Jr.[1], and C.F. Sirmans[3]

[1] Louisiana State University
[2] University of Alaska
[3] University of Connecticut

9.1 Introduction

Technological advances such as the global positioning system (GPS) and low-cost, high-quality geographic information systems (GIS) have led to an explosion in the volume of large data sets with locational coordinates for each observation. For example, the Census provides large amounts of data for over 250,000 locations in the US (block groups). Moreover, geographic information systems can often provide approximate locational coordinates for street addresses (geocoding). Given the volume of business information, which contains a street address field, this allows the creation of extremely large spatial data sets. Such data, as well as other types of spatial data, often exhibit spatial dependence and thus require spatial statistical methods for efficient estimation, valid inference, and optimal prediction.

Several barriers exist to performing spatial statistics with large data sets. Spatial statistical methods require the computation of determinants or inverses of n by n matrices. Allowing for space does not necessarily cure all of the problems encountered in typical data. For example, simple models fitted to housing and other economic data often exhibit heteroskedasticity, visible problems of misspecification for extreme observations, and non-normality (e.g., Goodman and Thibodeau, 1995; Subramanian and Carson, 1988; Belsley *et al.*, 1980). Simultaneously attacking these problems along with spatial dependence for large data sets presents a challenge.

Functional form transformations provide one technique, which can simultaneously ameliorate all of these problems. For example, better specification of the functional form could reduce spatial autocorrelation of errors given spatial clustering of similar observations. While not guaranteed, functional form transformations often simultaneously reduce heteroskedasticity and residual non-normality. Because of the potential interaction between the spatial transformation and the functional form transformation, it seems desirable to fit these simultaneously.

Accordingly, we wish to examine the following transformation of the dependent variable:

$$(I - \alpha D) Y(\theta),$$

where D represents n by n spatial weights matrix, α represents the spatial autoregressive parameter, and $Y(\theta)$ represents the dependent variable transformation parameterized by a vector of o parameters, θ. Least squares would not work for this problem, as it would reduce the sum-of-squared errors by reducing the range of the transformed variable. As an extreme case OLS could choose θ to make $Y(\theta)$ almost constant for a sufficiently flexible form and a regression with an intercept term would yield almost no error. Hence, this problem requires Maximum Likelihood with a Jacobian for the spatial transformation and a Jacobian for the functional form transformation.

The above form of the problem involves transformation of the functional form of the dependent variable first and the spatial transformation second. This seems a more natural formulation than transformation of the functional form of $(I-\alpha D)Y$ since the functional form of the dependent variable often has an interesting subject matter interpretation. However, spatial transformation first followed by functional form transformation is feasible and may offer some advantages.

The Box-Cox transformation is the most frequently used in regression. Recently, Griffith *et al.* (1998) discussed the importance of transformations for spatial data and examined bivariate Box-Cox/Box-Tidwell transformations of the dependent and independent variable in a spatial autoregression. The use of a parameter for the dependent variable as well as a parameter for the independent variable provided substantial flexibility in the potential transformation. Note, the Box-Cox/Box-Tidwell approach has an additional overhead in spatial problems, as one must compute the spatially lagged value of the new transformed variables at each iteration.

We take a different route in modeling the functional form of the variables in a spatial autoregression. Specifically, we use *B*-splines (de Boor, 1978; Ramsey, 1988) which are piecewise polynomials with conditions enforced among the pieces. The knots specify where each local polynomial begins and ends and the degree specifies the amount of smoothness among the pieces. A spline of degree 0 has no smoothness, a spline of degree 1 is piecewise linear, a spline of degree 2 is piecewise quadratic, and so forth.

Relative to the common Box-Cox transformation, the *B*-spline transformations do not require strictly positive untransformed variables and can assume more complicated shapes (Box and Cox, 1964). The standard one-parameter Box-Cox transformation either has a concave or convex shape. The *B*-spline transformation can yield convex shapes over part of the domain and concave shapes over the rest of the domain. Moreover, *B*-splines can yield more severe transformations of the dependent variable than the Box-Cox transformation. Burbidge *et al.* (1988) discusses the deficiencies of the Box-Cox transformation and the need for more severe transformations of the extreme values of the untransformed dependent variable. These beneficial features do have a price. Relative to transformations such as the Box-Cox, splines may require substantially more degrees-of-freedom.

This could create problems for small data sets or those with low amounts of signal-to-noise (i.e., low R^2).

Computationally, there are three components to the log-likelihood for this problem. These include: (1) a spatial Jacobian, (2) a functional form Jacobian, and (3) the log of the sum-of-squared errors term.

To address the spatial Jacobian part of the log-likelihood, we use the techniques proposed by Pace and Barry (1997a,b,c) to quickly compute the Jacobian of the spatial transformation ($\ln |I - \alpha D|$). This involves the computation of $\ln |I - \alpha D|$ across a grid of values of α. With sparse D, special techniques exist which make this computational tractable.

To address the functional form Jacobian part of the likelihood, we employ two additional techniques to greatly accelerate computational speed. First, we use an intermediate transformation of the dependent variable. Intermediate transformations are often used in nonparametric regression (regression with very flexible functional forms). By adopting a transformation, which partially models the nonlinearity, it requires less flexibility (fewer degrees-of-freedom) to model the remaining nonlinearity. The goal of our particular intermediate transformation is to make the dependent variable's histogram approximately symmetric.

Second, given an approximately symmetric dependent variable, we can employ evenly spaced knots. Equally spaced knots result in more observations between the central knots and fewer observations between the extreme knots. This makes the spline transformation more flexible in the tails and less flexible in the center, a desirable result. Such evenly spaced knots have often been used with B-splines (Hastie and Tibshirani, 1990, p. 24). Evenly spaced knots lead to a very simple functional form Jacobian (Eilers and Marx, 1996; Shikin and Plis, 1995, p. 44) suitable for rapid computation.

To address the log of sum-of-squared errors portion of the log-likelihood, we use the linearity of the B-spline and spatial transformations to write the overall sum-of-squared errors as a series of the sum-of-squared errors from regressions on the individual parts of the transformation. This allows us to recombine the sum-of-squared errors from a set of regressions rather than recompute the sum-of-squared errors fresh each iteration.

Cumulatively, these computational techniques accelerate the log-likelihood computations so that each iteration takes little time. Each estimate requires around 1,000 iterations. Yet, these could be computed in less than 10 seconds on a 200-megahertz Pentium Pro computer, even though the data set had 11,006 observations.

We apply this to a housing data set from Baton Rouge, Louisiana. The Real Estate Research Institute at Louisiana State University estimates regressions periodically to form an index of real estate prices over time. Since each house does not sell each quarter, the regression controls for the differences in sample composition over time by using a variety of independent variables such as age, living area, other area, number of bathrooms, number of bedrooms, and date of sale. In addition, variants of these data have been used to examine prediction accuracy of regression models (e.g., Knight *et al.*, 1994).

In real estate, predictions of the price of unsold homes have been extensively used for tax assessments. In fact, the majority of the districts in the country (and many foreign countries) use some form of statistical analysis to predict the prices of unsold homes (Eckert, 1990). In addition, the secondary mortgage markets have begun exploring the use of statistical appraisal for determining the value of collateral for loans (Gelfand *et al.*, 1998; Eckert and O'Connor, 1992). Note, both of these applications give rise to very large spatial data sets.

To handle these needs, we estimated a general model which includes the previously discussed transformations of the dependent variable, transformations of the independent variables, spatially lagged independent variables, time indicator, and miscellaneous variables. As an illustration of the ef˜ cacy of the proposed techniques, the general model reduced the interquartile range of the residuals by 38.38% relative to a simple model using the untransformed dependent variable. Moreover, the resulting dependent variable transformation greatly improved the pattern of the residuals.

Most estimates of the Box-Cox parameters yield a model somewhere between a linear and logarithmic transformation. The estimated dependent variable transformation also fell between a linear and a logarithmic transformation – it was close to a linear transformation for low-priced properties but approached a logarithmic transformation for the high-priced properties. In fact, it actually provided more damping than the logarithmic transformation for extremely high-priced properties. Finally, the estimated functional forms of the independent variables seemed plausible and of interest.

Section 9.2 develops the joint spatial and dependent variable transformation estimator while Sect. 9.3 applies the estimator to the Baton Rouge data. Section 9.4 concludes the chapter.

9.2 Simultaneous Spatial and Variable Transformations

This overall section presents the estimator and the various techniques facilitating computation. Section 9.2.1 sets up the log-likelihood, Sect. 9.2.2 discusses the application of splines to the problem, Sect. 9.2.3 shows how to simplify the *SSE*, Sect. 9.2.4 provides a computational simpli˜ cation of the spatial Jacobian, Sect. 9.2.5 gives a simple way of computing the functional form Jacobian, and Sect. 9.2.6 extends the model to transformations of the independent variables.

9.2.1 A Transformed Dependent Variable with Spatial Autoregression

Suppose the transformed variable follows a spatial autoregressive process:

$$\begin{aligned} Y(\theta) &= X\beta + u, \\ u &= \alpha D u + \varepsilon, \end{aligned} \tag{9.1}$$

where $Y(\theta)$ denotes the transformed dependent variable n element vector which depends upon the o element vector of parameters θ. In addition, X denotes an n by p matrix of the independent variables, D denotes an n by n spatial weights matrix, α represents the autoregressive parameter $(1 > \alpha \geq 0)$, β denotes the p element vector of regression parameters, u denotes the spatially autocorrelated error term, while ε denotes a normal *iid* error term.

The spatial weights matrix D has some special structure. First, it has zeros on the main diagonal which prevents an observation from predicting itself. Second, it is a non-negative matrix and positive entries in the jth column of the ith row means observation j directly affects observation i. We do not assume symmetry and so the converse does not necessarily hold. Third, we assume each observation is only directly affected by its m closest neighbors. This makes D very sparse (high proportion of zeros), which greatly aids computational performance. Fourth, D is row-stochastic and so each row sums to 1. This gives D a smoothing or linear ˇ l-ter interpretation (Davidson and MacKinnon, 1993). Intuitively, $DY(\theta)$ provides a construct similar to a lag in time series for $Y(\theta)$.

To estimate (9.1), we rewrite it as:

$$(I - \alpha D)\,Y(\theta) = X\beta + \varepsilon. \tag{9.2}$$

For a known α and θ, one could proceed to apply OLS to (9.2). Unfortunately, estimating α and θ via OLS results in biased estimates.

To motivate the defect in using OLS to estimate the parameters in this situation, consider momentarily the very simple model $(\xi I)\,Y = X\beta + \varepsilon$ where ξ represents a scalar parameter. If we employ OLS to estimate both ξ and β, the estimated value of the parameter ξ would equal 0 for any value of β. This would turn the dependent variable vector ξY into a vector of zeros that a model with an intercept would ˇ t perfectly.

To prevent this form of extreme behavior, one must employ Maximum Likelihood, which explicitly penalizes such pathological transformations using the Jacobian of the transformation. The Jacobian of the transformation measures the n-dimensional volume change caused by stretching or compressing any or all of the potential n dimensions. By premultiplying Y via the matrix ξI, we are performing a linear transformation. In this case we are compressing or stretching each of the n dimensions of Y by a factor ξ. Relative to a unit value for ξ, values of $\xi < 1$ correspond to more singular transformations. The Jacobian of the transformation is the determinant of the matrix of derivatives, which in this instance is ξ^n $(|\xi I| = \xi^n)$.[1] To make the example even simpler, we are dealing with a cube when n is 3. If we multiply each dimension of the cube by a factor of 2, we increase the volume of the cube by a factor of 8 (2^3). The need for the Jacobian is not speciˇ c to the normal Maximum Likelihood, but arises whenever making transformations with proper, continuous densities (Davidson and MacKinnon, 1993, p. 489; Freund and Walpole, 1980, pp. 230–252).

Assuming normality, the proˇ le log-likelihood for this example equals a constant plus the log of the Jacobian less $(n/2)\log(SSE(\xi))$. Taking as a reference point the sum-of-squared error when $\xi = 1$ $(SSE(\xi = 1))$, then:

$$SSE(\xi) = SSE(\xi = 1)\,\xi^2.$$

[1] Determinants measure the n-dimensional volume of the geometric object deˇ ned by its rows (or equivalently columns). See Lay 1997, pp. 199–204 for a nice discussion of this point.

As an example, multiplying Y by a constant of $1/2$ would multiply SSE by a constant of $1/4$. Hence, the pro˘ le log-likelihood becomes:

$$n\log(\xi) - (1/2)\, n\log\left(SSE(\xi = 1)\xi^2\right).$$

A simple expansion shows that the likelihood will be the same for any choice of ξ. Hence, the Maximum Likelihood choice for β does not depend upon ξ. Thus, one cannot affect the Maximum Likelihood estimate by a simple scaling of the dependent variable, a highly desirable result.[2]

In this simple case, the role of the Jacobian in Maximum Likelihood is clear. The Jacobian continues to play a similar role in more complicated transformations such as those arising from spatial transformations or from functional form transformations. Successive transformations result in Jacobians multiplied by each other in the multivariate density. Hence, for simultaneous transformations the log of the Jacobian of ABC would equal the sum of the logs of the individual Jacobians (e.g., $\ln(J_{ABC}) = \ln(J_A) + \ln(J_B) + \ln(J_C)$ where J denotes the relevant Jacobian).

Hence, the pro˘ le log-likelihood for estimation using a spatial and a functional form transformation equals:

$$L(\alpha, \theta) = C_{lik} + \ln\left(J(Y)_\alpha\right) + \ln\left(J(Y)_\theta\right) - (n/2)\ln\left(SSE(\alpha, \theta)\right), \tag{9.3}$$

where $J(Y)_\alpha$ and $J(Y)_\theta$ represent the Jacobians of the spatial and dependent variable transformations and C_{lik} represents an arbitrary constant.

Attacking the maximization of the above log-likelihood in the most straightforward way would likely result in very long execution times. We show methods for greatly accelerating the computation of each of these terms. We detail these computational accelerations in the sections below.

9.2.2 Linear Expansions of Non-Linear Functions

If one computed $Y(\theta)$ and subsequently $(I - \alpha D)Y(\theta)$ for every iteration of the maximization of the log-likelihood, this could greatly reduce the speed of the algorithm as $(I - \alpha D)$ is an n by n (albeit sparse) matrix. Hence, we ˘ rst seek ways of avoiding this step. Fortunately, if we can expand $Y(\theta)$ linearly, we can avoid this set of computations. A number of ways of linearly expanding a function exist. We could use indicator variables, polynomials, or splines. For our computations we chose B-splines (de Boor, 1978, 1999).

In fact, splines generalize both indicator variables and polynomials. Indicator variables provide a locally constant ˘ t to a function for their non-zero portions. B-splines of degree 0 yield indicator variables. The advantage of indicator variables or B-splines of degree 0 is their local ˘ t. Their disadvantage is that locally constant approximations are not necessarily continuous or smooth.

[2] Davidson and MacKinnon (1993) provide an excellent introduction to transformations in the context of Maximum Likelihood.

Polynomials, however, exhibit both continuity and smoothness. Polynomials attempt to approximate a function globally and gyrations of the function over parts of the domain can cause the polynomial to poorly fit other parts of the domain. A polynomial equates to a high degree *B*-spline with few knots.

Specifically, *B*-splines are piecewise polynomials with conditions enforced among the pieces. The knots specify where each local polynomial begins and ends and the degree specifies the amount of smoothness among the pieces. A spline of degree 0 has no smoothness, a spline of degree 1 is piecewise linear, a spline of degree 2 is piecewise quadratic, and so forth.

To provide some physical intuition, a spline was a thin strip of wood used in constructing ships. The spline attached to two points separated by less than its length would cause the spline to produce a curve. By introducing supports (ribs of the ship), the curve could be modified into many shapes. Hence, the spline knots act similar to the ship's ribs. Moreover, the flexibility of the strip of wood would determine the smoothness (affect the degree of the spline). The piecewise linear splines used here correspond to laying a string across the ribs of the ship.

Also, one can restrict *B*-splines to yield strictly monotonic transformations. One must have monotonic transformations of dependent variables for prediction in the original dependent variable space (Ramsey, 1988). Finally, *B*-splines can interpolate a given set of values (assuming satisfaction of the Schoenburg-Whitney conditions (de Boor, 1999, p. 1.10). The Schoenburg-Whitney conditions essentially require that each of the *B*-spline basis vectors have at least one non-zero value. Hence, given a set of values, some weighting of the associated *B*-spline basis vectors could return the same set of values.

To explain splines in detail is beyond the scope of this chapter. However, a specific example greatly aids in understanding some of their features. In example 1 we consider four values for the dependent variable Y of 1, 1.5, 2.25, and 3.0. Given knots of 1, 2, and 3 (with 1 and 3 being repeated), we used the SPCOL function in the Matlab Spline Toolbox 2.01 to produce the following matrix $B(Y)$ comprised of three basis vectors. The exact values of the basis vectors depend upon Y and hence we emphasize this by writing $B(Y)$.

Example 1			
Y	$B(Y)$		
1.00	1.00	0.00	0.00
1.50	0.50	0.50	0.00
2.25	0.00	0.75	0.25
3.00	0.00	0.00	1.00

In Example 1, $B(Y)$ illustrates a couple of B-spline features. First, $B(Y)$, the collection of basis vectors, contains only non-negative numbers. Second, each row sums to one. Third, the basis vectors have zero elements for elements of Y sufficiently far away from the knots. If we compute $B(Y)\theta$ for $\theta_{interpolate} = [1\ 2\ 3]$, we find it yields Y exactly. For other θ such that $\theta_1 < \theta_2 < \theta_3$, the plots of $B(Y)\theta$ versus Y show a monotonically increasing piecewise linear relations. Figures 9.1a–d show four such plots. In every case, the selected θ satisˇ ed the monotonicity constraints. Figure 9.1a shows how the function $B(Y)\theta$ exactly replicated the original Y (interpolated). Figure 9.1b shows a slightly concave transformation while Fig. 9.1c shows a more severe concave transformation. Figure 9.1d shows a convex transformation. With more points, one could generate combinations of convex and concave transformations (over different domains).

Assuming satisfaction of the Schoenburg-Whitney conditions, with *B*-splines our transformed dependent variable becomes:

$$Y(\theta) = B(Y)\theta, \tag{9.4}$$

where $B(Y)$ represents the n by o matrix containing the basis vectors and θ represents the o by 1 parameter vector. The number of basis vectors, o, depends upon the number of knots and the degree of smoothness required. As o rises, the transformed dependent variable $Y(\theta)$ can assume progressively more exible forms.

Substituting (9.4) into (9.2) yields:

$$(I-\alpha D)Y(\theta) = (I-\alpha D)B(Y)\theta = [\,B(Y)\ DB(Y)\,]\begin{bmatrix}\theta\\ -\alpha\theta\end{bmatrix}. \tag{9.5}$$

Hence, one can linearly expand the joint spatial and dependent variable into the product of a n by $2o$ matrix and a $2o$ by 1 parameter vector.

9.2.3 *SSE* Simpliˇ cations

Let M represent the idempotent least squares matrix $I - X(X'X)^{-1}X'$. We can write the residuals from the regression of $(I-\alpha D)Y(\theta)$ on X as follows:

$$\begin{aligned} e &= M\left(\,[\,B(Y)\ DB(Y)\,]\begin{bmatrix}\theta\\ -\alpha\theta\end{bmatrix}\right)\\ &= [\,MB(Y)\ MDB(Y)\,]\begin{bmatrix}\theta\\ -\alpha\theta\end{bmatrix}\\ &= E\rho, \end{aligned} \tag{9.6}$$

where the n by $2o$ matrix E contains all the residuals from the individual regressions and the vector ρ represents the $2o$ element parameter vector. The linearity of the problem means the least squares residuals e on the overall transformed variable $(I-\alpha D)Y(\theta)$ are simply a linear combination of the least squares residuals from regressing each basis vector in$B(Y)$ and their spatial lags $DB(Y)$ on X. Hence, forming parameterized sum-of-squared errors yields:

$$SSE(\alpha,\theta) = e'e = \rho'(E'E)\rho. \tag{9.7}$$

Note, the $2o$ by $2o$ error cross-product matrix $E'E$ is only computed once. Subsequent iterations of $\rho'(E'E)\rho$ involve only order of o^3 operations, a very small number which does not depend upon n, the number of observations or k, the number of regressors. Moreover, o is usually much less than k and strictly less than n. This reduction in the dimensionality of the sum-of-squared errors leads to an low dimensional proˇ le likelihood (Meeker and Escobar, 1995).

9.2.4 Spatial Jacobian Simpliˇ cations

Historically, the spatial Jacobian, $\ln|I-\alpha D|$, constituted the main barrier to fast computation of spatial estimators (e.g., Li, 1995). However, the use of a limited number of spatial neighbors lead to sparse matrices. (Pace and Barry, 1997a,b) show how various permutations of the rows and columns of such sparse matrices $(I-\alpha D)$ can vastly accelerate the computation of $\ln|I-\alpha D|$. Although computation of $\ln|I-\alpha D|$ is inexpensive for a particular value of α, one can further accelerate the computations by computing $\ln|I-\alpha D|$ for a large numbers of values of α (e.g., 100) and interpolating intermediate values. Insofar as α has a limited range (for stochastic D) and the function $\ln|I-\alpha D|$ is quite smooth, the interpolation exhibits very low error.

Moreover, these computations are performed only when changing the weight matrix D. Hence, one can reuse the grid of values (and interpolated points) when ˇ tting different models involving Y and X for a given D.

Pace and Barry have released a public domain Matlab-based package, "Spatial Toolbox 1.1", available at www.spatial-statistics.com, which implements these spatial Jacobian simpliˇ cations and contains copies of the articles which describe the implementation details.

9.2.5 Functional Form Jacobian Simpliˇ cations

The functional form log-Jacobian has a particularly simple form for piecewise linear splines with evenly spaced knots:

$$\begin{aligned}\ln(J(Y)_\theta) = C + n_2\ln(\theta_2-\theta_1) + n_3\ln(\theta_3-\theta_2) + \\ \ldots + n_{o-1}\ln(\theta_{(o-1)}-\theta_{(o-2)}),\end{aligned} \tag{9.8}$$

where $n_2, n_3, \ldots n_{(o-1)}$ represents the number of non-zero elements of all but the ˇ rst and last basis vectors and the distance between knots determines the constant C (Eilers and Marx, 1996; Shikin and Plis, 1995, p. 44). This very simple form lends itself to extremely rapid execution. Piecewise linear splines also facilitate enforcing strict monotonicity, provided $(\theta_{j+1}-\theta_j) > 0$, $J(Y)_\theta > 0$.

Unfortunately, an even placement of knots may not work well in many cases. However, transforming the original variable Y may result in a variable $g(Y)$ where an even knot placement will work better. In which case, the log of the Jacobian involving an intermediate transformation can be partitioned into the original log-Jacobian and a log-Jacobian for the intermediate transformation:

$$\ln(J(g(Y))_\theta) = \sum_{i=1}^{n}\ln\left(\frac{dg(Y)}{dY_i}\right) + \ln(J(Y)_\theta). \tag{9.9}$$

The intermediate transformation $g(\cdot)$ does not depend upon the parameters α or θ and hence these do not affect its contribution to the functional form Jacobian. However, the intermediate transformation $g(\cdot)$ does help adjust the placement of knots and therefore has some effect upon the ˇ nal ˇ t. Parameterizing knot placement within a Maximum Likelihood framework could make it easier to assess its statistical consequences.

Even knot placement results in nested models in some cases. For example, if the most exible model uses 12 knots, sub-models with six, four, three, and two knots correspond to parameter restrictions placed on the 12 knot model. Again, this aids the assessment of the statistical consequences of knot placement.

9.2.6 Extension to Functions of the Independent Variables

Naturally, one could include a spline expansion of the independent variables. In addition, one could include spatial lags of the independent variables. Let Z represent the untransformed independent variables. We could model X, the regressors as:

$$X = \left[B(Z) \; DB(Z) \right], \tag{9.10}$$

where $B(Z)$ represents the spline expansion of each one of the columns of Z. Note, without deletion of one basis vector for each column of Z, X would be linearly dependent as the sum of the rows of all the basis vectors always equals 1 for B-splines. Hence, if each basis function expansion takes o vectors, $B(Z)$ will have dimension of $p(o-1)$. Adding the spatial lags doubles the variable count. The spline expansion of each one of the core independent variables Z allows one to create a generalized additive model (Hastie and Tibshirani, 1990). In addition, this particular model allows the spatially lagged variables to follow a different functional form:

$$\begin{aligned}(I-\alpha D)Y(\theta) &= \left[B(Z) \; DB(Z) \right] \beta + \varepsilon \\ &= \sum_{i=1}^{p} f(Z_i) + \sum_{i=1}^{p} Dh(Z_i) + \varepsilon. \end{aligned} \tag{9.11}$$

This very general speciˇ cation subsumes the case of autocorrelated errors. This restriction would also make $f(\cdot) = h(\cdot)$. Imposing this restriction would substantially slow the speed of computing the estimates. However, the use of restricted least squares would still provide much more speed than a formulation which required computing $(X'X)$ each iteration. Moreover, this restriction will often be rejected by the data as n becomes large.

9.3 Baton Rouge Housing

This overall section presents the application of the techniques developed in the previous section to housing data from Baton Rouge. Section 9.3.1 discusses the data, Sect.9.3.2 gives details on the construction of the spatial weight matrix, Sect. 9.3.3 provides timing and other information on the determinant computations, Sect. 9.3.4 presents the general model, Sect. 9.3.5 discusses the estimated dependent variable transformation, Sect. 9.3.6 discusses the estimated independent variable transformations, Sect. 9.3.7 shows how to conduct the inference in this model, Sect. 9.3.8

discusses model performance in the untransformed variable space, and Sect. 9.3.9 conducts an experiment to document the uniqueness of the estimates and computation times.

9.3.1 Data

We selected observations from the Baton Rouge Multiple Listing Service which (1) could be geocoded (given a location in latitude and longitude based upon the house's address); and (2) had complete information on living area, total area, number of bedrooms, and number of full and half baths. In addition, we also discarded negative entries for these characteristics. In total, 11,006 observations survived these joint criteria.

9.3.2 Spatial Weight Matrix

To construct the spatial weights matrix D, compare the distance d_{ij} between every pair of observations i and j to d_i^m, the distance from observation i and its mth nearest neighbor. It seems reasonable to set to 0 the direct in uence of distant observations upon a particular observation. Accordingly, assign a weight of $1/m$ to observations whenever d_{ij} is greater than 0 and is less than or equal to d_i^m as:

$$0 < d_{ij} \leq d_i^m \quad \leftrightarrow \quad D_{ij} = \frac{1}{m}. \tag{9.12}$$

By construction D will be row-stochastic but not necessarily symmetric. For this particular problem, we set m equal to 4.

9.3.3 Determinant Computations

Following Pace and Barry (1997b) we computed $\ln|I - \alpha D|$ for:

$$\alpha = 0.01, 0.02, \ldots, 0.99.$$

The LU decomposition of $(I - \alpha D)$ results in the triangular matrices L and U, where the diagonal of U contains the pivots r_i. By construction, $(I - \alpha D)$ is strictly diagonally dominant and hence has bounded error sensitivity (Golub and van Loan, 1989). The magnitude of the determinant is determined by the product of the pivots r_i or the log-determinant by the sum of $\ln(r_i)$.

Computation of the 100 determinants took 57.6 seconds on an 200 megahertz Pentium Pro computer. By employing some of the permutation algorithms discussed in Pace and Barry (1997b) or by employing some devices to exploit symmetry as in Pace and Barry (1997a) we could further accelerate these times.

Given the grid of log-determinant values, we employed linear interpolation to arrive at intermediate values.

9.3.4 Model

We fitted the following model to the data. Each of the functions $f(\cdot), h(\cdot)$ for the independent variable's living area, other area, and age comes from piecewise linear *B*-splines with knots at the minimum value, the 1st, 5th, 10th, 25th, 50th, 75th, 90th, 95th, 99th quantiles, and the maximum value. Specifically, we used the Matlab Spline Toolbox (Version 1.1.3) function SPCOL to create the necessary basis vectors. Hence, applying SPCOL to a particular variable such as age would result in an n by 11 matrix whose columns contained the basis vectors. A particular linear combination of these basis vectors would create the function $f(\cdot)$ while a different linear combination of the same basis vectors would create $h(\cdot)$. De Boor wrote the Spline Toolbox and the functions in it closely resemble those described in de Boor (1978).

For the discrete full bath and beds variables, these functions are formed from indicator variables at each of the values these discrete variables assume. In addition, we used single indicator variables to control for age missing values, for age greater than 150 years, for the presence of half-baths, and for the year of sale. For both the spline and the sets of indicator variables, we deleted one column to prevent linear dependence, as the row-sum of *B*-splines equals 1, as does the sum of a complete set of indicator variables.

$$\begin{aligned}(I-\alpha D)g(\text{Price}) = {} & f_1(\text{living area}) + f_2(\text{other area}) + f_3(\text{age}) + f_4(\text{full baths}) \\ & + f_5(\text{beds}) + Dh_1(\text{living area}) + Dh_2(\text{other area}) \\ & + Dh_3(\text{age}) + Dh_4(\text{full baths}) + Dh_5(\text{beds}) \\ & + I_1(\text{age missing})\beta_1 + I_2(\text{age} > 150)\beta_2 \\ & + I_3(\text{half bath} > 0)\beta_3 + I_{1985-1992}\beta_{4-11} + \varepsilon \qquad (9.13)\end{aligned}$$

The full model involves 113 parameters. This very general model will hopefully span the true model. Moreover, the general model provides a way of investigating other potential problems and a starting point for subset selection. See Hendry *et al.* (1984) for more on the advantages of general to specific modeling.

9.3.5 Estimated Dependent Variable Transformations

As discussed in 9.2.5, the use of an intermediate transformation $g(\cdot)$ makes it possible to modify the effects of equal knot placements. We selected the Box-Cox transformation $g(Y) = (Y^{\varphi} - 1)/\varphi$ with log-Jacobian $(1-\varphi)\sum \ln(Y_i)$ for this step. We examined the transformation for a grid of φ and selected $\varphi = 0.25$ based upon maximizing the normality of Y as measured by the studentized range. This induced approximate symmetry, which made equal knot placement viable. We used 11 equally placed knots.

Based upon other work with transformations (e.g., Burbidge *et al.*, 1988) we expected most reasonable transformations would induce linearity for the bulk of the observations. The approximate normality of Y coupled with equal placement gave the desirable result of having a greater number of knots in the tails as opposed to the center of the density of Y. This gave the potential transformation more exibility in the tails where the differences among transformations emerge.

Figure 9.2 shows Y, ln (Y), and $Y(\theta)$, the optimal piecewise linear spline transformation of Y, plotted against ln (Y). The optimal transformation $Y(\theta)$ acts similar to a linear transformation for low-priced houses and acts more like the logarithmic transformation for high-priced houses.

Figure 9.3 shows the effects of this optimal transformation. Figure 9.3c shows the extreme heteroskedasticity (positively related to price) created by not using any transformation. Note the untransformed dependent variable model systematically underpredicts the high-priced properties as well.

Figure 9.3d shows the extreme heteroskedasticity (negatively related to price) created by using the logarithmic transformation. Note the logarithmically transformed dependent variable model overpredicts low-priced properties as well.

Figure 9.3b shows the intermediate transformation (Box-Cox with $\lambda = 0.25$) created heteroskedasticity for both low and high-priced properties and also created problems of systematic over and under prediction at the extremes of the price density.

Figure 9.3a shows how the spline transformation cures the problem of heteroskedasticity. Moreover, inspection of the low and high-priced properties does not reveal a systematic pattern of under or over prediction. Figure 9.4a shows the histogram of standardized residuals from the spatial regression on the transformed dependent variable with a normal curve superimposed. Similarly, Figure 9.4b shows the histogram of standardized residuals from the spatial regression on the untransformed dependent variable with a normal curve superimposed. Relative to the untransformed dependent variable spatial regression, the errors from the spatial regression on the transformed variable show substantially less leptokurtosis.

Previous work, such as Knight *et al.* (1994), avoided the problem of heteroskedasticity by truncating large portions of the sample based upon price.

9.3.6 Estimated Independent Variable Transformations

Figure 9.5 shows the optimal functions of the independent variables. Note, we did not enforce strict monotonicity with these optimal functions. Figure 9.5a depicts f_1 (living area), which apart from a decreasing section for very small houses not often observed in the sample, shows a positive, concave relation between $Y(\theta)$ and living area. Miscoding of observations, such as leaving out a digit in the living area ˇ eld, provides one possible explanation for this decreasing section. For example, if there are average-priced houses with 0 reported living area, the model might actually show a rise in price as living area goes to 0.

As depicted by Fig. 9.5b, age shows a decreasing relation up until about 40 years when it rises and declines again at 100 years. The Age variable confounds two phenomena. First, physical and hence economic depreciation rises with age. Second, age re ects the year of construction. If the year of construction proxies for features such as wood oors, high ceilings, or other desirable traits, one could see a non-monotonic relation between age and price. In addition, remodeling confuses the issue as the age of the improvements differs from the age of the original structure. Goodman and Thibodeau (1995) also found a non-monotonic relation between age and price. "Dwellings 20–40 years old appreciated slightly, while older dwellings depreciate."

As depicted by Fig. 9.5c, other area shows a very positive, concave relation between $Y(\theta)$ and other area. As depicted by Fig. 9.5d, baths shows a positive, concave relation between $Y(\theta)$ and baths up until four baths. Subsequently, it declines slightly. Again not many houses have five baths or more.

One would not necessarily expect a monotonic relation between bedrooms and price. Holding other variables constant, more bedrooms means smaller bedrooms. Hence, "bedrooms" is a design value with some optimal value. As depicted by Fig. 9.5e, this optimum is at three bedrooms, a plausible value. Finally, Fig. 9.5f shows the relation between $Y(\theta)$ and year-of-sale. This shows the precipitous drop in housing prices in 1988, which has been documented by others (e.g., Knight *et al.*, 1994).

We also examined the optimal independent variable transformations for the original untransformed dependent variable (no spatial or dependent variable transformations). For the most part, these arrived at qualitatively similar independent variable transformations. Some differences appeared. For example, the optimal transformation for living area was slightly convex instead of concave, baths showed a more precipitous drop for houses with more than five bathrooms, and age showed a rise after 20 years (as opposed to around 35 years for the model with spatial and dependent variable transformations).

9.3.7 Inference

Given the fast computation of the log-likelihood, it seems reasonable to conduct inference via Likelihood Ratio tests. Table 9.2 presents these Likelihood Ratios for a wide variety of hypotheses. In all cases these were significant at well beyond the 1% level. Hence, both the spatial and the transformation parts of the model seem highly significant. The spatial autoregressive parameter, α, equaled 0.5820 and had a deviance ($-2\log(LR)$) of 3936.62 with only one hypothesis. The transformation $Y(\theta)$ also proved quite significant with a deviance of 8114.82 with 10 hypotheses. Only 10 parameters vary independently due to the affine invariance of the regressand for linear regression. Note, deleting the transformation parameters equates to running a pure spatial model. For the pure spatial model α equaled 0.5099. Hence, rather than the transformation removing spatial autocorrelation through better specification, the model acted to transform the dependent variable to increase the use of the autocorrelation correction.

The individual variables were all significant with living area showing the greatest impact on the log-likelihood with a deviance of 3364.92. The general model dominated simpler models with fewer variables. Compared to running a regression with the untransformed dependent variable coupled with a simple set of independent variables ignoring space and transformations, the deviance was 14782.04 with 82 hypotheses.

Table 9.2. Likelihood Ratio Tests

Models	Log-likelihood	Deviance	Degrees of Freedom	Critical Values 1
Unrestricted Model	−154849.65			
Model Sans Beds Indicators	−154905.70	112.10	14	29.1
Model Sans Bath Indicators	−154936.91	174.52	12	26.2
Model Sans Age Spline	−154979.21	259.12	20	37.6
Model Sans Other Area Spline	−155131.04	562.78	20	37.6
Model Sans Time	−155317.68	936.06	11	24.7
Model Sans Living Area Spline	−156532.11	3364.92	20	37.6
Model Sans Lagged Dependent Variables ($a = 0$)	−156817.96	3936.62	1	6.6
Model Sans Spatial Lagged Independent Variables	−155095.00	490.70	56	83.5
Model Sans Transformation ($q = 0$)	−158907.06	8114.82	10	23.2
Log Dependent Variable Model with Spatial Lagged Independent Variables	−160313.48	10927.66	12	26.2
Linear Dependent Variable Model with Spatial Lagged Independent Variables	−160206.97	10714.64	12	26.2
Log Dependent Variable Model Sans Spatial Lagged Independent Variables	−162032.58	14365.87	82	114.7
Linear Dependent Variable Model Sans Spatial Lagged Independent Variables	−162240.67	14782.04	82	114.7

The use of restricted least squares, which avoids recomputing $(X'X)$, further aids in the speed of computing these likelihood ratio tests.

Finally, we do not account for the statistical consequences created by the monotonicity constraint. However, one could easily use a Bayesian inequality estimator as in Geweke (1986) to show how the prior associated with the monotonicity constraint affects the posterior distributions of the parameters of interest. See Gilley and Pace (1995) for an application of this estimator to another house price data set.

9.3.8 Performance in the Original Dependent Variable Space

Part of the goal of ˇ tting the general model was to improve upon prediction over simpler models in the original dependent variable space (Price). Given the Y and the strictly positive monotonic transformation $Y(\theta)$, we can take the prediction in the

Table 9.3. Sample Error Statistics Across Models For Prediction of the Untransformed Dependent Variable

	Model 1	Model 2	Model 3	Model 4	Model 5
Spatial *Y*	1	1	0	0	0
Spatial *X*	1	1	1	1	0
Transformed *Y*	1	0	1	0	0
Transformed *X*	1	1	1	1	0
Min	−173303.03	−228289.63	−220671.59	−241016.53	−252491.46
1st	−35807.31	−45655.02	−43785.12	−50025.25	−58528.35
5th	−20261.98	−23054.30	−25135.14	−28153.28	−33423.99
10th	−14270.14	−15912.10	−17853.04	−19654.08	−23087.08
25th	−6387.17	−6809.01	−8684.07	−9123.23	−10660.61
50th	42.30	348.76	−340.64	−15.06	−530.14
75th	6164.72	6927.99	7989.47	8762.24	9707.98
90th	13924.82	14010.39	18207.61	18189.98	21588.44
95th	21122.11	20214.30	27702.55	26686.41	32908.49
99th	52523.81	48008.43	63033.51	54432.72	73978.17
Max	328574.03	276177.59	409496.21	341369.79	389299.09
Interquartile Range	12,551.89	13,737.00	16,673.54	17,885.47	20,368.59

transformed space, $\hat{Y}(\theta)$ and with interpolation compute the prediction in the original space, $\hat{Y}$. Even if $\hat{Y}(\theta)$ comes from an unbiased estimator of $Y(\theta)$, $\hat{Y}$ does not unbiasedly estimate Y. To control for this bias, we allowed for it using the smearing estimator of Duan (1983).

We computed the predictions for a variety of models in the original dependent variable space. The performances of these models in the original dependent variable space appear in Table 9.3. We began with Model 5, a simple model in price space without transformation or spatial modeling of the independent or dependent variables. One could consider Model 5 as the standard model without using any transformations. The results from Model 5 closely match others in the literature. For example, Knight *et al.* (1994) examined the relation between list and transactions prices for the Baton Rouge data to investigate buyer search behavior. Their model uses a very similar speciˇ cation and has a R^2 of 0.72. The R^2 for Model 5 was a very similar 0.7299. This provides a benchmark for the subsequent models.

The residuals are asymmetric in Model 5 so while the mean error equals 0 by construction, the median error equals −530.14 dollars and the 25th and 75th quartiles are −10,660.61 and 9,707.98 dollars. Given the average price of the houses in the sample is \$75,597; this does not represent particularly good performance. Model 4, which includes spatial independent variables and transformed independent variables, improves considerably on Model 5. It shows more symmetric errors and dominates Model 5 for every order statistic. Similarly, Model 3 adds transformation of *Y*, and also improves on Model 4 for most order statistics. Model 2 does not use transformations of Y but does add spatially lagged Y. It shows a large reduction relative to previous models for all but the minimum and 1st quantiles of the empirical error density.

Model 1, the general model, displays considerable improvements over the previous models, except for the 95th quantile to the maximum of the empirical error density where the spatial model without dependent variable transformations (Model 2) displays lower error. Relative to the simple Model 5, Model 1 has a 38.38% lower interquartile range of the empirical error density. In addition, relative to Model 4, the next best performing model, it shows a 8.6% reduction in the interquartile range of the empirical error density. Hence, the improvements in the transformed space carry back to the untransformed space.

9.3.9 Timing and Uniqueness

Local maxima are the bane of complicated Maximum Likelihood models. To examine this problem in the context of this problem, we estimated the model 250 times with different random starting points. We picked α randomly from [0,1]. We picked θ_i from [0,1] with the restriction that $\theta_{i-1} > \theta_i$ to generate strictly positive monotonic starting points.

It took 493 iterations at minimum and 1642 iterations at maximum to ˇ nd the optimum. On average it took less than 10 seconds to arrive at the maximum likelihood estimates (given previous computation of $E'E$ and $\ln|I - \alpha D|$) using a computer with a 200Mhz Pentium Pro processor. All of the 250 estimates converged to the same log-likelihood value with a maximum error of 0.08 from the iteration, which took the longest to converge.

9.4 Conclusions

Locational data may suffer from both spatial dependence and a host of other problems such as heteroskedasticity, visible evidence of misspeciˇ cation for extreme values of the dependent variable, and non-normality. Functional form transformations of the dependent variable often jointly mitigate these problems. Moreover, the transformation to reduce spatial dependence and the transformation of the functional form of the dependent variable can interact. For example, a reduction in the degree of functional form misspeciˇ cation can also reduce the degree of spatial autocorrelation in the residuals. Alternatively, the functional form transformation may make the spatial transformation more effective. In fact, the latter occurred for the Baton Rouge data as the spatial autoregressive parameter rose from 0.5099 when using the untransformed variable to 0.5820 when using the transformed variable.

Application of the joint spatial and functional form transformations to the Baton Rouge data provided a number of gains relative to simpler models. First, the pattern of residuals in the transformed space improved dramatically. For example, unlike the residuals from simpler models, the general model's residuals seemed evenly divided by sign for all predicted values. Second, the magnitude of the sample residuals dropped dramatically even in the untransformed variable's space. Specifically, the interquartile range of the residuals from the general model using all the transformations when taken back into the untransformed variable's space fell by 38.38% relative to the residuals on a simple model with the untransformed variable. Third, the general model provided interesting insights into the functional form of the dependent and independent variables. The estimated functional form for the dependent variable followed an approximately linear transformation for low-priced properties, an approximately logarithmic transformation for high-priced properties, and a somewhat more severe than logarithmic transformation for the very highest-priced properties.

The computation of the model employs several innovations. First, it relies upon the sparse matrix techniques proposed by Pace and Barry (1997a,b,c) to compute 100 log-determinants of the 11,006 by 11,006 spatial transformation matrix in 57.6 seconds using a 200 megahertz Pentium Pro computer. Interpolation of this grid of log-determinants provides the spatial log-Jacobian, which greatly accelerates Maximum Likelihood maximization. Second, it uses an intermediate transformation to allow the use of evenly-spaced knots which have a particularly simple log-Jacobian for the functional form. Third, it expresses the overall sum-of-squared error as a linear combination of the sum-of-squared errors on individual parts of the transformations. Consequently, the actual maximization of the log-likelihood for the joint transformation takes less than 10 seconds on average (given prior computation of the spatial log-Jacobian and the individual sum-of-squared error computations). This part of the maximization of the log-likelihood does not directly depend upon the number of observations or the total number of regressors. The optimum appears unique as 250 iterations with different starting points returned the same log-likelihood value.

The computational speed of this model has at least two implications. First, inference can proceed by relatively straightforward likelihood ratio tests. The use of restricted least squares, which avoids recomputing $(X'X)$, further aids in the speed of computing the likelihood ratios. Second, the model becomes useful for exploratory work with large spatial data sets, an area which currently suffers from a lack of tools. By simultaneously fitting a generalized additive model and controlling for spatial dependence, it potentially provides a good first view of locational data. Such views can suggest simpler parametric specifications and the need for other adjustments such as reweighting. Naturally, the model could accommodate reweighting with an additional Jacobian for the weights.

While we primarily worked with economic data with this model, we suspect it could have applications to other fields. As the volume of spatial data continues to rise, methods, which simultaneously and quickly adapt to the problems, which arise in large data sets, should come into more common use.

Acknowledgments

We would like to thank Paul Eilers and Brian Marx for their comments, as well as the LSU Statistics Department Seminar participants. In addition, Pace and Barry would like to thank the University of Alaska for its generous research support. Pace and Sirmans would like to thank the Center for Real Estate and Urban Studies, University of Connecticut for their support. Pace and Slawson would like to thank Louisiana State University and the Greater Baton Rouge Association of Realtors for their support. All coauthors would like to thank Anton Andrenko at LSU Real Estate Research Institute for technical assistance and computer expertise.

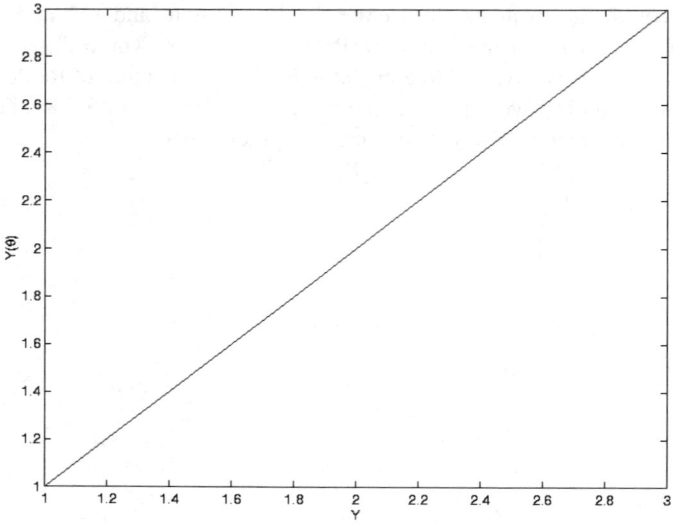

Fig. 9.1a. Linear piecewise linear transformation

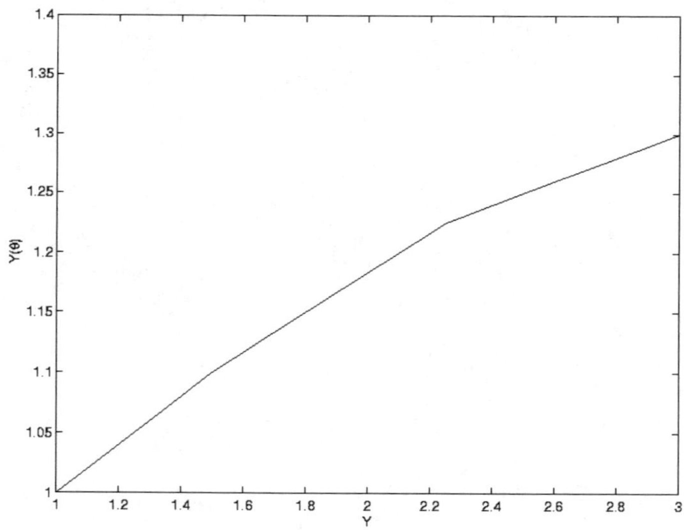

Fig. 9.1b. Slightly concave piecewise linear transformation

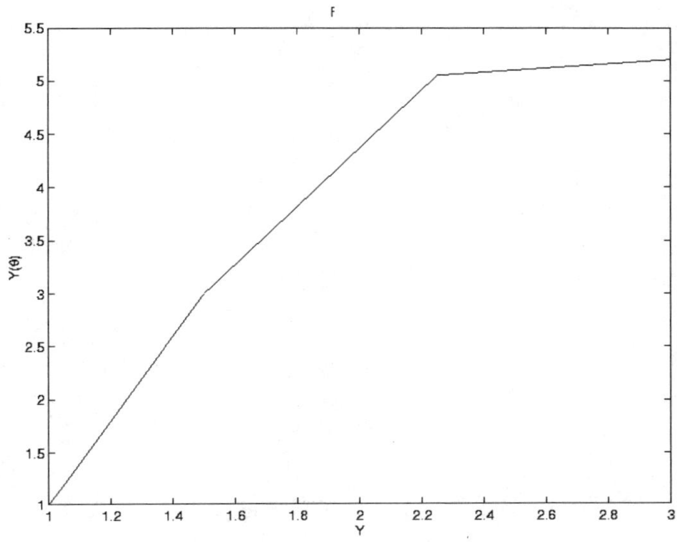

Fig. 9.1c. Severely concave piecewise linear transformation

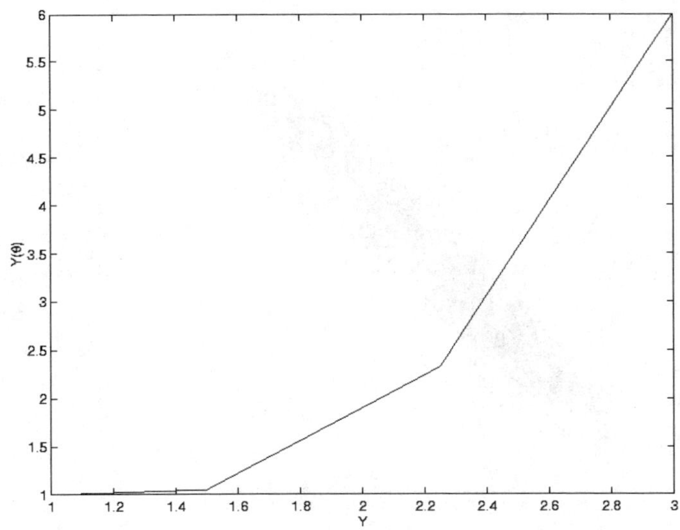

Fig. 9.1d. Convex piecewise linear transformation

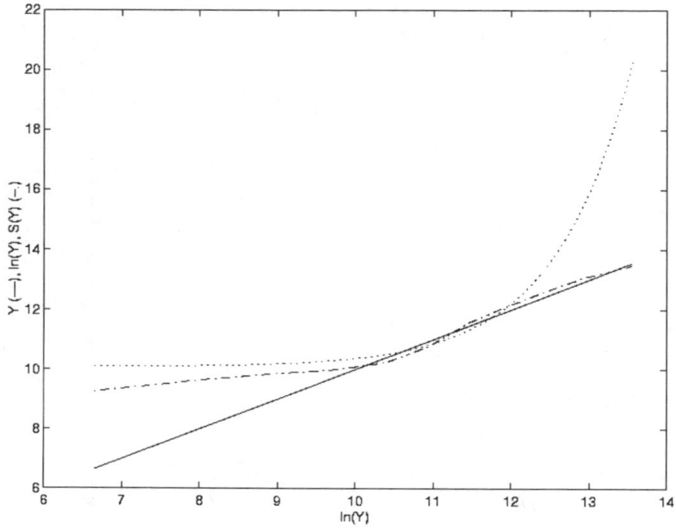

Fig. 9.2. Y, $\ln(Y)$, $S(Y)$

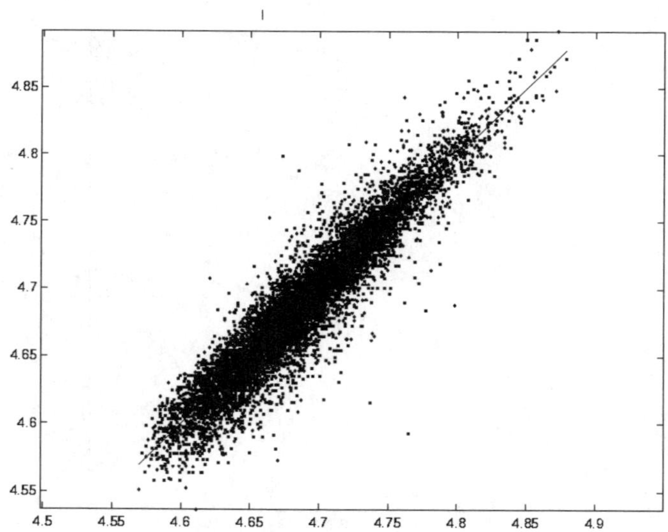

Fig. 9.3a. Predictions v $S(Y)$

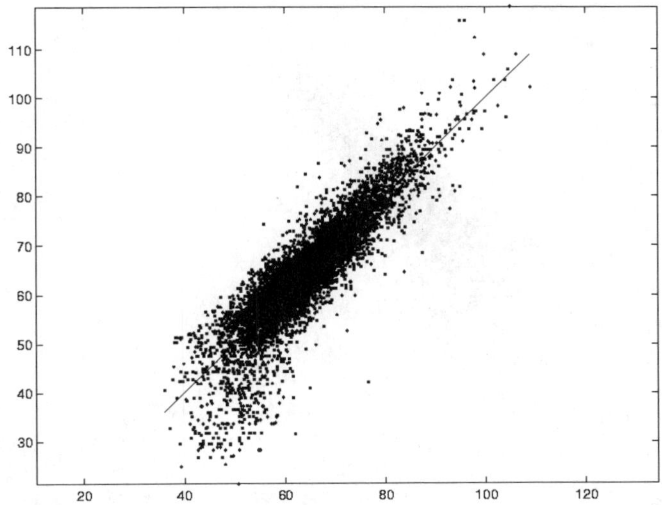

Fig. 9.3b. Predictions v $S(Y^{1/4})$

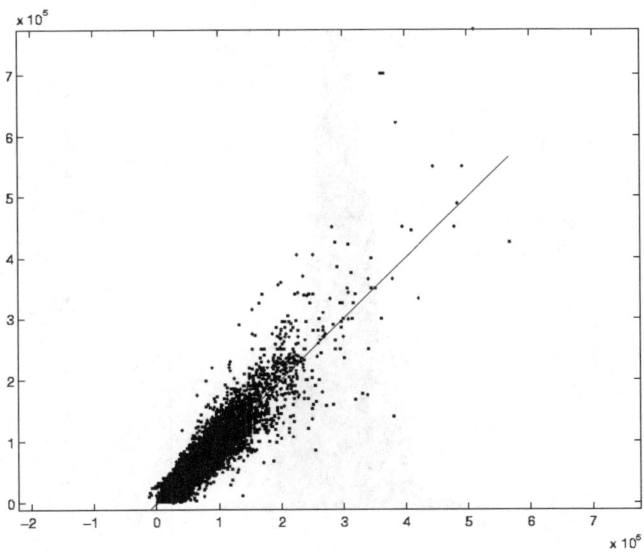

Fig. 9.3c. Predictions v $S(Y)$

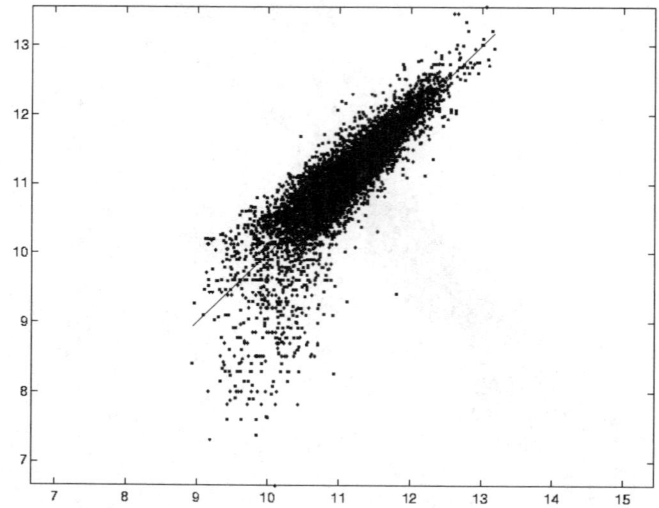

Fig. 9.3d. Predictions $v \ln(Y)$

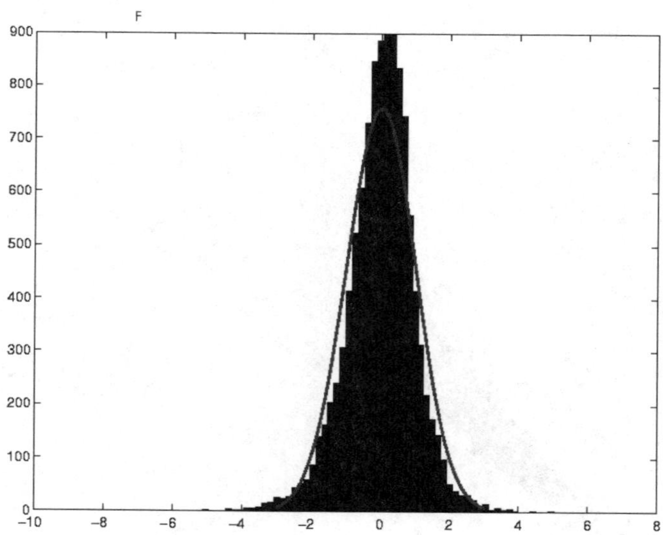

Fig. 9.4a. Histogram of spatial regression errors on transformed Y

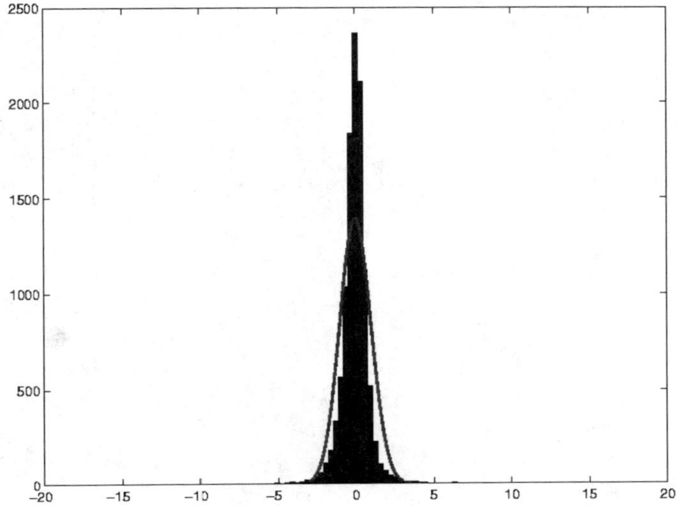

Fig. 9.4b. Histogram of spatial regression errors on untransformed Y

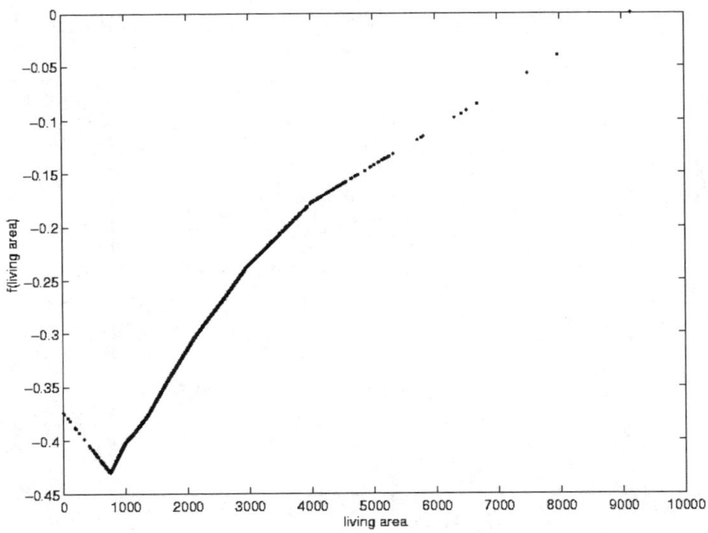

Fig. 9.5a. Living area transformation

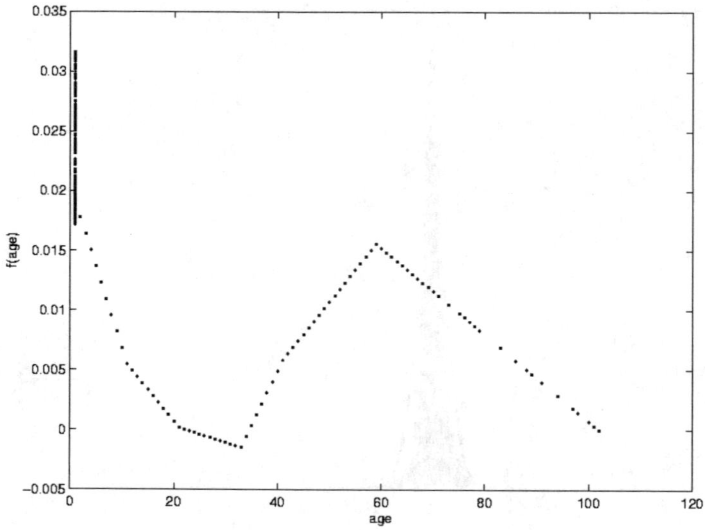

Fig. 9.5b. Age transformation

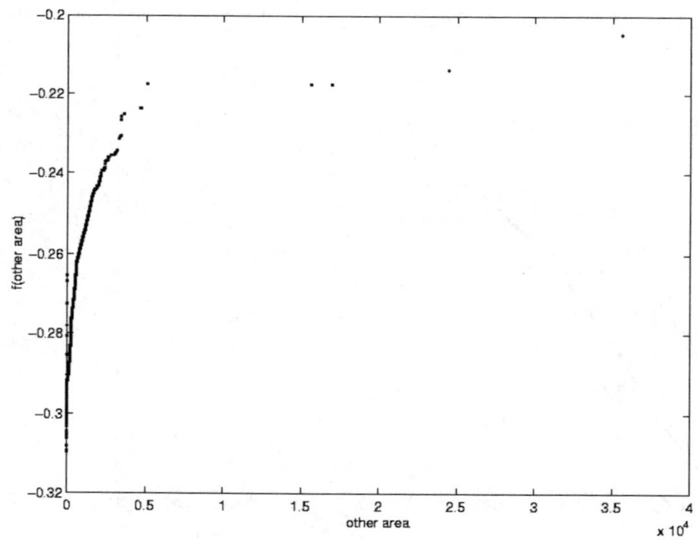

Fig. 9.5c. Other area transformation

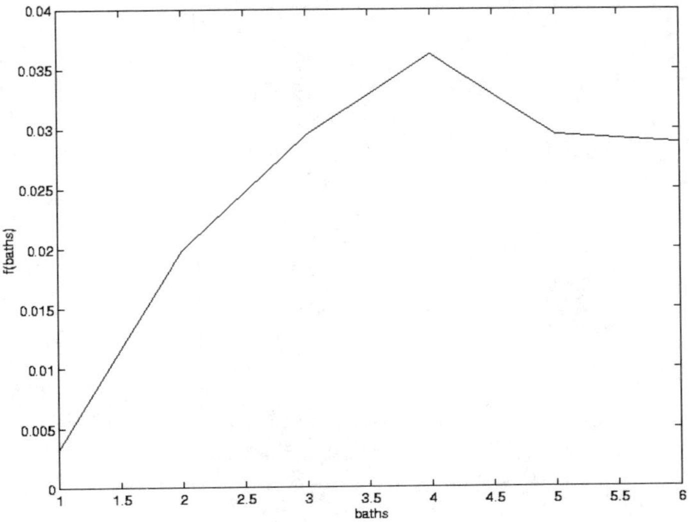

Fig. 9.5d. Baths transformation

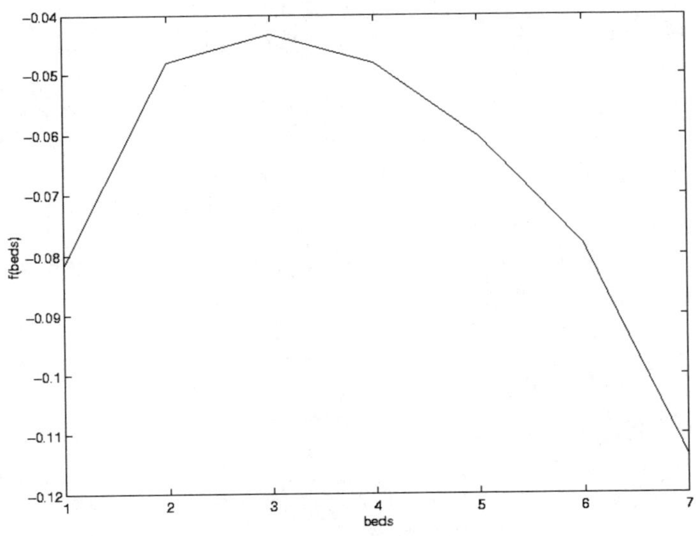

Fig. 9.5e. Beds transformation

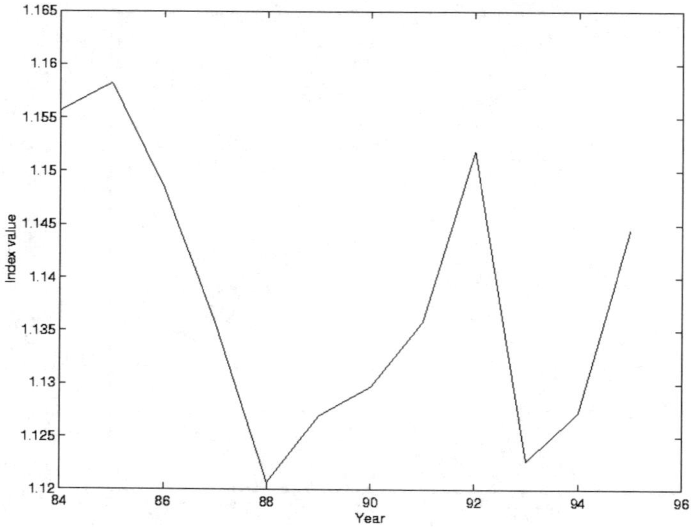

Fig. 9.5f. Time index

10 Locally Weighted Maximum Likelihood Estimation: Monte Carlo Evidence and an Application

Daniel P. McMillen and John F. McDonald

University of Illinois at Chicago

10.1 Introduction

Even small cities have complicated spatial patterns that are difficult to model adequately with a small number of explanatory variables. Shopping centers, parks, lakes, and the like have local effects on variables such as housing prices, land values, and population density. Proximity to such sites can be included as explanatory variables, but the number of potential sites is large and some may be unknown beforehand. Coefficient estimates are biased when relevant sites are omitted, but are inefficient when unimportant ones are included. Moreover, functional forms are often complex for urban spatial patterns even in the absence of local peaks and valleys.

Spatial econometric methods help to account for the effects of missing variables that are correlated over space. The starting point is usually a "spatial contiguity matrix", which specifies the relationship between neighboring observations. For example, we might have $\mu_i = \sum_{i \neq j} \omega_{ij}\mu_j$, where μ_i is an error term and ω_{ij} is the weight given to observation j's error term. Although this approach can be very useful, it has some disadvantages for urban modeling. It imposes restrictive structure that can bias the results when inappropriate. It can be difficult to implement for large data sets because existing estimation procedures typically require large matrices to be inverted. The approach accounts better for broad trends in spatial patterns than for local rises and falls. Finally, the standard approach starts with a simple functional form that may prove inadequate for complex spatial patterns even after controlling for spatial autocorrelation.

Nonparametric methods are a useful alternative for spatial modeling. The basic idea behind nonparametric modeling is to give nearby observations more weight when constructing an estimate for a target point. Whereas the measure of distance is often a general function of all of the explanatory models in many nonparametric models, distance has a natural geographic interpretation in spatial modeling. The central idea is that simple econometric models represent the data best in small geographic areas. When we estimate separate functions for several cities, we are recognizing that their structure is sufficiently different that the data should not be pooled. Enough variation exists within large cities that researchers often estimate separate functions for several areas. Nonparametric procedures simply formalize these heuristic approaches. They are amenable to large data sets, impose little structure, and can account for both broad nonlinear spatial trends and localized peaks and valleys.

Locally Weighted (LW) regression, which was proposed by Cleveland and Devlin (1988), has proved to be the most successful nonparametric procedure for spatial modeling. Applications include Brunsdon *et al.* (1996), Fotheringham *et al.* (1998), McMillen (1996), McMillen and McDonald (1997), and Meese and Wallace (1991). The estimation procedure simply involves repeated applications of Weighted Least Squares. LW regression produces separate coefficient estimates for each observation, but the procedure imposes enough smoothness to preserve degrees of freedom and to ensure that estimates are similar for nearby observations. Fotheringham *et al.* (1998) argue that LW regression is a natural evolution of the expansion method, which has enjoyed widespread use in geography (Casetti, 1972; Griffith, 1981; Jones and Casetti, 1992).

Spatial econometric methods have proved more difficult to develop for models with discrete dependent variables. Log-likelihood functions typically have multiple integrals, and the heteroskedasticity that is typical in spatial models produces inconsistent estimates when ignored in estimation. Existing estimation procedures either rely on restrictive specifications of the error structure (Case, 1992) or can be difficult to implement in practice (LeSage, 1997b, 2000; McMillen, 1992, 1995b)

Locally Weighted regression is readily adaptable to discrete dependent variable models (Tibshirani and Hastie, 1987; McMillen and McDonald, 1999). As in the continuous variable case, separate estimates are constructed for each observation, with more weight given to nearby sites. The weights are applied directly to the log-likelihood function. The estimates account for nonlinearity in the basic functional form as well as for local rises and falls in the function. The estimation procedure is easy to implement with existing software packages, and is suitable for large data sets.

McMillen and McDonald (1999) illustrate the feasibility of the LW approach for a multinomial logit model. In this chapter, we extend our earlier approach in two ways. First, we demonstrate by Monte Carlo procedures that the nonparametric approach provides an accurate alternative to Probit Estimation even when the assumptions behind the standard probit model are met. Importantly, Locally Weighted Probit continues to provide accurate estimates when the underlying functional form is misspecified. Second, we demonstrate the feasibility of the LW approach for the more complicated case of ordinal probit. We use the approach to analyze density zoning in 1920s Chicago. In 1923, all blocks in Chicago were zoned for one of five density categories. Standard ordinal Probit Estimates fit the data well and show that the same factors that influence land use zoning affect density zoning. LW ordinal probit provides a useful check on the estimates: most of the results are the same, but the apparently significant effects of two variables do not survive the scrutiny of the nonparametric estimator.

10.2 The Locally Weighted Log-Likelihood Function

The LW approach begins with the parametric function, $y_i = \beta' x_i + \mu_i$, for $i = 1, ..., n$. A simple linear function may fit well for observations near site i, but may be inappropriate when more distant observations are included. A simple weighting function makes this notion of proximity explicit. Let δ_{ij} be the Euclidean distance between observations i and j. The weight given to observation j in constructing the estimate for observation i is given by ω_{ij}. The tri-cube is a commonly used weighting function:

$$\omega_{ij} = \left[1 - \left(\frac{\delta_{ij}}{d_i}\right)^3\right]^3 I(\delta_{ij} < d_i), \tag{10.1}$$

where d_i is the distance of the qth nearest observation to i, and $I(\cdot)$ is an indicator function that equals one when the condition is true. The window size, q, determines which observations receive weight in constructing the estimate for observation i. The tri-cube was used in Cleveland and Devlin (1988), and has been used for locally weighted regression estimates by McMillen (1996), and McMillen and McDonald (1997).

Another common weighting scheme is the Gaussian function:

$$\omega_{ij} = \Phi\left(\frac{\rho_{ij}}{s_i b}\right), \tag{10.2}$$

where $\Phi(\cdot)$ is the standard normal density function, s_i is the standard deviation of the distances between observation i and all other observations, and b is the bandwidth.[1] The Gaussian weighting kernel has been used extensively in applications (examples include: Ahn and Powell, 1993; Horowitz and Härdle, 1996; McMillan *et al.*, 1989; Powell *et al.*, 1989; Thorsnes and McMillen, 1998; Ullah and Singh, 1989). The choice of weighting function is less important than the bandwidth or window size. For example, Thorsnes and McMillen (1998) present graphs of a function estimated with five different kernel weighting functions, and all five are virtually identical. All commonly-used functions are similar in that they place high weight on nearby observations and low weight on distant observations.

The bandwidth is similar to the window size in determining how rapidly the weights decrease with distance. Larger values of q or b put more weight on distant observations in forming the estimate for observation i. Either the bandwidth or window size can be chosen by the method of cross validation, which minimizes the overall residual sum of squares obtained when observation i is deleted in forming its own forecasted value (see McMillen and McDonald, 1997, for details). Highly nonlinear functions can be approximated adequately using small values of q or b even though the base function is linear, but small values produce a high variance. Cross validation formalizes the implicit tradeoff between bias and variance.

Nonparametric estimators provide estimates of both the dependent variable and the marginal effects of the explanatory variables. Under either weighting scheme in

[1] The search for the optimal bandwidth is simplified by removing the dependence of b on the scale of the distances. Note that the mean of δ does not affect the calculation because it cancels out when finding the distance between sites i and j. The calculation can be simplified by standardizing the distances.

equations (10.1) or (10.2), the LW estimate for observation i is obtained simply by Weighted Least Squares:

$$\hat{\beta}_i = \left(\sum_{j=1}^{n} \omega_{ij} x_j x_j' \right)^{-1} \left(\sum_{j=1}^{n} \omega_{ij} x_j y_j \right), \tag{10.3}$$

and $\hat{y}_i = \hat{\beta}_i' x_i$. The estimation procedure produces separate coef˜ cients for each observation, which are the marginal effect estimates. Analogs to standard F-tests are available to test whether variables have a signiˇ cant in uence on the dependent variable (McMillen, 1996; McMillen and McDonald, 1997).

LW regression captures the essential idea behind spatial econometrics – that nearby observations are more closely correlated than those farther away – without imposing an arbitrary, parametric weighting scheme. Small bandwidths and window sizes permit the base linear function to approximate overall nonlinear functions and also can account for local rises and falls in the regression surface. Limiting the estimation to a neighborhood of observation i while allowing for nonlinearity eliminates much of the heteroskedasticity and autocorrelation that is endemic to spatial data sets. [2] Bootstrap procedures that account for heteroskedasticity and autocorrelation can account for remaining violations of these classical assumptions.

The LW procedure is readily extended to more complicated nonlinear models that are estimated by Maximum-Likelihood methods. [3] In a typical Maximum Likelihood procedure, the log-likelihood function is $\sum_{i=1}^{n} \ln L_i$, which is maximized with respect to a parameter vector θ. The LW counterpart is to maximize separate pseudo log-likelihood functions for each observation in the data set, with more weight being given to nearby observations. For example, the base log-likelihood function for the standard regression model is:

$$\sum_{i=1}^{n} \left[\log \phi \left(\frac{y_i - \beta' X_i}{\sigma} \right) - \log \sigma \right].$$

The LW version of the model is obtained by maximizing the following pseudo log-likelihood function separately for each observation to obtain n different estimates of β_i and σ_i:

$$\sum_{j=1}^{n} \omega_{ij} \left[\log \phi \left(\frac{y_j - \beta_i' X_j}{\sigma_i} \right) - \log \sigma_i \right]. \tag{10.4}$$

[2] Of course, heteroskedasticity and autocorrelation may be intrinsic to the model, in which case they will still be present under nonparametric estimation. However, these problems are often caused by omitted explanatory variables that are correlated across space or by misspeciˇ ed functional forms. Errors will then be closer to being independent and homoskedastic in a small geographic than in the full sample.

[3] It is important to note that the LW Maximum Likelihood estimator does *not* produce Maximum Likelihood estimates, and it has no claim of ef˜ ciency. The pseudo log-likelihood function is a convenient basis for obtaining estimates in complicated settings where standard Maximum Likelihood is inappropriate. The point is to reduce bias, not to obtain ef˜ ciency.

Maximizing equation (10.4) with respect to β_i produces the LW estimator given by equation (10.3).

The Locally Weighted Maximum Likelihood (LWML) approach is adaptable to any standard Maximum Likelihood model. In general, the LW pseudo log-likelihood function is $\sum_{j=1}^{n} \omega_{ij} \ln L_{ij}$. The examples analyzed in the subsequent two sections of this chapter include probit and a three-choice ordinal probit model. For LW probit, the pseudo log-likelihood function for observation i is:

$$\sum_{j=1}^{n} \omega_{ij} \left[I_j \log \Phi(\beta_i' X_j) + (1 - I_j) \log \Phi(-\beta_i' X_j) \right], \tag{10.5}$$

where I_j is the discrete dependent variable and Φ is the standard normal cumulative density function. The LW ordinal probit pseudo log-likelihood function is:

$$\sum_{j=1}^{n} \omega_{ij} \Big[I_{0j} \log \Phi(-\beta_i' X_j) + I_{1j} \log \left(\Phi(\mu_i - \beta_i' X_j) \right) \\ + I_{2j} \log \Phi(-\mu_i + \beta_i' X_j) \Big], \tag{10.6}$$

where I_{0j}, I_{1j}, and I_{2j} are indicator variables for the three regimes, and μ_i is the threshold value for observation i. The same weighting schemes that are used for the regression case can be used for LWML. Cross validation can be used to choose the bandwidth or window size by estimating $\sum_{j=1}^{n} \ln L_{ij}$ separately for each observation i with that observation omitted, and choosing the value of b or q that maximizes $\sum_{i=1}^{n} \sum_{j=1}^{n} \ln L_{ij}$.

As in the continuous dependent variable case, LWML allows the data to determine the degree of nonlinearity. The estimation procedures are easy to implement with standard software packages, even for large data sets. Problems of heteroskedasticity and autocorrelation are potentially reduced by allowing for ample nonlinearity and by putting most weight on a neighborhood of observations where the base log-likelihood function is close to being correct. Bootstrap procedures can be used to construct hypothesis tests. The appendix presents a description of the computational steps needed to implement an LWML model, including bootstrap hypothesis tests.

10.3 Monte Carlo Experiments

This section contains the results of Monte Carlo experiments that demonstrate some of the benefits of LWML estimation for the probit model. We first generate an artificial data set that is based on a stylized urban model. We make two independent draws of n observations from uniform distributions with lower bounds of -10 and upper bounds of 10. These two variables, EAST and NORTH, are designed to measure distances from a city center. They are used to generate our first primary variable, x_1, which is straight-line distance from the center. The second variable, x_2, is drawn independently from a uniform distribution with a lower limit of 0 and an upper limit of 10.

The following model is used in estimation:

$$y_i^* = \beta_0 + \beta_1 x_{1i} + \beta_2 x_{2i} + \mu_i \tag{10.7}$$

where y_i^* is a latent variable that generates the observed dependent variable, $y_i = I(y_i^* > 0)$. The error term is drawn from a normal distribution with constant variance σ^2 and no autocorrelation, which implies that standard probit is consistent and efficient when equation (10.7) is the correct model specification. But we assume that the effect of distance from the city center is different on the north and south sides of the city. The true model is:

$$y_i^* = \beta_0 + \beta_1 x_{1i} \times N_i + \beta_2 x_{2i} \times S_i + \mu_i \tag{10.8}$$

where N and S are dummy variables indicating North > 0 and South ≤ 0. Having differential effects of x_{1i} on the north and south sides of the city introduces a very simple but realistic type of functional form misspecification that allows us to investigate the potential benefits and costs of LW probit estimation. Standard probit is consistent and efficient when $\beta_1 = \beta_3$; LW probit is consistent but has higher variance than standard probit in this case. The set of experiments with $\beta_1 = \beta_3$ allows us to determine the loss in efficiency from using LW probit when it is unnecessary. Standard probit applied to equation (10.7) is inconsistent when $\beta_1 \neq \beta_3$. LW probit can potentially reduce the bias by adapting locally to the change in functional form even when the model is misspecified.

The base coefficients for equation (10.8) are $\beta_0 = 5$, $\beta_1 = -0.5$, and $\beta_2 = 0.5$. We allow β_3 to vary from $-.5$ to -2.0 in increments of $-.5$. To ensure a similar base fit across experiments, we choose σ^2 to produce an average R^2 of 0.6:

$$\sigma^2 = \frac{2}{3} \text{Var}\,(\beta_0 + \beta_1 x_1 \times N + \beta_2 x_2 + \beta_3 x_1 \times S).$$

The variance on the right hand side of this expression increases as the absolute value of β_3 rises, which implies that σ^2 rises also. To ensure that $y_i = 1$ for about 50 percent of the observations, we subtract the mean value of the right hand side of equation (10.8) to obtain the final value of β_0 used in the experiments. Finally, note that Probit Estimates β/σ rather than β. To aid in keeping all of these transformations straight, we list the true value for each estimated coefficient in the tables. We replicate all experiments 500 times.

Standard probit is used to obtain the results reported in Table 10.1. We report the true coefficients, the average estimated coefficients, the standard deviation of the estimates, and the root mean squared error (RMSE) across the 500 replications. A constant, x_1, and x_2 are included as explanatory variables, but we do not distinguish between the north and south sides of the city in estimation. In contrast, the true model has different coefficients for x_1 on the north and south sides of the city except when $\beta_1 = \beta_3$. We report the RMSE for the estimated x_1 coefficient based on the true value on the south side of the city, β_3. As expected, standard Probit Estimates are very accurate when the true and estimated model are equivalent, which occurs when $\beta_3 = -0.5$. The RMSE rises substantially as the deviation between β_1 and β_3 rises.

Table 10.1. Standard Probit Monte Carlo Results

Sample Size, β_3(south side β_1),	β_0/σ true coef., estimate std. dev., RMSE	β_1/σ – south side true coef., estimate std. dev., RMSE	β_2/σ true coef., estimate std. dev., RMSE
$n = 250$,	0.677, 0.676	−0.282, −0.286	0.282, 0.288
$\beta_3 = -0.5$	0.301, 0.301	0.039, 0.039	0.040, 0.040
$n = 250$,	1.015, 0.983	−0.354, −0.221	0.177, 0.151
$\beta_3 = -1.0$	0.275, 0.277	0.031, 0.136	0.032, 0.041
$n = 250$,	1.038, 0.928	−0.343, −0.164	0.114, 0.085
$\beta_3 = -1.5$	0.243, 0.267	0.025, 0.180	0.026, 0.039
$n = 250$,	1.027, 0.886	−0.330, −0.140	0.083, 0.058
$\beta_3 = -2.0$	0.223, 0.264	0.023, 0.192	0.023, 0.034
$n = 750$,	0.646, 0.660	−0.290, −0.292	0.290, 0.292
$\beta_3 = -0.5$	0.162, 0.163	0.023, 0.023	0.022, 0.022
$n = 750$,	1.075, 0.984	−0.362, −0.224	0.181, 0.144
$\beta_3 = -1.0$	0.146, 0.172	0.017, 0.139	0.017, 0.041
$n = 750$,	1.122, 0.979	−0.348, −0.171	0.116, 0.076
$\beta_3 = -1.5$	0.130, 0.193	0.013, 0.178	0.014, 0.042
$n = 750$,	1.120, 0.933	−0.335, −0.147	0.084, 0.051
$\beta_3 = -2.0$	0.126, 0.225	0.013, 0.188	0.013, 0.036

The increased RMSE is entirely due to an increase in bias. The results for the LW probit model are reported in Tables 10.2 and 10.3. The results are harder to report because LW probit produces a different set of coefficients for each observation. We report average values of the coefficients across the south side observations, along with the standard deviations and RMSE of the average values. We use a Gaussian weighting function for all experiments, and vary the bandwidth from 0.4 to 1.0 in increments of 0.2. To avoid overwhelming the reader, we only report the results for $\beta_3 = -0.5$ and $\beta_3 = -1.5$. The average estimated coefficients under LW probit are about as accurate as standard probit when the true and estimated models are equivalent, i.e., when $\beta_3 = -0.5$. The standard deviation falls as the bandwidth increases, while the coefficient estimates do not change greatly. The RMSE's for all coefficients are nearly the same under LW and standard probit when $n = 750$ and $\beta_3 = -0.5$. There is little loss in efficiency from using LW probit relative to standard probit when focusing on average coefficient estimates.

LW probit is much more accurate than standard probit in identifying the true coefficient for x_1 when the estimated model is misspecified. For example, the RMSE is 0.041 for LW probit when $n = 750$, $\beta_3 = -1.5$, and $h = 0.4$, compared to 0.178 for standard probit. Smaller values of the bandwidth lead to lower RMSE when the estimated model is misspecified.

The Monte Carlo results illustrate the value of nonparametric procedures in a realistic setting. Our fictional researcher has imposed a nearly correct but still in-

Table 10.2. Locally Weighted Probit Monte Carlo Results: $n = 250$

Sample Size, β_3(south side β_1), bandwidth	β_0/σ – south side true coef., estimate std. dev., RMSE	β_1/σ – south side true coef., estimate std. dev., RMSE	β_2/σ – south side true coef., estimate std. dev., RMSE
$n = 250$,	0.677, 0.700	−0.282, −0.300	0.282, 0.304
$\beta_3 = -0.5$, $h = 0.4$	0.359, 0.360	0.055, 0.058	0.063, 0.067
$n = 250$,	0.677, 0.739	−0.282, −0.303	0.282, 0.300
$\beta_3 = -0.5$, $h = 0.6$	0.391, 0.396	0.053, 0.057	0.059, 0.061
$n = 250$,	0.677, 0.714	−0.282, −0.292	0.282, 0.290
$\beta_3 = -0.5$, $h = 0.8$	0.346, 0.348	0.044, 0.045	0.049, 0.050
$n = 250$,	0.677, 0.680	−0.282, −0.289	0.282, 0.292
$\beta_3 = -0.5$, $h = 1.0$	0.317, 0.317	0.044, 0.045	0.047, 0.048
$n = 250$,	1.038, 1.345	−0.343, −0.320	0.114, 0.083
$\beta_3 = -1.5$, $h = 0.4$	0.372, 0.482	0.049, 0.054	0.043, 0.053
$n = 250$,	1.038, 1.318	−0.343, −0.288	0.114, 0.076
$\beta_3 = -1.5$, $h = 0.6$	0.348, 0.446	0.040, 0.068	0.040, 0.056
$n = 250$,	1.038, 1.267	−0.343, −0.260	0.114, 0.072
$\beta_3 = -1.5$, $h = 0.8$	0.318, 0.391	0.035, 0.090	0.034, 0.055
$n = 250$,	1.038, 1.224	−0.343, −0.244	0.114, 0.076
$\beta_3 = -1.5$, $h = 1.0$	0.291, 0.345	0.032, 0.103	0.031, 0.049

accurate model on an almost symmetric city. As a consequence, standard Probit Estimates are inconsistent. By putting more weight on nearby observations in estimation, LW probit produces estimates with lower bias. On average, LW probit estimates of the coef˜ cient averages do not have substantially higher variance in large samples even when the assumptions behind standard probit are met. The Monte Carlo results suggest that there is little cost and much potential bene˘ t from using a nonparametric estimator as an alternative to standard probit.

10.4 Density Zoning in 1920s Chicago

Chicago adopted its ˘ rst zoning ordinance in 1923. As of April 23 of that year, every block in the city was zoned for one of four land use categories and one of ˘ ve density categories. We have analyzed land use zoning patterns in previous papers (McMillen and McDonald, 1999), but we have not yet analyzed density zoning. In this section, we present standard and LW ordinal probit models of the determinants of density zoning in the 1923 ordinance.

An ordinal model is appropriate for density zoning because density is clearly ordered from restrictive to unrestrictive. As described in the ordinance, city blocks designated for the "1st Volume District" must be developed at low density: "no building ... shall occupy more than 50 per cent of the area of a lot if an interior lot or 65 per cent if a corner lot ... ." In 2nd volume districts, the percentages are replaced

Table 10.3. Locally Weighted Probit Monte Carlo Results: $n = 750$

Sample Size, β_3(south side β_1), bandwidth	β_0/σ – south side true coef., estimate std. dev., RMSE	β_1/σ – south side true coef., estimate std. dev., RMSE	β_2/σ – south side true coef., estimate std. dev., RMSE
$n = 750$,	0.646, 0.666	−0.290, −0.295	0.290, 0.295
$\beta_3 = -0.5$, $h = 0.4$	0.189, 0.190	0.028, 0.028	0.028, 0.029
$n = 750$,	0.646, 0.659	−0.290, −0.295	0.290, 0.295
$\beta_3 = -0.5$, $h = 0.6$	0.195, 0.195	0.026, 0.026	0.027, 0.028
$n = 750$,	0.646, 0.661	−0.290, −0.293	0.290, 0.293
$\beta_3 = -0.5$, $h = 0.8$	0.173, 0.173	0.026, 0.026	0.025, 0.025
$n = 750$,	0.646, 0.656	−0.290, −0.293	0.290, 0.293
$\beta_3 = -0.5$, $h = 1.0$	0.182, 0.182	0.024, 0.024	0.025, 0.025
$n = 750$,	1.122, 1.301	−0.348, −0.317	0.116, 0.092
$\beta_3 = -1.5$, $h = 0.4$	0.196, 0.265	0.026, 0.041	0.023, 0.034
$n = 750$,	1.122, 1.300	−0.348, −0.296	0.116, 0.085
$\beta_3 = -1.5$, $h = 0.6$	0.189, 0.260	0.023, 0.058	0.021, 0.037
$n = 750$,	1.122, 1.287	−0.348, −0.278	0.116, 0.083
$\beta_3 = -1.5$, $h = 0.8$	0.171, 0.238	0.020, 0.074	0.018, 0.038
$n = 750$,	1.122, 1.256	−0.348, −0.260	0.116, 0.081
$\beta_3 = -1.5$, $h = 1.0$	0.167, 0.214	0.019, 0.091	0.018, 0.040

by 60 percent for an interior and 75 percent for a corner lot. They rise to 75 percent and 90 percent for 3rd volume districts; 4th and 5th volume districts have still higher densities, but such a small percentage of our sample falls in these categories (2.1 percent and 3.7 percent) that we combine them with the 3rd volume district, creating a single "high density district." In our sample of 1116 blocks, 239 are zoned for low density (1st volume districts), 593 for medium density (2nd volume districts), and 284 for high density (3rd, 4th, or 5th volume districts). Our dependent variable has a value of 0, 1, or 2 as the block is zoned for low, medium, or high density.

Explanatory variables include standard measures of access, which we have included in previous studies. They include distance from the city center, Lake Michigan, the nearest elevated train ("el") station, the nearest commuter train station, and the nearest navigable waterway. All distances are measured in straight-line miles. We define two dummy variables to represent highly localized effects. The first dummy variable equals one when a block is on a major street, and the second equals one when a block is near (within 1/8 of a mile, or 1 city block) a rail line. Finally, we define two dummy variables that control for the existing land use mix on the block. The first equals one when the block included commercial firms prior to the ordinance, and the second equals one when the block had residences.

Table 10.4. Ordered Probit Models for Density Zoning

Variable	Standard Ordered Probit[1]	Locally Weighted Ordered Probit[2]
Constant	4.821	5.679
	(0.288)	(0.355)
		[5.104, 6.525]
Distance to City Center	−0.573	−0.483
	(0.036)	(0.134)
		[-0.800, -0.328]
Distance to Lake Michigan	−0.372	−0.435
	(0.028)	(0.130)
		[-0.604, -0.023]
Distance to El Station	0.193	−0.096
	(0.063)	(0.311)
		[-0.621, 0.534]
Distance to Commuter Train Station	0.367	0.356
	(0.093)	(0.117)
		[0.086, 0.557]
Distance to River or Canal	0.274	−0.005
	(0.038)	(0.157)
		[-0.356, 0.333]
Near Rail Line	1.371	1.434
	(0.120)	(0.054)
		[1.335, 1.599]
Located on Major Street	0.799	0.874
	(0.107)	(0.113)
		[0.605, 1.052]
Block has Commercial Firms	−0.060	−0.176
	(0.092)	(0.098)
		[-0.475, -0.009]
Block has Residences	−0.273	−0.369
	(0.100)	(0.187)
		[-0.652, -0.058]
μ	2.968	3.538
	(0.125)	(0.182)
		[3.350, 4.105]
Log-likelihood	−620.635	−512.659

[1] (standard error)

[2] $h = 0.70$, (standard deviation), [minimum, maximum]

Although there is no previous historical evidence on the determinants of density zoning, standard bid-rent theory provides a useful framework for the analysis.

Our previous studies suggest that land use zoning closely followed the market in 1923. For instance, a block that had a relatively high land value in residential use was unlikely to be zoned manufacturing or commercial. Density zoning should follow a similar pattern. When land rents are high, builders will substitute capital for land, producing densely developed areas. If the zoning ordinance follows the market, high-rent areas will tend to be zoned for high densities. However, we also expect that non-residential areas will tend to be zoned for higher densities than residential areas even when land rents are the same in the two areas. The zoning ordinance was apparently motivated in large part by a desire to protect low-density residential areas from high-density non-residential development, which suggests that areas well suited to residences will tend to receive low-density zoning.

Following bid-rent theory, we expect blocks close to the city center, near Lake Michigan, near el stations, and along major streets to be zoned for high densities. We do not have an expectation for the effect of distance to commuter train stations because our previous studies suggest that they do not have reliably predictable effects on rents. Areas near commuter trains stations are often commercial, which tends to lead to high-density zoning. But planners may attempt to encourage residential development near the stations, which leads to low-density zoning. Sites close to navigable waterways, near rail lines, and along major streets are nearly always used for manufacturing or commercial enterprises, which leads to high-density zoning. However, our previous research suggests that proximity to waterways and rail lines lowers land values, which has the opposite effect on density zoning. A block with commercial firms should be more likely to be zoned for high densities, whereas the presence of residences should lead to low-density zoning.

Standard ordinal probit estimates are presented in the first column of results in Table 10.4. The results confirm most of our expectations. A block is estimated to have a higher probability of high-density zoning when it is closer to the city center or Lake Michigan, farther away from a navigable waterway, near a rail line, or along a major street. It is less likely to be zoned for high densities when it contains residential lots, but the presence of commercial land does not have a significant effect on density zoning patterns. Blocks closer to commuter trains stations are less likely to be zoned for high densities, which suggests that planners may have been attempting to encourage these areas to be residential. The positive coefficient on distance to the nearest el station is the only surprising result among those that are statistically significant. As with commuter train stations, it is possible that planners were attempting to encourage areas near el stations to be residential by zoning them for low densities.

LW ordinal probit results are presented in the last column of Table 10.4. We use a Gaussian weighting function. The bandwidth was chosen through cross validation. We report the average estimated coefficients across all 1116 estimates, along with the standard deviations and ranges. Although we do not formally test the significance of the coefficient means, the descriptive statistics reported in Table 10.4 provide measures of the robustness of the results. We have more confidence in estimates that have lower standard deviations and ranges that do not bracket zero.

Table 10.5. Predictions: Standard Probit Model

Actual Zoning	Predicted Zoning		
	0	1	2
0	130	109	0
1	64	471	58
2	1	71	212

By these measures, only two results undergo a substantive change. The effect of distance to the nearest el station is no longer estimated to be positive, a felicitous result because we had found the positive coefficient to be surprising. The positive effect of distance to a river or canal disappears, but we had no prior expectation for this coefficient. Overall, the LW results support the standard ordinal probit model, suggesting that the simpler model is not an overly restrictive specification.

Tables 10.5 and 10.6 present further evidence that the models fit the data well. Ordinal probit models often are unable to accurately predict middle categories, but all density zoning categories are identified accurately by both the standard and LW ordinal probit models. LW ordinal probit predicts better than the standard model, but the gains are not dramatic. The primary value of the nonparametric estimator in this application is its role as a diagnostic check. All important results survive the scrutiny of the nonparametric estimator.

10.5 Conclusions

Nonparametric models are useful alternatives for spatial econometric modeling. They directly incorporate the notion that nearby observations are more closely correlated than more distant sites. They can detect both local peaks and valleys and overall functional form nonlinearity. Although they are computer intensive, nonparametric estimators do not require large matrices to be inverted, and they do not require the specification of an arbitrary parametric structure. An important benefit of nonparametric estimation for discrete dependent variable models is that putting less weight on distant observations reduces the heteroskedasticity and autocorrelation problems that cause standard estimators to be inconsistent and inefficient.

Our Monte Carlo results demonstrate the value of nonparametric probit estimation in a stylized urban model. Standard probit does not have a large efficiency gain relative to LW probit when the restrictive assumptions of the standard model are met. The nonparametric procedure is much more accurate than standard probit when the standard model incorrectly assumes an absence of spatial variation in the coefficients. Our empirical application of LW ordinal probit to density zoning in 1920s Chicago illustrates the feasibility of nonparametric estimation for relatively complex Maximum Likelihood estimation. By demonstrating which results survive the application of a more flexible estimator, nonparametric estimation serves an important role as a diagnostic tool.

Table 10.6. Predictions: Locally Weighted Probit Model

Actual Zoning	Predicted Zoning		
	0	1	2
0	164	75	0
1	46	499	48
2	0	56	228

Appendix: Computational Steps for an LWML Model

In this appendix, we present the computational steps for an LWML model using a Gaussian weighting function. The models can be estimated easily with any computer software program that has do-loops and maximization routines. The models presented in this chapter were estimated using RATS.

A1.1 Algorithm for Maximizing the Pseudo Log-Likelihood Function

The objective is to maximize $L_i^* = \sum_{j=1}^n \omega_{ij} \ln L_{ij}(\theta_i)$ with respect to the k by 1 vector θ_i for each observation i. The steps are:

1. Initialize k variables to store the estimated values of θ: $K_1 = 0$, $K_2 = 0$, etc. Initialize a variable to store the estimated pseudo log-likelihood values: LOGL = 0. Each variable has n entries. Set the initial bandwidth, b.
2. Obtain initial estimates, θ_0, with the appropriate Maximum Likelihood procedure using all observations.
3. Begin a do-loop based on observations $i = 1, \ldots, n$.
   (a) Calculate δ_{ij}, the distance between observation i and observations $j = 1, \ldots,$ n. Calculate the standard deviation, s_i
   (b) Calculate the weighting function, $\omega_{ij} = \phi(\delta_{ij}/s_i b)$ for observations $j = 1, \ldots, n$.
   (c) Maximize L_i^* with respect to θ_i. The initial estimates are θ_0 for each $i = 1, \ldots, n$. Store the results in the ith entry of K_1, K_2, etc..
   (d) Calculate $\hat{\Lambda}_i = \sum_{j=1}^n \ln L_{ij}(\hat{\theta}_i)$, and store the result in the ith entry of LOGL.
   (e) Continue to $i = n$.
4. Calculate the pseudo log-likelihood value, $\hat{\Lambda} = \sum_{i=1}^n \hat{\Lambda}_i$.

The most dif˜ cult part of this procedure is step 3c. Standard maximization algorithms can be used, including those provided in such programs as RATS, TSP, Gauss, Stata, and Limdep. We did our own programming in RATS, based on a Newton-Raphson maximization procedure, because we found that the maximization procedure included in the program was slow.

A1.2 Cross Validation

We used the method of cross validation to choose the bandwidth. The steps are:

1. Choose a set of B bandwidths, $b = b_1, b_2, ..., b_B$.
2. Use the algorithm in A1.1 to estimate the model for each bandwidth, but set $\omega_{ii} = 0$. Thus, observation i gets zero weight in the estimation of θ_i. Only step 3b of the algorithm is altered.
3. The cross-validated bandwidth is the value of b that produces the highest value for $\hat{\Lambda}$.

The model is sometimes reestimated after the cross-validated bandwidth is determined, this time including all observations in estimation. Reestimation is not required, and may not be desirable because including observation i when estimating θ_i affects the asymptotic properties of the estimators. However, the model's ˇ t is improved when observation i is included.

A1.3 Using the Bootstrap to Calculate Standard Errors

Bootstrap resampling procedures can be used to calculate standard errors for any statistic of interest. Let τ represent the vector of statistics for which standard errors are desired. τ might be the mean value of the estimated θ_i, the estimated θ_i for an individual observation, or some function of the estimated coefˇ cients. Suppose that each observation i has data on a dependent variable, y_i, and a vector of explanatory variables, x_i. Draw randomly with replacement from the n values of y_i and x_i to form a new dependent variable, y_i^*, and a new set of explanatory variables, x_i^*, and reestimate the model using the new data set. The new value of the statistic of interest is τ_b, where b is now being used to denote an iteration of the bootstrap resampling procedure. The process is repeated B times, where again B is being used differently than in section A1.2.

At the end of this process, we have B estimates of τ. The bootstrap standard error for $\tau_i$is simply the standard deviation of the B values of $\hat{\tau}_b$:

$$s_\tau = \left(\frac{1}{B-1} \sum_{b=1}^{B} (\hat{\tau}_b - \hat{\tau}^*)^2 \right)^{1/2},$$

where $\hat{\tau}^* = B^{-1} \sum_{b=1}^{B} \hat{\tau}_b$. Bootstrap conˇ dence intervals can be constructed by assuming a standard normal distribution for τ. The 95 percent conˇ dence interval is $\hat{\tau} \pm 1.96 s_\tau$. Alternatively, the $\hat{\tau}_b$ can be ordered, and the bootstrap 95 percent conˇ - dence interval is the $0.025B$ to $0.975B$ entries of the vector of the ordered $\hat{\tau}_b$. Other versions of the bootstrap conˇ dence intervals can also be constructed (see Efron and Tibshirani (1986), for an excellent review), but these two versions are the most common.

Both nonparametric estimation and the bootstrap involve repeated applications of potentially time-consuming estimation procedures. Although the time involved may not be excessive for either one, the combination of the two may make the bootstrap impractical except for small values of B. The accuracy of the bootstrap improves as B increases, but it may be infeasible to apply the bootstrap repeatedly in large data sets. This problem arises when the nonparametric estimator is being applied to all n observations in the data set. The bootstrap is feasible even for large data sets if θ is calculated for only a few target observations, e.g., if τ is the estimated coef˜ cient vector at several representative sites instead of an average over all n observations.

11 A Family of Geographically Weighted Regression Models

James P. LeSage

University of Toledo

11.1 Introduction

A Bayesian approach to locally linear regression methods introduced in McMillen (1996) and labeled geographically weighted regressions (GWR) in Brunsdon *et al.* (1996) is set forth in this chapter. The main contribution of the GWR methodology is use of distance weighted sub-samples of the data to produce locally linear regression estimates for every point in space. Each set of parameter estimates is based on a distance-weighted sub-sample of "neighboring observations," which has a great deal of intuitive appeal in spatial econometrics. While this approach has a deˇ nite appeal, it also presents some problems. The Bayesian method introduced here can resolve some difˇ culties that arise in GWR models when the sample observations contain outliers or non-constant variance.

The distance-based weights used in GWR for data at observation i take the form of a vector W_i which can be determined based on a vector of distances d_i between observation i and all other observations in the sample. Note that the symbol W is used in this text to denote the spatial weights matrix in spatial autoregressive models, but here the symbol W_i is used to represent distance-based weights for observation i, consistent with other literature on GWR models. This distance vector along with a distance decay parameter are used to construct a weighting function that places relatively more weight on sample observations from neighboring observations in the spatial data sample.

A host of alternative approaches have been suggested for constructing the weight function. One approach suggested by Brunsdon *et al.* (1996) is:

$$W_i = \sqrt{\exp(-d_i/\theta)}. \tag{11.1}$$

The parameter θ is a decay or "bandwidth" parameter. Changing the bandwidth results in a different exponential decay proˇ le, which in turn produces estimates that vary more or less rapidly over space. Another weighting scheme is the tri-cube function proposed by McMillen and McDonald in Chapter 10 of this volume:

$$W_i = (1 - (d_i/q_i)^3)^3 \quad \mathrm{I}(d_i < q_i), \tag{11.2}$$

where q_i represents the distance of the qth nearest neighbor to observation i and I() is an indicator function that equals one when the condition is true and zero otherwise. Still another approach is to rely on a Gaussian function ϕ:

$$W_i = \phi(d_i/\sigma\theta), \tag{11.3}$$

where ϕ denotes the standard normal density and σ represents the standard deviation of the distance vector d_i.

The notation used here may be confusing since we usually rely on subscripted variables to denote scalar elements of a vector. Here, the subscripted variable d_i represents a vector of distances between observation i and all other sample data observations.

A single value of the bandwidth parameter θ is determined using a cross-validation procedure often used in locally linear regression methods. A score function taking the form:

$$\sum_{i=1}^{n} [y_i - \hat{y}_{\neq i}(\theta)]^2, \tag{11.4}$$

is minimized with respect to θ, where $\hat{y}_{\neq i}(\theta)$ denotes the ˇ tted value of y_i with the observations for point i omitted from the calibration process. Note that for the case of the tri-cube weighting function, we would compute an integer q (the number of nearest neighbors) using cross-validation. We focus on the exponential and Gaussian weighting methods for simplicity, ignoring the tri-cube weights.

The non-parametric GWR model relies on a sequence of locally linear regressions to produce estimates for every point in space using a sub-sample of data information from nearby observations. Let y denote an n by 1 vector of dependent variable observations collected at n points in space, X an n by k matrix of explanatory variables, and ε an n by 1 vector of normally distributed, constant variance disturbances. Letting W_i represent an n by n diagonal matrix containing the vector d_i of distance-based weights for observation i that re ect the distance between observation i and all other observations, we can write the GWR model as:

$$W_i y = W_i X \beta_i + \varepsilon_i. \tag{11.5}$$

The subscript i on β_i indicates that this k by 1 parameter vector is associated with observation i. The GWR model produces n such vectors of parameter estimates, one for each observation. These estimates are produced using:

$$\hat{\beta}_i = (X' W_i^2 X)^{-1} (X' W_i^2 y). \tag{11.6}$$

The GWR estimates for β_i are conditional on the parameter θ we select. That is, changing θ will produce a different set of GWR estimates. Our Bayesian approach relies on the same cross-validation estimate of θ, but adjusts the weights for outliers or aberrant observations. An area for future work would be devising a method to determine the bandwidth as part of the estimation problem, resulting in a posterior distribution that could be used to draw inferences regarding how sensitive the GWR estimates are to alternative values of this parameter. Posterior Bayesian estimates from this type of model would not be conditional on the value of the bandwidth, as this parameter would be "integrated out" during estimation.

One problem with GWR estimates is that valid inferences cannot be drawn for the regression parameters using traditional least squares approaches. To see this, consider that locally linear estimates use the same sample data observations (with different weights) to produce a sequence of estimates for all points in space. Given the conditional nature of the GWR on the bandwidth estimate and the lack of independence between estimates for each location, regression-based measures of dispersion for the estimates are incorrect.

Another problem is that the presence of aberrant observations due to spatial enclave effects or shifts in regime can exert undue influence on locally linear estimates. Consider that all nearby observations in a sub-sequence of the series of locally linear estimates may be "contaminated" by an outlier at a single point in space. The Bayesian approach introduced here solves this problem using robust estimates that are insensitive to aberrant observations. These observations are automatically detected and downweighted to lessen their influence on the estimates.

A third problem is that the locally linear estimates based on a distance weighted sub-sample of observations may suffer from "weak data" problems. The effective number of observations used to produce estimates for some points in space may be very small. This problem can be solved with the Bayesian approach by incorporating subjective prior information. We introduce some explicit parameter smoothing relationships in the Bayesian model that can be used to impose restrictions on the spatial nature of parameter variation. Stochastic restrictions based on subjective prior information represent a traditional Bayesian approach for overcoming weak data problems.

The Bayesian formulation can be implemented with or without the relationship for smoothing parameters over space, and we illustrate both uses in different applied settings. The Bayesian model subsumes the GWR method as part of a much broader class of spatial econometric models. For example, the Bayesian GWR can be implemented with a variety of parameter smoothing relationships. One relationship results in a locally linear variant of the spatial expansion method introduced by Casetti (1972, 1992). Another parameter smoothing relation is based on a monocentric city model where parameters vary systematically with distance from the center of the city, and still others are based on distance decay or contiguity relationships.

Section 11.2 sets forth the GWR and Bayesian GWR (BGWR) methods. Section 11.3 discusses the Markov Chain, Monte Carlo estimation method used to implement the BGWR, and Sect. 11.4 provides three examples that compare the GWR and BGWR methods.

11.2 The GWR and Bayesian GWR Models

The Bayesian approach, which we label BGWR is best described using matrix expressions shown in (11.7) and (11.8). First, note that (11.7) is the same as the GWR relationship, but the addition of (11.8) provides an explicit statement of the parameter smoothing that takes place across space. Parameter smoothing in (11.8) relies on a locally linear combination of neighboring areas, where neighbors are defined in terms of the GWR distance weighting function that decays over space. Other parameter smoothing relationships will be introduced later.

$$W_i y = W_i X \beta_i + \varepsilon_i \tag{11.7}$$

$$\beta_i = \begin{pmatrix} w_{i1} \otimes I_k \ \dots \ w_{in} \otimes I_k \end{pmatrix} \begin{pmatrix} \beta_1 \\ \vdots \\ \beta_n \end{pmatrix} + u_i \tag{11.8}$$

The terms w_{ij} in (11.8) represent normalized distance-based weights so the row-vector $(w_{i1}, \dots, w_{in})$ sums to unity, and we set $w_{ii} = 0$. That is:

$$w_{ij} = \exp(-d_{ij}/\theta) / \sum_{j=1}^{n} \exp(-d_{ij}\theta).$$

To complete our model speciˇ cation, we add distributions for the terms ε_i and u_i:

$$\varepsilon_i \sim N[0, \sigma^2 V_i], \quad V_i = \mathrm{diag}(v_1, v_2, \dots, v_n), \tag{11.9}$$

$$u_i \sim N[0, \sigma^2 \delta^2 (X' W_i^2 X)^{-1})]. \tag{11.10}$$

The $V_i = \mathrm{diag}(v_1, v_2, \dots, v_n)$, represent a set of n variance scaling parameters (to be estimated) that allow for non-constant variance as we move across space. Of course, the idea of estimating n terms $v_j, j = 1, \dots, n$ at each observation i for a total of n^2 parameters (and nk regression parameters β_i) with only n sample data observations may seem truly problematical! The way around this is to assign a prior distribution for the n^2 terms $V_i, i = 1, \dots, n$ that depends on a single *hyperparameter*. The V_i parameters are assumed to be *iid*. $\chi^2(r)$ distributed, where r is a hyperparameter that controls the amount of dispersion in the V_i estimates across observations. This allows us to introduce a single hyperparameter r to the estimation problem and receive in return n^2 parameter estimates.

This type of prior has been used by Lindley (1971) for cell variances in an analysis of variance problem, Geweke (1993) in modeling heteroscedasticity and outliers and LeSage (1997a) in a spatial autoregressive modeling context. The speciˇ cs regarding the prior assigned to the V_i terms can be motivated by considering that the mean of prior equals unity, and the prior variance is $2/r$. This implies that as r becomes very large, the prior imposes homoscedasticity on the BGWR model and the disturbance variance becomes $\sigma^2 I_n$ for all observations i.

The distribution for the stochastic parameter u_i in the parameter smoothing relationship is normal with mean zero and a variance based on Zellner's (1971) $g-$prior. This prior variance is proportional to the parameter variance-covariance matrix, $\sigma^2 (X' W_i^2 X)^{-1}$ with δ^2 acting as the scale factor. The use of this prior speciˇ cation allows individual parameters β_i to vary by different amounts depending on their magnitude.

The parameter δ^2 acts as a scale factor to impose tight or loose adherence to the parameter smoothing speciˇ cation. Consider a case where δ was very small, then the smoothing restriction would force β_i to look like a distance-weighted linear combination of other β_i from neighboring observations. On the other hand, as $\delta \rightarrow \infty$ (and $V_i = I_n$) we produce the GWR estimates. To see this, we rewrite the BGWR model in a more compact form:

$$\begin{aligned} \tilde{y}_i &= \tilde{X}_i \beta_i + \varepsilon_i, \\ \beta_i &= J_i \gamma + u_i. \end{aligned} \tag{11.11}$$

Where the deˇ nitions of the matrix expressions are:

$$\begin{aligned} \tilde{y}_i &= W_i y, \\ \tilde{X}_i &= W_i X, \\ J_i &= \left(w_{i1} \otimes I_k \; \ldots \; w_{in} \otimes I_k \right), \\ \gamma &= \begin{pmatrix} \beta_1 \\ \vdots \\ \beta_n \end{pmatrix}. \end{aligned}$$

As indicated earlier, the notation is somewhat confusing in that $\tilde{y}_i$ denotes an $n-$vector, not a scalar magnitude. Similarly, ε_i is an $n-$vector and $\tilde{X}_i$ is an n by k matrix. Note that (11.11) can be written in the form of a Theil-Goldberger (1961) estimation problem as shown in (11.12):

$$\begin{pmatrix} \tilde{y}_i \\ J_i \gamma \end{pmatrix} = \begin{pmatrix} \tilde{X}_i \\ -I_k \end{pmatrix} \beta_i + \begin{pmatrix} \varepsilon_i \\ u_i \end{pmatrix}. \tag{11.12}$$

Assuming $V_i = I_n$, the estimates β_i take the form:

$$\begin{aligned} \hat{\beta}_i &= R(\tilde{X}_i' \tilde{y}_i + \tilde{X}_i' \tilde{X}_i J_i \gamma / \delta^2), \\ R &= (\tilde{X}_i' \tilde{X}_i + \tilde{X}_i' \tilde{X}_i / \delta^2)^{-1}. \end{aligned}$$

As δ approaches ∞, the terms associated with the Theil-Goldberger "stochastic restriction", $\tilde{X}_i' \tilde{X}_i J_i \gamma / \delta^2$ and $\tilde{X}_i' \tilde{X}_i / \delta^2$ become zero, and we have the GWR estimates:

$$\hat{\beta}_i = (\tilde{X}_i' \tilde{X}_i)^{-1} (\tilde{X}_i' \tilde{y}_i). \tag{11.13}$$

In practice, we can use a diffuse prior for δ which allows the amount of parameter smoothing to be estimated from sample data information, rather than by subjective prior information. Details concerning estimation of the parameters in the BGWR model are taken up in the next section. Before turning to these issues, we consider some alternative spatial parameter smoothing relationships that might be used in lieu of (11.8) in the BGWR model.

One alternative smoothing speciˇ cation would be the "monocentric city smoothing" set forth in (11.14). This relation assumes that the data observations have been ordered by distance from the center of the spatial sample:

$$\begin{aligned} \beta_i &= \beta_{i-1} + u_i, \\ u_i &\sim N[0, \sigma^2 \delta^2 (X' W_i^2 X)^{-1}]. \end{aligned} \tag{11.14}$$

Given that the observations are ordered by distance from the center, the smoothing relation indicates that β_i should be similar to the coefˇ cient β_{i-1} from a neighboring concentric ring. Note that we rely on the same GWR distance-weighted data sub-samples, created by transforming the data using: $W_i y, W_i X$. This means that the estimates still have a "locally linear" interpretation as in the GWR. We rely on the same distributional assumption for the term u_i from the BGWR which allows us to estimate the parameters from this model by making minor changes to the approach used for the BGWR based on the smoothing relation in (11.8).

Another alternative is a "spatial expansion smoothing" based on the ideas introduced by Casetti (1972). This is shown in (11.15), where Z_{xi}, Z_{yi} denote latitude-longitude coordinates associated with observation i:

$$\beta_i = \begin{pmatrix} Z_{xi} \otimes I_k & Z_{yi} \otimes I_k \end{pmatrix} \begin{pmatrix} \beta_x \\ \beta_y \end{pmatrix} + u_i,$$
$$u_i \sim N[0, \sigma^2 \delta^2 (X' W_i^2 X)^{-1})]. \tag{11.15}$$

This parameter smoothing relation creates a locally linear combination based on the latitude-longitude coordinates of each observation. As in the case of the monocentric city speciˇ cation, we retain the same assumptions regarding the stochastic term u_i, making this model simple to estimate with only minor changes to the basic BGWR methodology.

Finally, we could adopt a "contiguity smoothing" relationship based on a ˇ rst-order spatial contiguity matrix as shown in (11.16). The terms c_{ij} represent the ith row of a row-standardized ˇ rst-order contiguity matrix. This creates a parameter smoothing relationship that averages over the parameters from observations that neighbor observation i:

$$\beta_i = \begin{pmatrix} c_{i1} \otimes I_k & \ldots & c_{in} \otimes I_k \end{pmatrix} \begin{pmatrix} \beta_1 \\ \vdots \\ \beta_n \end{pmatrix} + u_i,$$
$$u_i \sim N[0, \sigma^2 \delta^2 (X' W_i^2 X)^{-1})]. \tag{11.16}$$

These approaches to specifying a geographically weighted regression model suggest that researchers need to think about which type of spatial parameter smoothing relationship is most appropriate for their application. Additionally, where the nature of the problem does not clearly favor one approach over another, statistical tests of alternative models based on different smoothing relations might be carried out. Posterior probabilities can be constructed that will shed light on which smoothing relationship is most consistent with the sample data. This subject is taken up in Sect. 11.3.1 and illustrations are provided in Sect. 11.4.

11.3 Estimation of the BGWR Model

A recent methodology known as Markov Chain Monte Carlo is based on the idea that rather than compute a probability density, say $p(\theta|y)$, we would be just as happy to have a large random sample from $p(\theta|y)$ as to know the precise form of the density. Intuitively, if the sample were large enough, we could approximate the form of the probability density using kernel density estimators or histograms. In addition, we could compute accurate measures of central tendency and dispersion for the density, using the mean and standard deviation of the large sample. This insight leads to the question of how to efˇ ciently simulate a large number of random samples from $p(\theta|y)$.

Metropolis *et al.* (1953) demonstrated that one could construct a Markov chain stochastic process for $(\theta_t, t \geq 0)$ that unfolds over time such that: 1) it has the same state space (set of possible values) as θ, 2) it is easy to simulate, and 3) the equilibrium or stationary distribution which we use to draw samples is $p(\theta|y)$ after the Markov chain has been run for a long enough time. Given this result, we can construct and run a Markov chain for a very large number of iterations to produce a sample of $(\theta_t, t = 1, \ldots)$ from the posterior distribution and use simple descriptive statistics to examine any features of the posterior in which we are interested.

This approach, known as Markov Chain Monte Carlo, (MCMC) or Gibbs sampling has greatly reduced the computational problems that previously plagued application of the Bayesian methodology. Gelfand and Smith (1990), as well as a host of others, have popularized this methodology by demonstrating its use in a wide variety of statistical applications where intractable posterior distributions previously hindered Bayesian analysis. A simple introduction to the method can be found in Casella and George (1992) and an expository article dealing speciˇ cally with the normal linear model is Gelfand *et al.* (1990). Two recent books that deal in detail with all facets of these methods are Gelman *et al.* (1995), and Gilks *et al.* (1996).

We rely on Gibbs sampling to produce estimates for the BGWR model, which represent the multivariate posterior probability density for all of the parameters in our model. This approach is particularly attractive in this application because the conditional densities are simple and easy to obtain. LeSage (1997a) demonstrates this approach for Bayesian estimation of spatial autoregressive models, which represents a more complicated case.

To implement the Gibbs sampler we need to derive and draw samples from the conditional posterior distributions for each group of parameters, β_i, σ, δ, and V_i in the model. Let $P(\beta_i|\sigma, \delta, V_i, \gamma)$ denote the conditional density of β_i, where γ represents the values of other β_j for observations $j \neq i$. Using similar notation for the the other conditional densities, the Gibbs sampling process can be viewed as follows:

1. start with arbitrary values for the parameters $\beta_i^0, \sigma_i^0, \delta^0, V_i^0, \gamma^0$
2. for each observation $i = 1, \ldots, n$,
   (a) sample a value, β_i^1 from $P(\beta_i|\delta^0, \sigma_i^0, V_i^0, \gamma^0)$
   (b) sample a value, σ_i^1 from $P(\sigma_i|\delta^0, V_i^0, \beta_i^1, \gamma^0)$
   (c) sample a value, V_i^1 from $P(V_i|\delta^0, \beta_i^1, \sigma_i^1, \gamma^0)$
3. use the sampled values $\beta_i^1, i = 1, \ldots, n$ from each of the n draws above to update γ^0 to γ^1
4. sample a value, δ^1 from $P(\delta|\sigma_i^1, \beta_i^1 V_i^1, \gamma^1)$
5. go to step 1 using $\beta_i^1, \sigma_i^1, \delta^1, V_i^1, \gamma^1$ in place of the arbitrary starting values.

Steps 2 to 4 outlined above represents a single pass through the sampler, and we make a large number of passes to collect a sample of parameter values from which we construct our posterior distributions. Note that this is computationally intensive as it requires a loop over all observations for each draw. In one of our examples we implement a simpler version of the Gibbs sampler that can be used to produce robust estimates when no parameter smoothing relationship is in the model. This sampling routine involves a single loop over each of the n observations that carries out all draws, as shown below:

1. start with arbitrary values for the parameters $\beta_i^0, \sigma_i^0, V_i^0$
2. for each observation $i = 1, \ldots, n$, sample all draws using a sequence over:
3. Step 1: sample a value, β_i^1 from $P(\beta_i|\sigma_i^0, V_i^0)$
4. Step 2: sample a value, σ^1 from $P(\sigma_i|V_i^0, \beta_i^1)$
5. Step 3: sample a value, V_i^1 from $P(V_i|\beta_i^1, \sigma_i^1)$
6. go to Step 1 using $\beta_i^1, \sigma_i^1, V_i^1$ in place of the arbitrary starting values. Continue returning to Step 1 until all draws have been obtained.
7. Move to observation $i = i + 1$ and obtain all draws for this next observation.
8. When we reach observation n, we have sampled all draws for each observation.

This approach samples all draws for each observation, requiring a single pass through the n observation sample. The computational burden associated with the ˇ rst sampler arises from the need to update the parameters in γ for all observations before moving to the next draw. This is because these values are used in the distance and contiguity smoothing relationships.

The second sampler takes around 10 seconds to produce 1,000 draws for each observation, irrespective of the sample size. Sample size is irrelevant because we exclude distance weighted observations that have negligible weights. This reduces the size of the matrices that need be computed during sampling to a fairly constant size that does not depend on the number of observations. In contrast, the ˇ rst sampler takes around 2 seconds per draw for even moderate sample sizes of 100 observations, and computational time increases dramatically with the number of observations.

For the case of the monocentric city prior we could rely on the GWR estimate for the ˇ rst observation and proceed to carry out draws for the remaining observations using the second sampler presented above. The draw for observation 2 would rely on the posterior mean computed from the draws for observation 1. Note that we need the posterior from observation 1 to deˇ ne the parameter smoothing prior for observation 2. Assuming the observations are ordered by distance from a central observation, this would achieve our goal of stochastically restricting observations from nearby concentric rings to be similar. Observation 2 would be similar to 1, 3 would be similar to 2, and so on.

Another computationally efˇ cient way to implement these models with a parameter smoothing relationship would be to use the GWR estimates as elements in γ. This would allow us to use the second sampler that makes multiple draws for each observation, requiring only one pass over the observations. A drawback to this approach is that the parameter smoothing relationship doesn't evolve as part of the estimation process. It is stochastically restricted to the ˇ xed GWR estimates.

We rely on the compact statement of the BGWR model in (11.11) to facilitate presentation of the conditional distributions that we rely on during the sampling. The conditional posterior distribution of β_i given σ_i, δ, γ and V_i is a multivariate normal:

$$p(\beta_i|\ldots) \propto N(\hat{\beta}_i, \sigma_i^2 R), \tag{11.17}$$

where,

$$\begin{aligned}\hat{\beta}_i &= R(\tilde{X}_i' V_i^{-1} \tilde{y}_i + \tilde{X}_i' \tilde{X}_i J_i \gamma / \delta^2),\\ R &= (\tilde{X}_i' V_i^{-1} \tilde{X}_i + \tilde{X}_i' \tilde{X}_i / \delta^2)^{-1}.\end{aligned} \tag{11.18}$$

This result follows from the assumed variance-covariance structures for ε_i, u_i and the Theil-Goldberger (1961) representation shown in (11.12). The conditional posterior distribution for σ is a $\chi^2(m)$ distribution shown in (11.19), where m denotes the number of observations with non-negligible weights:

$$\begin{aligned}p(\sigma_i|\ldots) &\propto \sigma_i^{-(m+1)} \exp\{-\frac{1}{2\sigma_i^2}(\varepsilon_i' V_i^{-1} \varepsilon_i)\},\\ \varepsilon_i &= \tilde{y}_i - \tilde{X}_i \beta_i.\end{aligned} \tag{11.19}$$

The conditional posterior distribution for V_i is shown in (11.20), which indicates that we draw an m-vector based on a $\chi^2(r+1)$ distribution:

$$p\{[(e_i^2/\sigma_i^2) + r]/V_i \,|\ldots\} \propto \chi^2(r+1), \tag{11.20}$$

To see the role of the parameter v_{ij}, consider two cases. First, suppose (e_j^2/σ_i^2) is small (say zero), because the GWR distance-based weights work well to relate y and X for observation j. In this case, observation j is not an outlier. Assume that we use a small value of the hyperparameter r, say $r = 5$, which means our prior belief is that heterogeneity exits. The conditional posterior will have a mean and mode of:

$$\begin{aligned}\text{mean}(v_{ij}) &= (\sigma_i^{-2} e_j^2 + r)/(r+1) = r/(r+1) = (5/6),\\ \text{mode}(v_{ij}) &= (\sigma_i^{-2} e_j^2 + r)/(r-1) = r/(r-1) = (5/4),\end{aligned} \tag{11.21}$$

where the results in (11.21) follow from the fact that the mean of the prior distribution for V_{ij} is $r/(r-2)$ and the mode of the prior equals $r/(r+2)$.

In the case shown in (11.21), the impact of $v_{ij} \approx 1$ in the model is negligible, and the typical distance-based weighting scheme would dominate. For the case of exponential weights, a weight, $w_{ij} = \exp(-d_i)/\theta v_{ij}$ would be accorded to observation j. Note that a prior belief in homogeneity that assigns a large value of $r = 20$, would produce a similar weighting outcome. The conditional posterior mean of $r/(r+1) = 20/21$, is approximately unity, as is the mode of $(r+1)/r = 20/19$.

Second, consider the case where (e_j^2/σ_i^2) is large (say 20), because the GWR distance-based weights do not work well to relate y and X for observation j. Here, we have the case of an outlier for observation j. Using the same small value of the hyperparameter $r = 5$, the conditional posterior will have a mean and mode of:

$$\begin{aligned} \text{mean}(v_{ij}) &= (20+r)/(r+1) = (25/6), \\ \text{mode}(v_{ij}) &= (20+r)/(r-1) = (25/4). \end{aligned} \quad (11.22)$$

For this aberrant observation case, the role of $v_{ij} \approx 5$ will be to downweight the distance associated with this observation. The distance-based weight:

$$w_{ij} = \exp(-d_i)/\theta v_{ij},$$

would be de ated by a factor of approximately 5 for this aberrant observation. It is important to note that, a prior belief of homogeneity (expressed by a large value of $r = 20$) in this case would produce a conditional posterior mean of $(20+r)/(r+1) = (40/21)$. Downweighting of the distance-based weights would be only by a factor of 2, rather than 5 found for the smaller value of r.

It should be clear that as r becomes very large, say 50 or 100, the posterior mean and mode will be close to unity irrespective of the ˇ t measured by e_j^2/σ_i^2 This replicates the distance-based weighting scheme used in the non-Bayesian GWR model.

A graphical illustration of how this works in practice can be seen in Fig 11.1. The ˇ gure depicts the adjusted distance-based weights, $W_i V_i^{-1}$ alongside the GWR weights W_i for observations 31 to 36 in the Anselin (1988b) Columbus neighborhood crime data set. In Sect. 11.4.1 we motivate that observation #34 represents an outlier.

Beginning with observation 31, the aberrant observation #34 is downweighted when estimates are produced for observations 31 to 36 (excluding observation #34 itself). A symbol `o' has been placed on the BGWR weight in the ˇ gure to help distinguish observation 34. This downweighting of the distance-based weight for observation #34 occurs during estimation of β_i for observations 31 to 36, all of which are near #34 in terms of the GWR distance measure. It will be seen that this alternative weighting produces a divergence in the BGWR estimates and those from GWR for observations neighboring on #34.

Finally, the conditional distribution for δ is a $\chi^2(nk)$ distribution based on:

$$p(\delta|\ldots) \propto \delta^{-nk}\exp\{-\sum_{i=1}^{n}(\beta_i - J_i\gamma)'(\tilde{X}_i'\tilde{X}_i)^{-1}(\beta_i - J_i\gamma)/2\sigma_i^2\delta^2\}. \quad (11.23)$$

Now consider the modiˇ cations needed to the conditional distributions to implement the alternative spatial smoothing relationships set forth in Sect. 11.3. Because the same assumptions were used for the disturbances ε_i and u_i, we need only alter the conditional distributions for β_i and δ. First, consider the case of the monocentric city smoothing relationship. The conditional distribution for β_i is multivariate normal with mean $\hat{\beta}_i$ and variance-covariance $\sigma^2 R$ as shown in (11.24):

$$\begin{aligned} \hat{\beta}_i &= R(\tilde{X}_i' V_i^{-1}\tilde{y}_i + \tilde{X}_i'\tilde{X}_i\beta_{i-1}/\delta^2), \\ R &= (\tilde{X}_i' V_i^{-1}\tilde{X}_i + \tilde{X}_i'\tilde{X}_i/\delta^2)^{-1}. \end{aligned} \quad (11.24)$$

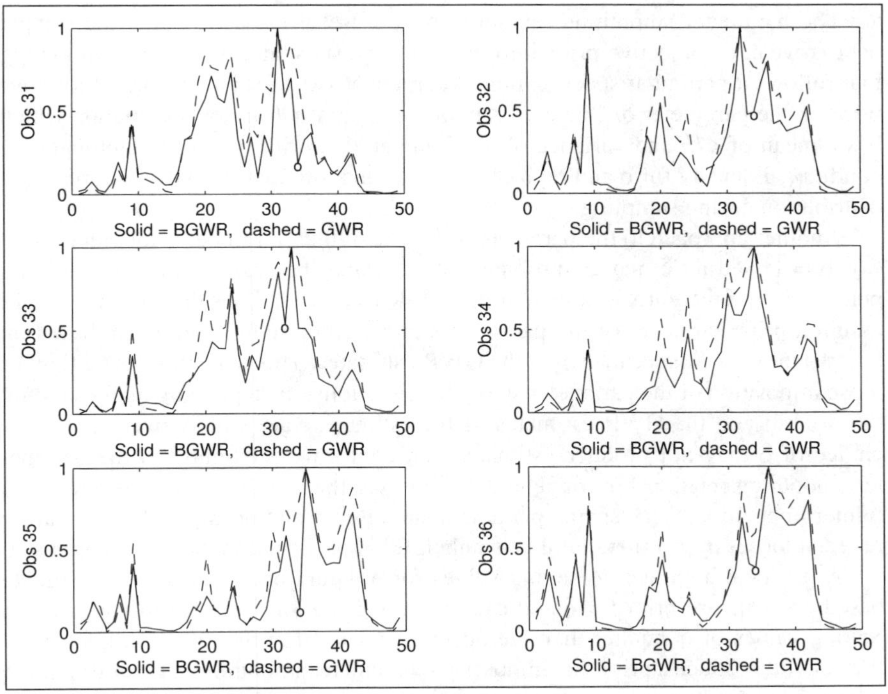

Fig. 11.1. Distance-based weights adjusted by V_i

The conditional distribution for δ is a $\chi^2(nk)$ based on the expression:

$$p(\delta|\ldots) \propto \delta^{-nk}\exp\{-\sum_{i=1}^{n}(\beta_i - \beta_{i-1})'(\tilde{X}'\tilde{X})^{-1}(\beta_i - \beta_{i-1})/\sigma_i^2\delta^2\}. \quad (11.25)$$

For the case of the spatial expansion and contiguity smoothing relationships, we can maintain the conditional expressions for β_i and δ from the case of the basic BGWR, and simply modify the deˇ nition of J, to be consistent with these smoothing relations.

11.3.1 Informative priors

Implementing the BGWR model with very large values for δ will essentially eliminate the parameter smoothing relationship from the model. The BGWR estimates will then collapse to the GWR estimates (in the case of a large value for the hyperparameter r that leads to $V_i = I_n$), and this represents a very computationally intensive way to obtain GWR estimates. If there is a desire to obtain robust BGWR estimates without imposing a parameter smoothing relationship in the model, the second sampling scheme presented in Sect. 11.3 can do this in a more computationally efˇ cient manner.

The parameter smoothing relationships are useful in cases where the sample data is weak or objective prior information suggests spatial parameter smoothing that follows a particular speciˇ cation. Alternatives exist for placing an informative prior on the parameter δ. One is to rely on a Gamma(a,b) prior distribution which has a mean of a/b and variance of a/b^2. Given this prior, we could eliminate the conditional density for δ and replace it with a random draw from the Gamma(a,b) distribution during sampling.

Another approach to the parameter δ is to assign an improper prior value using say, $\delta = 1$. Setting δ may be problematical because the scale is unknown and depends on the inherent variability in the GWR estimates. Consider that $\delta = 1$ will assign a prior variance for the parameters in the smoothing relationship based on the variance-covariance matrix of the GWR estimates. This may represent a tight or loose imposition of the parameter smoothing relationship, depending on the amount of variability in the GWR estimates. If the estimates vary widely over space, this choice of δ may not produce estimates that conform very tightly to the parameter smoothing relationship. In general we can say that smaller values of δ re ect a tighter imposition of the spatial parameter smoothing relationship and larger values re ect a looser imposition, but this is unhelpful in particular modeling situations.

A practical approach to setting values for δ would be to generate an estimate based on a diffuse prior for δ and examine the posterior mean for this parameter. Setting values of δ smaller than the posterior mean from the diffuse implementation should produce a prior that imposes the parameter smoothing relationship more tightly. One might use magnitudes for δ that scale down the diffuse δ estimate by 0.5, 0.25 and 0.1 to examine the impact of the parameter smoothing relationship on the BGWR estimates.

Posterior probabilities can be used as a guide for comparing alternative parameter smoothing relationships and various values for δ. These can be calculated using the log posterior for every observation divided by the sum of the log posterior over all models at each observation. Expression (11.26) shows the log posterior for a single observation of our BGWR model. Posterior probabilities based on these quantities provide an indication of which parameter smoothing relationship ˇ ts the sample data best as we range over observations:

$$\log P_i = \sum_{j=1}^{n} W_{ij}\{\log \phi([y_j - X_i\beta_i]/\sigma_i v_{ij}) - \log \sigma_i v_{ij}\}. \tag{11.26}$$

Keep in mind that these posterior probabilities re ect a measure of ˇ t to the sample data, as is clear from (11.26). In applications where robust estimates are desired, it is not clear that choice of models should be made using measures of ˇ t. Robust estimates require a trade-off between ˇ t and insensitivity to aberrant observations.

A similar Gamma prior for the hyperparameter r can be used, where values $a = 8, b = 2$ would indicate small values of r around 4. This should provide fairly robust estimates if there is spatial heterogeneity. In the absence of heterogeneity, the resulting V_i estimates will be near unity so the BGWR distance weights will be similar to those from GWR, even with a small value of r. We can also set an improper prior value for this hyperparameter, say $r = 4$ Additionally, a $\chi^2(c,d)$ natural conjugate prior for the parameter σ could be used in place of the diffuse prior set forth here. This would affect the conditional distribution used during Gibbs sampling in only a minor way.

Some other alternatives offer additional exibility when implementing the BGWR model. For example, one can restrict speciˇ c parameters to exhibit no variation over the spatial sample observations. This might be useful if we wish to restrict the constant term over space. Or, it may be that the constant term is the only parameter that we allow to vary over space.

These alternatives can be implemented by adjusting the prior variances in the parameter smoothing relationship:

$$\text{var - cov}(\beta_i) = \sigma^2\delta^2(\tilde{X}_i'\tilde{X}_i)^{-1}. \tag{11.27}$$

For example, assuming the constant term is in the ˇ rst column of the matrix $\tilde{X}_i$, setting the ˇ rst row and column elements of $(\tilde{X}_i'\tilde{X}_i)^{-1}$ to zero would restrict the intercept term to remain constant over all observations.

11.4 Examples

Section 11.4.1 provides two comparisons of the GWR and BGWR estimates without reliance on a parameter smoothing relationship. These illustrations demonstrate the sensitivity of GWR estimates to aberrant observations and show how outliers are downweighted by the V_i terms in the BGWR model.

An illustration that compares the GWR to the BGWR based on monocentric, distance and contiguity smoothing relations is provided in Sect. 11.4.2, along with the posterior probabilities for these alternative spatial smoothing approaches.

11.4.1 A comparison of GWR and BGWR

As an initial illustration of the problems created by outliers in GWR estimation, a generated data set containing 100 observations was used. A regression variable y was generated using coefˇ cients that vary over a regular grid according to the quadrant in which the observation falls. Coefˇ cients of 1 and -1 were used for two explanatory variables. A switch from 1 to -1 in the coefˇ cients occurs at observation 50, which is the type of spatial variation in relationships that the GWR model was devised to detect.

After producing GWR estimates based on this data set, we create a single outlier at observation 60 by multiplying the explanatory variables by 10. Another set of GWR estimates along with BGWR model estimates were produced using this outlier contaminated data set. If the BGWR model is producing robust estimates, we would expect to see estimates that are similar to those from the GWR model based on the data set with no outlier.

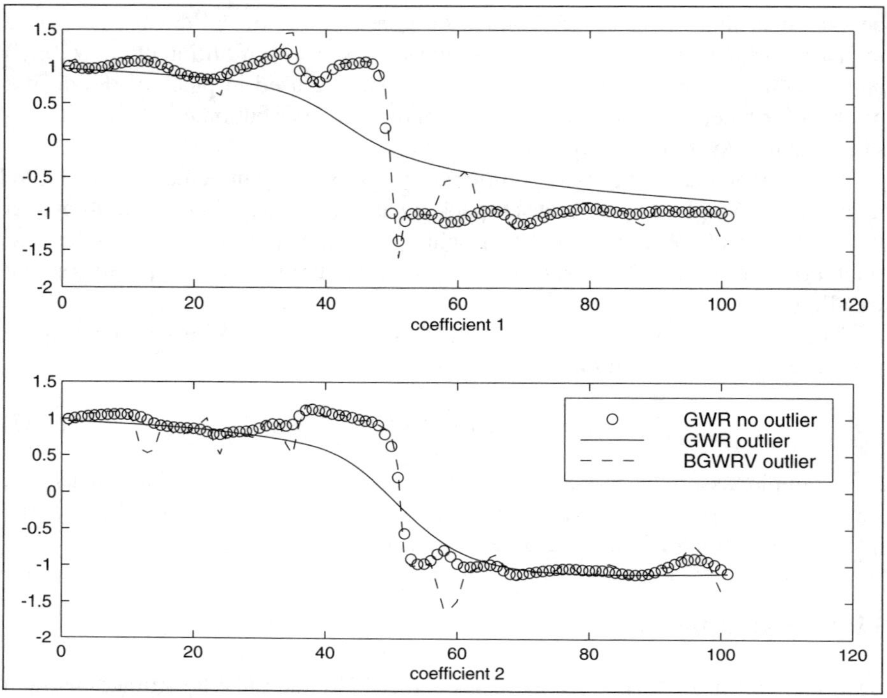

Fig. 11.2. β_i estimates for GWR and BGWRV with an outlier

The results from this experiment are shown in Fig. 11.2 where the adverse impact of the single outlier at observation 60 is clear. GWR estimates from the data set with no outlier captured the shift in relationship at observation 50 with a great deal of precision, as did the robust BGWR estimates based on the data set containing the outlier. In contrast, the GWR estimates based on the data set with a single outlier do not capture the abrupt shift in the relationship over space. It would be dif˜ cult to infer the abrupt shift in regime at the appropriate point in space based on these GWR estimates.

In addition to adversely impacting the coef˜ cient trajectories over space, the single outlier also affects the $t-$statistics that would be used to draw inferences regarding shifts in regime as we move over space. Figure 11.3 shows $t-$statistics from the GWR model based on both data sets as well as the BGWR $t-$statistics for the data set containing the outlier. Here again, we see that the BGWR estimates are close to those from the GWR model based on no outliers. A closer examination of the $t-$statistic from the GWR model in the case of the outlier data set indicated that the estimate of the noise variance, σ^2 which enter into calculation of the $t-$statistics was the source of the problem.

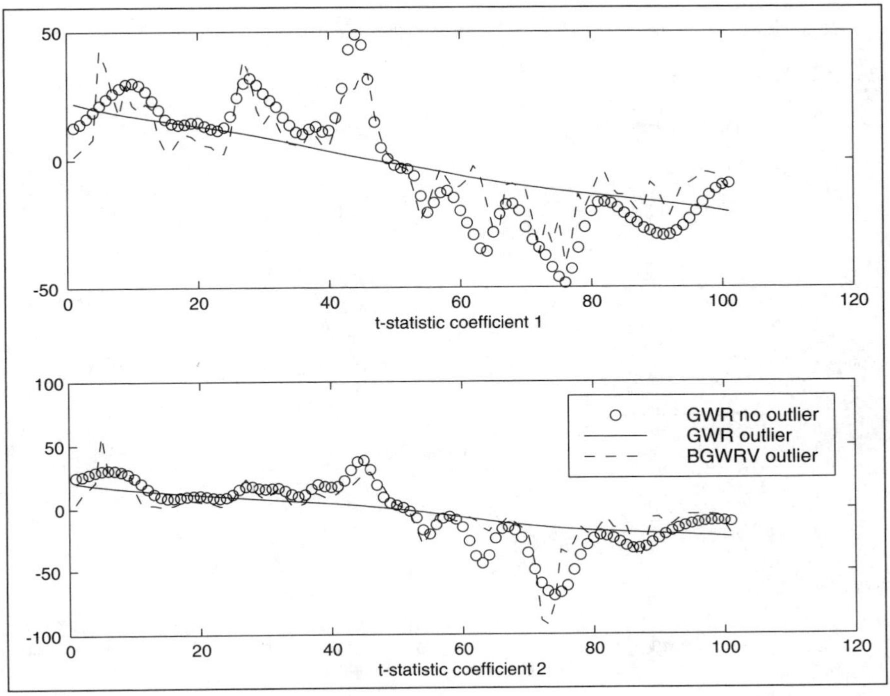

Fig. 11.3. $t-$statistics for the GWR and BGWRV with an outlier

As an applied illustration of the BGWR model we used a spatial data set from Anselin (1988b) on neighborhood crime in Columbus, Ohio. A model was estimated using neighborhood crime incidents as the dependent variable, household income and house values along with a constant term as explanatory variables, that is:

$$\text{Crime}_i = \beta_{1i} + \beta_{2i}(\text{Household Income})_i + \beta_{3i}(\text{House Value})_i + \varepsilon_i. \quad (11.28)$$

Estimates from a GWR model are compared to those from a BGWR model based on $r = 4$ representing a heteroscedastic prior, and a Gaussian weighting approach. For this sample of 49 observations and 3 explanatory variables, it took around 250 seconds to produce 1,250 draws, and 120 seconds for 550 draws on an Apple 266 Mhz. G3 Powerbook. The posterior means of the parameter estimates were virtually identical for the sample of 550 and 1,250 draws, suggesting no problems with convergence of the Gibbs sampler.

Figure 11.4 shows the comparison of GWR and BGWR estimates from the heteroscedastic version of the model. We see deˇ nite evidence of a departure between the GWR and BGWR estimates. The large V_i estimates presented in Fig. 11.5 point to non-constant variance as we move over the spatial sample.

An interesting question is – are these differences signiˇ cant in a statistical sense? We can answer this question using the 1,000 draws produced by the Gibbs sampler

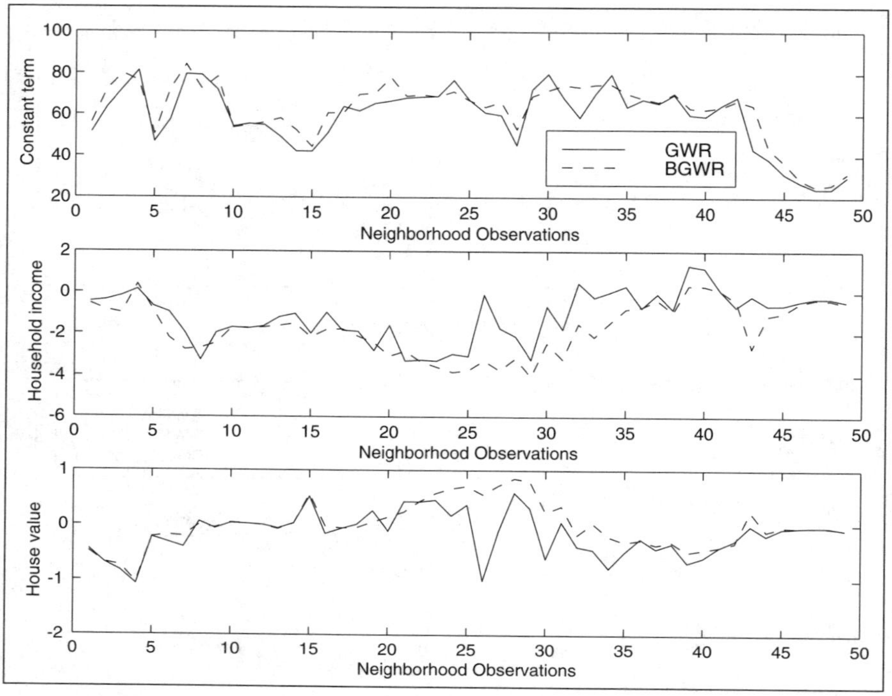

Fig. 11.4. GWR versus BGWR estimates for Columbus data set

to compute a two standard deviation band around the BGWR estimates. If the GWR estimates fall within this conˇ dence interval, we would conclude the estimates are not signiˇ cantly different. Figure 11.6 shows the GWR estimates and the conˇ dence bands for the BGWR estimates. The actual BGWR estimates were omitted from the graph for clarity. We see that the GWR estimates are near the two standard deviation conˇ dence intervals for sample observations in the range from 20 to 44, which implies we might draw different inferences from the GWR and BGWR estimates.

Another way to visualize the impact of non-constant variance over space is to examine a map of the absolute differences between the GWR and BGWR estimates. Neighborhoods surrounding areas with large V_i values should exhibit differences in the GWR and BGWR estimates. A change in the noise variance for a single observation tends to produce different trajectories for the estimates in all surrounding neighborhoods because the GWR relies on a sequence of sub-samples of the data.

Figures 11.7 and 11.8 show maps of the absolute differences between the GWR and BGWR coefˇ cient estimates for household income and housing values in the 49 Columbus neighborhoods. Darker areas re ect larger differences between the GWR and BGWR estimates.

In the case of the income coefˇ cient shown in Fig. 11.7, we see a pattern where the absolute differences between the GWR and BGWR estimates are largest around

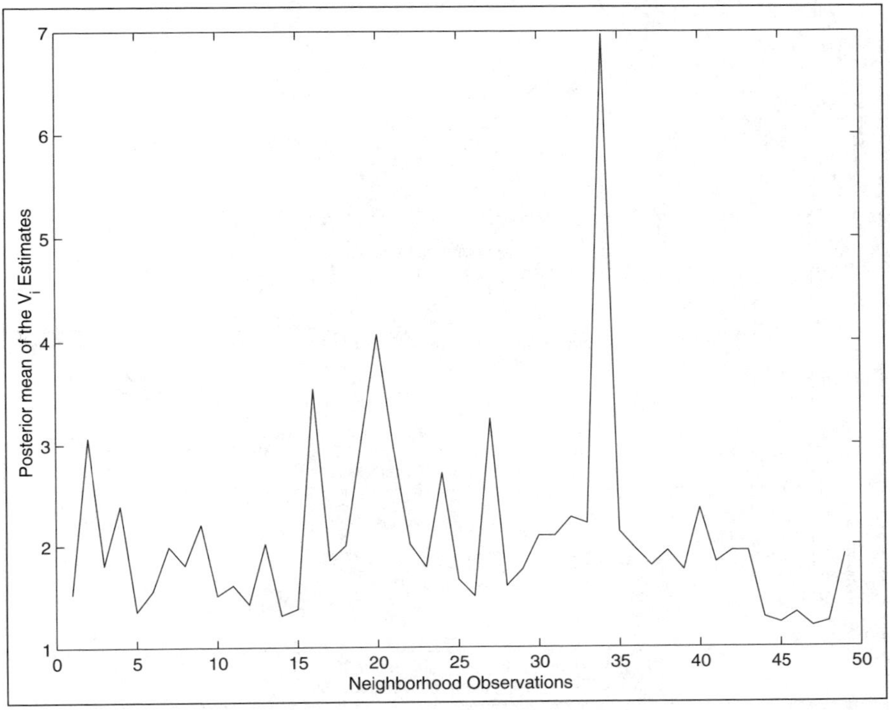

Fig. 11.5. Average V_i estimates over all draws and observations

neighborhoods bordering on observations 2 in the west, 16 and 27 in the north, 20 and 24 near the center and observation 34 in the south. Note that large V_i estimates for these observations shown in Fig. 11.5 produced large differences between GWR and BGWR estimates for surrounding neighborhoods, not just the observations containing large V_i values. A similar pattern exists in Fig. 11.8 showing absolute differences between the GWR and BGWR estimates for housing values.

The mean of the V_i estimates averaged over all observations in the spatial sample can be used as a diagnostic measure to detect aberrant observations. These V_i values re ect observations that consistently produced large residuals during estimation of each β_i parameter. The average V_i draws in Fig. 11.5 indicate that observations 2, 16 and 27, 20 and 24 as well as observation 34 were consistently downweighted during estimation of the β_i for all 49 observations. This is desirable if we wish to keep these aberrant observations from contaminating the estimates produced for neighbors.

Ultimately, the role of the parameters V_i in the BGWR model and the prior assigned to these parameters re ect our prior knowledge that distance alone may not be reliable as the basis for spatial relationships between variables. If distance-based weights are used in the presence of aberrant observations, inferences will be contaminated for whole neighborhoods and regions in our analysis. Incorporating this

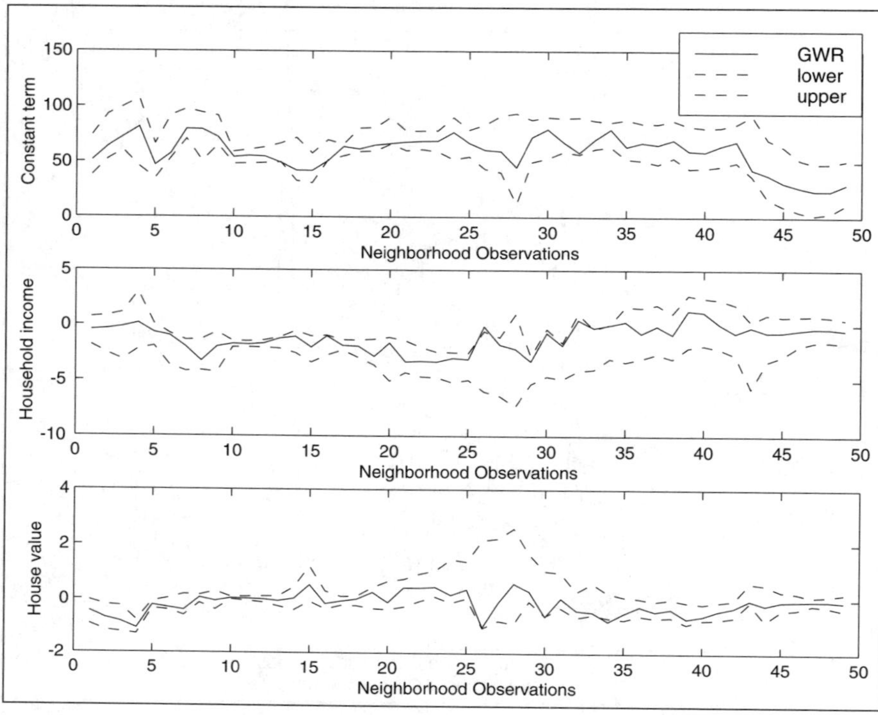

Fig. 11.6. GWR versus BGWR conˇ dence intervals

prior knowledge turns out to be relatively simple in the Bayesian framework, and it appears to effectively robustify estimates against the presence of spatial outliers.

11.4.2 Alternative spatial smoothing relations

To illustrate alternative parameter smoothing relationships we use a data set consisting of employment, payroll earnings and the number of establishments in all ˇ fty zip (postal) codes from Cuyahoga county Ohio during the ˇ rst quarter of 1989. The data set was created by aggregating establishment level data used by the State of Ohio for unemployment insurance purposes. It represents employment for workers covered by the state unemployment insurance program. The regression model used was:

$$\ln(E_i/F_i) = \beta_{0i} + \beta_{1i}\ln(P_i/E_i) + \beta_{2i}\ln(F_i) + \varepsilon_i, \tag{11.29}$$

where E_i is employment in zip code i, P_i represents payroll earnings and F_i denotes the number of establishments. The relationship indicates that employment per ˇ rm is a function of earnings per worker and the number of ˇ rms in the zip code area. For presentation purposes we sorted the sample of 50 observations by the dependent

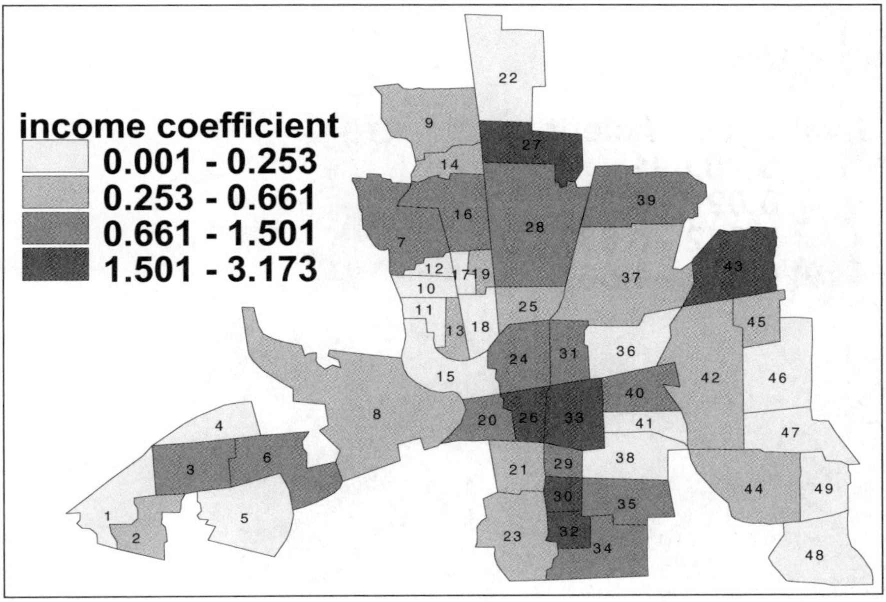

Fig. 11.7. Absolute differences between GWR and BGWR household income estimates

variable from low to high, so observation #1 represents the zip code district with the smallest level of employment per firm.

Three alternative parameter smoothing relationships were used, the monocentric city prior centered on the central business district, the distance decay prior and the contiguity prior. We would expect the monocentric city prior to work well in this application. An initial set of estimates based on a diffuse prior for δ are discussed below and would typically be generated to calibrate the tightness of alternative settings for the prior on the parameter smoothing relations.

A Gaussian distance weighting method was used, but estimates based on the exponential weighting method were quite similar. All three BGWR models were based on a hyperparameter $r = 4$ reflecting a heteroscedastic prior.

A graph of the three sets of estimates is shown in Fig. 11.9, where it should be kept in mind that the observations are sorted by employment per firm from low to high. This helps when interpreting variation in the estimates over the observations.

The first thing to note is the relatively unstable GWR estimates for the constant term and earnings per worker when compared to the BGWR estimates. Evidence of parameter smoothing is clearly present. Bayesian methods attempt to introduce a small amount of bias in an effort to produce a substantial increase in precision. This seems a reasonable trade-off if it allows clearer inferences. The diffuse prior for the smoothing relationships produced estimates for δ^2 equal to 138 for the monocentric city prior, 142 and 113 for the distance and contiguity priors. These large values

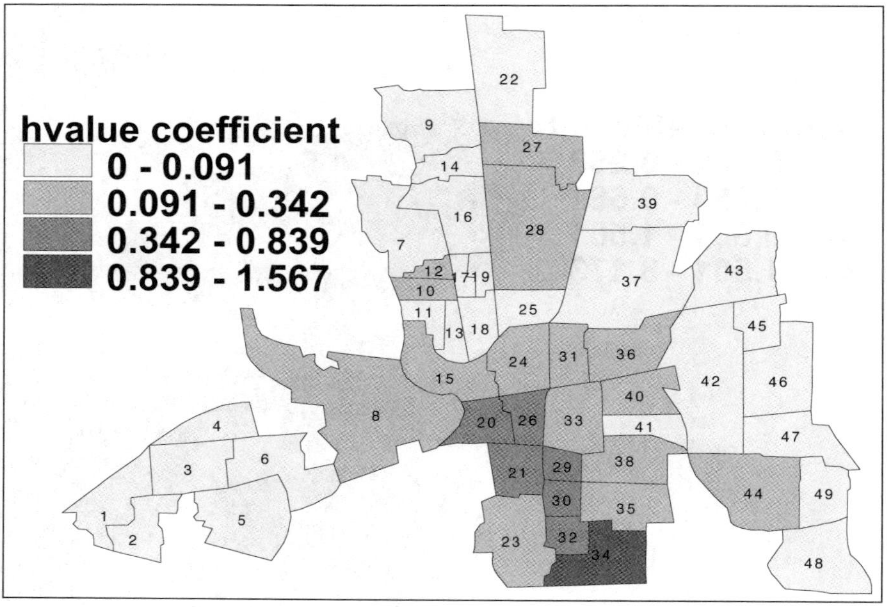

Fig. 11.8. Absolute differences between GWR and BGWR house value estimates

indicate that the sample data are inconsistent with these parameter smoothing relationships, so their use would likely introduce some bias in the estimates. From the plot of the coefˇ cients it is clear that no systematic bias is introduced, rather we see evidence of smoothing that impacts only volatile GWR estimates that take rapid jumps from one observation to the next.

Note that the GWR and BGWR estimates for the coefˇ cients on the number of ˇ rms are remarkably similar. There are two factors at work to create a divergence between the GWR and BGWR estimates. One is the introduction of v_i parameters to capture non-constant variance over space and the other is the parameter smoothing relationship. The GWR coefˇ cient on the ˇ rm variable is apparently insensitive to any non-constant variance in this data set. In addition, the BGWR estimates are not affected by the parameter smoothing relationships we introduced. An explanation for this is that a least-squares estimate for this coefˇ cient produced a $t-$statistic of 1.5, signiˇ cant at only the 15 percent level. Since our parameter smoothing prior relies on the variance-covariance matrix from least-squares (adjusted by the distance weights), it is likely that the parameter smoothing relationships are imposed very loosely for this coefˇ cient. Of course, this will result in estimates equivalent to the GWR estimates.

A ˇ nal point is that all three parameter smoothing relations produced relatively similar estimates. The monocentric city prior was most divergent with the distance and contiguity priors very similar. We would expect this since the latter priors rely

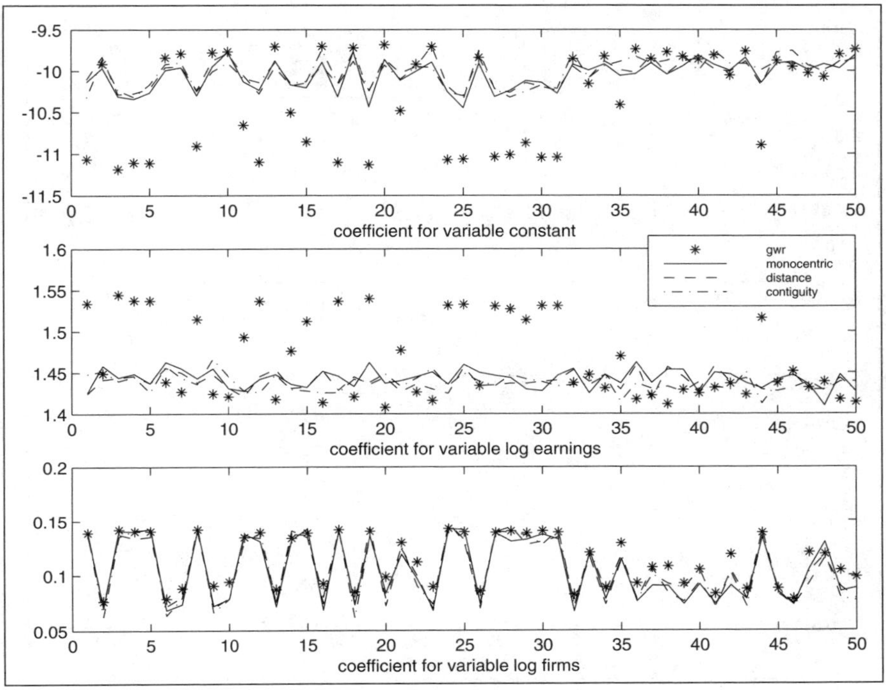

Fig. 11.9. Ohio GWR versus BGWR estimates

on the entire sample of estimates whereas the monocentric city prior relies only on the estimate from a neighboring observation.

The times required for 550 draws with these models were: 320 seconds for the monocentric city prior, 324 seconds for the distance-based prior, and 331 seconds for the contiguity prior.

Turning attention to the question of which parameter smoothing relation is most consistent with the sample data, a graph of the posterior probabilities for each of the three models is shown in the top panel of Fig. 11.10. It seems quite clear that the monocentric smoothing relation is most consistent with the data as it receives slightly higher posterior probability values for all observations. There is however no dominating evidence in favor of a single model, since the other two models receive substantial posterior probability weight over all observations, summing to over 60 percent.

For purposes of inference, a single set of parameters can be generated using these posterior probabilities to weight the three sets of parameters. This represents a Bayesian solution to the model speciˇ cation issue (see Leamer, 1983a). In this application, the parameters averaged using the posterior probabilities would look very similar to those in Fig. 11.9, since the weights are roughly equal and the coefˇ cients are very similar.

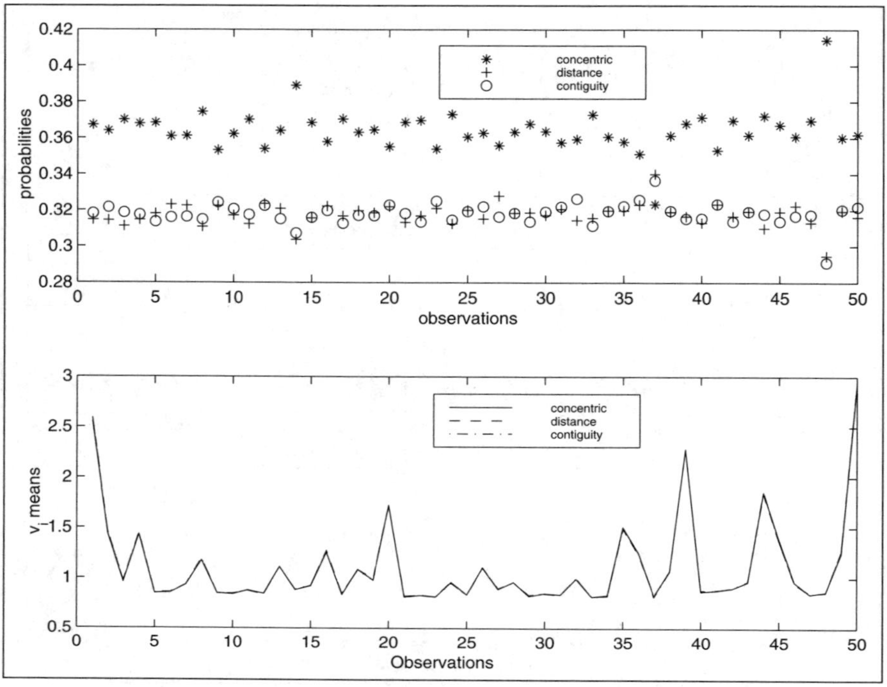

Fig. 11.10. Posterior probabilities and v_i estimates

Figure 11.10 also shows a graph of the estimated v_i parameters from all three versions of the BGWR model. These are nearly identical and point to observations at the beginning and end of the sample as regions of non-constant variance as well as observations around 17, 20, 35, 38 and 44 as perhaps outliers. Because the observations are sorted from small to large, the large v_i estimates at the beginning and end of the sample indicate our model is not working well for these extremes in ˇ rm size. It is interesting to note that outlying GWR estimates by comparison with the smoothed BGWR estimates correlate highly with observations where the v_i estimates are large. As we saw in the generated data example, the GWR model tends to "chase" after the outliers, and we see evidence of this here as well.

A ˇ nal question is – how sensitive are these inferences regarding the three models to the diffuse prior used for the parameter δ? To test alternative smoothing priors in an attempt to ˇ nd a single best model we impose the priors in a relatively tight fashion. In the face of a very strict implementation of the smoothing relationship, the posterior probabilities will tend to concentrate on the model that is most consistent with the data. To illustrate this, we constructed another set of estimates and posterior probabilities based on scaling δ to 0.1 times the estimate of δ from the diffuse prior. This should re ect a fairly tight imposition of the prior for the parameter smoothing relationships.

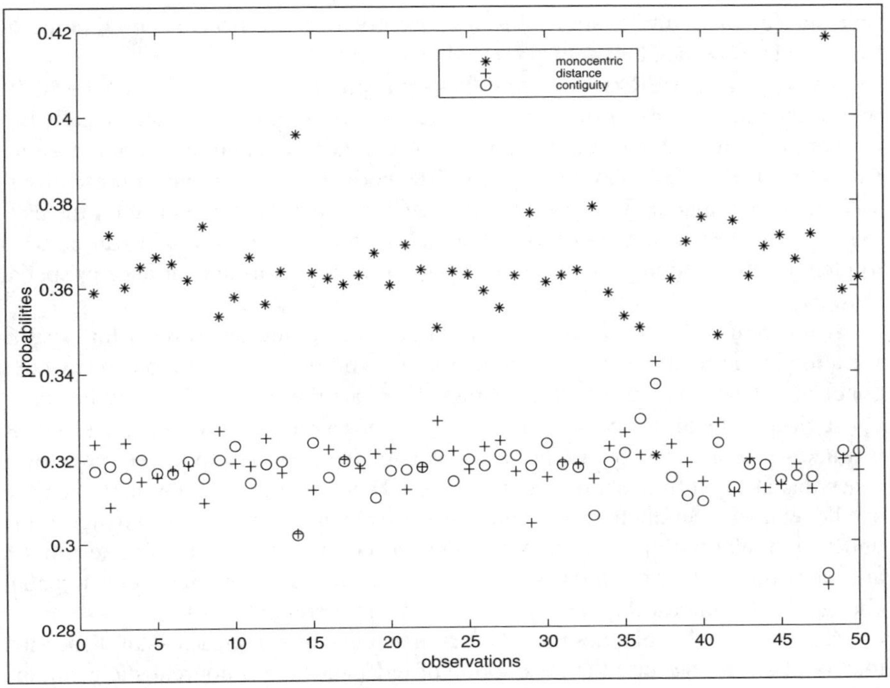

Fig. 11.11. Estimates based on a tight imposition of the prior

The posterior probabilities and estimates from these three models were very similar to those from the diffuse prior implementation. This suggests that even with this tighter imposition of the prior, all three parameter smoothing relationships are relatively compatible with the sample data. No smoothing relationship obtains a distinctive advantage over the others.

We need to keep the trade-off between bias and efˇ ciency in mind when implementing tight versions of the parameter smoothing relationships. For this application, the fact that both diffuse and tight implementation of the parameter smoothing relationships produced similar estimates indicates our inferences would be robust with respect to relatively large changes in the smoothing priors.

11.5 Conclusions

We have demonstrated that GWR models can be subsumed as a special case of a broader set of Bayesian models. This was accomplished by adding a parameter smoothing relationship to the GWR model that stochastically restricts the estimates based on spatial relationships.

In addition to replicating the GWR estimates, the Bayesian model presented here can produce estimates based on parameter smoothing speciˇ cations that rely on distance, contiguity relationships, monocentric distance from a central point, or the latitude-longitude locations proposed by Casetti (1972).

The Bayesian GWR model also solves some problems that arise when the GWR model encounters non-constant variance over space or outliers. Given the locally linear nature of the GWR estimates, aberrant observations tend to contaminate entire sub-sequences of the estimates. The BGWR model robustifies against these observations by automatically detecting and downweighting their influence on the estimates. A further advantage of this approach is that a diagnostic plot can be used to identity observations associated with regions of non-constant variance or spatial outliers.

If the goal of locally linear estimation is to make inferences regarding spatial variation in the relationship, contamination from outliers may lead to an erroneous conclusion that the relationship is changing. In fact the relationship may be stable but subject to the influence of a single outlying observation. In contrast, the BGWR estimates indicate changes in the parameters of the relationship as we move over space that abstract from aberrant observations. From the standpoint of inference, we can be relatively certain that changing BGWR estimates truly reflect a change in the underlying relationship as we move through space. In contrast, the GWR estimates are more difficult to interpret, since changes in the estimates may reflect spatial changes in the relationship, or the presence of an aberrant observation.

A final issue that plagues the GWR is that conventional measures of dispersion may not be valid because the assumption of independence is not realistic given the reuse of sample observations. Bayesian estimates produced using the Gibbs sampler overcome these problems using measures of dispersion based on the posterior distributions derived from the Gibbs sampler that are not affected by a lack of sample independence.

Part III

Spatial Externalities

12 Hedonic Price Functions and Spatial Dependence: Implications for the Demand for Urban Air Quality

Kurt J. Beron[1], Yaw Hanson[2], James C. Murdoch[1], and Mark A. Thayer[3]

[1] University of Texas at Dallas
[2] Fannie Mae
[3] San Diego State University

12.1 Introduction

In 1967, Ronald Ridker and John Henning conducted the first study that linked air pollution to property values. Using census level data, they found that, for St. Louis, air pollution had a negative and significant affect on median housing prices. Research since has verified, modified, and redefined the economic interpretation of this relationship. In summarizing twenty-five years of property value/air pollution literature, Smith and Huang (1993, 1995) reported that approximately 74 percent of the studies found at least one significant air pollution variable. Even allowing for a publication bias toward significant findings, there seems to be a preponderance of evidence that air pollution is negatively related to housing prices. This is important because it reveals information about the willingness to pay for air quality – a non-market commodity. Moreover, to the extent that policymakers use the results from air pollution/property value studies, the findings are socially relevant. The South Coast Air Quality Management District, for example, uses a property value based model in formulating their Air Quality Management Plans.

In this paper, our goal is to re-examine the air pollution-property value relationship using a large, detailed data set that we specifically constructed for this purpose. Ultimately, we wish to present estimates for the demand for air quality. However, much of the analysis focuses on the hedonic regressions, wherein some measure of house price is the dependent variable and measures of the characteristics of housing; e.g., living area, existence of a pool, neighborhood quality, school district, etc., as well as measures of pollution are the independent variables. Like Can (1992) and Dubin (1988, 1992), we are worried that the potential for misspecifying the role of neighborhood quality as a determinant of housing prices is high. For us, however, this is relevant to the extent that it may significantly alter the estimate of the air pollution effect. We are also concerned that, even if we correctly specify the neighborhood influence, the measurement error in neighborhood level variables could affect the estimates on the air pollution variable.

To analyze these issues, we use the tools of spatial econometrics as defined by Anselin (1988b); i.e., tools for handling spatial dependence and spatial heterogeneity. Since, by definition, homes close to each other are "neighbors," problems measuring and modeling the neighborhood characteristics likely cause the errors in the hedonic regression model to be spatially dependent. By hypothesizing a structure for the spatial dependence, we can test for it and where appropriate use the information about the dependence to improve the efficiency of the estimators. Hence, our concerns regarding neighborhood effects, to the extent that they are captured by spatial dependence, can be analyzed with the tools from spatial econometrics. In a recent re-analysis of the Harrison and Rubinfeld (1978) data, Pace and Gilley (1997) demonstrated that the air pollution effect changed rather substantially after incorporating spatial dependence in the model. Therefore, it seems clear that a systematic study is warranted.

The study area is the South Coast Air Basin (SCAB), which provides the life-sustaining atmosphere for approximately 14 million people in four counties in Southern California: Los Angeles, Orange, Riverside, and San Bernardino. Urban air pollution is a significant problem. From 1983 to 1992, there were 2052 days (56 percent) where the Pollutant Standards Index (PSI) exceeded 100 ("unhealthful"), a rate more than triple that of the next worse US air shed – that covering the New York MSA (USEPA, 1993). The pollution problem has been addressed with a large dose of regulatory action. Incredibly, the regulatory polity, the South Coast Air Quality Management District (SCAQMD), employed more than 900 people at its Diamond Bar, CA facility in 1990. By some measures, the regulatory action appears to be working. For example, the maximum hourly readings for ozone have declined at most monitoring stations over the last 15 years. The extent to which the regulation is efficient remains an open question. The answer, of course, depends on numerous factors, one of which is the social valuation of improvements in air quality – the subject of this chapter.

The chapter is organized as follows. In the next section, we review some of the literature regarding the estimation of hedonic price functions and the demands for the characteristics. Then, we present the econometric issues followed by the estimations. Brief remarks are presented in the last section. An appendix contains complete descriptions of the data sources and variable names.

12.2 Hedonic Functions and Benefit Estimation

Given that the purpose of this paper is to present some estimates of the willingness to pay for air quality, the relevant take-off point is Ridker and Henning (1967) who interpreted their estimate on the air pollution term as a measure of the willingness to pay (WTP) for air quality improvements. Rosen (1974) and Freeman III (1974, 1979) noted that this interpretation was incorrect, stressing that the coefficients measured marginal willingness to pay (MWTP). They outlined a multi-step method for estimating the demand for a characteristic from which benefits (WTP) could be estimated. In the first step, the hedonic price function is estimated using data on home prices (e.g., sales price, rental price, or appraised price) and the characteristics of the home that are believed to influence the price (e.g., living area, school district, air pollution, etc.). Let p denote the price and Z a vector of characteristics. Then, the first step is to estimate $p(Z)$ which, assuming hedonic market equilibrium, describes the equilibrium prices. With an estimate for $p(Z)$ in hand, the MWTP for a particular characteristic, (z_i), is the partial derivative of the hedonic price function with respect to z_i: $\text{MWTP}_i = \partial p(Z)/\partial z_i = p_i(Z)$.

Following earlier work by Halvorsen and Pollakowski (1981), Atkinson and Crocker (1987), Leamer (1983b), Klepper and Leamer (1984), Spitzer (1984), and others, Graves *et al.* (1988) examined the robustness of hedonic MWTP estimates for air pollution using a systematic comparative analysis on a single data set. The relative impact of four specific sources of inaccuracy were studied: variable selection and treatment, functional form, measurement error, and error distribution. The primary result of this inquiry was that hedonic-based MWTP estimates could vary widely, dependent upon these various influences. From a policy perspective this is an uncomfortable situation as it implies that a wide range of willingness to pay estimates can be empirically "justified." Additionally, many of the issues remain confusing. For example, Graves *et al.* (1988) found that the functional forms generally used in hedonic studies (linear, log-linear, semi-log) were consistently outperformed by more flexible forms using the criteria of goodness of fit (see also Halvorsen and Pollakowski, 1981). However, Cassell and Mendelsohn (1985) and Cropper *et al.* (1988) argue that emphasis on goodness of fit measures was misplaced since this criterion does not guarantee the correct relationship between the focus and dependent variables. Graves *et al.* (1988) and Cropper *et al.* (1988) both suggest that part of the problem can be attributed to poor measurement and missing measures of the neighborhood variables. Thus, the tests and corrections for spatial dependence are particularly relevant in the context of this literature.

The second step in the Rosen-Freeman hedonic method involves estimating the underlying demand and supply functions for the characteristic of interest, using the previously estimated $p_i(Z)$. Initially, Rosen suggested that the identification of the demand and supply parameters represented a standard identification problem.[1] Follain and Jimenez (1985), Bartik (1987), Epple (1987), and Kahn and Lang (1988) however, noted that because consumers and firms choose the level of the characteristic (z_i) and $p_i(Z)$ simultaneously, the identification of the demand and supply functions was more complicated. The essential problem is that unmeasured individual (consumer or firm) tastes and preferences are correlated with the Z, making some of the independent variables in the second step correlated with the error terms. Hence, OLS estimates of the underlying demand and supply parameters are inconsistent and any inferences drawn from them (i.e., benefit estimates) highly suspect. The standard econometric approach in this situation is to use Instrumental Variables that are correlated with the Z yet uncorrelated with the error terms. However, the traditional method of using the exogenous variables from the supply equation as in-

[1] Brown and Rosen (1982) recognized that, within a single market, some functional forms for the hedonic (e.g., quadratic) could not be used to identify other functional forms of the demand (e.g., linear). They suggested that multiple market data would avoid this problem.

struments for the demand equation does not work in this case. The instruments need to be exogenous to the demand and supply.[2]

How can we find instruments? One way to proceed (Bartik, 1987; Follain and Jimenez, 1985; Palmquist, 1984) is to use multi-market data (determined by time or space) and estimate the hedonic price functions for each market. Then, measures of the markets (market dummy variables and interactions of the dummies with other demand variables) can be used as instruments for the Z. While this approach is recognized in the literature, very few multimarket hedonic studies have actually been performed, especially with respect to air pollution. In fact, we have found no recent studies that actually estimate the demand for air quality using the two-step procedure.

12.3 Econometric Issues

The point of departure is the data generating process that is assumed in most hedonic studies of environmental attributes:

$$y = S\beta + N\zeta + E\gamma + \varepsilon. \tag{12.1}$$

In (12.1), y denotes an n by 1 vector of the housing prices as measured by sales transactions, S is an n by j matrix of site specific characteristics (plus the constant), N is an n by k matrix of neighborhood characteristics, E is an n by l matrix of ambient environmental characteristics, β, ζ, and γ are, respectively, j, k, and l length vectors of unknown parameters, and ε is a random error vector. Estimation of the unknown parameters in equation (12.1) constitutes the first step in the hedonic methodology with the primary focus on γ.[3]

Our concern is with the impact of spatial dependence and spatial heterogeneity on the estimates of the γ parameters. In terms of spatial dependence, we follow the spatial econometrics literature and specify a spatial lag (LAG) and a spatial autoregressive error (SAR) model and then test them against the traditional model. The spatial dependence is described by a n by n spatial weights matrix, wherein each nonzero element represents the strength of the dependence between the observations with the row, column indices. For heterogeneity, we specify a model with spatial trend in housing prices and test it against a traditional model with fixed effects based on geographic areas. The spatial trend is modeled as a quadratic function of the latitude and longitude for each observation.

Let W denote a row standardized spatial weights matrix (zeros on the diagonal) that describes the spatial dependence. Then, the spatial lag model is:

$$y = \rho Wy + S\beta + N\zeta + E\gamma + \varepsilon, \tag{12.2}$$

[2] Follain and Jimenez (1985) note that the traditional simultaneity fails to obtain when using microlevel data; hence, it is not even necessary to incorporate the supply side variables into the demand estimation.

[3] The linear form is assumed for exposition. Other functional forms are often employed in practice.

while the spatial autoregressive error model is,

$$y = S\beta + N\zeta + E\gamma + \alpha W\varepsilon + \mu, \tag{12.3}$$

and μ is a random error vector. In (12.2), the estimate of ρ measures the spatial dependence, while in (12.3), the spatial parameter is α. The consequences of ignoring spatial dependence vary by speciˇ cation. If (12.2) is the true model and (12.1) is estimated with OLS, the estimates are biased and inconsistent. If (12.3) is the true model, then the OLS estimates are unbiased but inefˇ cient (Anselin, 1988b).

The parameters of both models can be estimated with the method of Maximum Likelihood (ML) and tested against (12.1) (Anselin, 1988b). In the case that the estimates for both ρ and α are signiˇ cant, Anselin and Bera (1998) offer useful Lagrange Multiplier (LM) tests that may help determine the type of dependence. In hedonic studies, both speciˇ cations seem possible, *a priori*. For example, the lack of adequate neighborhood measures in many studies suggests the SAR model; i.e., the errors of neighbors would tend to be spatially autocorrelated. The appraisal process (formal or informal), on the other hand, suggests the LAG speciˇ cation because the prices of neighboring properties in uence the price of the observation under consideration. Of course, neither model may be correct. Pace *et al.* (1998a) point out that the appraisal process usually means that the *previous* prices of neighboring houses actually in uence the price of the property under consideration. Moreover, we may have very rich measures of neighborhood and observe no spatial autocorrelation in the errors.

Turning to the second stage model, we wish to specify a statistical model for the MWTP for the components of E in (12.1). Let t denote the market. Then:

$$\text{MWTP}_t = \lambda G_t + \psi_t, \tag{12.4}$$

where G_t includes the environmental characteristic and "demand shifters" like income (net of housing expenditures) and education. As discussed above, the parameters of (12.4) need to be estimated with Instrumental Variables. With multimarket data, the instruments can be market dummy variables and interactions of other variables with the market dummies (Kahn and Lang, 1988).

The calculation of the MWTP is in uenced by the type of spatial dependence. We need the derivative of $\hat{y}$ with respect to E. In in the spatial lag model, (12.2), $\hat{y} = \hat{\rho} Wy + S\hat{\beta} + N\hat{\zeta} + E\hat{\gamma}$, so the derivative at a particular location depends on the prices of neighboring houses. To see the calculation in the spatial error model, (12.3), let $\hat{\varepsilon}$ denote the residuals deˇ ned by $y - S\hat{\beta} - N\hat{\zeta} - E\hat{\gamma}$. Then, the prediction of the dependent variable is $\hat{y} = S\hat{\beta} + N\hat{\zeta} + E\hat{\gamma} + \hat{\alpha} W\hat{\varepsilon}$. In this case the MWTP depends on neighboring residuals.

12.4 Estimates

Our empirical strategy is as follows. First, we employ an almost ideal data set for hedonic property value analysis. The list of variables included is given in Table 12.1,

Table 12.1. Variable description

Variable	Description
PRICE	sales price ($1,000)
LIV	living area (× 100 sq ft)
BATHS	bathrooms
FIRE	ˇ replaces
AIR	central air
HEAT	central heat
POOL	existence of pool dummy
LAND	land area (× 10,000 sq ft)
VIEW	existence of view dummy
TWORK	mean travel time to work
BDUM	within 5 miles of beach dummy
WHITE	percentage white
CRIME	FBI index of major crimes
BPOV	percentage below poverty
SCHOOL	district average assessment
ORANGE	Orange county dummy
RIVSIDE	Riverside county dummy
SANB	San Bernadino county dummy
AQ	120 less average PM10
NETINC	mean income less housing expenses ($1,000)
COLLEGE	percentage with college degree

and the mean values in the six years covered by our analysis are listed in Table 12.2. A detailed description of the data set and the steps taken to construct the speciˇ c variables is given in the Appendix. We feel that it is one of the largest and most detailed data sets ever used to look at the relationship between property values and air pollution. It contains numerous variables that measure the site-speciˇ c, neighborhood, and ambient air quality characteristics. Second, we use this data to produce estimates of a "traditional" hedonic price function, (12.1), and estimates of the WTP (demand) function for air quality, (12.4). Third, we employ the LM tests for the LAG and SAR models. This leads to the last step in the analysis, "introducing" the spatial dependence and comparing to the benchmark hedonic and WTP equations.

In order to highlight the in uence of the neighborhood variables, the spatial dependence, and the spatial heterogeneity on the WTP for air quality, we look at sets of three models. In Model 1, we include all of the neighborhood variables, while in Model 2, we drop the county dummies. Thus, Model 2 highlights the in uence of large scale heterogeneity. Then, in Model 3, we drop all of the city, school district and census tract level variables in order to focus on the role of the localized variables. Model 2 is nested within 1 and Model 3 is nested within 2 and, therefore, within 1 as well. Each of these models is then estimated with the quadratic

Table 12.2. Descriptive statistics

Variable	1980	1983	1986	1989	1992	1995
PRICE	103.03	119.62	151.70	236.20	227.51	198.47
LIV	16.12	16.13	16.12	15.76	15.79	15.72
BATHS	1.88	1.86	1.89	1.84	1.84	1.81
FIRE	0.68	0.69	0.66	0.64	0.64	0.62
AIR	0.28	0.26	0.30	0.27	0.25	0.24
HEAT	0.19	0.20	0.21	0.20	0.19	0.21
POOL	0.15	0.16	0.16	0.14	0.15	0.16
LAND	0.88	0.91	0.87	0.85	0.85	0.96
VIEW	0.04	0.03	0.03	0.03	0.03	0.03
TWORK	28.31	28.61	28.75	29.45	28.91	28.86
BDUM	0.02	0.03	0.02	0.02	0.02	0.02
WHITE	79.01	70.56	70.52	58.24	59.28	57.98
CRIME	68.59	68.47	68.68	72.51	69.22	70.53
BPOV	9.01	8.98	9.01	10.03	9.53	10.26
SCHOOL	251.35	251.64	252.16	257.99	259.08	255.37
ORANGE	0.21	0.19	0.19	0.17	0.18	0.16
RIVSIDE	0.03	0.05	0.06	0.07	0.05	0.05
SANB	0.12	0.11	0.11	0.11	0.08	0.08
AQ	60.91	74.26	67.63	63.15	80.30	77.82
NETINC	48.59	49.22	49.32	47.16	48.54	47.32
COLLEGE	19.61	22.24	22.35	22.73	23.91	23.23

expansion of the X, Y coordinates in order to model the spatial trend. These two sets of estimates are referred to as OLS and OLS XY, respectively. The estimates for the semilog form of the hedonic functions in 1992 are presented in Table 12.3.[4] While minor differences appear in the other years, the results in Table 12.3 offer a good representation of the full set of estimates. Generally, the estimates on the site-specific and neighborhood characteristics are significant and of the anticipated sign. The notable exceptions are coefficient estimates on CRIME and AIR. Turning to the XY specifications (OLS XY), we see some important changes in the estimates. First, notice how much closer the log-likelihoods are for Models 1 and 2. In fact, with the OLS XY model we can not reject the restriction that sets the coefficients on the county dummies equal to zero (0.025 level of significance). Had we started with

[4] The semilog form was selected on the basis of some Box-Cox estimations. We looked at the Box-Cox linear form (the right-hand side is linear, while the dependent variable is transformed) and the Box-Cox quadratic form (the right-hand side is quadratic, while the dependent variable is transformed). In both specifications the transformation parameter, albeit significant, was close to zero. The highest value for the transformation parameter was less than 0.25. Thus, we felt that the semilog form offered an adequate representation of the model.

Table 12.3. OLS estimates of the semilog hedonic price functions (1992)

	OLS			OLS XY		
Variable	Model 1	Model 2	Model 3	Model 1	Model 2	Model 3
LIV	0.02952	0.0316	0.03362	0.02911	0.02924	0.03241
BATHS	0.08058	0.05442	0.08291	0.0862	0.08556	0.09336
FIRE	0.07641	0.07764	0.09265	0.07373	0.07284	0.09614
AIR	0.0157*	−0.0054*	−0.0002*	0.0269	0.0275	0.0323
HEAT	0.04614	0.05728	0.04929	0.04843	0.04507	0.05855
POOL	0.03373	0.05913	0.07777	0.03533	0.03843	0.06263
LAND	0.01519	0.01271	0.01349	0.01623	0.01633	0.01808
VIEW	0.06663	0.09612	0.09552	0.07303	0.07617	0.08654
TWORK	−0.00701	−0.00921		−0.0071	−0.00739	
BDUM	0.16108	0.13452	0.17071	0.17484	0.17652	0.22039
WHITE	0.00362	0.00193		0.00374	0.00356	
CRIME	−0.0006*	−0.001		−0.0006*	−0.00047	
BPOV	−0.00433	−0.00709		−0.00419	−0.00461	
SCHOOL	0.00109	0.00086		0.00112	0.00114	
ORANGE	−0.11346			−0.036*		
RIVSIDE	−0.36872			−0.09176		
SANB	−0.31504			−0.10079		
AQ	0.01155	0.02022	0.0242	0.01094	0.01152	0.02067
X				−0.0036*	−0.5097*	20.9587
Y				151.537*	153.7677*	−422.6631
X^2				−1.19588	−1.50337	−1.663
Y^2				−41.257*	−42.204*	117.0434
XY				1.226*	1.675*	−3.793
INT	10.3783	9.9133	9.4303	−276.572*	−280.571*	758.108*
LOGLIK	−30833.2	−31017.9	−31383.0	−30783.6	−30789.6	−31230.0
LM-ERR	2029.8	3134.6	5219.0	1700.2	1725.4	3752.6
LM-LAG	224.1	327.6	635.7	199.7	202.6	506.5
RLM-ERR	1828.4	2834.3	4649.8	1523.3	1546.0	3311.9
RLM-LAG	22.8	23.2	65.8	22.9	23.2	65.8

All estimates are statistically signiˇ cant at $p = 0.05$ except for those indicated by *

the XY speciˇ cation, we would have dropped the county dummies on the basis of a statistical test, concluding that the county dummies duplicated the spatial trend captured by the X, Y coordinates. Second, consider the estimates on the AIR variable. In the OLS XY models, they are signiˇ cant and of the expected sign. Evidently, central air conditioning is spatially correlated, probably re ecting the relationship between distance to the beach and weather. Interestingly, BDUM is not seriously affected by the inclusion/exclusion of the X, Y coordinates.

Of particular interest are the estimates on the AQ measure, which are positive and significant in every estimation in every year. Within any particular year, the AQ estimates are rather stable between the OLS and OLS XY specifications, especially when compared to the estimates on AIR. As shown in Models 2 and 3, the AQ estimates seem more sensitive to inclusion/exclusion of the neighborhood variables. Hence, our initial concern about correctly measuring and modeling the neighborhood appears justified.

The Lagrange Multiplier tests (Anselin, 1988b) for spatial dependence in the error (LM-ERR), spatial lagged dependent variable (LM-LAG) and their robust counterparts (Anselin and Bera, 1998), RLM-ERR and RLM-LAG are also displayed in Table 12.3. The LM tests are based on the OLS estimates and a hypothesized spatial weights matrix, W. The specification of W is somewhat ad hoc and alternative specifications should be considered in future research (Bell and Bockstael, 2000). Here, we give a weight equal to 1 for observations within 1.5 miles and 0 for observations beyond 1.5 miles. This gives a n by n matrix with zeros on the diagonal and either zeros or ones in the off-diagonal elements. For (say) the first row, a 1 in the 2000th column would indicate that house 1 and house 2000 are within 1.5 miles of each other. The actual W matrix used in the analysis is row standardized. Thus, if for house 1 there are 30 other houses within 1.5 miles, then each weight will be 1/30.[5]

Both the LM-ERR and LM-LAG indicate nonzero α and ρ. Unfortunately, the robust versions fail to rule out one of the models. However, both the LM-ERR and the RLM-ERR are much larger than the LAG statistics. Following Anselin and Rey (1991), we suggest that the SAR structure like that in equation (12.3) is more likely than the lagged dependent variable structure, and we proceed to estimate the SAR models.

The SAR estimates corresponding to those in Table 12.3 are presented in Table 12.4. Looking at Table 12.4, we see significant estimates of the autocorrelation parameters in every model.[6] Not surprisingly, as the neighborhood variables are dropped from the model, the autocorrelation generally strengthens; i.e., $\hat{\alpha}$ approaches one.

Comparing the AQ estimates in Table 12.4 with those in Table 12.3, we see, in contrast to Pace and Gilley (1997), very minor differences. As noted above, AIR is rather unstable between the OLS specifications. Moving to the SAR estimates, however, we see that the site-specific characteristics estimates are basically invariant with respect to the model. Apparently, AIR is partially measuring a localized variable (perhaps vintage) that is effectively filtered by the SAR model. Similarly, VIEW and TWORK are significantly altered in the SAR model. In both cases, the point estimates

[5] All of the estimations were performed in Matlab, which takes advantage of the sparseness of the W matrices. We benefited greatly from the set of Matlab functions written by Pace and Barry (1998).

[6] Significance of α is tested by comparing the log-likelihoods from Table 12.3 to their corresponding value in Table 12.4. For example, the Model 1 log-likelihood from the OLS model is -30833.2, while from Table 12.4 the corresponding value is 30469.8. Minus two times the difference is distributed χ^2 with one degree of freedom under the null hypothesis that $\alpha = 0$. The value of 726.8 indicates rejecting the null hypothesis.

Table 12.4. Maximum Likelihood estimates of the semilog hedonic price functions (1992)

	SAR			SAR XY		
Variable	Model 1	Model 2	Model 3	Model 1	Model 2	Model 3
LIV	0.02492	0.02491	0.02547	0.02487	0.02493	0.02549
BATHS	0.08916	0.08657	0.08988	0.09033	0.09044	0.0911
FIRE	0.05751	0.05576	0.0614	0.057	0.05625	0.06191
AIR	0.03067	0.02753	0.03306	0.03267	0.0317	0.03562
HEAT	0.0457	0.04565	0.04899	0.04692	0.04435	0.05022
POOL	0.05069	0.05457	0.05981	0.05073	0.05163	0.05895
LAND	0.01496	0.01443	0.01454	0.01515	0.01523	0.01508
VIEW	0.0228*	0.0226*	0.022*	0.0246*	0.025*3	0.023*
TWORK	−0.0027*	−0.0037*		−0.0027*	−0.0032*	
BDUM	0.0766*	0.055*	0.0753*	0.08426	0.08596	0.09701
WHITE	0.00403	0.00326		0.00405	0.00397	
CRIME	−0.0008*	−0.00095		−0.0008*	−0.0008*	
BPOV	−0.00356	−0.0042		−0.00369	−0.00377	
SCHOOL	0.00091	0.00078		0.00096	0.00098	
ORANGE	−0.08257			0.02123		
RIVSIDE	−0.36159			−0.09738		
SANB	−0.32645			−0.16057		
AQ	0.01294	0.02215	0.02481	0.01037	0.01237	0.0206
X				−12.304*	0.8907*	26.727*
Y				127.225*	159.062*	−320.425*
X^2				−0.88892	−1.41131	−1.3812
Y^2				−37.838*	−43.174*	91.759*
XY				4.149*	1.195*	−5.637*
INT	10.23957	9.57162	9.51989	−206.53*	−292.83*	554.99*
α	0.63	0.69	0.75	0.62	0.62	0.73
LOGLIK	−30469.8	−30499.7	−30602.3	−30459.6	−30463.1	−30590.6

All estimates are statistically signiˇ cant at $p = 0.05$ except for those indicated by *

are much less in the SAR, perhaps indicating that these variables are measuring additional localized characteristics.

Four sets of demand functions are presented in Tables 12.5 and 12.6, corresponding to hedonic model estimates illustrated in Tables 12.3 and 12.4. Table 12.5 shows the estimates from the OLS models (i.e., from models like those displayed in the ˇ rst three columns of Table 12.3), and from the OLS XY models. The corresponding results for the SAR models are given in Table 12.6. The demand estimations follow the procedures outlined by Epple (1987), Bartik (1987), and Kahn and Lang (1988) and are based on all six years of data. First, the AQ and the hedonic price of AQ ($\partial\hat{y}_i/\partial \text{AQ}_i$) for each observation in each year are merged with their corresponding census tract average household income net of housing expenditures (NETINC)

Table 12.5. Estimates of the demand for air quality – OLS-based[a]

	OLS			OLS XY		
Variable	Model 1	Model 2	Model 3	Model 1	Model 2	Model 3
AQ	91.401	−86.578	−103.124	−61.024	−63.252	−156.853
NETINC	0.045	0.045	0.048	0.026	0.023	0.033
COLLEGE	11.111	78.900	85.702	52.541	49.149	96.418
Y80	1043.053	−3659.585	−4016.066	−1952.407	−1955.320	−5169.665
Y83	288.733	−1689.666	−1588.902	229.632	345.399	−1310.104
Y86	846.014	−1573.715	−1843.249	−1195.880	−838.479	−2726.102
Y89	3979.607	1415.894	1861.874	1024.280	1012.716	1363.560
Y92	−233.461	613.329	1187.725	316.477	924.292	1208.375
INT	−6978.279	6789.801	8123.154	4505.679	4449.646	12112.933
R^2	0.36	0.34	0.23	0.34	0.34	0.29
Mean HP	2902	3796	4342	2376	2106	3659
WTP 10 percent	15639	30885	33803	17334	14223	30489

[a] Estimation by 2SLS; all estimates are statistically signiˇ cant at $p = 0.05$

Table 12.6. Estimates of the demand for air quality – SAR-based[a]

	SAR			SAR XY		
Variable	Model 1	Model 2	Model 3	Model 1	Model 2	Model 3
AQ	120.916	−73.986	−72.820	−87.251	−116.866	−179.354
NETINC	0.047	0.044	0.045	0.022	0.018	0.027
COLLEGE	13.751	96.932	105.857	70.686	77.663	126.987
Y80	1528.625	−3293.408	−3362.188	−2689.792	−3353.222	−5908.976
Y83	416.906	−1428.680	−1191.416	86.585	−148.146	−1322.349
Y86	1342.152	−1434.068	−1433.395	−1665.807	−1688.406	−2901.287
Y89	4604.529	1883.338	2491.671	528.587	−0.437	416.860
Y92	−115.314	738.099	963.782	−62.273	711.419	748.401
INT	−9294.241	5708.577	5723.483	6636.142	8724.706	13918.270
R^2	0.40	0.41	0.39	0.38	0.36	0.37
Mean HP	3114	4185	4676	2476	2315	3874
WTP 10 percent	15719	32222	34650	20148	19205	34154

[a] Estimation by 2SLS; all estimates are statistically signiˇ cant at $p = 0.05$

and percentage of the population with a college degree (COLLEGE). Then, a linear speciˇ cation of the implicit demand for AQ is estimated using Two-Stage Least Squares (2SLS). The instruments for AQ are the year dummies and the interaction of the dummies with the exogenous variables NETINC and COLLEGE (Kahn and Lang, 1988). At a minimum, the estimates in Tables 12.5 and 12.6 provide a mechanism for analyzing the empirical consequences of the alternative hedonic models. Ideally, they provide relevant information on the WTP for air quality. The "bottom line" for each set of estimates gives the estimated household WTP for a 10 percent change in AQ and offers a uniform measure for comparing models.[7]

Substantial differences are evident in the OLS estimates. First, the slope of the demand curve is actually positive for Model 1. Second, the WTP estimates essentially double from Model 1 to Model 2 and, somewhat surprisingly, third, the coefˇ cients on the dummies vary dramatically from Model 1 to Model 2. Returning to Table 12.3, the restrictions imposed in Model 2 (Model 3) can be tested by the standard likelihood ratio test; i.e., minus two times the difference in the log-likelihoods. Scanning the log-likelihood values it is clear that the restricted models can not be statistically justiˇ ed, implying that Model 1 should be maintained.

Looking at the OLS XY demand estimations, we see very little difference between the Model 1 and Model 2 estimates. As noted above, the spatial expansion terms effectively remove the in uence of the county dummies. This highlights an important issue for beneˇ t analysis from hedonic price functions. We are not sure about the speciˇ cation of the hedonic function and the choices that we make regarding inclusion/exclusion of the uncertain variables signiˇ cantly alter the beneˇ t assessment. While we can often rely on a statistical test to select among speciˇ cations, it is never obvious where to start; i.e., it is difˇ cult to choose the unrestricted model.

While the variability in the WTP estimates between Model 1 and Model 2 is greatly reduced in the OLS XY estimations, when compared to the OLS estimations, the addition of the X,Y coordinates does little to reduce the impact of the neighborhood variables (Model 3). *A priori* we expected that the SAR would capture these effects. Looking at the SAR demand estimations, however, we see that in terms of the beneˇ ts of improving air quality, the SAR speciˇ cation actually has very little empirical impact.

12.5 Conclusions

From a policy analysis point-of-view, large ranges in beneˇ t estimates are a source of uncertainty concerning the economic consequences of a particular policy action. We have illustrated that, in the case of urban air pollution, the beneˇ ts estimates from hedonic studies depend on ad hoc choices about the speciˇ cation of the model. Ideally, we would like to identify a speciˇ cation or set of speciˇ cations that offer less variability yet accurately re ect the property value market. Introducing localized spatial dependence (within 1.5 miles), while providing a statistically superior speciˇ cation did little to help reduce the beneˇ t variability. Clearly, we need to expand and explore other structures of spatial dependence. In particular, a look at models with dependence out to 3 and 5 miles and some models with weights that decline with distance appears warranted. On the other hand, by speciˇ cally modeling

[7] For this calculation, we use NETINC = 50000, COLLEGE = 22, all dummies equal to zero, and AQ = 70. Thus, the estimated function is integrated over AQ from 70 to 77, a 10 percent change.

the spatial trend in the property value market, we did “remove” the county dummies as a source of variability. Thus, it seems worthwhile to more fully consider characterizations of the trend. This suggests that hedonic studies could beneˇ t from three dimensional exploratory spatial data analysis of the residuals and dependent variable.

Acknowledgments

This research was supported by grants from NSF/EPA and the South Coast Air Quality Management District.

Appendix: Data Sources

The property value and site-speciˇ c characteristics data were purchased from the Experian Company (formerly TRW-REDI). Each data record represents a single family home in the South Coast Air Basin counties of Los Angeles, Orange, Riverside, and San Bernardino in 1996. We geocoded the data using the 1995 TIGER Line ˇ les and ARC/INFO.[8] The records contain the last two previous sales transactions, facilitating a temporal organization of the sales data. From 1980 through 1995, we have over 1.6 million sales transactions.[9]

The site-speciˇ c characteristics are organized into quantity and quality measures. House size or quantity is described through such variables as square footage of living space (LIV), number of bathrooms (BATHS), and lot size (LAND). The quality variables are number of ˇ replaces (FIRE), existence of centralized heating (HEAT), air conditioning (AIR), existence of a pool (POOL), and existence of a view (VIEW).[10]

The neighborhood characteristics are measured by census tract, school district, and city variables. From the TIGER ˇ les, we attached the 1990 census tract, city, and school district to the sales transactions records, facilitating the addition of census

[8] The success rate in the geocoding process was about 80 percent.

[9] We admit to some ignored selectivity issues. We do not have the earlier transactions for properties that were sold more than twice over the 1980–1995 period. Thus, the size of the annual transactions tails off as we get further away from 1995. One danger in using these previous sale transactions is that the characteristics recorded from the last sale may re ect improvements to the property that were not re ected in the previous sale. For example, consider a home that sold in 1982 for $100,000. In 1984, a new bathroom was added and then the home sold in 1987 for $200,000. We do not have a mechanism for knowing that the bathroom was added after the 1982 transaction, although there is an improvement code that may facilitate a crude analysis of the magnitude of the problem.

[10] We should point out that the data set contains many other measures of quantity and quality; i.e., garage characteristics, number of bedrooms, locational in uences, type of roof, etc. We feel that the set presented here fairly represent the site-speciˇ c effects. We have run numerous other speciˇ cations and have not noticed that the environmental variables are impacted by alternative sets of site-speciˇ c variables.

tract data, city data and school district data to the sales data. Given that the census geography changed from 1980 to 1990, we used the Census Bureau's published relationship between 1980 and 1990 in order to attach 1980 census tract data. A problem that we encountered here is that the relationship is many-to-many. Thus, we were forced to use census tract measures that were standardized by population rather than aggregates or measures standardized by area. Once again, the 1980 and 1990 census files offer numerous options for constructing measures of neighborhood characteristics. Here, we use the percent of the population below the poverty level (BPOV), the percent of the population that is white (WHITE), and the average travel time to work (TWORK). Because the census tract data are only available for 1980 and 1990, we must arbitrarily assign some data to the other years. We decided to use the 1990 data for all years from 1989–1995, the simple average for 1986 and 1983, and the 1980 data for 1980. Clearly, we could devise numerous other schemes for assigning the data so we are particularly interested in how important the census tract variables are to the findings regarding air quality.

Using the city codes, we attached the FBI index of seven major crimes (CRIME). While the crime data are available each year, we have several city codes without a corresponding FBI crime index. To assign the data, we used the county average of the index and assigned this to the cities with missing values. The school district code allowed us to merge California's Assessment Test results (SCHOOL) as a measure of school district quality. Unfortunately, the school district data are incomplete. California apparently changed the make-up of the test after 1993. Moreover, some of the tests were not administered in every district in every year. We assigned the composite 1981 score to 1980 and 1983 data, the 1986 math score to 1986, 1989 math score to 1989, and the 1992 math score to 1992 and 1995 data. Once again this is an arbitrary process that could affect the air quality results.

For air pollution, we used measures of particulate matter that we obtained from the SCAQMD for approximately 40 sites scattered around the study area. The raw particulate matter data are every sixth day readings, giving either 60 or 61 observations per station per year. During our study period (1980–1995), the policy emphasis shifted from total suspended particulates (TSP) to particulate matter of size 10 microns or less (PM10). Hence, in the early years we have few (none before 1984) stations recording PM10, while in the later years the number of TSP readings drops off. Others have used a conversion factor of 0.55 to convert TSP to PM10. We had enough overlap in PM10 and TSP readings that we were able to run a series of regressions to estimate the conversion. We found a conversion of approximately 0.57, which we employed to scale all of the TSP data. In this fashion, we were able to get comparable PM10 data over the 1980–1995 period. Our variable, PM10, is the annual average of the PM10 data.

The PM10 measure provides a summary for a particular location in the study area; i.e., that of the associated monitor. In order to assign these data to the site-specific and neighborhood data, we used a kriging routine. The kriged data were merged with the geocoded property value data using the X, Y coordinates. For the hedonic functions, we define a new variable, AQ, as 120 minus PM10 so that the expected sign on the AQ term is positive.

In addition to the site-specific, neighborhood, and air quality variables, we considered three measures that address spatial heterogeneity. The first is the influence of the beach on property values in Southern California. Because properties close to the beach sell at a premium, all else equal, we have included a dummy variable (BDUM) that is equal to one for homes within 5 miles of the beach and zero otherwise. The second is a set of county dummies, reflecting the four counties in the study area (LA, the left out dummy, ORANGE, RIVSIDE for Riverside, and SANB for San Bernadino). The county dummies were the source of the instability in the air quality effects reported in Graves *et al.* (1988). Thus, we are particularly interested in how these influence the estimates. The third measure is spatial trend. Following Dubin (1992) and Pace and Gilley (1997), some of our specifications include a quadratic expansion of the coordinates of the homes. The variables X, Y are the coordinates, X^2 and Y^2 are one-half times the coordinates squared, and XY is the interaction.

Given our concern with spatial dependence, we reduced the size of the data set by random sampling from six years of data. For temporal coverage, we selected 1980, 1983, 1986, 1989, 1992, and 1995 as the years for analysis. Within each of these years, we drew a random sample of 10,000 observations from the main dataset, giving us 60,000 observations. After removing the observations with obvious coding errors (like lot size less than interior living area) and missing values, we were left with 51,110 observations, 8478 in 1980, 8696 in 1983, 8516 in 1986, 8364 in 1989, 8587 in 1992, and 8469 in 1995. Variable names, a brief description, and the annual means are presented in Tables 12.1 and 12.2.

13 Prediction in the Panel Data Model with Spatial Correlation

Badi H. Baltagi[1] and Dong Li[2]

[1] Texas A&M University
[2] Kansas State University

13.1 Introduction

The econometrics of spatial models have focused mainly on estimation and testing of hypotheses, see Anselin (1988b), Anselin *et al.* (1996) and Anselin and Bera (1998) to mention a few. In this chapter we focus on *prediction* in spatial models based on panel data. In particular, we consider a simple demand equation for cigarettes based on a panel of 46 states over the period 1963–1992. The spatial autocorrelation due to neighboring states and the individual heterogeneity across states is taken explicitly into account. In order to explain how spatial autocorrelation may arise in the demand for cigarettes, we note that cigarette prices vary among states, primarily due to variation in state taxes on cigarettes. For example, in 1988, state excise taxes ranged from 2 cents per pack in a producing state like North Carolina, to 38 cents per pack in the state of Minnesota. In 1997, these state taxes varied from a low of 2.5 cents per pack for Virginia to $1.00 per pack in Alaska and Hawaii. Since cigarettes can be stored and are easy to transport, these varying taxes result in casual smuggling across neighboring states. For example, while New Hampshire had a 12 cents per pack tax on cigarettes in 1988, neighboring Massachusetts and Maine had a 26 and 28 cents per pack tax. Border effect purchases not explained in the demand equation can cause spatial autocorrelation among the disturbances.[1]

In this chapter, we model the demand for cigarettes as follows:

$$y_{it} = x'_{it}\beta + \varepsilon_{it} \qquad i = 1,...,n;\ t = 1,...,T, \tag{13.1}$$

where y_{it} denotes the real per capita sales of cigarettes by persons of smoking age (14 years and older) measured in packs per head. The explanatory variables include the average retail price of a pack of cigarettes measured in real terms, and the real per capita disposable income of each state. All variables are expressed in logarithms and the estimated coefˇ cients represent elasticities. The dimensions of the panel are $n = 46$ states and $T = 30$ years. We only use the ˇ rst 25 years for estimation and reserve the last 5 years for out of sample forecasts. Details on data sources are given in Baltagi and Levin (1986). Here, we update the data 12 years from 1981

[1] Alternatively, one can model this using spatially lagged regressors like population density of neighboring states and prices and incomes of neighboring states. In fact, Baltagi and Levin (1986) used the minimum price in neighboring states to capture border effects purchases.

to 1992. The disturbance term follows an error component model with spatially autocorrelated residuals (see Anselin, 1988b, p. 152). The disturbance vector for time t is given by:

$$\varepsilon_t = \mu + \phi_t, \tag{13.2}$$

where $\varepsilon_t = (\varepsilon_{1t}, ..., \varepsilon_{nt})'$, $\mu = (\mu_1, ..., \mu_n)'$ denotes the vector of state effects and $\phi_t = (\phi_{1t}, ..., \phi_{nt})'$ are the remainder disturbances which are independent of μ. The ϕ_t's follow the spatial autoregressive error dependence model:

$$\phi_t = \lambda W \phi_t + \nu_t, \tag{13.3}$$

where W is the matrix of known spatial weights of dimension n by n and λ is the spatial autoregressive coefˇ cient. The elements of the error vector $\nu_t = (\nu_{1t}, ..., \nu_{nt})'$ are iid$(0,\sigma_\nu^2)$ and are independent of the elements of ϕ_t and μ. The spatial matrix W is constructed as follows: a neighboring state takes the value 1, otherwise it is zero. The rows of this matrix are standardized so that they sum to one. The μ_i's are the unobserved state speciˇ c effects which can be ˇ xed or random (see Hsiao, 1986; Baltagi, 1995). State speciˇ c effects include but are not limited to the following:

- Indian reservations sell tax-exempt cigarettes. States with Indian reservations like Montana, New Mexico and Arizona are among the biggest losers of tax revenues from these tax exempt sales. The Advisory Commission on Intergovernmental Relations (1985) estimated a loss of $309 million from tax exemption or tax evasion in 1983.
- States with tax exempt military bases like Florida, Texas, Washington and Georgia also lose revenues from these tax exempt sales.
- Utah, a state with a high percentage of Mormons (a religion which forbids smoking) had a per capita sales of cigarettes in 1988 of 55 packs, a little less than half the national average of 113 packs.
- Nevada, a highly touristic state, has per capita sales of cigarettes above the national average.

Not accounting for these state speciˇ c effects may lead to biased estimates.

13.2 Estimation

13.2.1 Pooled Model

Table 13.1 reports the estimates of a simple, albeit naive demand model for cigarettes using pooled OLS.[2] These estimates ignore the states heterogeneity and the spatial autocorrelation. The price elasticity estimate is -0.62, while the income elasticity estimate is 0.11 and both are statistically signiˇ cant. Next, we take into account the spatial autocorrelation, and estimate the model using Maximum Likelihood (MLE)

[2] For a dynamic demand model of cigarettes, see Baltagi and Levin (1986) and for a rational addiction model, see Becker *et al.* (1994).

Table 13.1. Pooled estimates of cigarette demand[a]

	Price	Income
Pooled OLS	−0.618	0.114
	(−13.7)	(4.00)
Pooled spatial	−0.882	0.285
	(−16.4)	(8.29)
Average heterogeneous OLS	−1.193	0.476
	(−37.6)	(26.4)
Average spatial MLE	−1.235	0.505
	(−39.1)	(27.2)
Fixed Effects	−0.474	−0.259
	(−17.7)	(−12.6)
Fixed Effects−spatial	−0.775	−0.131
	(−20.7)	(−3.45)
Random Effects	−0.474	−0.251
	(−17.8)	(−12.3)
Random Effects−spatial	−0.803	−0.070
	(−20.8)	(−1.77)

[a] The F-statistic for $H_0 : \mu = 0$ yields a value of 88.95, which is statistically signiˇ cant. The one-side Breusch-Pagan test for $H_0 : \sigma_\mu^2 = 0$ yields a N(0,1) test statistic of 81.1 which is statistically signiˇ cant. Hausman's test based on ˇ xed andrandom effects yields a χ_2^2 of 26.8 which is statistically signiˇ cant.

described in Anselin (1988b), but ignoring the heterogeneity across states. This is reported as "Pooled spatial" in Table 13.1. This yields slightly higher price (-0.88) and income elasticities (0.29) than OLS ignoring the spatial correlation. Both elasticities are signiˇ cant. The estimate of λ is 0.41.[3] In addition, we conducted a grid search procedure over λ to ensure a global maximum. The likelihood ratio test for $\lambda = 0$ yields a value of 120.8 which is asymptotically distributed as χ_1^2 under the null hypothesis. The null is rejected justifying concern for spatial autocorrelation.

13.2.2 Time-Wise Heterogeneity

Table 13.2 allows for different parameter (heterogeneous) estimates for each year. The ˇ rst set of estimates gives the cross-sectional demand equation estimates using OLS for each year. The price elasticity estimates varied between −0.66 in 1963 to −1.44 in 1967, while the income elasticity estimates varied between 0.16 in 1980 to a high of 0.83 in 1968. Pesaran and Smith (1995) suggested averaging these heterogeneous estimates to obtain a pooled estimator. This yields a price elasticity estimate of −1.19 and an income elasticity estimate of 0.48, both of which are signiˇ cant. These are reported as "average heterogeneous OLS" in Table 13.1.

[3] This was obtained using the OPTMUM procedure of GAUSS version 3.2.37.

Table 13.2. Heterogeneous estimates of cigarette demand

	Heterogeneous OLS		Heterogeneous Spatial			
	Price	Income	Price	Income	λ	LM*
1963	−0.663	0.718	−0.625	0.730	0.097	0.278
	(−1.925)	(5.621)	(−1.841)	(5.600)	(0.517)	(0.597)
1964	−1.215	0.619	−1.210	0.622	0.039	0.044
	(−3.368)	(4.629)	(−3.450)	(4.712)	(0.206)	(0.834)
1965	−1.204	0.634	−1.203	0.635	0.003	0.000
	(−3.465)	(4.525)	(−3.563)	(4.575)	(0.021)	(0.986)
1966	−1.429	0.736	−1.435	0.740	0.070	0.218
	(−4.438)	(4.710)	(−4.526)	(4.743)	(0.411)	(0.641)
1967	−1.438	0.791	−1.455	0.797	0.081	0.331
	(−4.494)	(5.426)	(−4.571)	(5.452)	(0.489)	(0.565)
1968	−1.411	0.831	−1.417	0.833	0.030	0.040
	(−4.478)	(5.861)	(−4.526)	(5.969)	(0.175)	(0.842)
1969	−1.155	0.787	−1.164	0.790	0.044	0.080
	(−4.609)	(5.502)	(−4.669)	(5.583)	(0.251)	(0.777)
1970	−0.998	0.779	−1.010	0.786	0.067	0.209
	(−4.078)	(4.929)	(−4.135)	(4.960)	(0.395)	(0.648)
1971	−0.882	0.661	−0.882	0.667	0.062	0.200
	(−3.129)	(3.669)	(−3.195)	(3.710)	(0.377)	(0.655)
1972	−1.003	0.573	−1.028	0.600	0.148	1.191
	(−3.955)	(2.872)	(−4.078)	(2.905)	(0.923)	(0.275)
1973	−1.022	0.394	−1.072	0.442	0.195	1.966
	(−3.980)	(1.964)	(−4.093)	(2.097)	(1.213)	(0.161)
1974	−1.048	0.432	−1.102	0.463	0.189	1.820
	(−4.353)	(2.179)	(−4.440)	(2.261)	(1.169)	(0.177)
1975	−1.142	0.400	−1.207	0.435	0.179	1.576
	(−4.681)	(2.096)	(−4.763)	(2.198)	(1.091)	(0.209)
1976	−1.245	0.443	−1.450	0.510	0.298	4.056
	(−4.666)	(2.189)	(−4.921)	(2.402)	(1.859)	(0.044)
1977	−1.278	0.381	−1.448	0.456	0.291	3.769
	(−4.638)	(1.913)	(−4.899)	(2.176)	(1.782)	(0.052)
1978	−1.308	0.298	−1.482	0.419	0.287	3.092
	(−4.482)	(1.528)	(−4.758)	(1.963)	(1.671)	(0.078)
1979	−1.253	0.270	−1.314	0.319	0.140	0.803
	(−4.217)	(1.484)	(−4.296)	(1.657)	(0.802)	(0.370)
1980	−1.267	0.164	−1.289	0.191	0.089	0.341
	(−3.903)	(0.920)	(−4.017)	(1.037)	(0.516)	(0.560)
1981	−1.275	0.300	−1.493	0.432	0.336	4.083
	(−4.733)	(1.890)	(−5.262)	(2.512)	(2.000)	(0.043)
1982	−1.263	0.316	−1.280	0.344	0.160	1.258

continued on next page

Table 13.2. Continued

	Heterogeneous OLS		Heterogeneous Spatial			
	Price	Income	Price	Income	λ	LM*
	(−4.212)	(1.867)	(−4.375)	(2.016)	(0.973)	(0.262)
1983	−1.433	0.295	−1.480	0.340	0.281	3.963
	(−5.086)	(1.971)	(−5.593)	(2.239)	(1.777)	(0.047)
1984	−1.263	0.327	−1.253	0.316	0.301	5.510
	(−4.407)	(2.205)	(−4.670)	(2.180)	(2.046)	(0.019)
1985	−1.235	0.260	−1.231	0.256	0.222	2.115
	(−4.681)	(1.955)	(−4.757)	(1.920)	(1.336)	(0.146)
1986	−1.328	0.289	−1.317	0.300	0.254	3.220
	(−4.338)	(2.047)	(−4.509)	(2.098)	(1.600)	(0.073)
1987	−1.064	0.209	−1.040	0.208	0.329	4.922
	(−3.584)	(1.556)	(−3.698)	(1.519)	(2.099)	(0.026)

* LM statistic for $\lambda = 0$ with p-value in parentheses.

These individual cross-section regressions and their average do not take the spatial autocorrelation into account. Using the normality assumption, we re-estimate these cross-sectional demand equations using the Maximum Likelihood estimators (MLE) described in Anselin (1988b) which account for spatial autocorrelation in the disturbances. These heterogeneous spatial estimates are reported in Table 13.2 along with the corresponding estimate of λ. We also report for each year the LM test for a null hypothesis of $\lambda = 0$, given by equation (59) of Anselin and Bera (1998). Most of the spatial coef˜ cients estimates are insigni˜ cant at the 5 percent level, except for ˜ ve out of the 25 years used for estimation. These are 1976, 1981, 1983, 1984 and 1987. The heterogeneous MLE estimates accounting for spatial autocorrelation do not differ much from the heterogeneous OLS estimates ignoring spatial autocorrelation. The price elasticity estimates varied from a low of −0.63 in 1963 to a high of −1.49 in 1981, while the income elasticity estimates varied from a low of 0.19 in 1980 to a high of 0.83 in 1968. The average pooled spatial heterogeneous MLE estimator yields a price elasticity estimate of −1.24 and an income elasticity estimate of 0.51 with a spatial autocorrelation parameter estimate of λ of 0.17, all of which are signi˜ cant. These are reported in Table 13.1 as "average spatial MLE." Note that these estimates are slightly higher than the average heterogeneous OLS estimates that ignore spatial autocorrelation.

13.2.3 Fixed Effects Estimators

Next, we account for heterogeneity across states by using the ˜ xed effects (FE) estimator. This model assumes that the μ_i 's are ˜ xed parameters to be estimated. The F-statistic for testing the signi˜ cance of the state dummies (see equation (2.12) of Baltagi, 1995), yields a value of 88.9, which is statistically signi˜ cant. Note that if these state effects are ignored, the OLS estimates and their standard errors in Ta-

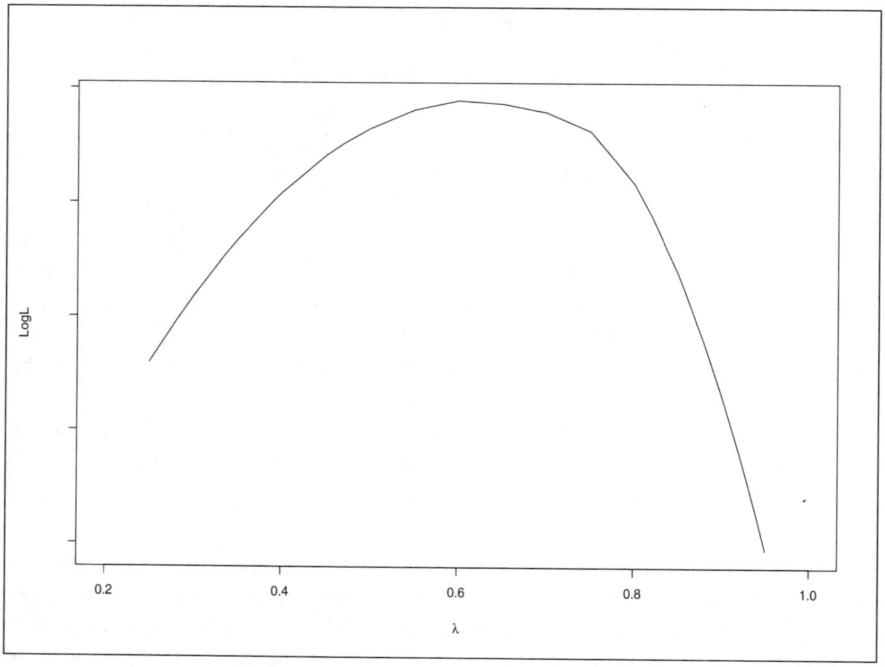

Fig. 13.1. Log-likelihood for the FE-spatial model

ble 13.1 would be biased and inconsistent (see Moulton, 1986).[4] Ignoring the spatial effects, the FE estimator can be obtained by running the regression with state dummy variables or by performing the within transformation and then running OLS (see Hsiao, 1986). We denote these estimates by $\hat{\beta}_{FE}$. They are reported in Table 13.1 as "FE." Compared to the OLS estimates, the price elasticity estimate drops to -0.47. In addition, the income elasticity estimate becomes negative as well, at -0.26. Both coefficients are significant. While surprising, the negative sign for income is not unlikely, since income can be a proxy for education levels and smoking is known to decrease with higher education levels.

This FE estimator still does not take into account the spatial autocorrelation. We therefore proceed to estimate a fixed effects model with spatial error autocorrelation using MLE.[5] In order to make sure a global optimum was obtained, we used a grid search procedure over the autoregressive parameter λ. As illustrated in Fig. 13.1, the likelihood function is well behaved for values of λ around the global maximum.

[4] Note that prices vary across states mainly due to tax changes across states. To the extent that endogeneity in prices is due to its correlation with the state effects, the fixed effects estimator becomes a viable estimator which controls for endogeneity by wiping out the state effects.

[5] This was obtained using the OPTMUM procedure of GAUSS version 3.2.37.

The corresponding estimates are reported in Table 13.1 as "FE-spatial." The results yield a slightly higher price elasticity estimate of -0.78 and a slightly lower income elasticity estimate of -0.13 than the non-spatial FE estimator. Both estimates are also statistically signi˜ cant. The estimate for the spatial autoregressive coef˜ cient λ is 0.61. A likelihood ratio test for $\lambda = 0$, yields a χ_1^2 test statistic of 251.4. This is statistically signi˜ cant and rejects the null of $\lambda = 0$ in the FE model.

13.2.4 Random Effects Estimators

In a random effects model, the μ_i's are $iid(0, \sigma_\mu^2)$ and are independent of the ϕ_{it}'s (see Anselin, 1988b). For this model, we need to derive the variance-covariance matrix. Let $B = I_n - \lambda W$, then the disturbances in equation (13.3) can be written as follows:

$$\begin{aligned}\phi_t &= (I_n - \lambda W)^{-1} \nu_t \\ &= B^{-1} \nu_t. \end{aligned} \tag{13.4}$$

Substituting ϕ_t from (13.4) in (13.2), we get:

$$\varepsilon = (\iota_T \otimes I_n)\mu + (I_T \otimes B^{-1})\nu, \tag{13.5}$$

where ι_T is a vector of ones of dimension T, and I_n is an identity matrix of dimension n. The corresponding variance covariance matrix is obtained as:

$$\begin{aligned}\Omega &= \mathrm{E}(\varepsilon\varepsilon') \\ &= \sigma_\mu^2(\iota_T\iota_T' \otimes I_n) + \sigma_\nu^2(I_T \otimes (B'B)^{-1}). \end{aligned} \tag{13.6}$$

In order to simplify notation, let $\Omega = \sigma_\nu^2 \Psi$, such that:

$$\Psi = \frac{\sigma_\mu^2}{\sigma_\nu^2}(\iota_T\iota_T' \otimes I_n) + (I_T \otimes (B'B)^{-1}), \tag{13.7}$$

or, with $\theta = \sigma_\mu^2/\sigma_\nu^2$, $\bar{J}_T = (1/T)\iota_T\iota_T'$, $V = T\theta I_n + (B'B)^{-1}$, and $E_T = I_T - \bar{J}_T$, we obtain the following expression for Ψ (see Anselin, 1988b, p. 154),

$$\begin{aligned}\Psi &= \bar{J}_T \otimes (T\theta I_n) + I_T \otimes (B'B)^{-1} \\ &= \bar{J}_T \otimes V + E_T \otimes (B'B)^{-1}. \end{aligned} \tag{13.8}$$

Using the result from (Anselin, 1988b, p. 154) or a similar trick in Wansbeek and Kapteyn (1983) for the classical error component model without spatial autocorrelation, the inverse of Ψ can be obtained as:

$$\Psi^{-1} = \bar{J}_T \otimes V^{-1} + E_T \otimes (B'B). \tag{13.9}$$

In the random effects model, a Generalized Least Squares procedure carried out on (13.1) that uses the inverse Ψ^{-1} yields the estimates β_{GLS}. Note that the computation is simpli˜ ed, since the nT by nT matrix Ψ^{-1} is reduced to inverting two lower order matrices, V and B, both of dimension n by n.

When $\lambda = 0$, so that there is no spatial autocorrelation, then $B = I_n$, and Ω from (13.6) becomes the usual error component variance-covariance matrix:

$$\begin{aligned}\Omega_{RE} &= \mathrm{E}(\varepsilon\varepsilon') \\ &= \sigma_\mu^2(\iota_T\iota_T' \otimes I_n) + \sigma_\nu^2(I_T \otimes I_n).\end{aligned} \tag{13.10}$$

In this case, a number of simplifying results are obtained, since:

$$\begin{aligned}V &= (T\theta + 1)I_n \\ &= \left(\frac{T\sigma_\mu^2 + \sigma_\nu^2}{\sigma_\nu^2}\right) I_n,\end{aligned} \tag{13.11}$$

such that,

$$\Psi_{RE}^{-1} = \frac{\sigma_\nu^2}{\sigma_1^2}(\bar{J}_T \otimes I_n) + E_T \otimes I_n, \tag{13.12}$$

where $\sigma_1^2 = T\sigma_\mu^2 + \sigma_\nu^2$.

Applying Generalized Least Squares to (13.1) while using (13.12) in the transformation yields the Random Effects (RE) estimator, which we will denote by $\hat{\beta}_{RE}$.

For our data, a one-sided Breusch and Pagan (1980) test for the null hypothesis of no random effects, $\sigma_\mu^2 = 0$, yields a $\mathrm{N}(0,1)$ test statistic of 81.1, which is statistically signi˘ cant. Feasible GLS can be based on a method for estimating the variance components due to Amemiya (1971). This is an analysis of variance method that uses the FE residuals in place of the true disturbances (see Baltagi, 1995). The results are reported as "FE" in Table 13.1. We ˘ nd the corresponding estimate for the price elasticity to be -0.47 and the income elasticity estimate to be -0.25. As for the FE estimator, both coef˘ cients are negative and signi˘ cant. The RE estimates are also close in value to those of the FE estimator. However, we are concerned about possible correlation between the regressors and the random effects. To this effect, we compute a Hausman (1978) test statistic for misspeci˘ cation, based on the difference between the FE and RE estimators for β. This yields a χ_2^2 test statistic of 26.8 which is statistically signi˘ cant. Therefore, the null hypothesis of exogeneity is rejected and the RE estimator cannot be considered to be consistent.

When $\lambda \neq 0$, estimation may be based on MLE, assuming normality of the disturbances. Such an estimator for the error component model with spatial autocorrelation is derived Anselin (1988b).[6] We apply this method by implementing a grid search procedure over λ and $\rho = \sigma_\mu^2/(\sigma_\mu^2 + \sigma_\nu^2)$ and check for a global maximum. Since the ρ term is a positive fraction, this allows a grid search over values of ρ between zero and one. As illustrated in Fig. 13.2, the likelihood function is well behaved for values of λ and ϕ around the global maximum. The results are reported in Table 13.1 as "RE-spatial." These results yield a higher price elasticity estimate of -0.80 and a lower income elasticity estimate of -0.07 than the non-spatial RE estimator. The price elasticity is statistically signi˘ cant, but the income elasticity is not.

[6] We apply this MLE using the OPTMUM procedure of GAUSS version 3.2.37.

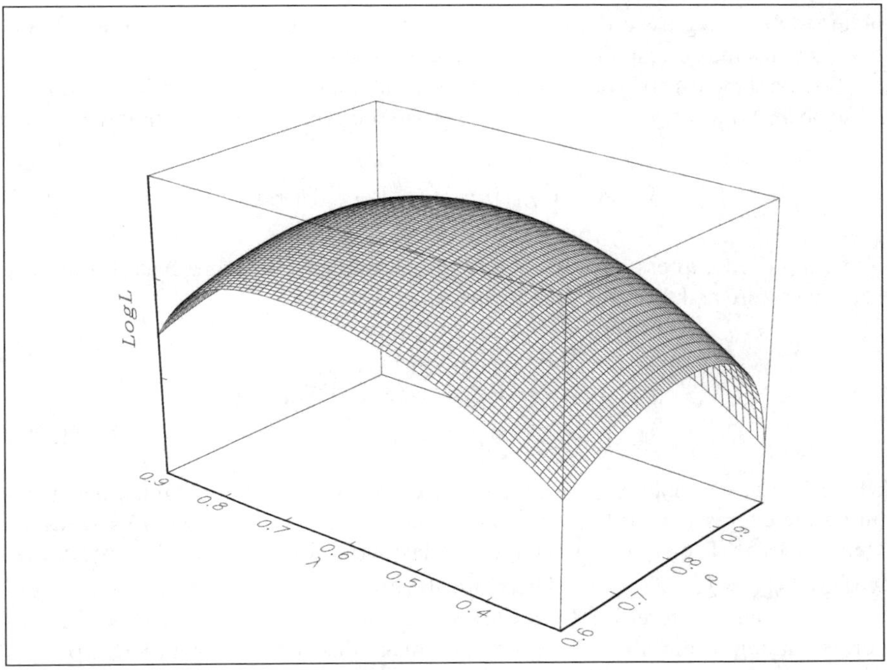

Fig. 13.2. Log-likelihood for the RE-spatial model

The λ estimate is 0.65 which is close to that of the FE-spatial model. A likelihood ratio test for the null hypothesis of $\lambda = 0$ yields a χ_1^2 test statistic of 249.4. This is statistically signiˇ cant and conˇ rms the importance of a spatial autoregressive disturbance in the RE model.

We now turn to comparing the performance of the various estimators based on ˇ ve years ahead forecasts. These are the out of sample predictions for 1988, 1989, ..., and 1992.

13.3 Prediction

13.3.1 Standard Case

A classic result of Goldberger (1962) shows that, for a given error variance covariance matrix Ω, the best linear unbiased predictor (BLUP) for the ith state at a future period $T+S$ is given by:

$$\hat{y}_{i,T+S} = x'_{i,T+S}\hat{\beta}_{GLS} + \omega'\Omega^{-1}\hat{\varepsilon}_{GLS}, \tag{13.13}$$

where $\omega = E(\varepsilon_{i,T+S}\varepsilon)$ is the covariance between the future disturbance $\varepsilon_{i,T+S}$ and the sample disturbances ε. The term $\hat{\beta}_{GLS}$ corresponds to the GLS estimator of β

obtained in the regression (13.1), based on the appropriate speciˇ cation for Ω, and $\hat{\varepsilon}_{GLS}$ denotes the corresponding GLS residual vector.

For the standard error component model, without spatial autocorrelation ($\lambda = 0$), Wansbeek and Kapteyn (1978) and Taub (1979) derived the corresponding BLUP as:

$$\hat{y}_{i,T+S} = x'_{i,T+S}\hat{\beta}_{GLS} + \frac{\sigma_\mu^2}{\sigma_1^2}(\iota'_T \otimes l'_i)\hat{\varepsilon}_{GLS}. \tag{13.14}$$

The simpliˇ ed expression follows due to the structure of the relevant covariance between future and "current" disturbances:

$$\begin{aligned}\omega &= E(\varepsilon_{i,T+S}\varepsilon)\\ &= E[(\mu_i + \nu_{i,T+S})\varepsilon]\\ &= \sigma_\mu^2(\iota_T \otimes l_i),\end{aligned} \tag{13.15}$$

where l_i is the ith column of I_n. We can see how this result is obtained by substituting the expression for Ψ_{RE}^{-1} from equation (13.12) into equation (13.13), which yields (13.15). The typical element of the last term of (13.14) is $\left(T\sigma_\mu^2/\sigma_1^2\right)\bar{\varepsilon}_{i.,GLS}$, where $\bar{\varepsilon}_{i.,GLS} = \sum_{t=1}^T \hat{\varepsilon}_{ti,GLS}/T$. Therefore, the BLUP of $y_{i,T+S}$ for the RE model modiˇ es the usual GLS forecasts by adding a fraction of the mean of the GLS residuals corresponding to the ith state. In order to make this forecast operational, $\hat{\beta}_{GLS}$ is replaced by its feasible GLS estimate $\hat{\beta}_{RE}$ (reported in Table 13.1), and the variance components are replaced by their feasible estimates. The results for the predictors are given in Table 13.3, where they are labeled "RE."

For the ˇ xed effects model without spatial autocorrelation ($\lambda = 0$), the BLUP can be obtained along the same lines as (see Baillie and Baltagi, 1998):

$$\tilde{y}_{i,T+S} = x'_{i,T+S}\tilde{\beta}_{FE} + \tilde{\mu}_i, \tag{13.16}$$

where μ_i is estimated as,

$$\tilde{\mu}_i = \bar{y}_{i.} - \bar{x}'_{i.}\hat{\beta}_{FE}, \tag{13.17}$$

and $\bar{y}_{i.} = \sum_{t=1}^T y_{it}/T$, with $\bar{x}_{i.}$ similarly deˇ ned. Note that in this case, since $\lambda = 0$, the term ϕ_{it} in equation (13.3) reduces to ν_{it} and the latter are not serially correlated over time. Therefore, $\omega = E(\nu_{i,T+S}\nu) = 0$, and the last term of equation (13.13) becomes zero for the FE model. However, the $\tilde{\mu}_i$ do appear in the predictions, as shown in equation (13.16). The results for the corresponding predictor are labeled "FE" in Table 13.3.

13.3.2 Spatial Case

So far, the results presented to obtain predicted values correspond to the classic case of no spatial autocorrelation. In this chapter, we also derive the BLUP correction term for both the RE and the FE models in the presence of spatial error autocorrelation.

First, we consider the random effects speci˘ cation. In this case:

$$\begin{aligned}\omega &= E(\varepsilon_{i,T+S}\varepsilon)\\ &= E[(\mu_i + \phi_{i,T+S})\varepsilon]\\ &= \sigma_\mu^2(\iota_T \otimes l_i),\end{aligned} \tag{13.18}$$

since the ϕ's are not correlated over time. Using $\Omega^{-1} = \sigma_\nu^{-2}\Psi^{-1}$ from (13.7), and with the inverse from equation (13.9), we obtain:

$$\begin{aligned}\omega'\Omega^{-1} &= \frac{\sigma_\mu^2}{\sigma_\nu^2}(\iota_T' \otimes l_i')[(\bar{J}_T \otimes V^{-1}) + (E_T \otimes (B'B))]\\ &= \theta(\iota_T' \otimes l_i'V^{-1}),\end{aligned} \tag{13.19}$$

since $\iota_T' E_T = 0$. Therefore:

$$\begin{aligned}\omega'\Omega^{-1}\hat{\varepsilon}_{GLS} &= \theta(\iota_T' \otimes l_i'V^{-1})\hat{\varepsilon}_{GLS}\\ &= \theta\, l_i'V^{-1}\sum_{t=1}^{T}\hat{\varepsilon}_{t,GLS}\\ &= T\theta\sum_{j=1}^{n}\delta_j\bar{\varepsilon}_{j.,GLS},\end{aligned} \tag{13.20}$$

with δ_j as the jth element of the ith row of V^{-1}, and,

$$\bar{\varepsilon}_{j.,GLS} = \sum_{t=1}^{T}\hat{\varepsilon}_{tj,GLS}/T. \tag{13.21}$$

In other words, the BLUP adds to $x_{i,T+S}'\hat{\beta}_{GLS}$ a weighted average of the GLS residuals for the n regions, averaged over time. The weights depend upon the spatial matrix W and the spatial autocorrelation coef˘ cient λ. To make this predictor operational, we replace $\hat{\beta}_{GLS}$, θ and λ by their Maximum Likelihood estimates, reported as RE-spatial in Table 13.1. The corresponding predictor is labeled "RE-spatial" in Table 13.3.

When there is no spatial autocorrelation, i.e., $\lambda = 0$, the BLUP correction term given in (13.20) reduces to the Wansbeek and Kapteyn (1978) and Taub (1979) predictor term given in equation (13.14). Also, when there are no random state effects, so that $\sigma_\mu^2 = 0$, then $\theta = 0$ and the BLUP prediction term in (13.20) drops out completely from equation (13.13). In this case, Ω in (13.6) reduces to $\sigma_\nu^2(I_T \otimes (B'B)^{-1})$ and feasible GLS on this model, based on the MLE of λ, yields the pooled spatial estimator reported in Table 13.1. The corresponding predictor is labelled "pooled spatial" in Table 13.3.

The results for the ˘ xed effecs model with spatial autocorrelation are slightly different. The problem is to predict:

$$y_{i,T+S} = x_{i,T+S}'\beta + \mu_i + \phi_{i,T+S}, \tag{13.22}$$

Table 13.3. Out of sample forecast – RMSE performance

	1988	1989	1990	1991	1992	5 years
Pooled OLS	0.1947	0.2022	0.2239	0.2226	0.2016	0.2093
Pooled spatial	0.1862	0.1888	0.2072	0.2002	0.1769	0.1922
Average heterogeneous OLS	0.1927	0.1896	0.2029	0.1913	0.1674	0.1892
Average spatial MLE	0.1901	0.1862	0.1990	0.1867	0.1666	0.1860
FE	0.1152	0.1241	0.1595	0.1739	0.1680	0.1501
FE-spatial	0.1027	0.1051	0.1360	0.1404	0.1478	0.1278
RE	0.1158	0.1249	0.1604	0.1749	0.1687	0.1509
RE-spatial	0.1042	0.1070	0.1371	0.1407	0.1444	0.1279

with $\phi_{T+S} = \lambda W \phi_{T+S} + \nu_{T+S}$ obtained from (13.3). Unlike the standard ˇ xed effects case, $\lambda \neq 0$ and the μ_i's and β have to be estimated by means of Maximum Likelihood, i.e., using the FE-spatial estimates. The disturbance vector from (13.3) can be written as $\phi = (I_T \otimes B^{-1})\nu$, so that $\omega = E(\phi_{i,T+S}\phi) = 0$ since the ν's are not serially correlated over time. So the BLUP for this model looks like that for the FE model without spatial correlation given in (13.16) except that the μ_i's and β are estimated assuming $\lambda \neq 0$. The corresponding predictor is labeled "FE-spatial" in Table 13.3.

13.3.3 Results

Table 13.3 gives the root mean squared error (RMSE) for the one year, two year,..., and ˇ ve year ahead forecasts along with the RMSE for all 5 years. These are out of sample forecasts from 1987 to 1992. Each year's RMSE is obtained from 46 state by state predictions. We compare the forecasts for all 5 years. The pooled OLS predictor in Table 13.3 is computed as $\hat{y}_{i,T+S} = x'_{i,T+S}\hat{\beta}_{OLS}$. Pooled OLS, which ignores spatial autocorrelation and heterogeneity across the states gives the highest RMSE of 0.2093. Accounting for spatial autocorrelation using the pooled spatial estimator lowers the RMSE to 0.1922. This predictor replaces the OLS estimator of β by that of pooled spatial MLE, reported in Table 13.1. The average heterogeneous OLS estimator (which ignores spatial autocorrelation but allows for parameter heterogeneity across time) lowers the RMSE to 0.1892. The forecast performance is slightly improved by accounting for spatial autocorrelation. The average heterogeneous spatial MLE yields an RMSE of 0.1860. A substantial improvement in the forecast performance occurs when one takes into account the state heterogeneity. The simple FE estimator without spatial autocorrelation yields an RMSE of 0.1501, followed closely by the RE estimator without spatial autocorrelation, with an RMSE of 0.1509. These predictors were described in (13.16) and (13.14), respectively.

Additional reduction in the forecast RMSE is obtained by taking into account both heterogeneity and spatial autocorrelation. The best forecast performance for all ˇ ve years is obtained by the FE estimator with spatial autocorrelation, which yields an RMSE of 0.1278, followed closely by the RE with spatial autocorrelation, with an RMSE of 0.1279.

13.4 Conclusions

For the simple cigarette demand model chosen to illustrate our forecasts, the speciˇ cation that takes into account both the heterogeneity across states and the spatial autocorrelation yields the best out of sample forecasts, as measured by their RMSE. The FE-spatial estimator gives the lowest RMSE for the ˇ rst four years and is only surpassed by the RE-spatial model in the ˇ fth year. Overall, both the RE-spatial and FE-spatial estimators perform well in predicting cigarette demand.

Some of the limitations of our study are that we used a simple static model of cigarette demand when a dynamic or a rational addiction model of cigarette demand may be more appropriate. However, the latter models introduce additional econometric complications for our forecasting illustrations, which are beyond the scope of this chapter. Despite these limitations, this chapter lays out a simple methodology for forecasting with panel data models that are spatially autocorrelated. These methods will hopefully prove useful to researchers forecasting with these models.

14 External Effects and Cost of Production

Rosina Moreno, Enrique López-Bazo, Esther Vayá, and Manuel Art´s

University of Barcelona

> "The whole is more than the sum of the parts, in that, not only does the interrelation of parts bring out latent characteristics in each, as in any complex, but the complex as a whole takes on a new character not explainable out of the parts." (Hartshorne, 1939).

14.1 Introduction

Recent studies (Romer, 1986; Lucas, 1988) have stressed the importance of factors external to the ˇ rm in the production process. Such externalities are assumed to have a direct effect on the level of production or to enhance the productivity of traditional inputs. Broadly speaking, we can identify two types of externalities. First, inputs that are not explicitly taken into account in the ˇ rm's decision-making process although they contribute to the production process (for instance, the availability of human capital, public capital or infrastructure, and social capital). We will refer to these external effects as "external inputs." Second, externalities that are relevant outside the economies giving rise to the externality, regardless whether these economies are understood as the economy of a speciˇ c industry or a speciˇ c country or region. This type of externality has recently been considered theoretically in growth models dealing with open economies.

Several papers report empirical results with respect to the relevance of both abovementioned types of externalities. During the last two decades, the empirical economic growth literature starts to consider external inputs as main engines for increases in total factor productivity. For instance, Aschauer (1989), Munnell (1990b), and Garc´a-Mila and McGuire (1992) analyze the contribution of the stock of public capital to the economy's performance. Kiriacou (1991), and Benhabib and Spiegel (1994) devote special attention to the role of human capital as a factor in the process of economic growth. Although some results tend to support the relevance of external inputs, lacking robustness of the results seems to be a main characteristic of these analyses.

With respect to externalities across economies, several studies followed up on Caballero and Lyons' (1990) seminal paper in trying to empirically test for the existence and magnitude of spillovers across industries within an economy (Caballero and Lyons, 1992; Burnside, 1996). It is being argued that using national aggregates instead of data disaggregated by industry makes it impossible to identify returns to scale external to an industry, as these returns are internalized at the national level. Internal returns to scale and returns to scale external to a speciˇ c industry are thus confounded in one parameter. An obvious practical implication is that, unless properly speciˇ ed, external economies may cause the estimated internal returns to scale to be biased. A similar problem occurs when externalities across geographical barriers of an economy are considered. When a country is considered in isolation, across-economy linkages are mixed up with a country's own returns, and the effect that is due to contagiousness (an economy grows because neighbors are growing as well) is disregarded. The latter is not at all unlikely given the nexus of economic growth, international trade and the diffusion of knowledge across economies (Grossman and Helpman, 1991b; Coe and Helpman, 1995; Park, 1995). Trade makes available products and services that embody foreign knowledge and therefore provide technologies that would otherwise be unavailable or very costly to acquire. It has even been suggested that there may be reasons beyond trade flows that can account for the occurrence of technology flows among countries (see, for instance, Verspagen, 1997; Keller, 1998).

Regardless of the exact reason for the occurrence of externalities, both across industries and across geographical units, we set out to introduce three new ideas in this chapter.

First, we proxy across-industry spillovers by a measure accounting for forward and backward linkages across sectors, rather than a raw measure for thick-market effects that is frequently used in this literature. With respect to spillovers across aggregate economies, we focus on the regional level given the higher incidence of the relevance of externalities. In addition, we also subscribe to the common assertion that geographical clustering among regions induces both pecuniary and technological external effects.

Second, in the empirical analyses we use spatial econometric techniques. To date, most empirical analyses have not used econometric methods that can robustly test and estimate externalities of this kind. Our empirical exercise directly addresses this issue. Specifically, we assess the adequacy of traditional spatial statistics for detecting externalities and adapt them to the specific features of our empirical model. In particular, we deal with nonlinearity in some of the parameters, and with the cross-section and time-series dimension of the data. In the case of spatial externalities, we consider spatial dependence based on the interaction between contiguous regions. For sectoral spillovers, we suggest the application of a similar setup to sectors by modeling interdependence among industries on the basis of input-output relationships.

Finally, most studies analyzing external inputs or across-economy linkages have focused on production functions. We, however, use a production function that explicitly considers externalities, and subsequently derive a cost function using duality theory. In this framework, the impact of external effects on the costs of production can potentially be broken down to assess the effect of externalities through each private input and the level of output, while we separate such effects from input utilization.

The remainder of this chapter is organized as follows. Section 14.2 reviews the literature dealing with the sources of regional and industrial externalities. The role of what we call external inputs is well documented in the literature on economic growth, and we therefore do not further discuss this topic here. Section 14.3 presents the conceptual model based on duality theory including external effects. In Sect. 14.4, we suggest an empirical framework that can be used to test for the existence of external effects and subsequently estimate their impact. Section 14.5 describes the database. In Sect. 14.6, we apply the theoretical and empirical framework to the case of the manufacturing industries in Spanish regions, covering the time-period from 1980 to 1991. Finally, Sect. 14.7 concludes.

14.2 Sources of Regional and Industrial Externalities

Evidence of the spatial concentration of economic activity has been widely reported (Krugman, 1991b; Glaeser *et al.*, 1992; Henderson, 1992). A simple look at a map depicting the density of economic activity reveals that the spatial distribution of economic activity is neither random nor homogeneous. Rather, firms tend to cluster spatially, depending on the previous location of other firms in the same or in different industries. The former is referred to as Marshall-Arrow-Romer and Porter externalities, the latter as Jacobs externalities. Although a firm can freely select its geographical location, the probability of selection for each possible location in a given territory is not equally distributed (Ellison and Glaeser, 1997). There is a tendency to concentrate economic activity in locations offering advantages due to the existence of large, specialized markets. Marshall (1920) explained the concentration of industries in a territory through the concept of external economies operating as a centripetal force. Specifically, Marshallian externalities explain the geographical concentration of economic activity due to the presence of highly specialized markets for labor and intermediate inputs, forward and backward linkages in the production process, and the facilitation of the diffusion of ideas, technology and information. Scitovsky (1954) classifies the first two factors as pecuniary externalities, and they have been incorporated in the new theories of industrial location and trade as engines for agglomeration. Krugman (1991b), Krugman and Venables (1995), Puga and Venables (1996), and Martin and Ottaviano (1999) provide numerous examples. They explicitly address the role of agglomeration economies as the main engine for endogenous growth. In a similar fashion, the diffusion of technological innovations is facilitated by geographical closeness, and constitutes an important determinant of growth of individual firms as well as aggregate economic growth.

While recognizing the importance of geographical proximity, we argue that two types of externalities should be considered: externalities across industries and across geographical units (regions or countries).

14.2.1 Industrial Externalities

Several mechanisms justify the existence of externalities across firms within a geographical area. For instance, by investing in physical capital, a firm accumulates knowledge from which other firms might benefit, increasing their productivity without incurring the associated costs (Arrow, 1962). In other words, when one firm buys intermediate goods from another, it is paying a price that is smaller than the value corresponding to the information embodied in these goods, because the innovative firm is unable to internalize the whole benefit associated with the innovation. This phenomenon is known as knowledge spillover. In addition, yet another externality mechanism linked to physical capital has been distinguished. The complementarities between activities and firms may result in advantages related to within-industry specialization (Durlauf, 1991).

Even though most of the abovementioned externalities focus on economies that are external to the firm and at the same time internal to the industry (i.e., so-called industry-specific externalities), we are particularly interested in spillovers across industries, as reported in studies by Chang (1981), Diamond (1982) and Herberg *et al.* (1982). In these papers, externalities correspond to transaction or thick-market effects arising from easier matching between agents during expansions. Firms in the industrial sector are linked by input-output relationships that create forward and backward linkages. If transport costs are assumed to exist, proximity to suppliers allows costs to be reduced, thereby generating forward linkages. Similarly, proximity to customers generates backward linkages. Bartelsman *et al.* (1994) find a clear prevalence of the customer-driven externality in the short run, whereas linkages with suppliers are dominant in the long run. Keller (1997) provides further evidence. He estimates the elasticity of total factor productivity with respect to own-industry R&D investments and other industries' R&D investments. His results show that the elasticity for R&D investments of other industries is strongly significant in explaining total factor productivity, and amounts to 0.2–0.5 times the elasticity to a firm's own R&D investments. Assuming R&D investments adequately proxy for the improvement in technology levels, it is obvious that it is worthwhile to consider externalities across industries.

We end this brief summary on the available evidence for the relevance of spillovers across industries by referring to two papers by Caballero and Lyons (1990, 1992). They proxy across-industry externalities by including the output of industry defined at a higher level of sectoral aggregation in the production function for each industry.[1] They show that, for a given input level, an industry's output is on average significantly higher when the output of the industry defined at a higher aggregation level is higher. They also present evidence for estimated returns to scale being larger for the manufacturing sector as a whole than for two-digit industries. This difference is due to the externality only being internalized at the higher level of aggregation. When considering aggregate data, returns to scale external to the industry cannot be identified because external economies become internal as the aggregation level rises. However, this line of reasoning shows that for the case where aggregate economies (such as regions or countries) are the unit of analysis, considering spillovers is potentially important as well. We therefore discuss empirical evidence for the existence of externalities across regions or countries in the next subsection.

[1] The authors acknowledge that this type of specification of external effects may induce endogeneity problems.

14.2.2 Spatial Externalities

Up to this point, we have focused on externalities in terms of spillovers across industries within one economy. However, the world economy has undergone a major globalization process during recent decades. Inventions and innovations generated in a particular location are easily and quickly absorbed and adapted in other locations. Undoubtedly, among other things, direct foreign investments and trade of intermediate and final goods play an important role in such a process. The importance of trade relationships has increased, particularly between countries belonging to integrated trade areas such as the European Union and the North American Free Trade Agreement. Countries trade, establish links, and learn from each other more so than before. In empirical research, however, each economy has been treated as an island so that economic growth in effect solely depends on factors internal to the economy. It is of course logical to hypothesize the existence of sources of growth going beyond the scope of a single economy. The increasing exchange of goods and knowledge at an international level has led to an increasing interdependence in growth in different countries (Coe and Helpman, 1995; Ciccone, 1996).

When we consider regional (instead of national) economies, these interdependence mechanisms can be expected to be even more relevant. The existence of common output and input markets is more likely at a regional level within a single country than across countries. Another reason why externalities are likely to exhibit geographical spillovers is the existence of similar social conditions, which may play a significant role in the way regional economies incorporate and adapt innovations (Rodriguez-Pose, 1999). If the regions of a country share similar local conditions, knowledge spillovers between them may be more intense. Kollmann (1995), to give an example, shows that productivity growth is more strongly correlated across the states of the U.S. than across the G7 countries.

Several authors consider external effects, and particularly innovation diffusion, to be more important among homogeneous groups or "clubs" of economies. Durlauf and Quah (1999) maintain that for "naturally generated" groups of economies the average income to which they converge will change in a groupwise fashion. This is in accordance with the greater intensity of trade flows and technological diffusion among such clusters of regions. The importance of geographical proximity of units of production for innovation transmission is also demonstrated in Henderson (1992) and Glaeser *et al.* (1992).

Even though theoretical and empirical evidence seems to support the existence of externalities across industries and regions, it is not clear as to what type of externalities are stronger. Costello (1993) shows how total factor productivity growth is more strongly correlated across industries within one country than across countries within one industry. Conversely, Kollmann (1995) concludes that correlations across industries within a region are weaker than across regions within an industry. López-Bazo *et al.* (1998) observe how both sources of externalities are similar in magnitude for the Spanish economy. These results support the relevance of transfers of technology across regions. The higher degree of integration among the U.S.

states and regions in Spain may explain why technology and growth spread more easily among those regions than across heterogeneous countries.

Despite these arguments, studies explicitly considering such externalities across economic areas are few. Barro and Sala-i-Martin (1995, chapter 12), Ciccone (1996), and Ades and Chua (1997) incorporate externalities across countries. Quah (1996), López-Bazo *et al.* (1998), Fingleton and McCombie (1998), Vayá *et al.* (1998), and Rey and Montouri (1999) do so for regions.

14.3 Theoretical Framework: Duality Theory and External Effects

The analysis in this chapter develops along the basic line of reasoning adopted in studies incorporating industry and spatial externalities in order to simultaneously assess the signiˇ cance and strength of both types of spillovers in the cost of production. In this section, we incorporate external inputs and spillovers across economies in a cost function. Given the main objective of this chapter, we focus on the development of the elasticities measuring the externalities, omitting the derivation of the traditional elasticities regarding private inputs and output.[2]

We consider an aggregate production function, where Y_{it} is the output in the ith economy (region or industry) at time t, and $X_j (j = 1,\ 2,\ ...,\ r)$ the jth input:

$$Y_{it} = f(X_{1it}, X_{2it}, ..., X_{rit}). \tag{14.1}$$

Taking into account the potential role of external inputs and the existence of across-region and across-industry externalities implies that the output in an economy also depends on the stock of the external inputs and the amount of inputs and output in neighboring economies.[3] As a result, internal and external returns to scale can be separated, and we obtain the following expression:

$$Y_{it} = f(X_{1it}, X_{2it}, ..., X_{rit}, E_{it}, E_{\rho it}), \tag{14.2}$$

where E_{it} is a measure of the external input under consideration, and $E_{\rho it}$ the externalities across regions and industries.

We assume the ˇ rm is constrained to accept a vector of input prices, $P_1, ..., P_r$, so that the optimization problem of ˇ rms consists of determining the amount of inputs that minimizes the cost for producing a given level of output, $\bar{Y}$. As a result, the technology of the ˇ rm depicted by (14.2) can be represented by a variable cost function that also includes external effects:

$$VC_{it} = \textstyle\sum_j P_{jit} \cdot X_{jit} = g(P_{1it}, ..., P_{rit}, \bar{Y}_{it}, E_{it}, E_{\rho it}), \tag{14.3}$$

[2] See Berndt (1991) and Morrison and Schwartz (1996) for the derivation of the traditional elasticities.

[3] We use a broad concept of neighborhood in the analysis. In the regional case, it refers to geographical proximity, while in the sectoral case it is based on trade ows across industries.

where VC is the level of variable costs, and X_j the amount of the jth input at the optimum.[4]

Specifically, taking into account the presence of externalities, the variable cost function used in this chapter can be specified as follows:

$$VC_{it} = VC(P_{Lit}, P_{Mit}, \bar{Y}_{it}, Kp_{it}, E_{it}, E_{\rho it}), \tag{14.4}$$

where we consider two variable private inputs, labor (L) and intermediates (M), appearing in the cost function through their prices, P_L and P_M, respectively, and a quasi-fixed input, private capital (Kp).[5] This cost function allows for the combination of internal scale economies in the production process due to private inputs (both variable and quasi-fixed), and external scale economies caused by differing types of external inputs and/or across-economy spillovers. The cost function also overcomes one of the criticisms raised against empirical evidence on across-industry spillovers based on analyses using a production function approach, where significant externalities can be due to variations in the use of internal inputs. In (14.4), we consider Kp as an input that may not be at its optimum level in each time-period, and we can therefore isolate the external effects on production from the over- or underutilization of capacity.

Assuming that variable input prices are exogenous to the producer, Shephard's Lemma (Shephard, 1953) states that it is possible to obtain the unique vector of the different variable inputs that minimize costs (cost-minimizing demands), and hence, their factor share (z_j), that is, the percentage of the cost implied by the jth input, is:

$$z_j = \frac{P_{ij} \cdot X_j}{VC} = \frac{\partial \ln VC}{\partial \ln P_j} = f(P_l, P_M, \bar{Y}, Kp, E, E_\rho), \text{ for } j = L, M. \tag{14.5}$$

For ease of notation, the variables in (14.5) and in subsequent equations do not carry indices for the period of time or the economy. Equations (14.4) and (14.5) constitute the solution to the equilibrium related to variable factors. Testing the validity of Shephard's Lemma is therefore equivalent to testing the validity of the restrictions on the parameters of the cost function and the share equations for variable inputs.

Once an empirical specification for the variable cost function has been estimated, the usual cost-private input elasticities and the elasticities of substitution between inputs can be obtained. However, here we focus on computing the effect on costs of the external input, E, and the spillovers across economies, E_ρ. It is important to note that, despite imposing constant parameters for all individuals and time-periods, the use of a flexible functional form allows for a separate elasticity for each region/sector and time-period.

In order to find out whether a marginal addition to the stock of an external factor decreases the cost per unit of output, the elasticity of production cost with respect

4 See Chambers (1988) for a detailed description of cost function properties.

5 In order to test the assumption that private capital is a quasi-fixed input, the test developed in Schankerman and Nadiri (1986) can be used. The assumption is not rejected for the Spanish economy (Moreno *et al.*, 1998).

to this input can be obtained as:

$$\varepsilon_{VCE} = \frac{\partial \ln VC}{\partial \ln E} = \frac{\partial VC}{\partial E}\frac{E}{VC}. \tag{14.6}$$

This elasticity will be negative as long as the external factor represents efficiency changes in terms of decreases in variable input utilization, and therefore in costs. These effects can be computed as the elasticity of the conditional demand for private inputs with respect to E:

$$\varepsilon_{X_jE} = \frac{\partial \ln X_j}{\partial \ln E} = \frac{\partial X_j}{\partial E}\frac{E}{X_j}, \text{ for } j = L, M. \tag{14.7}$$

We are also interested in the quantification of changes in manufacturing costs due to the presence of spillovers, in other words, due to the manufacturing performance in neighboring economies (regions or industries). This elasticity is obtained as:

$$\varepsilon_{VCE_\rho} = \frac{\partial \ln VC}{\partial \ln E_\rho} = \frac{\partial VC}{\partial E_\rho}\frac{E_\rho}{VC}. \tag{14.8}$$

According to the literature on externalities, we can expect ε_{VCE_ρ} to be negative in the case of externalities enhancing production, indicating that the greater the interdependencies across economies, the greater the efficiency and hence, the lower the costs.

14.4 Spatial and Sectoral Externalities

Several external inputs can affect the production process. Given that our empirical exercise illustrates the results obtained for the stock of publicly provided capital (Kg), we will from now on use Kg instead of E. There are also several ways of accounting empirically for spillovers across economies. We present evidence for the case of spillovers specified by the level of output in neighboring regions or sectors (Y_ρ), measuring “thick markets.” In addition, when analyzing the regional case we also include public capital of the neighbors (Kg_ρ), providing another source of spillovers. Obviously, the method described can be applied to other measures in a straightforward manner.

14.4.1 Empirical Cost Function

In order to implement duality theory, we assume a translog cost function of the following form:[6]

$$\begin{aligned}\ln \frac{VC}{P_M} = {} & \beta_0 + \beta_L \ln \frac{P_L}{P_M} + \beta_Y \ln Y + \beta_{Kp} \ln Kp + \beta_T t \\ & +0.5\left(\beta_{LL} \ln^2 \frac{P_L}{P_M} + \beta_{YY} \ln^2 Y + \beta_{KpKp} \ln^2 Kp + \beta_{TT} t^2\right) \\ & +\beta_{LY} \ln \frac{P_L}{P_M} \ln Y + \beta_{LKp} \ln \frac{P_L}{P_M} \ln Kp + \beta_{LT} \ln \frac{P_L}{P_M} t \\ & +\beta_{YKp} \ln Y \ln Kp + \beta_{YT} \ln Y t + \beta_{KpT} \ln Kp t,\end{aligned} \tag{14.9}$$

where t denotes a time-trend that captures exogenous technical change. When applying Shephard's Lemma to (14.9), we obtain the share equations for variable inputs associated with the variable cost function above.

Following the reasoning in Sect. 14.3, the variable cost function in (14.9) should be modified in order to include both external input and across-economy spillovers. For the external input in the regional case, we consider the stock of public capital in the region itself. In the sectoral case, the stock of infrastructure is computed for each industry according to its importance in the total manufacturing sector. The aggregate public capital stock is thus weighted according to the proportion of the output of an industry in total output. In this way, the potential utilization of the national public infrastructure endowment by an industry is accounted for.

Concerning the across-economy spillovers, we introduce output in the nearest economies (regions or industries) both linear and squared, as well as the cross-product with private capital. The former allows for a marginal effect of the externality, whereas the latter picks up the fact that capital intensity is positively correlated with the potential to benefit from spillovers. Alternatively, the cross-product can be taken as an indication of private capital being more profitable as externalities increase (see Azariadis and Drazen, 1990). In the regional case, we also consider the effect of the infrastructure stock in neighboring regions as a potential source for spillovers. This type of effect is considered as an additional production input in Mas *et al.* (1996), and Kelejian and Robinson (1997). However, we propose a specification that allows for a global effect of public capital specified as a geometric mean of the own and the neighboring regions' capital, that is, $G_{it} = Kg_{it}^{\theta} \cdot Kg_{\rho_i t}^{1-\theta}$, where $\theta \in [0,1]$.[7] In this specification, the weight of the region's own public capital stock, θ, is parameterized and estimated simultaneously with the other parameters in the model. The parameter θ measures the contribution of the region's own public capital

[6] This functional form implies a large range of substitution possibilities and can be fitted to any production technology. We introduce intermediates price as a relative factor to ensure that the function is homogeneous of degree one in factor prices. No a priori returns to scale are imposed.

[7] To the best of our knowledge this specification has only been used in the literature concerning R&D spillovers (see Jovanovic *et al.*, 1992; Nadiri and Kim, 1996).

stock on manufacturing costs in the region, and $(1-\theta)$ measures the importance of public capital in neighboring regions in the costs in this region.

This type of speciˇ cation has two distinct advantages. First, it implies a complementary relation between a region's own capital and that of its neighbors, re ecting the network characteristics of most transport and communication infrastructure. Consequently, we in fact include a composite variable re ecting the infrastructure in the region in which the ˇ rm is located and in neighboring regions, rather than two separate variables. This has the additional advantage that adding new regressors resulting from the interaction of each argument in the function with other arguments can be avoided. This is attractive because it reduces the multicollinearity problem endemic to the translog functional form, the downside being that the inclusion of G_{it} now necessitates the application of nonlinear estimation techniques. We note that in the industrial case, the inclusion of the stock of public capital in the closest industries is of course pointless.

Including the external input as well as the two sources of across-economy externalities in the variable cost function turns the speciˇ cation given in (14.9) into the following expression:

$$
\begin{aligned}
\ln \frac{VC}{P_M} = {} & \beta_0 + \beta_L \ln \frac{P_L}{P_M} + \beta_Y \ln Y + \beta_{Kp} \ln Kp + \beta_{Kg} \ln\,(Kg^{\theta} \cdot Kg_{\rho}^{1-\theta}) \\
& + \beta_T t + 0.5 \left[\beta_{LL} \ln^2 \frac{P_L}{P_M} + \beta_{YY} \ln^2 Y + \beta_{KpKp} \ln^2 Kp \right. \\
& \left. + \beta_{KgKg} \ln^2 (Kg^{\theta} \cdot Kg_{\rho}^{1-\theta}) + \beta_{TT} t \right] + \beta_{LY} \ln \frac{P_L}{P_M} \ln Y \\
& + \beta_{LKp} \ln \frac{P_L}{P_M} \ln Kp + \beta_{LKg} \ln \frac{P_L}{P_M} \ln\,(Kg^{\theta} \cdot Kg_{\rho}^{1-\theta}) \\
& + \beta_{LT} \ln \frac{P_L}{P_M} t \beta_{YKp} \ln Y \ln Kp + \beta_{YKg} \ln Y \ln\,(Kg^{\theta} \cdot Kg_{\rho}^{1-\theta}) + \beta_{YT} \ln Y t \\
& + \beta_{KpKg} \ln Kp \ln\,(Kg^{\theta} \cdot Kg_{\rho}^{1-\theta}) + \beta_{KpT} \ln Kp t + \beta_{KgT} \ln\,(Kg^{\theta} \cdot Kg_{\rho}^{1-\theta}) t \\
& + \beta_{Y_{\rho}} \ln Y_{\rho} + \beta_{Y_{\rho}Y_{rho}} \ln^2 Y_{\rho} + \beta_{Y_{\rho}Kp} \ln Y_{\rho} \ln Kp. \qquad (14.10)
\end{aligned}
$$

The estimation of (14.10) in the regional case must be carried out by Nonlinear Least Squares (NLLS) because of the nonlinearity caused by the interaction of θ with the parameters measuring the effect of public capital. Given that the columns in the matrix of pseudo-regressors are linearly independent, the identiˇ cation is guaranteed, although a high degree of collinearity endemic to the translog approach may be prevalent (see Berndt and Hanson, 1992, for a discussion). As mentioned above, in the sectoral case (14.10) simpliˇ es because of the assumption that $\theta = 1$.

14.4.2 Spatial Externalities

The estimation of (14.9) suffers from spatial dependence, affecting standard estimation and inference procedures, when the external effects are erroneously omitted. The spatial econometric toolbox provides the necessary equipment to deal with this problem (Anselin, 1988b). Using the concept of a spatial lag, we can rewrite the terms representing the spillovers in (14.10): $\ln Y_\rho$ can be expressed as $W \ln Y$, where W is a matrix defining across-regional linkages. Assuming the data form a panel data set with n regions and T time-periods, and only contemporaneous spatial dependence (i.e., the effect of the externality is exhausted within the period in which it is generated) is relevant, we can define a weight matrix W as a nT by nT block-diagonal matrix:

$$W = I_T \otimes C, \tag{14.11}$$

where I_T is the T by T identity matrix and C is a n by n row-standardized weight matrix specified according to the physical contiguity criterion (1 for contiguous regions and 0 otherwise). The spatially lagged dependent variable, $W \ln Y$, is the weighted average of output in the regions contiguous to a specific region, as defined by W. Similarly, $W \ln Kg$ can be obtained as an operationalization of Kg_ρ. One should note that the parameters referring to spatial effects are assumed constant over time. As for the other parameters of the model, we estimate an average parameter over T time-periods.

Now that the spillovers are specified using a weight matrix, it is easy to see that the empirical model is a mixed regressive-spatial cross-regressive model (Anselin, 1988b), in which only a subset of the regressors enters with corresponding spatial lags. Taking into account that this model is derived on the basis of an explicit theory, we prefer to estimate (14.10) prior to checking for the occurrence of any remaining spatial effects in the residuals. The nonlinearity of the empirical model should be taken into consideration when deriving the expressions for the spatial Lagrange Multiplier statistics. It is not difficult to prove that the expression of the Lagrange multiplier test for spatial error dependence (LM-ERR) is not affected by the nonlinearity in the parameters of the exogenous variables. In contrast, the test for spatial lag dependence (LM-LAG) for this case is:

$$LM - LAG = \frac{\left(e_*' WY/\hat{\sigma}^2\right)^2}{RJ^0_{\rho-\beta}}, \tag{14.12}$$

where $e_* = Y - h(X, \hat{\beta}_*$ is the vector of residuals in the nonlinear estimation under the null hypothesis, and $RJ^0_{\rho-\beta} = T_1 + \frac{1}{\hat{\sigma}^2}(WX_0\hat{\beta}_*)' M_0 (WX_0\hat{\beta}_*)$, with $T_1 = tr(W^2 + W'W)$ and $M = I - X_0(X_0'X_0)^{-1}X_0'$, where $X_0 = \partial(X,\beta_*)/\partial\beta_*'$ is the (n by T) by k matrix of pseudo-regressors. Equation (14.12) only differs from that of the linear case in the use of the matrix of pseudo-regressors rather than the regressors themselves, and the residuals from the nonlinear model under the null. In the case where the spatial Lagrange Multiplier tests point to the existence of remaining spatial dependence, we consider the estimation of a spatial model that incorporates either a substantive or a nuisance spatial process (see Florax and Folmer, 1992; Anselin and Florax, 1995c).

It is important to bear in mind that most empirical studies in this field directly estimate expressions such as (14.9), without considering external effects. Given that the erroneous omission of externalities affects the statistical inference, it is imperative that the model is checked for the occurrence of spatial dependence.

14.4.3 Across-Industry Externalities

As in the spatial case where regions are related to contiguous neighbors, industries are related to each other through input-output links. The use of cross-sections of industries can therefore result in sectoral autocorrelation as well. Sectoral dependence has the same detrimental econometric consequences as spatial dependence, and should therefore be checked and, when appropriate, corrected.

As pointed out in Sect. 14.2, several studies include spillovers across industries within an economy in order to identify returns to scale that are external to the industry. However, the studies differ in the way external effects are modeled. For four and two-digit SIC-level manufacturing industry data, Caballero and Lyons (1989) and Burnside (1996) use aggregate manufacturing inputs as an index for the external effect, whereas Caballero and Lyons (1992) use output. There are, however, a number of limitations to these studies. First, they do not explicitly test for the existence of external effects. Second, they use output at an aggregation level that is higher than the aggregation level applying to the externality, without considering the strength of the dependence across industries. Finally, the standard methods used for estimation, suffer from endogeneity problems that may cause the estimators to be biased.

The use of spatial econometric techniques applied to the sectoral context can help to overcome these limitations. Speciˇ cally, we can explicitly test for the presence of across-industry externalities, while the use of the dual approach avoids the problem of endogeneity when including Y_ρ. For the speciˇ cation of the sectoral interaction structure, we suggest the use of input-output linkages between industries (for a similar reasoning, see Bartelsman *et al.*, 1994; Keller, 1997). The general expression for the direct-requirements matrix of an input-output table is:

$$\Lambda = \begin{bmatrix} \alpha_{11} & \alpha_{12} & \cdots & \cdots & \alpha_{1J} \\ \alpha_{21} & \alpha_{22} & \cdots & \cdots & \alpha_{2J} \\ \vdots & \vdots & \ddots & & \vdots \\ \vdots & \vdots & & \ddots & \vdots \\ \alpha_{J1} & \alpha_{J2} & \cdots & \cdots & \alpha_{JJ} \end{bmatrix}, \tag{14.13}$$

where the element α_{lm} re ects the value of products from industry l used as an intermediate input in industry m.

For externalities concerned with technology diffusion through purchases of intermediates (supplier-driven externalities), the weights for the industrial linkages can be measured as the rates of purchases from all other industries. This weight is the lmth element of Λ divided by the sum of the lth column:

$$c_{lm} = \frac{\alpha_{lm}}{\sum_{l=1}^{J} \alpha_{lm}}, \quad l \neq m. \tag{14.14}$$

For externalities derived from sales to other industries (customer-driven externalities), the appropriate weights can be defined as the rates of sales to all other industries. This weight is the lmth element of Λ divided by the sum of the mth row:

$$c_{lm} = \frac{\alpha_{lm}}{\sum_{m=1}^{J} \alpha_{lm}}, \quad l \neq m. \tag{14.15}$$

Because we are interested in evaluating the significance and size of sectoral linkages affecting cost levels in each industry, supplier-driven externalities are a priori the most relevant. We therefore consider how industries supplying industry l exert an influence on its cost level and structure through a weight representing the importance of the purchases industry l makes from each industry.[8] The resulting sectoral matrix is as in (14.11), where I is a (J by T) by (J by T) block-diagonal matrix, J the total number of industries, T the number of time-periods, and the characteristic element for row l and column m is c_{lm}, as in (14.14).

Now that the sectoral weight matrix is defined, it is possible to use the concept of sectoral dependence in the same way as spatial dependence, and to test for the presence of sectoral externalities. The testing and estimation procedure is similar to the one presented for the spatial case, with the exception of the nonlinearity aspect.

14.5 Data

We use data for 12 manufacturing industries in 15 regions in Spain (NUTS II regions, excluding the islands) over the time-period 1980–1991 in the empirical analysis. The data are obtained from two main sources. Data on output, intermediate deliveries, labor cost and the number of workers are obtained from the *Encuesta Industrial* (Industrial Survey) produced by the *Instituto Nacional de Estadística* (INE, Spanish Statistical Office).[9] Data series for private and public capital stocks are taken from *El Stock de Capital en la Economía Española* (The Capital Stock in the Spanish Economy, Fundación BBV 1995). Table 14.1 provides an overview of the twelve manufacturing sectors considered in the analysis.

The price for employment (P_L) is obtained by dividing labor costs by the number of jobs. The index price of intermediate inputs (P_M) is measured by dividing the nominal intermediate input series by the constructed real intermediate input series. Private capital is measured by the total net capital stocks of the manufacturing industry. Public capital stock includes the net monetary stock of core infrastructure, such as roads and highways, railway, harbors and maritime signaling, airports, water and sewage facilities, and urban structures.[10] Since public infrastructure is not as-

[8] Compare this idea with to Coe and Helpman (1995) where the relevance of international spillovers in R&D investments depends on the trade volumes between economies.

[9] Data provided by the *Encuesta Industrial* are given in nominal values. Sector and region-specific producer price indices, supplied by the *Programa de Investigaciones Económicas* (Economic Research Program), are used to deflate the regional sectoral data.

[10] It has been shown that basic public infrastructure has a positive impact on regional productivity in the Spanish regions. The effects for social public are less clear (see, for instance, Mas *et al.*, 1996; Moreno *et al.*, 1997).

Table 14.1. Description of the industrial sectors

Sector ID	Sector name
1	Metallic minerals and first transformation of metals
2	Non metallic minerals and products
3	Chemistry
4	Metallic products and metalwork
5	Agricultural and industrial machinery and equipment
6	Electric machinery and material
7	Transport materials
8	Food products, alcohol, drinks and tobacco
9	Textiles, leather and shoes
10	Paper and derivatives and printing
11	Rubber and plastic derivatives
12	Wood, cork and derivatives and remaining industries

sumed to have an immediate effect on industrial activity, we use a one-year lag. The data needed for compiling the sectoral weight matrix are taken from the input-output table for the Spanish economy in 1990.

14.6 Empirical Results

We present the main results for both the regional and the sectoral case in this section. In both cases, the restrictions between the parameters implied by Shephard's Lemma do not fit our data. We therefore compute the relevant elasticities by estimating the variable cost function. The results are generated using a program written in Gauss v3.2.8.

14.6.1 The Regional Case

Although our primary concern is to test for the significance of externalities and to estimate their effect on costs, we first estimate (14.9) and check for spatial dependence in the residuals. This gives an indication of the potential bias of the traditional specification in which externalities are neglected.

The model is estimated using pooled data. Several studies (for instance, Seitz and Licht, 1995; Morrison and Schwartz, 1996) have estimated a fixed effect model to account for unobservable economy effects on the cost level. The fixed effects model is chosen because random effects are assumed to be correlated with the arguments in the cost function, which is not appropriate. The use of fixed effects, however, causes an incidental parameter problem when the maximum likelihood (ML) principle is applied in the spatial context (both for the tests and the estimation procedure in the presence of spatial dependence).[11] Given that spatial effects are

[11] We are grateful to the editors for pointing this out.

Table 14.2. Spatial dependence tests in the regional case with p-values in parentheses

	Moran's I	LM-LAG	LM-ERR
Without external effects	0.895	16.759	0.161
	(0.371)	(0.000)	(0.688)
Including external input (Kg)	1.507	9.339	1.119
	(0.132)	(0.002)	(0.275)
Including external input (Kg) and	0.941	0.327	0.364
across-region externalities (Kg and Y_ρ)	(0.347)	(0.567)	(0.546)

our main concern, we therefore consider exogenous economy-wide heterogeneity by means of a dummy variable that separates regions with a high share of manufacturing in total output from regions specialized in other sectors. This variable is signiˇ cant in all regressions and, as expected, indicates lower exogenous cost levels in regions specialized in manufacturing. The results for the spatial autocorrelation tests are shown in Table 14.2. The LM-LAG test clearly rejects the null hypothesis, and points to the relevance of spatial externalities in explaining the cost level of the manufacturing industry. However, neither Moran's I nor LM-ERR rejects the null hypothesis of no spatial autocorrelation in the residuals or the errors, respectively.

In order to obtain the effect of the external input on manufacturing cost, we start by introducing the stock of infrastructure.[12] A Likelihood Ratio (LR) test for the joint signiˇ cance of all terms related to Kg rejects the null hypothesis of the effects being zero (LR $= 35.82$, $p < 0.001$), revealing the necessity of including such a variable.

Since it is difˇ cult to analyze the plausibility of signs and signiˇ cance of the estimates given that the speciˇ cation contains quadratic and cross-product terms referring to each variable, the results for the initial estimation are not presented and instead the relevant effects are summarized by the elasticities.[13] The elasticities regarding the effects of public capital are shown in Table 14.3. The results show that the elasticity of costs with respect to public capital is on average (weighted by the share of regional output in total national) negative. The value of -0.034 indicates that the Spanish industries beneˇ ted only slightly from cost reductions related to increasing public capital during the 1980s. The negative average sign for the infrastructure elasticity implies a global net substitution relationship between public capital and private inputs. The elasticities of the conditional demand for labor and intermediates show that, on average, infrastructure is labor using (0.179) and intermediates saving (-0.064). The results also indicate that returns to scale ($RTS = 1/\varepsilon_{VCY}$) are nearly constant (1.073).[14] This is in accordance with earlier results, including Suárez (1992) and Velázquez (1993).

[12] This implies the restrictions $\theta = 1$ and $\beta_{Y_\rho} = \beta_{Y_\rho Y_\rho} = \beta_{Y_\rho Kp} = 0$ in (14.10).

[13] The results of the estimations are available upon request.

[14] Testing the average elasticities requires knowledge about individual standard errors, which are a complicated function of the estimated standard errors for the parameters in the model.

Table 14.3. Elasticities from the speciˇ cations with the external input in the regional case

	ε_{VCKg}	ε_{LKg}	ε_{MKg}	RTS
Andaluc´a	−0.069	0.312	−0.069	1.035
Aragón	−0.016	0.064	−0.060	1.140
Asturias	0.010	0.190	−0.066	1.021
Cantabria	0.056	0.216	−0.062	1.065
Castilla-Leon	−0.034	0.143	−0.058	1.148
Castilla-Mancha	−0.027	1.150	−0.066	1.109
Cataluña	−0.061	0.233	−0.066	1.027
Valencia	−0.049	0.338	−0.068	1.044
Extremadura	−0.001	0.318	−0.067	1.131
Galicia	−0.028	−0.150	−0.064	1.117
Madrid	−0.011	0.352	−0.055	1.143
Murcia	0.020	0.031	−0.061	1.155
Navarra	0.032	0.539	−0.053	1.217
Pais Vasco	−0.026	−0.790	−0.067	0.992
Rioja	0.040	1.538	−0.054	1.280
Average	−0.034	0.179	−0.064	1.073

It can be seen in Table 14.2 that inclusion of the external input reduces the magnitude of the LM-LAG statistic, although it continues to be signiˇ cantly different from zero. Therefore, following the "classical" speciˇ cation search approach frequently used in the spatial econometric literature, we estimate the spatial lag model. When estimated by ML, the spatial lag of the endogenous variable is signiˇ cant (LR $= 9.689$, $p = 0.002$), indicating the adequacy of considering variable costs in neighboring regions.[15] In this speciˇ cation there does not seem to be any remaining spatial dependence. However, although the consideration of the spatial lag model results in an econometric solution for the spatial dependence problem, it does not identify the sources of these spatial externalities. Considering our theoretical sources of externalities across economies as described in Sect. 14.4, we therefore proceed estimating (14.10) by NLLS in order to deal with the nonlinearity caused by the functional form for the composite of public capital. The LR test rejects the hypothesis of all the terms related to Y_ρ jointly being zero (LR $= 22.859$, $p < 0.001$), so the thick-market externality needs to be considered. Table 14.2 shows that the inclusion of these externalities completely removes spatial autocorrelation.

Alternatively, one can use the dispersion in the values for each sector and time period. Constant returns to scale cannot be rejected in the latter case.

[15] For reasons of space, the elasticities concerning the effect of the external input when including a spatial lag of the endogenous variable are not presented. However, it is worth noting that expressing the spatial lag model as a reduced form shows that the explanatory variables are multiplied by $(I-\rho W)^{-1}$ in the spatial lag speciˇ cation.

Table 14.4. Elasticities from the speciˇ cation with the external input and the across-region externality in the regional case

	ε_{VCKg}	ε_{LKg}	ε_{MKg}	RTS	ε_{VCY_p}
Andaluc´a	0.287	−0.091	0.024	1.063	0.037
Aragón	0.253	0.219	0.021	1.121	0.058
Asturias	0.306	0.129	0.023	0.931	0.050
Cantabria	0.298	−0.088	0.023	0.928	0.056
Castilla-Leon	0.252	−0.143	0.021	1.165	0.049
Castilla-Mancha	0.245	−0.144	0.023	1.113	0.055
Cataluña	0.298	−0.084	0.025	1.097	0.053
Valencia	0.278	−0.078	0.025	1.095	0.057
Extremadura	0.236	−0.244	0.022	1.028	0.058
Galicia	0.259	−0.133	0.023	1.105	0.051
Madrid	0.276	−0.105	0.021	1.125	0.050
Murcia	0.241	−0.171	0.022	1.086	0.058
Navarra	0.251	−0.241	0.020	1.101	0.053
Pais Vasco	0.319	−0.103	0.024	0.979	0.043
Rioja	0.221	0.052	0.020	1.127	0.057
Average	0.282	−0.084	0.023	1.084	0.050

The estimated parameter θ is equal to 0.58, indicating that, although the public capital endowment in the region under consideration is the most relevant, the endowments in neighboring regions also play an important role. This is likely the result of the network characteristic of most public infrastructure (Rietveld, 1995). The elasticity of costs with respect to the composite of public capital (ˇ rst column in Table 14.4) now equals 0.282 on average. Apparently, public capital did not induce a reduction in manufacturing costs in the Spanish regions during the eighties. This result is in accordance with other studies dealing with the effectiveness of public capital investment in enhancing productivity in modern industrialized economies (Holtz-Eakin (1994), and Garc´a-Mila *et al.* (1996), for the U.S., and de la Fuente (1996), for Spain). This result can be explained by the fact that, during the 1980s, the Spanish regions already had a substantial stock of public capital, suggesting the threshold level for infrastructure had already been reached.[16] Even more important is the observation that the positive cost elasticity of infrastructure contradicts the estimated sign for the model without externalities across economies. It is obvious that spatial misspeciˇ cations may evoke erroneous conclusions. What is also sur-

[16] In several studies analyzing the effect of public capital stock on economic growth in the Spanish regions (e.g., Mas *et al.*, 1996; Moreno, 1998), it has been shown that the impact of infrastructure decreased during the eighties. This was partly due to the existence of decreasing returns to scale for public capital, indicating that it is a factor with a threshold level that once reached reduces its effect.

Table 14.5. Spatial dependence tests in the sectoral case with p-values in parentheses

	Moran's I	LM-LAG	LM-ERR
Without external effects	2.813	17.645	5.789
	(0.005)	(0.000)	(0.016)
Including external input (Kg)	2.564	16.321	2.790
	(0.010)	(0.000)	(0.095)
Including external input (Kg) and	1.139	0.160	0.108
across-region externalities (Y_ρ)	(0.255)	(0.689)	(0.742)

prising is that in our sample the output of neighboring regions increases the cost in the region under consideration. Although small in magnitude, it seems that Spanish regions suffer in the case of proximity to regions with a high manufacturing output. This may be indicative of a certain competition during a period of major restructuring in the manufacturing industry in Spain. It is obvious that these conclusions may depend on the speciˇ cation of the weight matrix.

14.6.2 The Sectoral Case

As for the regional case, we estimate (14.9) by OLS using dummy variables in order to allow for distinct levels of exogenous costs in a group of sectors characterized by higher technology levels[17] and another by included mature activities. As shown in Table 14.5, the spatial statistics reveal the existence of sectoral dependence. In contrast to the regional case, all tests are signiˇ cantly different from zero. In order to analyze the effect of public capital on manufacturing costs, we ˇ rst only introduce the public capital stock variable. The LR test shows the joint signiˇ cance of all the new terms (LR $= 16.459$, $p = 0.011$). The cost elasticity with respect to public capital is shown in Table 14.6, and reveals a positive industry-weighted average (0.305). There is, however, a strong cross-industry variation both in the sign and the magnitude of this elasticity. This may re ect differences in the capacity of industries to take advantage of available public capital in the Spanish economy, as proposed in a number of theoretical models (see, e.g., Holtz-Eakin and Lovely, 1996). As for the relationship between public capital and each variable factor, we again obtain the result that infrastructure capital is labor using (2.289) and intermediates saving (−1.197). Finally, average returns to scale are increasing, 1.849, with large cross-industry variability. This is similar to the results obtained in studies analyzing returns to scale at the industry level (for instance, Caballero and Lyons, 1990; Burnside, 1996).

As depicted in Table 14.5, the introduction of the external input does not eliminate sectoral dependence. The LM-LAG test is the most signiˇ cant test. As in the regional case, if we follow the classical strategy in seeking the best model, we have to resort to the spatial lag model. Although the lag of the endogenous variable is

[17] This group includes sectors 3, 4, 6, 7 and 10 (see Table 14.1).

Table 14.6. Elasticities from the speciˇ cation with the external input in the sectoral case

Sector ID	ε_{VCKg}	ε_{LKg}	ε_{MKg}	RTS
1	0.720	0.898	−2.146	5.724
2	0.428	1.321	−1.213	1.680
3	0.540	1.091	−1.606	2.721
4	0.274	1.774	−0.992	1.206
5	0.012	10.170	−0.594	0.855
6	0.098	9.185	−0.686	0.927
7	0.434	1.470	−1.220	2.513
8	0.288	1.142	−1.415	1.569
9	−0.005	3.606	−0.836	0.865
10	0.337	2.090	−0.915	1.390
11	0.301	3.521	−0.790	1.183
12	−0.084	−1.620	−0.682	0.766
Average	0.305	2.289	−1.197	1.849

signiˇ cant (the value for the LR test is 24.747, $p < 0.001$) and completely removes sectoral autocorrelation, it does not provide us with an explanation of the origin of the externalities across industries. We therefore estimate our empirical model given in (14.10) for the sectoral case. As shown in Table 14.5, introducing the spillover across industries also completely removes sectoral dependence. This result, as well as the global signiˇ cance of all parameters including the supplier-weighted product (LR $= 20.900$, $p < 0.001$) supports our hypothesis that the source of the externality is thick markets.

Finally, Table 14.7 displays the results of the elasticities including the cross-industry externality in the speciˇ cation. The value for the cost elasticity with respect to public capital changes to an average of −0.341, with both positive and negative effects. This implies that the manufacturing industry in Spain beneˇ ted from the increase in infrastructure provision during the 1980s. However, the variability across industries is large, which is in line with the differing effects of public capital on economic activities as shown in Holtz-Eakin and Lovely (1996). It is worth noting that this cost-reduction effect of public capital appears when we allow for externalities across economies. Similarly, returns to scale seem more reasonable in this speciˇ cation as well. The results reported in Table 14.6 may be strongly biased due to the omission of the externality. This conclusion is in line with studies applying the primal model. The cost elasticity with respect to the cross-industry externality is negative on average (−0.325). Consequently, the higher the output in the supplier industries, the greater the technological diffusion embodied in goods, and the higher the supplier-driven externalities, with correspondingly lower manufacturing costs.

Table 14.7. Elasticities from the speciˇ cation with the external input and the across-industry externality in the sectoral case

Sector ID	ε_{VCKg}	ε_{LKg}	ε_{MKg}	RTS	$\varepsilon_{VCY_{\rho}}$
1	0.175	0.415	−1.102	1.046	−0.591
2	0.042	0.672	−0.538	0.944	−0.375
3	−0.244	0.565	−0.647	0.841	−0.490
4	−0.202	1.060	−0.415	0.822	−0.280
5	0.186	0.377	−0.283	1.134	0.279
6	0.031	7.140	−0.316	0.989	0.076
7	−0.492	0.883	−0.504	0.781	−0.413
8	−1.086	0.958	−0.436	0.569	−0.612
9	−0.368	−0.617	−0.320	0.713	−0.098
10	0.096	1.002	−0.434	1.033	−0.134
11	0.427	1.174	−0.403	1.293	0.030
12	0.098	1.767	−0.292	0.893	0.082
Average	−0.341	1.148	−0.481	0.828	−0.325

14.7 Conclusions

This chapter addresses the relevance of external effects for the economic performance of ˇ rms. We distinguish two sources for the occurrence of externalities: externalities derived from inputs within the same economy but external to the ˇ rm, and externalities related to spillovers from industries or geographical areas generating the effects. The latter case contributes to the debate on the scope of externalities across economies. While one strand of the literature argues that externalities with respect to ˇ rms exist, albeit only for ˇ rms within the same industry, others emphasize linkages between ˇ rms from different sectors. The same reasoning can be applied to ˇ rms located in different geographical areas (i.e., regions or countries). Unlike most studies conducted in this area, our analysis is carried out within the duality framework. This overcomes some of the shortcomings of an analysis based on the production function. First, it allows internal and external returns to be disentangled from variations in input utilization. Second, measures of externalities, such as those from thick-markets proxied by output, do not cause problems of endogeneity. Finally, this approach should give more information about the effects of the externalities through the different substitution effects with internal inputs.

We utilize spatial econometric techniques to model the external effects, and show that the concept of spatial dependence can be extended to account for linkages between different sectors of an economy. While traditional deˇ nitions of weight matrices can be used in the case of externalities across regions, proper counterparts for the sectoral case need to be deˇ ned. We propose the use of input-output linkages as a measure of contiguity or neighborhood for the empirical analysis of externalities across industries. In addition, given the nonlinearity affecting the parameters of certain exogenous variables of the proposed empirical model, we obtain the expressions of the Lagrange Multiplier tests for spatial dependence in the context of such a nonlinear model.

We apply this theoretical and conceptual framework to the case of the stock of publicly provided capital (external input) and to the output in the neighboring economies (cross-economy spillovers) for manufacturing sectors in Spanish regions. Although the results may be sensitive to multicollinearity problems endemic to the use of unrestrictive cost functions, we can conclude that externalities have a signiˇ -cant impact in terms of cost reduction in the sectoral case. This effect is opposite (in sign) to the effect of spatial externalities. The effect of public capital is not homogeneous, and remains unclear. Finally, we show that the omission of external effects evokes biased parameters for the effect of traditional inputs and internal returns to scale.

Acknowledgments

We would like to thank the editors for helpful comments and suggestions, and the *Fundación Empresa Pública* for providing the data of the Spanish Industrial Survey. Part of this research was ˇ nanced by the *Plan Nacional de I+D*, Spanish Ministry of Education, Project 2FD97-1004-C03-01.

Part IV

Urban Growth and Agglomeration Economies

15 Identifying Urban-Rural Linkages: Tests for Spatial Effects in the Carlino-Mills Model

Shuming Bao[1], Mark Henry[2], and David Barkley[2]

[1] University of Michigan
[2] Clemson University

15.1 Introduction

A continuing interest of regional scientists is the development of econometric models for the identiˇ cation of local characteristics associated with regional growth (e.g., Carlino and Mills, 1987; Thurston and Yezer, 1994; Boarnet, 1994a). Recent advances in spatial econometrics and geographic information systems (GIS) enhance the reliability of small region growth models by incorporating the in uences of spatial linkages on the local development process (e.g., Anselin, 1988b; Anselin and Florax, 1995b). Modeling the in uence of spatial linkages along with local characteristics appears most beneˇ cial in studies of small area economic change where inter-area spillovers may be extensive. For example, economic and population change in the "edge cities" of urban complexes may affect development of nearby rural areas.

The purpose of this chapter is to evaluate the role that both space and local amenities play in the rural development process. We construct a model of small area employment and population change that weaves together local characteristics with several facets of the spatial dimensions of rural development within a Functional Economic Area (Fox, 1974). Spatial linkage effects are examined by construction of models that include both spatial lags and intra-regional interaction terms.[1]

In this chapter, we modify the Carlino and Mills (1987) model of local (county) change by incorporating a spatial autoregressive process in the econometric models. While the incorporation of temporal lags in time series analysis is commonplace, an analysis of the role of spatial lags in rural development models is still rare. Few studies assemble the spatial weights matrices needed in small area spatial econometric models despite the recognition of the possible effects that spatial processes can have on estimators in cross sectional studies (for example, Anselin, 1988b). However, recent advances in the use of GIS to construct spatially lagged variables have greatly facilitated the use of spatial econometrics by practitioners. Moreover, the results of our analysis reveal that parameter estimates in commonly used models of local development can be sensitive to the inclusion of a spatial autoregressive process in the model.

---

[1] In earlier work we have carried out tests for linkages between urban centers and nearby rural areas using population density functions within an FEA (see Barkley *et al.*, 1994). While we found change in the density functions between 1980 and 1990, we were not able to explain why some areas near the urban centers grew in density while others declined.

In the next section, we describe the areal dimension of the problem and related hypotheses about the regional development process. In Sect. 15.3, the modiˇ ed Carlino-Mills model of regional change in employment and population is developed. In Sect. 15.4, empirical results for our region and rural policy implications are examined. A summary ofˇ ndings is provided in Sect. 15.5.

15.2 Spatial Context of the Analysis

Gaile (1980) proposes that economic development in an urban core impacts the surrounding region through a complex set of spatial processes. These processes include intra-regional ows of private capital, private and public expenditures for goods and services, information and technology, residents and commuters, political in uence, and public investments. The ows generally provide both beneˇ cial and detrimental impacts on the peripheral region, with the net result depending on the relative magnitude of the positive and negative forces. If the cumulative process results in an increase in the absolute level of activity in the periphery, the resulting impact is urban *spread* or spillover. A decline in the absolute level of economic activity in the rural periphery in conjunction with core expansion is evidence of a *backwash* effect.

However, the geographical dimension in much of the small region econometric work may be too crude to reveal the spread and backwash processes between urban centers and the hinterlands. For example, Carlino and Mills (1987) used U.S. counties as *independent* observations in a semi-structural model of employment and population change.

While the Carlino-Mills model provides an empirical structure for reconciling the jobs-people direction of causality issue in county-level development processes, it lacks the spatial structural detail needed to assess the role that "place in the FEA (Functional Economic Areas)" and urban core-fringe-hinterland linkages may play in development processes within a FEA.[2] Speciˇ cally, potentially important spatial dependence in the error term between the counties (the units of observation) is not considered by Carlino-Mills.

Boarnet (1994a) provides an extension of the Carlino-Mills model by explicitly recognizing that city labor market areas are larger than a municipality. Recognizing that place within a region matters, he adjusts for this spatial mismatch by incorporating labor market potential variables that are a function of distance from a given city and number of employees/population in each city – essentially a simple gravity model concept.

15.2.1 Modiˇ cations to the Carlino-Mills Model

We modify the Carlino-Mills model in a direction that re ects our interest in intra-FEA linkages between the urban core, urban fringe, and rural hinterland. Accordingly, we identify potential spatial linkages between the core, fringe and hinterland

[2] Fox (1974) deˇ ned the Functional Economic Area (FEA) as a relatively self-contained labor market that contains a metropolitan central city and hinterlands within commuting distance.

areas explicitly in the model. We do this by including two *interaction variables* on the right hand side of the equations to test for the presence of core-hinterland (CORE) and fringe-hinterland (FRINGE) linkages. This procedure allows for explicit tests on hypothesized spillover and backwash effects on the rural hinterland. We also distinguish between urban core-fringe *size* and urban core-fringe *growth* effects. This results in two different model specifications, which we will refer to as RU(I) for the growth model, and RU(II) for the size model. This approach fits in the tradition of testing for spread and backwash effects employed in monocentric models of metropolitan area development (see Hughes and Holland, 1994; Schmitt, 1996, for recent tests of these effects).

We also address the issue of spatial processes in the model by specifying a spatial autoregression and estimating a spatial lag parameter. If the population and employment change process in a FEA yields a spread or backwash from the urban areas to the rural hinterland, then change in rural areas near urban areas should be affected by the growth of the urban areas and by the location of the rural area relative to the urban area. Spatial lags, if important, should then be recognized in the econometric models of rural area employment and population change (see Anselin and Florax, 1995b).

Our analytical approach to evaluate these hypotheses proceeds as follows. We define a series of Functional Economic Areas based on observed 1990 commuting patterns in three southern states (South Carolina, parts of Georgia and North Carolina). Our study region consists of the 46 South Carolina counties, 21 adjacent North Carolina counties, and 18 Georgia counties adjacent to South Carolina. We use an algorithm that groups counties together to maximize within-region commuting (see Bao *et al.*, 1995). Within each of the FEAs we use census tract observations and define urban core, urban fringe and rural hinterland tracts based on population density and distance from the urban cores. Specifically, the urban core is defined as the Census Urbanized Area in 1990 together with the surrounding tracts that have a population density of over 1000 persons per square mile. The urban fringe is defined as the area outside of the core but within a 30 minute travel distance to the urban core. The travel distance is calculated in a GIS and takes into account the speed limit imposed on each route. Finally, the rural hinterland is composed of the remaining tracts in each FEA.

A three-part process is used to accomplish the operational delineation of FEAs. First, the MSA county with the maximum in-commuter flow of all the counties in the set is identified. Next, using ARCINFO GIS software (ESRI, 1992) and the Census TIGER files, non-metro counties are assigned to a MSA based on the minimum distance from each non-metro county's centroid to the centroid of the MSA core county. This establishes a temporary FEA of MSA core and proximate MSA and non-metro counties. Finally, journey-to-work data from the 1990 Census of Population are used to reallocate non-metro and metro counties to FEAs such that commuter flows are maximized within an FEA.

As illustrated in Fig. 15.1, eight FEAs were eventually identified in our data set. These include four that are relatively large, each containing at least 11 counties

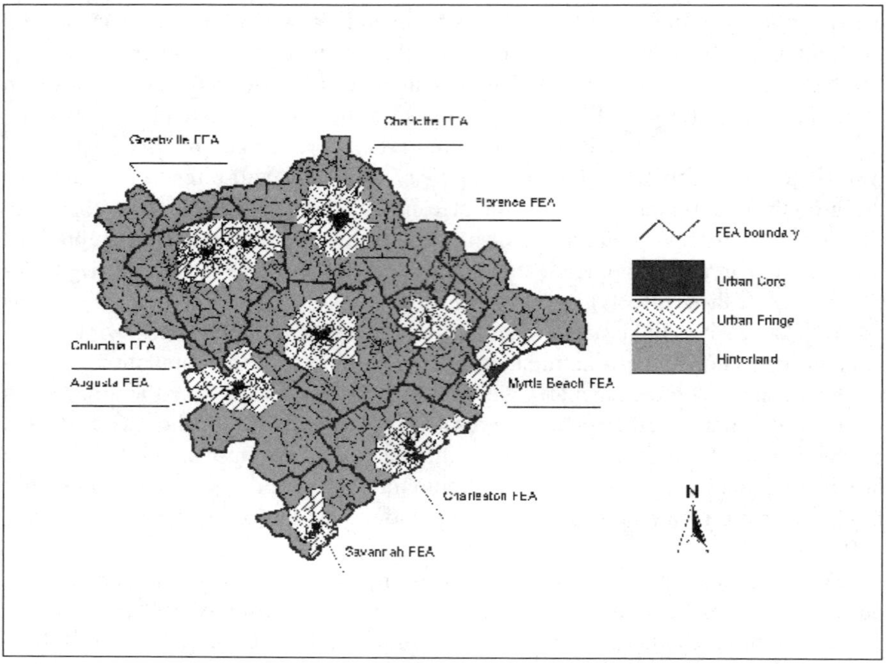

Fig. 15.1. Functional economic areas with classification of urban core, fringe and hinterland

(Augusta, GA; Charlotte, NC; Columbia, SC; and Greenville, SC). Three FEAs border the Atlantic coast (Charleston, SC; Myrtle Beach, SC; and Savannah, GA), each with six or fewer counties. The small number of counties in these FEAs results from their inability to expand east of their core counties. The remaining FEA with a South Carolina core county is a five county area surrounding the small MSA of Florence, SC.

In this context, we are particularly interested in the presence and role of spatial autocorrelation. Such spatial dependence can be caused by spatial interactions of people and firm's choices of location, by core-fringe-hinterland spillover effects, and by the demand for and supply of local amenities.[3]

At this juncture, we adopt a framework for analysis that is derived from Carlino and Mills (1987), with a focus on rural hinterland development and intra-region relationships. We modify the analysis of Carlino-Mills in three ways. First, we use observations at the census tract level for 1980 and 1990 for core, fringe and hinterland areas. Thus we incorporate the Functional Economic Area concept into the model and test explicitly for linkages between the three zones. Second, we employ

[3] In this study, we found significant spatial autocorrelation. This indicates significant cross-tract effects of spatial interactions between population and employment. Details of the tests and the Moran *I* results are available in Henry *et al.* (1994).

the estimation techniques in SpaceStat (Anselin, 1992) to account for potential spatial autocorrelation. Third, we focus on the rural hinterland areas in our empirical analyses and evaluate how their population and employment growth are affected by local amenities as well as the activities in nearby urban core and fringe areas.

15.3 Econometric Model

15.3.1 Carlino-Mills Model

Tiebout and monocentric models of the regional development process indicate that firm and household location decisions within a region will be influenced by transportation costs and community characteristics. Thus, the bids for sites within a region (or the revealed location decisions) of firms and households can be expressed as functions of the beginning period spatial distribution of activity, local transport conditions, and community characteristics.

Empirical models of regional development reflect the interdependencies between household residential choices and firm location decisions. This view is well established as a result of work on identification of the direction of causality in the "jobs follow people" or "people follow jobs" literature.

Following Steinnes and Fisher (1974), Carlino and Mills defined a simultaneous interaction model for population and employment at equilibrium as:

$$P^\star = f\left(E^\star|\Omega^p\right), \quad (15.1)$$

$$E^\star = g\left(P^\star|\Omega^e\right), \quad (15.2)$$

where $P^\star$ and $E^\star$ are equilibrium population and employment, and Ω^p and Ω^e are the sets of the historical information and the spatial structural information.

To account for the interdependency between population and employment, Carlino and Mills suggest the following linear simultaneous equations for reconciling the direction of causality issue in county-level development processes:

$$P^\star = A_P E + B_P T, \quad (15.3)$$

$$E^\star = A_E P + B_E S. \quad (15.4)$$

where S and T are vectors of exogenous variables for local amenities, A_P and A_E are coefficients of the endogenous variables, and B_P and B_E are vectors of coefficients of exogenous variables.

Starting from a general spatial equilibrium model, both households and firms are assumed mobile over space and adjustable towards equilibrium. Since the equilibrium values, $P^\star$ and $E^\star$ in equations (15.3–15.4), are not observable, Mills and Price (1984) suggest the relationship between equilibrium population and employment and their observed values to be as:

$$P = P_{-1} + \lambda_P(P^\star - P_{-1}), \quad (15.5)$$

$$E = E_{-1} + \lambda_E(E^\star - E_{-1}), \quad (15.6)$$

where the subscript -1 refers to the variable lagged one period in time, and λ_P and λ_E are parameters for the speed of adjustment, with $0 \leq \lambda_P, \lambda_E \leq 1$.

Substituting equations (15.3) and (15.4) in (15.5)–(15.6), a system of simultaneous equations in the observable P and E follows as:

$$P = \lambda_P A_P E + \lambda_P B_P T + (1 - \lambda_P) P_{-1}, \tag{15.7}$$

$$E = \lambda_E A_E P + \lambda_E B_E S + (1 - \lambda_E) E_{-1}. \tag{15.8}$$

Alternatively, equations (15.7) and (15.7) can be represented in difference form, as:

$$\begin{aligned} \Delta P &= P - P_{-1}, \\ &= \lambda_P A_P E + \lambda_P B_P T + \lambda_P P_{-1}, \end{aligned} \tag{15.9}$$

and,

$$\begin{aligned} \Delta E &= E - E_{-1}, \\ &= \lambda_E A_E P + \lambda_E B_E S + \lambda_E A_E E_{-1}. \end{aligned} \tag{15.10}$$

Equations (15.7) and (15.8), suggested by Carlino and Mills, are straightforward to verify empirically. However, in this form, control for potentially important spatial dependence between counties is ignored, and the model lacks the spatial structural information that can provide the basis for assessing the role that core-periphery-hinterland linkages may play in local development. Moreover, the estimates from an econometric model without spatial autocorrelation correction may be biased and inconsistent if the speciˇ cation fails to properly capture spatial structural information (Anselin, 1988b).

In this chapter, a spatial autoregressive speciˇ cation is constructed as a solution to the problem of spatial dependence, and the appropriate spatial econometric estimators are applied (Anselin, 1988b, 1993; Upton and Fingleton, 1985; Henry *et al.*, 1994).

15.3.2 Spatial Structure in the Models

In order to obtain appropriate estimates when spatial autocorrelation is present, spatial process models can be constructed to capture the information on the spatial structure. A typical spatial autoregressive model can be deˇ ned as:

$$y = \rho W y + X\beta + \varepsilon, \tag{15.11}$$

where ρ is the spatial autoregressive parameter, y is a random variable in vector form, with a spatial autoregressive structure, W is the pre-deˇ ned spatial weights matrix, with Wy as the spatial lag of the dependent variable, X is a matrix of explanatory variables that are assumed to be uncorrelated with the error term, β is a matching vector of regression parameters, and ε is the random error term.

Before we consider spatial effects, we specify the following baseline model:

$$\begin{aligned} P_{90} &= a_0 + a_1 E_{90} + a_2 P_{80} + a_3 \mathrm{PHL}_{80} + a_4 \mathrm{PDH} + a_5 \mathrm{RMILL} \\ &\quad + a_6 \mathrm{PUPTEA}_{80} + a_7 \mathrm{RSEW}_{80} + a_8 \mathrm{POV}_{80} + \varepsilon_P, \qquad (15.12) \\ E_{90} &= b_0 + b_1 P_{90} + b_2 E_{80} + b_3 \mathrm{WSL}_{80} + b_4 \mathrm{RMILL} \\ &\quad + b_5 \mathrm{PUPTEA}_{80} + b_6 \mathrm{RSEW}_{80} + b_7 \mathrm{POV}_{80} + \varepsilon_E, \qquad (15.13) \end{aligned}$$

where $P_{90/80}$ is the 1990/80 population in persons per square mile, $E_{90/80}$ is the 1990/80 on site employment in employees per square miles, PHL_{80} is the 1980 highway density in miles or highway per square mile, WSL_{80} is the 1980 water and sewer line density in miles of sewer line per square miles, PDH is the 1980 average distance for each census tract to its nearest hospital in miles, RSEW_{80} is the 1980 percentage of occupied houses with public sewer utilities, RMILL is the 1990 aggregated mill rate (city and county mill rates), PUPTEA_{80} is the 1980 per student-teacher ratio in high schools, and POV_{80} is the 1980 percentage of persons living below the poverty level. All the densities are calculated on the basis of land area with water area excluded.

In spatial econometric models, a key issue is how to deˇ ne a proper spatial weights matrix to construct the spatial lags of variables. Different deˇ nitions of spatial weights matrices may re ect different spatial structure. We consider three different weights: re ecting spatial contiguity linkage, spatial distance linkage, and commuter linkage.

- Contiguity linkage is deˇ ned by the topological information on census tract boundaries, such that the elements of the weights matrix $w_{ij} = 1$ if tract i and tract j are contiguous and zero otherwise. Also, by convention, $w_{ii} = 0$, such that the weights matrix has a zero diagonal.
- Spatial distance linkage is deˇ ned in terms of the distance between tract centroids: $w_{ij} = 1$ for those tract pairs i, j that are within a given threshold distance of each other. Since census tracts are deˇ ned to capture about 4,000 persons, the area and spatial shape of census tracts are not regular. For example, the area of a rural tract may be more than ten times the area of an urban tract. A distance criterion may be more appropriate to capture association in terms of geographical distribution. Since the daily activities of most people are carried out within 30 miles radius of their residence, this distance is used in our empirical analysis. It is interesting to note that this same threshold is used by analysts at the South Carolina Department of Commerce to deˇ ne labor sheds for new ˇ rms that express an interest in a South Carolina location (Henry *et al.*, 1997).
- Both contiguity and distance linkages preclude directional effects in the speciˇ cation of spatial association. However, people's activities are usually affected by geographical characteristics such as highways, lakes and mountains. For example, the access to a highway may make the spatial linkage between two tracts stronger, while a lake may isolate two tracts even if they are geographical neighbors. The notion of a commuter linkage is deˇ ned according to people's working activities, which re ect a directional spatial relationship. It is formalized

as:

$$w_{ij} = \frac{C_{i \to j} + C_{j \to i}}{\sum_k^n (C_{i \to k} + C_{k \to i})}, \tag{15.14}$$

where $C_{i \to j}$ is the number of persons living in tract i, but working in tract j, and $C_{j \to i}$ is the number of persons living in tract j but working in tract i.

- We selected the spatial weights that will be used in the modified Carlino-Mills model on an empirical basis. To this effect, we estimate equations (15.12–15.13) by means of two stage least squares to control for standard endogeneity (no spatial lag terms are included) and test for spatial autocorrelation in the associated residuals by means of Moran's I. The results indicated that the distance-based weights (for a 30 mile cut-off) were the most appropriate for our data.[4]

15.3.3 Modified Carlino-Mills Models

Since our focus is on the hinterland, we construct a spatial form of the Carlino-Mills model that hypothesizes a link from both the urban core and the urban fringe to the rural hinterland. In this model, values for the hinterland tracts are endogenous, but variables with urban core and fringe observations (as aggregations over core or fringe tracts within each FEA) are exogenous.

The point of departure is the linear change specification from equations 15.9–15.10. A spatially lagged dependent variable (respectively, $W\Delta P$ or $W\Delta E$) is introduced on the right hand side of these equations. In addition, two interaction variables are included that take the form P*CORE and P*FRINGE (and the corresponding expressions for interaction with E). Two different models result, depending on whether the interaction terms are based on the change in population or employment in the core/fringe tracts, model RU(I), or on the absolute size of these variables, model RU(II).[5]

The final specification also includes two vectors of amenity variables to control for the effect on residential locational choices (F_1 in the equation for ΔP) or on firm location (F_2 in the equation for ΔE). A problem with such variables is a high degree of multicollinearity. In order to avoid this problem, we carried out a factor analysis on a candidate set of amenity variables and identified how they were grouped into the resulting factors. For each of the groups, selected amenity variables were included in the model that were representative of each group. The results of this analysis are summarized in Table 15.1, with the selected variables marked by an asterisk.[6]

---

[4] Results not reported here, but available upon request.

[5] A positive sign for these interaction variables would suggest a spillover effect from urban places to the rural hinterland. Conversely, preferences for smaller centers should yield negative coefficients.

[6] The detailed results, data sources and coefficients used in the factor analysis are not included here, but may be obtained from the authors.

Table 15.1. Selected amenity variables from factor analysis

	Access to urban areas for job, commercial and recreational opportunities
1.*	DISTH, distance from the rural tract centroid to the nearest hospital (km)
2.	PDH, 1980 average distance from census tract to nearest hospital (miles)
3.*	PHL80, 1980 highway density (four lane miles per square mile)
	Local government services, infrastructure and costs
1.	EXP2, 1990 per capita local government spending (county and city)
2.	RMLL, 1990 aggregate mull rates (city and county)
3.*	WSL80, 1980 water and sewer line density (miles per square miles)
4.*	RSEW80, 1980 percent of occupied houses with public sewer utilities
	Human captial and educational quality
1.*	PUPTEA80, 1980 students per teacher in tract high school(s)
2.	T11, 1980 test score mean on standard BASP for high school juniors
3.	DEH80, quality of local labor (high school graduates per working age population)
4.*	DEC80, quality of local labor (persons with at least some college per working age population)
	General quality of socioeconomic milieu
1.	PCI90, 1990 tract per capita income
2.*	POV80, 1980 percent persons below poverty level
3.	CR90, 1990 crimes per 1000 residents
4.*	RHOU78, houses build in 1970–80 per total houses 1980
5.	P80, 1980 tract population per square mile

In sum, the model used in the empirical veriˇ cation is:

$$\Delta P_{89} = a_0 + \rho_1 W \Delta P_{89} + a_1 \Delta E_{89} + a_2 P_{80} + a_3 E_{80} + a_4 P_{80} * \text{CORE} + a_5 P_{80} * \text{FRINGE} + a_6 F_1 + u_1, \tag{15.15}$$

$$\Delta E_{89} = b_0 + \rho_2 W \Delta E_{89} + b_1 \Delta P_{89} + b_2 E_{80} + b_3 P_{80} + b_4 E_{80} * \text{CORE} + b_5 E_{80} * \text{FRINGE} + b_6 F_2 + u_2, \tag{15.16}$$

where the subscript 89 refers to the change between 1980 and 1990 and the subscript 80 refers to the value at 1980.

15.4 Empirical Results

The empirical analysis is based on a data set with 268 observations at the tract level, derived from a spatially consistent 1980–90 geography constructed by means of a GIS.[7] The parameter estimates obtained for the linkage models are presented in Table 15.2, for both the change deˇ nition of the interaction variables CORE and FRINGE, model RU(I), as well as for the size deˇ nition, model RU(II).

[7] For details on the procedure used, see Bao *et al.* (1995).

15.4.1 Spatial Linkages

A comparison of the spatial spillover effect of the growth rates of urban clusters to the effect of their size can be based on an analysis of the estimation results for models RU(I) and RU(II). Of specific interest are the differences between the CORE and FRINGE interaction terms in these specifications.

In the model for population change using RU(I), the CORE interaction term is significant with a negative sign. This suggests a possible backwash effect for an urban core that experiences higher population growth, and a spread-through-decentralization effect for an urban core having a slower population growth. In other words, more rapid population growth in an urban core is associated with slower growth in rural hinterland population. The spatial lag term is significant and positive, suggesting positive spatial externalities for population growth in the rural hinterland.

The employment equation for RU(I) also shows a significant and negative sign for the CORE interaction term. Thus, here too there is evidence that more rapid employment growth in an urban core is associated with slower growth in rural hinterland employment, supporting an employment backwash effect. However, in the employment equation, the spatial lag parameter is negative (and significant), which, in contrast to the population model, does not support a notion of positive spatial externalities for hinterland employment growth.

The effect of the FRINGE interaction term is quite different. There is strong evidence of positive spillovers between the fringe population growth and the population change in nearby rural areas. The same phenomenon holds for FRINGE employment as well.

In the models RU(II), the interaction term is specified in terms of the size of the core or fringe populations. In the population equation, there is no significant interaction effect of the CORE population, but a negative and significant effect of the FRINGE. This suggests a population backwash effect of large fringe areas on nearby rural communities. The effect in the employment equation works in the opposite direction. The CORE interaction term is negative and significant, while the FRINGE interaction term is positive and significant. As in model RU(II), there is evidence of positive spatial externalities for population change in the rural hinterland, but negative externalities for employment change

In sum, there is evidence of both backwash and spread effects occurring in the southern FEAs considered here. Rural area population grew faster than average between 1980 and 1990 if it was in an FEA with a large urban core population and a rapidly growing population in the urban fringe. Rural employment grew faster than average in FEAs that had larger populations in the urban fringe.

Table 15.2. Parameter estimates for the rural/urban linkage models[a]

Variable[b]	RU(I) ΔP_{89}	RU(I) ΔE_{89}	RU(II) ΔP_{89}	RU(II) ΔE_{89}
Constant	−110.082	−2.4282	−247.993	−4.0861
	(585.681)	(46.7471)	(588.911)	(46.8491)
$W\Delta P_{89}$	0.0082		0.0157	
$W\Delta E_{89}$		−0.0070		−0.0040
P_{80}	−0.2018	−0.0004	0.1031	−0.0005
	(0.1463)	(0.0011)	(0.0310)	(0.0011)
E_{80}	3.5834	5.1726	3.7198	1.0055
	(1.4858)	(0.7850)	(1.5063)	(0.1614)
ΔP_{89}		0.0027		0.0036
		(0.0082)		(0.0082)
ΔE_{89}	−7.1581		−7.4185	
	(3.1659)		(3.2090)	
DISTH	−8.1642	0.6417	2.5720	0.7890
	(12.0871)	(0.8858)	(12.2820)	(0.8950)
RSEW80	−43.0293		−39.7680	
	(24.8418)		(25.1048)	
WSL80		−0.1163		−0.5001
		(1.5810)		(1.5563)
POV80	−35.0323	−0.4565	−40.5758	−0.4011
	(6.3940)	(0.5496)	(6.7670)	(0.5536)
PHL80	228.0240		282.8450	
	(347.3120)		(352.3230)	
RHOU78	45.4995		46.2797	
	(5.8438)		(5.9357)	
DEC80		−0.0003		0.0935
		(0.4362)		(0.4369)
PUPTEA80	−34.4359	0.2365	−32.4743	−0.0297
	(21.8745)	(1.6316)	(22.1597)	(1.6355)
CORE*P_{80}	−1.9484		0.0220	
	(0.7392)		(0.1680)	
FRINGE*P_{80}	3.6540		−0.1527	
	(1.1566)		(0.0749)	
CORE*E_{80}		−16.6340		−3.9845
		(4.7124)		(0.7401)
FRINGE*E_{80}		25.8659		0.5109
		(5.9327)		(0.1912)

continued on next page

Table 15.2. Continued[a]

Variable[b]	RU(I) ΔP_{89}	RU(I) ΔE_{89}	RU(II) ΔP_{89}	RU(II) ΔE_{89}
AIC	4429.16	4436.56	3032.94	3034.15
P_{LR}	0.0981	0.0761	0.0875	0.0892

[a] Estimated standard errors in parentheses

[b] Variable deˇ nitions as in Table 15.1

In contrast, rural area population grew more slowly than average if it was in an FEA with a fast growing population in the urban core and a larger than average population in the urban fringe. On the employment side, rural area employment grew slower than average in FEAs with large urban employment in urban core and rapid urban population growth in the fringe. The sign of the spatial lag parameter suggests the presence of a backwash of employment growth between adjacent rural tracts.

15.4.2 Amenity Effects

In addition to the spatial spillover effects, it is of interest to assess the sign and significance of the various amenity factors that were incorporated in the model speciˇ cation. Public service, proxied by RSEW80 in the population equation and by WSL80 in the employment equation, has a negative effect in both, but is only signiˇ cant for the population equation. Since the density of water and sewer lines is usually highly associated with the population density, this may suggest a spread effect of rural population and employment.[8]

Community wealth, as proxied by POV80 shows a similar pattern. The estimated coefˇ cient is signiˇ cant and negative for the population eqations, suggesting that new residents avoid rural hinterland tracts that are poor. However, while negative, the corresponding coefˇ cients in the employment equations are not signiˇ cant.

Local infrastructure as proxied by PHL80 is not signiˇ cant in the regression results. This could indicate that new residents seek tracts that had relatively small public capital stocks in 1980. This suggests that land developed for new residential development may have been in tracts that required additions to their capital stock. This is consistent with residential choices pushing land development into rural areas that were dominated by farm or forest land uses. It also means that it is probably not prudent for rural leaders to invest in roads or water and sewer capacity in the hope that development will be attracted to it. Distance to hospitals (DISTH) is not signiˇ cant in either equation.

[8] Theoretically, rural governments that provide basic public services for housing units should be more competitive in attracting residents than poor public service areas. However, beginning stocks of water and sewer capital seem to have no pull for new residents and ˇ rms and may repel them if they re ect higher millage rate for debt service.

Several indicators of "quality of life" were included as regressors as well. Age of housing stock (RHOU78) and school quality (PUPTEA80) are signiˇ cant in the population equations, with the expected sign.[9] In other words, a greater availability of new housing stock and good schools are strong attractors of rural population growth in the hinterland. Surprisingly, the effect of school quality is not signiˇ cant in the employment equation. Similarly, the variable re ecting human capital (DEC80) is not signiˇ cant in the employment equation, suggesting that other factors may be more important in attracting job growth to the rural hinterland.

15.4.3 Causality in Population-Employment Dynamics

One of the strengths of the FEA approach to the analysis of regional development processes is the recognition of commuter ows as an important force affecting ˇ rm and residential location choices. Firms want to locate in a place that has access to a labor force of the size and quality needed for their production. Households want to locate in places with good access to employment opportunities. In our RU models, there are several variables that capture these access dimensions of the location decision.

Consider the residential location decision ˇ rst. Do people choose to locate near jobs? The coefˇ cient for the employment base in 1980 (E_{80}) in the population equations is positive and signiˇ cant throughout. However, the coefˇ cient of change in employment (ΔE_{89}) is negative and signiˇ cant. Therefore, there is no strong evidence that people choose to move to where the new jobs are. In contrast, there might be a tendency for clusters of new jobs and clusters of new residents to form in separate locations. Most new residents move to new exclusively residential subdivisions, while most of the new jobs are concentrated in a few industrial districts.

In terms of new ˇ rm and employment location, do ˇ rms indeed locate near large and highly qualiˇ ed labor force pools? The only signˇ cant coefˇ cient relevant to this question pertains to the beginning rural employment (E_{80}), neither population size nor population change are signiˇ cant. This may suggest some type of cummulative effect on employment growth in the rural hinterland but does not indicate a strong link with population dynamics.

15.5 Conclusions

In this chapter, we used a spatial econometric extension of the Carlino-Mills model to study the development process of selected southern Functional Economic Areas. Our results suggest a mix of spillover and backwash effects from urban core and fringe areas to their rural hinterlands. As urban complexes grow in the South, they have impacts on the rural hinterland. These impacts include spillovers or urban spread effects. The population in a rural area grew faster than average between 1980

---

[9] A negative sign of PUPTEA80 is expected since lower student/teacher ratios indicate higher quality.

and 1990 if it was in an FEA with a large urban core population and with a rapidly growing population in the urban fringe. Rural employment on the other hand grew faster than average in FEAs with larger populations in the urban fringe.

We also found evidence of backwash impacts by urban areas on their rural hinterlands. Rural employment grew faster than average in FEAs that had larger populations in the urban fringe. This scenario is consistent with small rural town business losing markets in FEAs where there has been a rapid growth in shopping malls, service and manufacturing in the urban fringe of large cities.

Greater availability of basic public services does not appear to be a strong attraction to new residents in rural communities. There is evidence that new jobs are accumulated in those areas with a large base employment. The quality of the local high schools was important to new residents, but not for new ˇ rm location. New residents also seem to avoid rural areas that are poorer than average and that are most remote from urban centers.

Finally, we ˇ nd that spatial autoregressive models are useful in research that looks at development within FEAs. The spatial autoregressive parameters in each of the Rural/Urban linkage models are statistically signiˇ cant for both the population and the employment equations. This means that ignoring these spatial effects may lead to misleading inference in empirical models of the development process that are based on data for counties or census tracts.

Acknowledgments

Funding for this research was supported by grant 92-37401-8251 from the United States Department of Agriculture, National Research Initiative.

16 Economic Geography and the Spatial Evolution of Wages in the United States

Yannis M. Ioannides

Tufts University

16.1 Introduction

Questions pertaining to the location of economic activity, to the relative sizes of cities in different countries, and to changing roles for different geographical areas in the process of economic growth have attracted considerable interest recently. Work by several theorists who developed the so-called new economic geography, including recent contributions by several researchers, but in particular by Masahisa Fujita, Paul Krugman and Anthony Venables have added important new spatial insight to the established system of cities literature, represented most notably by research by Henderson (1974, 1988). The system of cities approach features powerful models of the intrametropolitan spatial structure, but lacks an explicit model of intermetropolitan spatial structure. Certain aspects of the intermetropolitan spatial structure have played a key role in the new economic geography literature, as, for example, in Krugman (1991b). Krugman (1998) provides an excellent overview of this literature. Tabuchi (1998) proposes a step towards a synthesis of the older system of cities literature with the newer economic geography based theories by incorporating intrametropolitan commuting costs in addition to intermetropolitan transport costs.

This chapter attempts to examine empirically some of the consequences of new economic geography theories for the spatial evolution of wages. Dobkins and Ioannides (1998) examines the basic dynamics of spatial interactions among U.S. cities and its impact on their populations. It uses a data set on U.S. metro areas, spanning this century from 1900 to 1990, to look at patterns of city growth and the distribution of city sizes as new cities enter the distribution. Entry of new cities is an important characteristic of the United States system of cities. The present chapter looks in more detail at the impact upon wages of spatial aspects of that expanding system, considering the presence of neighboring cities, regional in uence, distance between cities, and the time since ˇ rst settlement, "age," of cities in the system.

It is organized as follows. Section 16.2 outlines the theoretical points that guide the questions we ask. Section 16.3 provides a theoretical framework and uses to obtain empirical predictions. Section 16.4 describes the data set. Section 16.5 develops the estimation models and presents the empirical results. Section 16.6 concludes.

16.2 Theoretical Strands

It is well known that the distribution of economic activity over space is not uniform. Its clustering over space is held as evidence in favor of the importance of increasing returns to scale and information spillovers and other external effects from proximity. According to Henderson-type system of cities theories, such agglomeration effects vary across industries, a fact that also serves as key reason for cities to specialize (Henderson, 1974). The beneˇ ts from agglomeration cause workers to want to concentrate where their skills are most valued, which because of increasing returns is also where many workers concentrate. The resulting congestion removes some of the attractiveness for workers, making it economical for similar industries to concentrate in different locations, "cities." They thus avail themselves of economies of scale and of the beneˇ ts from external effects of proximity. However, as the diseconomies from congestion depend primarily on city size, it makes sense for economic activities to cluster only if they confer mutual beneˇ ts to one another. The resulting cities have optimum sizes, where such sizes may differ because of specialization. This basic theory confers an important role to "large" agents, land developers who foresee the advantages from developing different types of cities. Theories of this type have found a fair amount of empirical support, when tested with U.S. and other data, by Henderson as well as others. Henderson (1988) presents the international evidence as well.

Henderson-type system of cities theories do account for spatial differentiation within metropolitan areas, but unfortunately neglect intermetropolitan space. It is there where modern Krugman-type economic geography has made major contributions to our understanding of how economic activity locates itself in space. Roughly speaking, broad geographical features and historical accidents are credited for determining the location of new centers. However, once a city has established itself, its own size affects critically its further development through its own "agglomeration shadow." In spite of its great popularity, Krugman-type economic geography remains relatively unexplored empirically. Hanson (1998) and Thomas (1996) are arguably the only exceptions. They use Krugman (1991b) as a starting point and modify it in order to allow for diseconomies from congestion and to develop an estimable model.

With relatively few impediments to labor mobility and non-existent barriers to interregional and intermetropolitan trade in the United States, we would expect that wage differences would steadily, but perhaps slowly, disappear over time. While we do observe a fair amount of convergence of wages across cities, their evolution exhibits a number of interesting features. For instance, Table 16.3 shows that cities with neighbors are more "homogeneous," as measured by the coefˇ cient of variation of the logarithm of wages, than cities without neighbors, as we shall study in more detail below.

16.3 The Model

We start with the deˇ nition of geography. Let there be a set of geographical *sites* $\varsigma = \{1, \ldots, s, \ldots, S\}$ deˇ ned within particular geography, such as the real line (or an interval thereof), a circle, a one-dimensional or a two-dimensional lattice, or simply the North American landscape. For each appropriate geography, we consider a notion of neighborhood. Among alternative possibilities we shall restrict attention to two cities' being neighbors if they occupy adjacent sites: the neighbors of city i, $\nu(i)$, are all other sites with which a city has a boundary. The particular geographic characteristics of all sites are in principle available in detail. With this geography we associate below a set of intercity transport costs of the "iceberg" type. For convenience, we deˇ ne the set of all cities that are neighbors of city i including itself, as $N_i = \{i, \nu(i)\}$. This is helpful in deˇ ning the set of all cities that have neighbors in time t as $\mathcal{N}_t = \bigcup_{i=1}^{I_t} N_i$, $\mathcal{N}_t \subseteq I_t$, and its size as N_t.

Let I denote a set of names of cities, i.e., $i = 1$, denotes Abilene, TX, $i = 206$ denotes New York, NY, etc. Let I_t denote the set of *cities extant* at time $t : I_t \subseteq I$, Let $I_t = |I_t|$, and let L_{it} denote the size, in terms of population (or employment), of city i at time t, $i \in I_t$, $1 \leq i \leq I_t$, and time periods $t = 1, \ldots, T$. Let L_t denote the vector of sizes of the I_t cities in existence in the economy at time $t : P_t \in R_{+}^{I_t}$. We shall assume that not all potential urban sites need be occupied, and there is plenty of space for new urban development: $\max_t : I_t < |\varsigma|$.

Next we deˇ ne a *settlement mapping*, $g_t : \varsigma \rightarrow \{0, I_t\}$, where $g_t(s) = 0$, denotes that site s, $s \in \varsigma$, is not settled at time t, and $g_t(s)$, if site s is settled, $g_t(s) \neq 0$, denotes the site occupied by city $g_t(s) \in I_t$. We keep track of the evolution of settlement sites by means of the vector $G_t = (g_t(1), \ldots, g_t(s), \ldots, g_t(S))$.

A site once settled is indistinguishable from the city which occupies it, for as long as it remains settled by the same city. We would like to explicitly model sites within the particular geography which have not yet been settled. We keep track of the time site s of city $i = g_t(s)$ was ˇ rst settle, t_i^s, the settlement date. The settlement date is different from the time a city enters the data, e_i. This is, of course, due to our deˇ nition of cities. A city may disappear, that is, ghost-towns are possible, though relatively rare in the United States during the twentieth century.

We borrow the basic features of the adaptation of Krugman (1991b, 1992) by Hanson (1998), Helpman (1998), and Thomas (1996) in order to deˇ ne a new economic geography system of cities in a dynamic setting. We modify the model in order to account for the impact of different types of spatial interactions upon the dynamic evolution of wages. We assume that consumers are inˇ nitely lived but cannot save, that is, they are assumed to maximize utility in each period. In each city i, individuals are assumed to be identical in terms of human capital and preferences. Their preferences over consumption of housing services, C_{hit}, and of manufacturing goods, represented by a composite, C_{mit}, are deˇ ned by:

$$U_{it} = C_{mit}^{\mu} C_{hit}^{1-\mu}. \tag{16.1}$$

The manufacturing composite C_{mit} is deˇ ned as a function of the quantities of symmetric manufacturing product varieties, given by:

$$C_{mit} = \left(\sum_{j=1}^{n_t} c_{jit}^{1-\frac{1}{\sigma}} \right)^{\frac{\sigma}{\sigma-1}}, \tag{16.2}$$

where σ is the direct partial elasticity of substitution between any two varieties, and n_t the number of varieties available in the entire economy. Each individual variety is produced with labor only by means of an increasing returns to scale technology: $L_{ji} = \alpha + \beta_i c_{ji}$, where we allow for the marginal productivity of labor to depend upon the city where production takes place. It is well known (Dixit and Stiglitz, 1977; Krugman, 1991b) that in equilibrium, each variety is produced by a single monopolistically competitive ˇ rm. With free entry, each ˇ rm produces so as proˇ ts are zero. Its optimum output in city i is equal to $(\sigma - 1)\alpha/\beta_i$, and its price is equal to $\sigma/(\sigma-1)\beta_i W_{it}$.

The supply of housing in each city, H_{it}, is ˇ xed and owned communally by all members of the economy. That is, each individual owns $1/L_t$ of the housing sock in each city. They supply all housing inelastically in perfectly competitive markets. The economy's total population, L_t, is assumed to be equal to the total labor force at time t. Workers are assumed to be perfectly mobile between any two adjacent cities. This expresses the notion that neighboring cities have integrated housing markets. We allow for the possibility that workers may have different levels of human capital across cities. We assume that neighboring cities have different labor markets. However, we introduce some friction in the mobility of labor among all other cities, that is cities without neighbors.

Goods shipped between cities incur transport costs in the form of iceberg costs: for each unit shipped between locations s' and s'', $s', s'' \in \varsigma$, which are not necessarily settled, only a fraction $v_{s's''} = e^{-\tau D_{s's''}}$, where $D_{s's''}$ denotes the "effective" distance between those two locations. The transport cost parameter τ may change over time, which affects transport costs between any two sites. Changes in the national transportation system may affect all effective distances.

Let us deˇ ne the following variables, associated with city i : W_{it}, T_{it}, R_{it}, Y_{it}, and λ_{it}, denote, respectively, the nominal wage rate, the price index of manufactures, the rental rate of housing, total income, and population as a share of total population.

The ˇ *rst* set of equilibrium conditions pertains to labor mobility and is deˇ ned differently for cities with neighbors from those without neighbors. For all cities, that is with or without neighbors we assume that their population growth rate γ_{it}:

$$\gamma_{it} \equiv \frac{\lambda_{it} L_t - \lambda_{it-1} L_{t-1}}{\lambda_{it-1} L_{t-1}}$$

is proportional to the gap, in logarithmic terms, between actual real wages and average real wages across the entire economy in the previous period,

$$\frac{\lambda_{it} L_t - \lambda_{it-1} L_{t-1}}{\lambda_{it-1} L_{t-1}} \rho \left[\ln W_{it-1} - (1-\mu) \ln R_{it-1} - \mu \ln T_{it-1} - \bar{u}_{t-1}\right],$$
$$\forall i \in I_t,\ t = 1, \ldots, T, \tag{16.3}$$

where $\bar{u}_{t-1}$ is deˇ ned as the geometric average real wage across the economy at time $t-1$,

$$\bar{u}_{t-1} \equiv \sum_{j=1}^{I_t} \lambda_{jt-1} \left[\ln W_{jt-1} - (1-\mu) \ln R_{jt-1} - \mu \ln T_{jt-1}\right]. \tag{16.4}$$

We assume that *neighboring* cities are always in locational equilibrium. That is, the real wage, deˇ ned as the value of indirect utility enjoyed by each resident, is equalized across any two cities that are neighbors with one another:

$$\frac{W_{it}}{R_{it}^{1-\mu}T_{it}^{\mu}} = \frac{W_{kt}}{R_{kt}^{1-\mu}T_{kt}^{\mu}}, \forall i \mathcal{N}_t, \forall k \in \nu(i), \; t = 1, \ldots, T, \tag{16.5}$$

where $\nu(i)$ denotes the cities of cities that are neighbors of city i. Since adjacent cities are assumed to be neighbors, immediate equalization of utility is an appropriate assumption. Moving across cities that are not neighbors could cause individuals to suffer utility which does depend upon distance between them. As we do not model such moves, we think it appropriate to highlight a sharp difference in equilibrium adjustment between cities that are neighbors and those that are not.

By using (16.5) in (16.3), it follows that the dynamics of population adjustment over time are different for cities with neighbors than for cities without neighbors. That is, for a city with neighbors, by using (16.5) and (16.4) in (16.3), we have:

$$\begin{aligned} \frac{\lambda_{it}L_t - \lambda_{it-1}L_{t-1}}{\lambda_{it-1}L_{t-1}} = \rho \sum_{j=1,\, j \ni \nu(i)}^{I_t} \lambda_{jt-1} [&\ln W_{it-1} - (1-\mu)\ln R_{it-1} - \mu \ln T_{it-1} \\ &- \ln W_{jt-1} - (1-\mu)\ln R_{jt-1} - \mu \ln T_{jt-1}], \\ &\forall i \in I_t, \; t = 1, \ldots, T. \end{aligned} \tag{16.6}$$

The *second* equilibrium condition deˇ nes total income in terms of labor income in each city:

$$Y_{it} = \lambda_{it}L_t W_{it} + \frac{1-\mu}{\mu}\lambda_{it}\sum_{j}^{I_t}\lambda_{jt}L_t W_{jt}, \; \forall i \in I_t, \; t = 1, \ldots, T. \tag{16.7}$$

The *third* set of equilibrium conditions expresses equilibrium in the housing market in each city:

$$R_{it}H_{it} = (1-\mu)Y_{it}, \; \forall i \in I_t, \; t = 1, \ldots, T. \tag{16.8}$$

Cities that are neighbors of one another share the same housing market:

$$R_{it} = R_{i't}, \forall i \in I_t, \forall i' \in \nu(i). \tag{16.9}$$

We note that the Thomas modiˇ cation of the Krugman model allows for effects of congestion: cities that are attractive from the viewpoint of geography may sustain higher housing costs.

The *fourth* equilibrium condition expresses equilibrium in the labor market in each city (Krugman, 1992). Spending by all cities on each city's products must equal to that city's labor income:

$$W_{it} = \left[\sum_{k=1}^{I_t} Y_{kt}\left(T_{kt}\mathrm{e}^{-\tau D_{ik}}\right)^{\sigma-1}\right]^{\frac{1}{\sigma}}. \tag{16.10}$$

We note that this condition holds even for sites that are not yet settled, or that they have no manufacturing.

As Krugman notes, *ibid.*, condition (16.10) resembles closely the condition for equilibrium in market-potential type models. It differs from the market potential model only because the true price indices for all locations also enter the deˇ nition of each index. This re ects the effects of competition from producers in other locations, a feature which is absent from market-potential models. The value for labor is higher in locations which are "closer" in terms of transport costs to areas with high consumer demand: this expresses a notion of *backward linkages.*

The ˇ*fth* equilibrium condition is the deˇ nition of price index for manufactures in each city (Krugman, 1992):

$$T_{it} = \frac{\sigma}{\sigma - 1}\left[\sum_{k=1}^{I_t} \lambda_{kt} \left(\beta_k W_{kt} e^{\tau D_{ik}}\right)^{1-\sigma}\right]^{\frac{1}{1-\sigma}}. \qquad (16.11)$$

As Krugman noted, condition (16.10) expresses a notion of *forward linkages:* the price index will be lower, the higher the share of manufacturing that is located in cities with low transport costs. However, advantageous forward linkages are in part offset by congestion costs.

General equilibrium is deˇ ned in this economy in terms of the values of of the $5I_t - 1$ unknowns at time t, $\{W_{it}, \lambda_{it}, Y_{it}, P_{it}, T_{it}\}_{i\in I_t}$, which satisfy the $5I_t - 1$ independent equations (16.5), (16.3), (16.7), (16.8), (16.9), (16.10), and (16.11), and given the lagged values of all endogenous variables. Key properties of the model have been explored by means of simulations by Krugman and others. In general, the highly nonlinear nature of the model suggests multiplicity of solutions, which in our dynamic setting implies a dramatic increase in the richness of the dynamics.

16.3.1 The Workings of the Model

It is an essential feature of the model that as neither the price markup nor optimum production of each manufacturing variety are affected by the size of the market, all scale effects work themselves through the range of varieties produced. This is a well known source of enormous simpliˇ cation of the Dixit-Stiglitz model. Therefore, more population means that more sites may be occupied and thus more varieties may be produced. This increases welfare, both because of the increased number of varieties and the reduced pressure on the housing market. However, a tradeoff emerges: while agglomeration is checked by congestion costs, settlement of distant sites causes an increase in the price index of manufactures for all sites. Therefore, the spatial expansion of the system of cities is checked by congestion costs relative to the impact of geography on transportation costs.

As Krugman notes, existence of a settlement may skew further development in its favor via its own "agglomeration shadow." Therefore, the availability of data on the time of settlement may help anchor the original location of economic activity in an otherwise homogeneous setting. Of course, the fact that a particular site has been settled before itself may convey advantages of location. Various shocks may cause reallocations of economic activity, which may bring into the picture the full force of the asymmetric nature of nonlinear dynamics. Symmetry breaking will demonstrate itself in the form of new urban developments. Existing settlements will be able to survive as long as they are subject to forces which although would not have caused them to appear in the ˇ rst place, are nonetheless above a critical value for existence (Krugman, 1996a).

In this basic model, as long as population growth continues, welfare increases via the increase in the range of manufacturing varieties (Ioannides, 1994). If population growth ceases, there is no further force for welfare improvement. The availability of suitable sites for urban development is re ected in housing supply. The improvement in welfare that is associated with a greater number of varieties may be strengthened if we were to to adopt a modelling "trick" from Krugman and Venables (1995), also adapted by Puga (1996), and assume that each variety-producing ˇ rm utilizes a composite of raw labor and an aggregate of the differentiated goods available in the economy. This provides an additional force for increase in welfare through reduction in the cost of producing each variety that follows from availability of a greater range of varieties available in the entire economy.

Since neither housing supply nor housing rents are observed in our data, we assume that at a ˇ rst level of approximation, all cities have equal housing stocks: $H_{it} = \bar{H}_{it}$. We justify this by referring to empirical evidence that population densities across U.S. states vary with population. Thomas (1996) shows that state population explains 40% of the variation in population density across U.S. states. Therefore, by using (16.8) in (16.5) and (16.3) we may eliminate R_{it}. Next, we may eliminate income, Y_{it}, which is unobservable in our data, from the model by solving from (16.7). We are thus left with with two sets of simultaneous equations for wages and price indices, W_{it} and T_{it}, the condition for perfect mobility among neighbors, which involves wages and price indices only.

The spatial evolution of the economy over time in this model is driven by population growth. As population grows, individuals migrate from one city to another in response to differences between real wage in the city of their residence and the average real wage in the economy. This model, as it stands, lacks an engine of growth, in the absence of population growth. In contrast, Eaton and Eckstein (1997) assume a Lucas-type human capital externality, that generates city-speciˇ c human capital. The presence of migration costs in the Eaton-Eckstein model dampens potential moves across cities by individuals who wish to take advantage of better opportunities and a nontrivial distribution of city sizes emerges.

We augment the model by assuming, as Eaton and Eckstein do, that residents of a city are accorded a city-speciˇ c productivity in employment regardless of how long they have lived there. That is, we assume that the marginal labor requirements in each city, β_{kt}, steadily decline over time. Therefore, we generalize (16.11) as follows:

$$T_{it} = \frac{\sigma}{\sigma - 1} \left[\sum_{k=1}^{I_t} \lambda_{kt} \left(\beta_{kt} W_{kt} e^{\tau D_{ik}} \right)^{1-\sigma} \right]^{\frac{1}{1-\sigma}}. \qquad (16.12)$$

Therefore, steady decreases in the β_{kt}'s over time could cause the price indices to decrease and through (16.11) also cause nominal wages to decrease. Both those types of changes cause the real wage to increase. To see this, let ω_T, ω_W, ω_β denote the rates at which the price indices, wage rates and the β's decrease over time. From (16.11) and (16.10), it follows that the price indices decrease faster than wages: $\omega_T > \omega_W$. However, larger values for W_{kt}'s are compatible with larger values for incomes. This is an application of the notion of forward linkages. It would be consistent with the spirit of new economic geography to consider improvements in labor productivity coming from Romer-type pecuniary externalities associated with the range of intermediate inputs (Romer, 1990) available in each city. This could be considered as an effect of agglomeration and be related to city size. Combining the role of city-speciˇ c human capital, measured through lagged schooling, S_{it-1}, with agglomeration effects as determinants of marginal labor productivity, we assume that $\beta_{it} = \beta(L_{it}, S_{it-1})$, where $\beta_1 < 0$, $\beta_2 < 0$.

16.3.2 Empirical Implications of Theories

The key empirical implication of the theoretical framework is a prediction the dynamic evolution of wages re ects spatial considerations. The spatial evolution of the economy affects those dynamics through geographical distances among sites as well as proximity. The above analysis of the workings of the model implies that the dynamics may be different depending upon whether or not a city has neighbors.

We summarize here key results from Dobkins and Ioannides (1998), which emphasizes the dynamics of city sizes. It is shown there that spatial considerations are important in urban growth. The likelihood that an entering city will locate so as to have neighbors is increasing with its own size and its age. Distance is a very important determinant of size and growth and has nonlinear effects. However, among cities that have neighbors in the sense of adjacency, the average growth rate among a city's neighbors is a very important determinant of a city's own growth rate, while distance from the nearest higher-tier city is insigniˇ cant. The opposite is true for cities without neighbors. Dobkins and Ioannides conclude that overall geography is important in urban growth. For cities with neighbors, growth rates are closely interdependent. If a city is outside a major agglomeration, its growth rate is subject to an impact by the nearest center. However, the marginal impact of distance is maximized at around 460 miles. We interpret this as evidence in favor of the notion advanced by Krugman that a city once created generates its own agglomeration shadow.

In the remainder of the chapter we study the impact of spatial considerations upon the dynamic evolution of wages. We take as given the presence or absence of neighbors and estimate a reduced-from model that roughly re ects the qualitative predictions of the above theoretical model.

16.4 Data

There are a variety of ways to define cities.[1] In this study we primarily use contemporaneous Census Bureau definitions of metropolitan areas, with adaptations for availability. From 1900 to 1950, we have metropolitan areas defined by the 1950 census. That is, for years previous to 1950, we use Bogue (1953) reconstructions of what populations would have been in each metropolitan area in each year if the cities had been defined as they were in 1950. For each decennial year from 1950 to 1980, we use the metropolitan area definitions that were in effect for those years. Between 1980 and 1990, the Census Bureau redefined metropolitan areas in such a way that the largest U.S. cities would seem to have taken a huge jump in size, and several major cities would have been lost. While this might be appropriate for some uses of the data, we want to be able to track cities as neighbors. Therefore, we reconstructed the metro areas for 1990, based on the 1980 definitions, much as Bogue did earlier. We believe that this gives us the most consistent definitions of U.S. cities (metropolitan areas) that we are likely to find.

The method also raises a question as to which cities, as defined or reconstructed, should be included. In the years from 1950 to 1980, we use the Census Bureau's listing of metropolitan areas. Although the wording of the definitions of metropolitan areas has changed slightly over the years, the number 50,000 is minimum requirement for a core area within the metropolitan area. Therefore, we used 50,000 as the cutoff for including metropolitan areas as defined by Bogue prior to 1950. Consequently we have a changing number of cities over time, from 112 in 1900 to 334 in 1990. While it is often difficult to deal with an increasing number of cities econometrically, we think that this is a key aspect of the U.S. system of cities, and is worthy of being factored into our studies.

We also have data on earnings in all cities in the sample for all years, drawn from Census reports, although the data set is not ideal because the Census Bureau changed the categories it reported over the years. We have data on schooling in each city over the century, reported as the percentage of the population in the 15 to 20 year old category who are in school.

As noted above, spatial expansion over geographical regions is an important feature of the US experience. The Census Bureau divides the country into nine regions (see Fig. 16.1), which we recombine into five regions, when necessary. The east-west movement that is at the heart of mercantile theory would predict a steady increase of cities in the Midwest, Mountain, and Pacific Coast areas.

16.4.1 Spatial Measures

We measure *distance* in terms of driving distances from each city in the sample to the nearest (larger) city in a higher tier. In order to construct the tiers, we took as our basic classification a listing of U.S. cities by "function" (nodal centers) from Knox (1994). We amended the top tier slightly to include New York City, Chicago,

[1] This section draws extensively from Dobkins and Ioannides (1998).

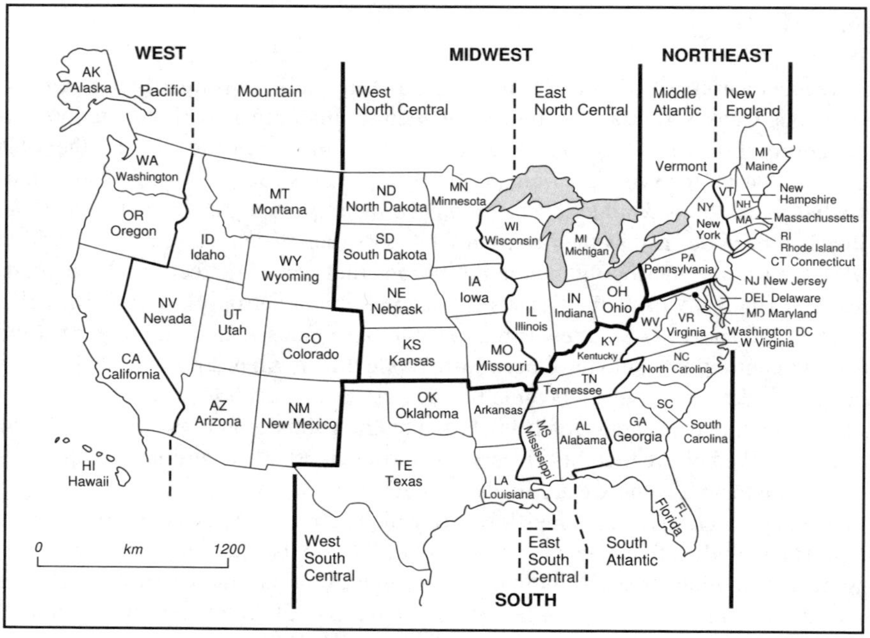

Fig. 16.1. U.S. States and Census Regions

Los Angeles, Houston, Miami, San Francisco, Washington D.C., Atlanta, Denver and Seattle. The data entry for each of these cities is to the nearest city in the set. The next classiˇ cation is the regional nodal centers, which includes fourteen large cities: Baltimore, Boston, Cincinnati, Cleveland, Columbus OH, Dallas, Indianapolis, Kansas City MO, Minneapolis, New Orleans, Philadelphia, Phoenix, Portland OR, and St. Louis MO. The entry for these cities is the mileage to the nearest city in the top tier. The third classiˇ cation is the subregional nodal centers, nineteen cities whose entry is the distance to the nearest city in either of the larger tiers. The third tier cities are: Birmingham, Charlotte, Des Moines, Jackson MS, Little Rock, Memphis, Mobile, Nashville, Oklahoma City, Omaha, Salt Lake City, Shreveport, Syracuse, Richmond, Detroit, Hartford, Milwaukee, Tampa, and Pittsburgh. Distances for all other of the 334 cities present in 1990 are the mileages to the nearest city in any of the three top tiers. The only exceptions are Honolulu and Anchorage, for which we used arbitrary ˇ gures of 1200 and 1100, respectively, as driving distances are irrelevant. These numbers are simply larger than any of the other mileages we record. (It is 1,029 miles from Denver to Los Angeles.)[2] We note that in a key sense

[2] It would be possible to reˇ ne our measures of the spacing of cities in order to account for an overall constraint of geography. That is, just as a point within a canonical triangle has distances from all edges that sum up to a constant, the measures of distance for all cities

Table 16.1. Descriptive statistics, decennial data (1900 – 1990)

1 Year	2 U.S. Pop. (1,000)	3 U.S.Pop.: Urban (1,000)	4 Mean Size	5 Median Size	6 GNP billion $
1900	75,995	29,215	259952	121830	71.2
1910	91,972	39,944	286861	121900	107.5
1920	105,711	50,444	338954	144130	135.9
1930	122,775	64,586	411641	167140	184.8
1940	131,669	70,149	432911	181490	229.2
1950	150,697	85,572	526422	234720	354.9
1960	179,323	112,593	534936	238340	497.0
1970	203,302	139,419	574628	259919	747.6
1980	226,542	169,429	526997	232000	963.0
1990	248,710	192,512	577359	243000	1277.8

All ˇ gures are taken from *Historical Statistics of the United States from Colonial Times to 1970*, Volumes 1 and 2, and *Statistical Abstract of the United States, 1993*. Column 6: GNP adjusted by the implicit price de ator, constructed from sources above; 1958 = 100.

our empirical measure of distance anchors all city locations relative to those in the higher tier.

Another measure of proximity that we employ is whether or not two cities are adjacent. We consider cities to be adjacent if the Census Bureau has ever grouped them together in various extended, but pertinent, deˇ nitions. For example, the Census Bureau's consolidated metropolitan area for Los Angeles includes San Bernardino/Riverside, Anaheim, and Oxnard. We consider each of those as separate cities in our sample. When they enter the data set on their own, they are denoted as neighbors to Los Angeles and to each other. The average number of neighbors (of cities with neighbors) uctuates around 1.00, until the 1960s, after which time it starts varying between a low of 1.383 to a high of 2.111.

Neighbors "happen" in several ways. In some cases, cities simply grew up near each geographically, as in the case of Dallas and Fort Worth. In other cases, neighboring cities may have been a part of a city's hinterland and simply grow with the core city until they reach a population threshold and enter the distribution. An example of this is Rock Hill SC, which enters in 1980 as a neighbor to Charlotte NC. In other cases, neighbors enter and in so doing separate from an existing city. The most dramatic case is Nassau and Suffolk counties in New York state, which enter in 1980 at more than two million population, lowering the population of New York City, of which they were previously a part, by that amount.

Descriptive statistics for our data, given in Tables 16.1–16.3, but especially in Table 16.3, reveal important features of the force of agglomeration in US eco-

are not independent from one another. We think that this problem is not as important for the U.S. case, whose geomass still has a substantial hinterland.

Table 16.2. Descriptive statistics for all cities, 1900 – 1990, 1990 observations

Variable	Mean	Std. Dev.	Skewness	Kurtosis	Min.	Max.
Population (000)	479.5	1001.5	6.6	58.8	50.7	9,372.0
Log(Population)	12.4028	0.9895	1.0	4.1	10.8343	16.374
Growth Rate (%)	10.62	41.98	−1.1	5.8	−0.999	1.8752
New England	0.0879	0.2833	2.9	9.5	0.00	1.00
Mid Atlantic	0.1276	0.3338	2.2	6.0	0.00	1.00
South Atlantic	0.1673	0.3734	1.8	4.2	0.00	1.00
East North Central	0.2030	0.4023	1.5	3.2	0.00	1.00
East South Central	0.0663	0.2489	3.5	13.1	0.00	1.00
West North Central	0.0910	0.2876	2.8	9.1	0.00	1.00
West South Central	0.1221	0.3275	2.3	6.3	0.00	1.00
Mountain	0.0462	0.2100	4.3	19.7	0.00	1.00
Paciˇ c	0.0884	0.2840	2.9	9.4	0.00	1.00
Education (%)	57.1085	20.9284	−0.4	1.8	11.80	92.73
Real Wage ($)	3197.92	1132.37	0.2	2.3	1020.00	7311.00

Data on education and real wage are taken from *Historical Statistics of the United States from Colonial Times to 1970*, Volumes 1 and 2, and *Statistical Abstract of the United States, 1993*. Educational percentage refers to the mean percent of 15 to 20 age cohort in school. Mean real annual earnings, by city proper or metro area, are in dollars, de ated by the Consumer Price Index, 1967 = 100.

nomic geography. When we compare earnings among all cities and among cities with neighbors, we conˇ rm a prediction implied by agglomeration. The variance of wages in all cities, normalized by the mean of wages in all cities, is always larger than the variance of wages in cities relative to their neighbors. If we alter the test slightly, we also note that the variance of wages in all cities without neighbors, again normalized by the mean of wages in those cities, is greater than the variance of wages of cities with neighbors relative to those of their neighbors. The latter formulation is more robust to precise testing and we ˇ nd that the variance is significantly larger (at the 10% level) in 1900, 1910, 1930, 1950, 1970, 1980, and 1990.[3] We see evidence in Table 16.3 of some of the enduring facts of U.S. economic geography, and an interesting spatial interpretation. The population "boom" of the 1950s resulted in 48 new cities entering the system, almost a third of them as neighbors to either existing cities and/or to each other.[4] The so-called "rural renaissance" of the 1970s, however, resulted in 79 new cities entering the system, with less than 10 percent of those being neighbors.

[3] This ˇ nding is similar to Quah (1996) for European regions.

[4] The so-called population boom of the 1950s is, of course, relative and modern. The 19 percent increase in population would have rated as the smallest increase in the period from 1790 up through the ˇ rst decade of this century.

Table 16.3 columns 7 and 8 suggest that cities with neighbors and the neighbors themselves tend to be larger than isolated cities. Whereas column 7 shows that average city sizes generally grow over time, the opposite is true for city sizes relative to total urbanized population, reported in column 8, of Dobkins and Ioannides (1998). Column 7, Table 16.3, indicates that the average size of a city with no neighbors in 1900 was 192,000. The average size of a city with neighbors was 487,000 and the average size of the neighbors was 571,000. These numbers differ because some cities have more than one neighbor and because not all neighbors to a central city are neighbors to each other. This pattern continues through the century.

In our set of 78 cities that have neighbors over the years from 1900 to 1990, 56 are involved in either entering as a new neighbor or being the existing neighbor to a new entrant. The other 22 are cities that co-exist as a neighbor in 1900, and do not overlap the previous set. (For example, Bridgeport CT is a neighbor to New York City in 1900 and is tallied among the 22. New York City is counted among the 56 with its ten other neighbors that enter over the century.) Among the 56, all entering neighbors are smaller than their existing neighbors except for Greensboro, NC, which enters as a neighbor to Winston-Salem. These cities are an exception to the rule throughout the years, as the Greensboro-Winston-Salem-High Point area grows together quite quickly. Among the 56, excepting Greensboro and Winston-Salem, the average percentage of the size of the entering city to the size of the existing city is 18 percent. This does include such large concentrations as Nassau and Suffolk counties, noted above.

Interestingly enough, of the neighbors that coexist in 1900, the smaller neighbors are, on average, 32 percent of the size of their larger neighbors. This may highlight a feature of the data set, in that cities are designated as neighbors if they are ever grouped together by the Census Bureau. These groupings were published relatively late in the century. Perhaps, with less ef˜ cient transportation, these cities were actually further apart in a real sense in 1900. To check this, we note the average percentage of the same group of neighbors in 1990. This averages turns out to be 28 percent; it would be 21 percent if we were to leave out Scranton, PA and Wilkes-Barre, PA. This is another problematic set of cities (which the Census Bureau simply calls "Northeast Pennsylvania" in 1980); and which reverses dominant size, with Scranton the smaller city in 1900 and the larger in 1990. Although these numbers deal with a small set of cities, the analysis does seem to bear out some of the theoretical predictions. Cities tend to be smaller than the core city in an "agglomeration shadow," although the entire agglomeration is larger than isolated cities. Furthermore, cities with some initial advantage (in 1900, for instance), may "lock-in" and remain relatively large even as a neighbor grows more rapidly.

Table 16.3. Earnings, schooling and size of cities and their neighbors

1 Year/nei's	2 Number of cities	3 Number with > 1 neighbor	4 New Cities	5 Wage mean, $	6 Schooling percent	7 Size (1,000s) w/ nei's / of nei's	8 SD $\ln W$	9 CV
1900/no nei's	86			1725.63	28.43	192	0.1994	0.0268
1900/nei's	26	2:2		1918.77	25.76	487/571	0.1359	0.0180
1900/all	112			1770.46	27.82	261/133	0.1924	0.0257
1910/no nei's	109		25	1908.67	28.03	202	0.1954	0.0259
1910/nei's	30	2:2	2	2049.67	26.64	597/687	0.1771	0.0233
1910/all	139		27	1939.10	27.73	287/148	0.1935	0.0256
1920/no nei's	113		7	1848.91	25.79	215	0.1670	0.0222
1920/nei's	36	2:2	3	1958.56	24.06	726/818	0.1819	0.0240
1920/all	149		10	1875.40	25.37	339/198	0.1717	0.0228
1930/no nei's	117		6	2526.96	37.91	260	0.1854	0.0237
1930/nei's	40	2:2	2	2584.80	39.31	855/954	0.1864	0.0238
1930/all	157		8	2541.69	38.27	412/243	0.1853	0.0237
1940/no nei's	120		3	1946.14	43.84	279	0.2274	0.0301
1940/nei's	40	2:2	0	2094.70	46.13	916/1,017	0.1913	0.0251
1940/all	160		3	1983.28	44.41	438/254	0.2213	0.0292
1950/no nei's	122		2	2765.75	53.18	342	0.1896	0.0240
1950/nei's	40	2:2	0	3014.00	56.15	1,096/1,211	0.1451	0.0181
1950/all	162		2	2827.05	53.91	526/299	0.1837	0.0232
1960/no nei's	150		33	3956.03	62.60	365	0.1386	0.0167
1960/nei's	60	4:2; 7:3; 1:6	15	4487.50	64.82	964/1,425	0.1499	0.0179
1960/all	210		48	4107.88	63.23	535/407	0.1518	0.0183

continued on next page

Table 16.3. Continued

1	2	3	4	5	6	7	8	9
Year/nei's	Number of cities	Number with > 1 neighbor	New Cities	Wage mean, $	Schooling percent	Size (1,000's) w/ nei's / of nei's	SD ln$W$	CV
1970/no nei's	173		24	4598.29	73.06	374	0.1256	0.0149
1970/nei's	70	11:2; 9:3; 5:4; 1:7	7	5170.74	75.78	1,067/1560	0.0991	0.0114
1970/all	243		31	4763.19	74.22	575/449	0.1303	0.0154
1980/no nei's	244		72	3859.96	71.54	356	0.1529	0.0188
1980/nei's	78	16:2; 7:3; 8:4; 1:11	7	3411.91	69.62	1,060/1,706	0.1244	0.0151
1980/all	322		79	3520.44	70.08	527/413	0.1562	0.0192
1990/no nei's	256		12	4518.99	81.86	391	0.1510	0.0184
1990/nei's	78	16:2; 7:3; 8:4; 1:11	0	3636.13	80.91	1,184/1,860	0.1599	0.0190
1990/all	334		12	3842.31	81.13	577/434	0.1781	0.0216
Total	1988							
Total w/ nei's	498							

Initial advantage is a challenging concept for operationalization. We use the date of settlement for each city, reasoning that good sites were selected first. At first glance, one would suppose that the east to west settlement of the country would determine settlement dates, but we find early settlement dates in the west and late ones along the east coast when we consider so many cities. Settlement here refers to historical references to settlement in a location, and our variable is compiled by sifting through historical texts. In a number of cases, the dates are references to military forts. We use those dates because often the site of the fort determined the site of the city that grew up nearby. The earliest date is that of Jacksonville, Florida, in 1564, and the latest is Richland, Washington, originally the site of a nuclear facility settled in 1944. It is an interesting statistic in and of itself to see how age of settlement correlates with city size. If older age (a better site) makes a city larger, which indicates importance in the system, then we would expect the "date" variable to have a negative sign. Of course, this variable also has implications for Marshall's prediction, as noted above.

So, in summary, we use the distance variable as well as population to explain the role of central place considerations. We use the date variable to explain initial advantage. We expect the distance variable to correlate positively with population if central place theory as we have interpreted it is to be valid. However, as we indicated earlier, the threshold effects that Krugman (and his coauthors) has emphasized would imply that the marginal effect of distance would not be monotonic. In fact, this prediction is confirmed by the data. We expect the date variable to have a negative impact on population; that is, the older the city, the larger we would expect its population to be, indicating that it was a good site and obtained initial advantage.

16.5 Econometric Analysis

The economic model presented above may be expressed through a spatial autoregressive model, whose linear version is usually written as:

$$W_t = \eta \Pi_t W_t + \zeta I W_{t-1} + ZX + \varepsilon_t, \tag{16.13}$$

where W denote the I_t-vector of city wages at time t, I denotes an I_t by I_t identity matrix, Π_t denotes a possibly time-varying I_t by I_t proximity matrix, X is a I_t-vector of city-specific time invariant characteristics, η, ζ, Z are unknown parameters, and ε_t an I_t-vector of random shocks. Equation (16.14) follows from (16.13) if we assume that the interaction between a city i and its neighbors operates through the average wage among its neighbors, in which case row i of Π contains the terms $1/|\nu(i)|$ for each column associated with the elements of of $\nu(i)$. Such models have been extensively analyzed in spatial econometrics (see Anselin, 1988b, for a standard reference). In view of the endogeneity of W_t in the right hand side of models like (16.13) instrumental estimation methods or Maximum Likelihood estimation methods are typically used.

Taking a broad interpretation of the above economic model, we assume that a city's wage rate is related to its own lagged value, to the city's contemporaneous size, which we understand as an agglomeration effect, and to the quality of the city's labor force, which we measure through our schooling variable. The impact of geography is expressed through the discrete event of whether or not a city has neighbors, the contemporaneous average wage rate among a city's neighbors when it has neighbors, and the distance from the nearest center.

The model we estimate is deˇ ned by:

$$\ln W_{it} = \zeta_0^n + \zeta_W^n \ln W_{it-1} + \zeta_P^n \ln P_{it} + \zeta_S^n \ln S_{it-1} + \zeta_\nu^n \ln W_{\nu(i)t} + ZX_i + \varepsilon_{it}, \quad (16.14)$$

where n indicates sample separation with $n = 0$ denoting that city i has no neighbors and $n = 1$ that city i has neighbors, P_{it} denotes population, S_{it-1} (lagged) schooling, and $\ln W_{\nu(i)t}$ the (geometric) average value of wages among city i's neighbors (if any, $\zeta_\nu^0 = 0$). X_i is a vector of time-invariant characteristics, such as regional dummies, a polynomial structure for distance [5] D_i, and date of settlement e_i. Finally, the error term ε_{it} re ects a time-invariant idiosyncratic component, a time effect and a random component which is independently and identically distributed across observations.

The regression system we estimate resembles Hanson's estimation models. We test in effect whether the in uence of the entire system of cities upon each individual city, the notion of market potential, as reˇ ned by Krugman, is reducible in terms of distances from the nearest higher-tier city and of the characteristics of neighboring cities.

Equation (16.14) clearly resembles the standard spatial autoregressive model (16.13). However, our particular setting is such that standard packages may not readily be used to estimate the model in a most general full-information setting. We explain why by referring ˇ rst to Table 16.3 to note that the data as a panel data set are very unbalanced. The cross-sectional size in 1990 is three times as much as that in 1900. About one-fourth of the cities in the sample have neighbors, 78 out of 334, while about one fourth of all observations pertain to cities with neighbors, 498 out of 1988.

The econometric model calls for estimation of a simultaneous equations model with the following technical characteristics. First, the evolution of wages is different for cities with neighbors and for those without. This requires a switching-regressions model. Second, endogeneity of the right hand side of (16.14) must be accounted for. Third, the panel data are very unbalanced, which also implies that the proximity matrix changes every year. Fourth, the construction of the data spans the entire twentieth century, and the panel aspect is intended to represent continuity in the identiˇ cation of different metro areas as economic units.

[5] We recognize that distance need not be time-invariant, as the urban system may realign itself over time. However, our attempts to treat it as time-varying did not produce any signiˇ cant differences.

Table 16.4. Wages and Spatial Evolution

	1	2	3	4	5	6	7	8
Regression	$\ln W_t$	$\ln W_t$	$\ln W_t$	$\ln W_t$	$\ln W_t$	FE, C.3	FE, C.4	FE, C.5
Sample	All	w/ nei's	all	w/ nei's	w/o nei's	all	w/ nei's	w/o nei's
Constant	2.54	1.05				4.81	4.99	4.16
	(18.62)	(4.58)				(136.78)	(55.33)	(96.23)
$\ln W_{t-1}$	0.609	0.224	0.2598	0.1351	0.2638			
	(31.93)	(6.17)	(10.08)	(2.41)	(9.10)			
$\ln P_t$	0.029	0.020	0.0535	0.1520	0.0938			
	(9.23)	(3.07)	(5.89)	(1.24)	(6.86)			
$\ln S_{t-1}$	0.034	0.127	0.0654	0.0493	0.0763			
	(2.13)	(4.02)	(3.15)	(1.37)	(8.84)			
$\ln W_{\nu(i),t}$		0.068		0.1843				
		(0.78)		(4.81)				
Date							-0.0005	0.0003
							(2.74)	(2.26)
Dist						-0.6710^{-3}	-0.001	0.210^{-4}
						(3.20)	(2.27)	(0.09)
Dist^2						0.1410^{-5}	0.3210^{-5}	-0.410^{-6}
						(2.99)	(1.82)	(0.61)
Dist^3						-0.7010^{-11}	-0.2210^{-10}	0.410^{-11}
						(2.24)	(1.48)	(0.92)
North East	-0.003							
	(0.37)							
South East	-0.042							

continued on next page

Table 16.4. Continued

	1	2	3	4	5	6	7	8
Regression	$\ln W_t$	$\ln W_t$	$\ln W_t$	$\ln W_t$	$\ln W_t$	FE, C.3	FE, C.4	FE, C.5
Sample	All	w/ nei's	all	w/ nei's	w/o nei's	all	w/ nei's	w/o nei's
	(5.13)							
South West /Mountain	−0.055							
	(6.51)							
Paci˘ c	−0.012							
	(1.13)							
New England						0.068	0.026	0.086
						(2.29)	(0.32)	(2.40)
Mid Atlantic						0.004	−0.007	−0.007
						(0.16)	(0.09)	(0.23)
South Atlantic						−0.047	−0.083	−0.036
						(2.09)	(1.09)	(1.50)
East North Central						0.029	0.048	0.038
						(1.31)	(0.65)	(1.57)
East South Central						−0.081		−0.073
						(2.91)		(2.56)
West South Central						−0.072	0.002	−0.071
						(3.10)	(0.30)	(3.00)
Mountain						−0.010		−0.022
						(0.35)		(0.18)
Paci˘ c						0.011	0.074	−0.022
						(0.44)	(0.99)	(0.79)

continued on next page

Table 16.4. Continued

	1	2	3	4	5	6	7	8
Regression	$\ln W_t$	$\ln W_t$	$\ln W_t$	$\ln W_t$	$\ln W_t$	FE, C.3	FE, C.4	FE, C.5
Sample	All	w/ nei's	all	w/ nei's	w/o nei's	all	w/ nei's	w/o nei's
Fixed effects	no	yes	yes	yes	yes	no	no	no
Observations	1654	418	1654	434	1220	322	77	243
R^2	0.914	0.923	0.9395	0.9504	0.9384	0.2624	0.272	0.167
F	1164	327	31258	35636	28672	10.52	3.88	5.06

Columns 1–5 report regressions that include wave dummies. Column 2 reports results after correcting for endogeneity of neighbor wage. Columns 2–5 report panel regressions with ˇ xed effects. Columns 6–8 report regressions of the ˇ xed effects from Cdumns 3–5, respectively, against time-invariant variables.

After a lot of experimentation and in view of available econometric packages, we concluded that the best possible picture of what our data suggest would be given by the following two sets of regressions. First, we report estimation results in Table 16.4 with a pair of switching regressions, with *exogenous*[6] sample separation: the samples of those with neighbors and those without neighbors are used as a panel to estimate the law of the evolution of wages, where the average wage rate among a city's neighbors is treated as exogenous. Second, we discuss estimations results for the evolution of wages, separately for each pair of two consecutive waves of the data. The actual estimation results are not reported for reasons of brevity.

16.5.1 Wages and Spatial Evolution: Panel Data Estimation

Table 16.4 reports our estimation results with model (16.14). Columns 1 – 5 report regressions with time effects estimated by means of census dummies. Columns 1 and 2 report results with time-invariant characteristics other than regional dummies being excluded, and with the error term assumed to be iid across all observations. The regression coefficients reported in Column 1 do not reveal any unusual dynamics. The autoregression coefficient for the lagged value of wages implies convergent dynamics, the coefficient of population implies small but highly significant agglomeration effects, and the coefficient of schooling confirms a productive role for education. Looking across the other columns reveals that the significance of population and of lagged schooling diminishes when the average wage rate among a city's neighbors is also a regressor. So, we see that when we estimate (16.14), for $n = 1$, the coefficient of the average wage rate among a city's neighbors is very significant, although numerically small, in the case of no fixed effects. When fixed effects are accounted for, as in the regressions reported in Columns 3 – 5, the role of economies of scale, as reflected in the coefficient of population, is strengthened. When we account for the endogeneity of the average wage rate among a city's neighbors, however, by instrumenting it, we find that it is no longer significant.

The most noteworthy feature of the results for the full switching regressions model with fixed effects and census dummies, reported in Table 16.4, Columns 4 and 5, is the statistical significance and numerical importance of the average wage rate among a city's neighbors relative to the own lagged wage rate. When we instrument the average wage rate among a city's neighbors by means of all exogenous variables of the model, its coefficient becomes negative and insignificant. These results are reported in Column 2, Table 16.4. This continues to be the case after we control for sample selection, by estimating a probit model for the event of whether or not a city has neighbors.

We examine further the role of time-invariant individual characteristics as determinants of the dynamic evolution of wages. We report in Columns 6 – 8 regression results for the estimated fixed effects, from Columns 3 – 5, respectively, against regional dummies, settlement date and a cubic function for distance. Several regional

[6] We eschew here an analysis of the endogeneity of sample separation and refer instead to Dobkins and Ioannides (1998).

dummies are significant, implying that some of the unobserved components in the evolution of wages are of regional geographic origin. Yet, the most interesting feature is the role of the settlement date, which implies that wages in older settlements are more likely to grow faster if they have neighbors than if they have not. Therefore, older settlements that remain isolated are likely to be associated with slower growth of wages. This result confirms the role of early advantage in the dynamics of wages. Finally, the role of distance is reflected in a nonlinear fashion, that implies that distance from the nearest center confers an advantage, which decreases with distance less than 240 miles, and increases with distance more than 240 miles, from the nearest higher-tier city. Therefore, in this precise sense, the individual effect's dependence on distance defines a higher-tier city's agglomeration shadow.

16.5.2 Wages and Spatial Evolution: Repeated Cross-Sections Estimation

For a number of reasons, estimations with our data being utilized in the form of repeated cross sections gives an arguably superior view of the underlying economic phenomena. The panel structure of the data for cities with neighbors is rather short, and after selection for non-missing values and the like, there are only 322 out of 498 observations left that are associated with more than one observation. That remaining subsample continues to be very unbalanced.

We have therefore performed two groups of regressions with data for nine pairs of two consecutive periods. The first group uses data for cities with neighbors only and instruments the average lagged wage rate of neighbors, by regressing it first against all exogenous variables in the model. The second group involves a selection model with endogenous sample separation for the evolution of wage rates by using all observation for each pair of consecutive periods. For cities with neighbors, Equation (16.14) is estimated, where in the place of the average wage for neighbors we use its predicted value from the first stage regression. For cities without neighbors, that same equation is estimated, but without the term of the lagged wage rate of neighbors. We use a city's lagged population as the only explanatory variable in the discrete choice part.

The regressions with the sample of cities with neighbors only show that the fitted value for the average wage among a city's neighbors as a instrument is significant before, but *insignificant* after, correcting for endogeneity. This pattern persists in the case of the selection model. The selection model is always significant, and the correlation coefficient between the discrete choice part and the continuous part is generally insignificant for the case of cities with neighbors, but significant and most often negative with a large absolute value, for the case of cities without neighbors. This finding suggests that the unobserved factors that determine whether or not a city has neighbors are associated with slower dynamic evolution of wages for cities without neighbors.

16.6 Conclusions

We test implications of the new economic geography for the impact of spatial evolution of US cities on wages. We use a data set consisting of 1900–1990 metro area populations, and spatial measures including distance from the nearest larger city in a higher-tier, adjacency, and location within U.S. regions. We also date cities from their time of settlement.

The data show that the dispersion of wages is smaller among cities with neighbors than without neighbors, and generally decrease over time, although the pattern is different for cities with neighbors and without neighbors. We ˇ nd that panel regressions of real wages against their own lagged values, contemporaneous population and lagged schooling give very good ˇ ts, especially when ˇ xed effects are included. When the wages of neighbors are included, for cities with neighbors, agglomeration effects are not signiˇ cant, but the effect of the contemporaneous average wage among a city's neighbors is stronger than that of the own lagged wage. This ˇ nding appears only after we control for individual effects. Agglomeration effects are signiˇ cant, for cities without neighbors. The individual effects exhibit strong regional effects and nonlinear dependence on distance from the nearest larger city. The individual effects are larger for older cities with neighbors and for younger cities without neighbors.

The statistical signiˇ cance of the average wage among a city's neighbors disappears after we correct for its endogeneity. Nonetheless, the fact that presence of neighbors is a statistically important feature of the spatial evolution of wages strengthens the importance of the forces of spatial clustering. The notion, emphasized especially by Krugman, that a city once created generates its own agglomeration shadow continues to be an important challenge for empirical work.

Acknowledgments

An earlier version was presented at the Regional Science Association International Conference, Santa Fe, NM, November 1998. I thank Dan McMillen for useful comments. I am grateful for comments from the editors of the volume and two referees. I thank Linda Dobkins for her suggestions during earlier work with these data, Vernon Henderson and Stuart Rosenthal for insightful discussions and valuable comments, and Tracey Seslen for exceptional research assistance. I am grateful to the John D. and Catherine T. MacArthur Foundation and the National Science Foundation for generous research support.

17 Endogenous Spatial Externalities: Empirical Evidence and Implications for the Evolution of Exurban Residential Land Use Patterns

Elena Irwin[1] and Nancy Bockstael[2]

[1] The Ohio State University
[2] University of Maryland

17.1 Introduction

The notion that "neighbors" may generate spatial externalities is well established in economics. In addition to textbook examples of externalities among firms, a significant body of empirical work in urban and environmental economics has provided evidence of the effects of neighboring, undesirable land uses on residential location decisions and housing values. The goal of this chapter is not to challenge or augment this literature, but rather to use it as a starting point in asking whether spatial externalities may influence actual land use conversion decisions by landowning agents. The basic thesis proposed here is that agents' consideration of these spatial externalities may influence their land use decisions if the resulting change in a parcel's relative values in alternative land uses is sufficiently strong. If so, then the presence of such spatial externalities creates an interdependence among neighboring agents' land use decisions, which implies that land use conversion may be partially driven by a process of endogenous change.

The intuition behind the agent interaction hypothesis is straightforward. If spatial externalities from neighboring land uses influence agents' land use conversion decisions, then the local neighborhood around an individual parcel matters. As a result, the regional land use pattern, which is the cumulative outcome of agents' individual conversion decisions, will depend on the nature and timing of these spatial externalities. A variety of externality effects are possible. Positive externalities between developed parcels may include various "community" spillover effects; people may find it desirable to live in close enough proximity to others so as to have the social benefits of neighbors. In addition, there may be positive effects associated with a critical density of residents in an area, which may be necessary to attract public and private services to the area. Negative spillover effects may occur between neighboring developments, however, due to congestion and aesthetic considerations. This, in conjunction with positive externalities associated with undeveloped parcels (e.g., open space amenities), could result in a "repelling" effect among residential development, such that a scattered pattern of development could result.

This term has been used by some researchers to distinguish areas that are neither fully suburban nor rural, but that have elements of both.

This chapter investigates whether the "agent interaction" hypothesis can explain residential land use pattern changes in rural-urban fringe areas. These exurban[1] areas are of particular interest, as they have experienced the highest levels of population growth and rates of land conversion over the last decade and are projected to receive the majority of development in coming years. Figure 17.1 illustrates recent changes in the exurban development pattern in Calvert County, MD. This increasing fragmentation of the landscape by residential development illustrated by Fig. 17.1 is typical of how exurban landscapes are changing across the U.S. Recent results from hedonic models of residential exurban land values provide evidence of signiˇ cant spillover effects from both neighboring open space and neighboring higher density development (Bockstael and Geoghegan, 1999; Bell and Bockstael, 2000; Geoghegan *et al.*, 1997). This evidence is consistent with the hypothesis that the land use conversion process is at least partially driven by neighboring agents' actions. If so, an alternative conception of the underlying spatial process to that assumed by standard urban economic models is implied. Rather than an underlying spatial process based on the assumption that accessibility to one or more centrally located employment centers drives land rents, a more general conception of the underlying spatial process is suggested – one in which endogenous interactions among neighboring land use agents also matter.

17.2 Spatial Externalities and Residential Location

Recognizing that urban areas are prime generators of externalities, urban economists have long been concerned with the spatial consequences of externalities. Indeed, external economies among ˇ rms are commonly cited as the underlying force of city formation. The literature contains discussions of many different types of externalities, e.g., externalities from producers to households; externalities among households; externalities among producers; and externalities associated with other elements of urban structure, e.g., transportation networks (see Miyao and Kanemoto, 1987, for an overview). Spatial externalities have been incorporated into the monocentric framework by assuming that proximity to the exogenous city center is desirable because of lower commuting costs but undesirable because of higher pollution. For example, the effects of industrial pollution on households' residential location choice are considered by assuming that industrial plants are located in the city center and that negative externalities from pollution decrease with distance from the city center. As a result, if the negative externalities are sufˇ ciently large, they may offset the negative land rent gradient created by transportation costs and generate a positive land rent gradient.

The case of externalities among households has also been addressed in the context of racial preferences. If the racial composition of one's neighborhood matters, then the location of households relative to each other in uences subsequent households' location decisions. In the so-called "global externality models" (e.g.,

[1] This term has been used by some researchers to distinguish areas that are neither fully suburban nor rural, but that have elements of both.

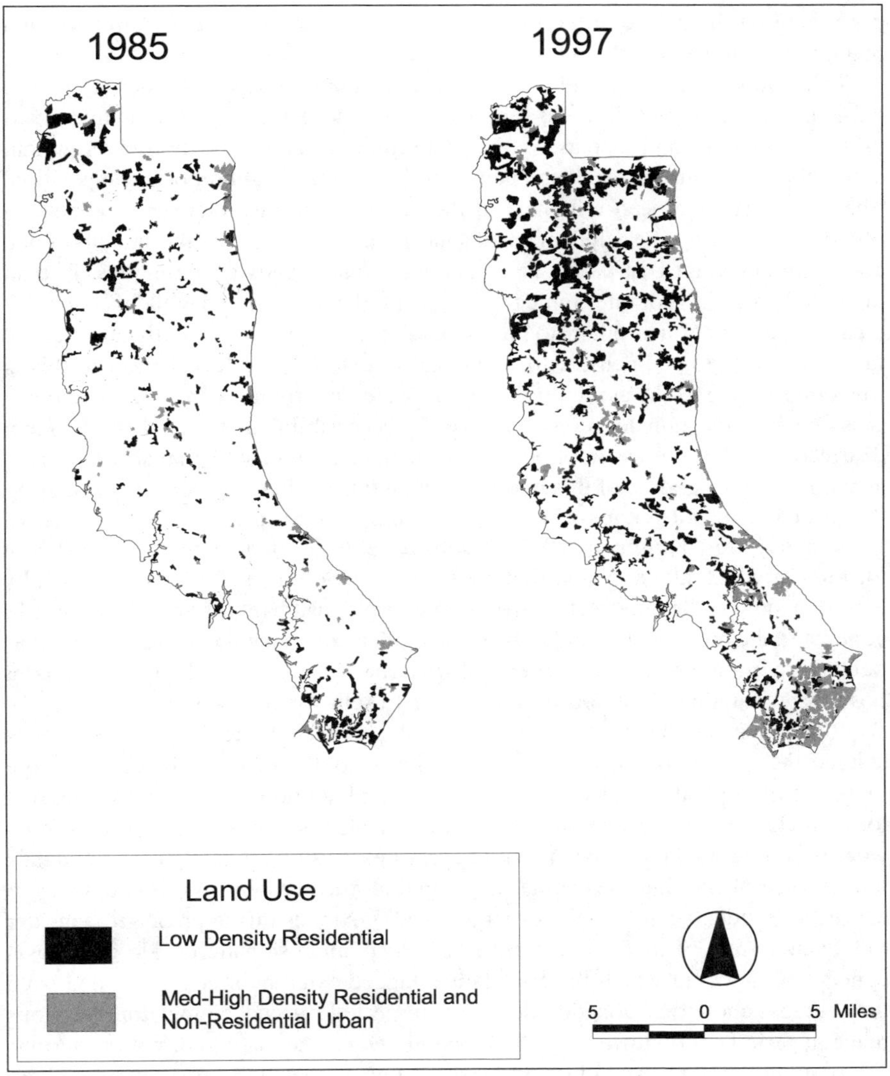

Fig. 17.1. Changes in land use pattern in Calvert County, MD

Kanemoto, 1980), utility is a function of neighborhood racial composition and therefore bid rent gradients that make each type of household indifferent to location can be derived. The degree of spatial segregation in equilibrium and the stability of equilibrium residential patterns are determined by examining the relative slopes of the bid rent functions for both types of households. One of the most interesting results of these models is that a completely segregated pattern may or may not result, depending on the nature of the racial preferences (and therefore the direction and magnitude of the externalities).

Other models have considered the dynamic consequences of externalities. In particular, Schelling (1971) and others examine the possibility of dynamic instability in which a small increase in the proportion of one type of household within a neighborhood causes a sudden change in the overall neighborhood composition. These models have been criticized in the urban economics literature for not explicitly considering the adjustment of land rents as a result of the neighborhood composition changes. In particular, it is argued that negative externalities from an increased proportion of one ethnic type in a central neighborhood will only cause increases in emigration of others to the suburbs, if land rents in the central area do not fall sufficiently to compensate for the negative externality effect. However, this is true only so long as the assumption regarding the underlying economic spatial process, in which location decisions are driven by accessibility to a central employment district, is valid. If this assumption is not valid, then a framework that abstracts from an explicit consideration of this notion of land rents could be useful for considering the role of such endogenous effects in the formation of residential land use patterns. Recognizing this, Krugman (1995, 1996b) develops a model of endogenous firm location in which all exogenous heterogeneity is ignored and firm location is fully determined by spillover effects generated by firm densities at different locations. In contrast to the monocentric and policentric models, in which the evolution of a spatial pattern is the result of an exogenously determined city center, Krugman's model posit the evolution of firm location patterns as a self-organizing process.

Regional economic models of residential location across regions have also considered the cumulative influence of many households' location choices, e.g., the influence of population density in a region on the likelihood that an individual will locate in that region. For example, Anas (1981, 1983) uses the random utility framework to link individual decisions with aggregate flows by modeling an individual's joint choice of housing, residential location, and travel mode and linking these to aggregate residential zones. Werczberger (1987) extends this approach to consider the dynamic role of spatial externalities in urban land use patterns. He develops a conceptual model in which the cumulative, lagged external effects generated from households' and firms' location decisions in period t influence location decisions made in period $t+1$. However, as Werczberger notes, the magnitudes of the various externalities among households and firms are unknown and therefore, empirical evidence of these effects is necessary in order to simulate the model he develops using realistic parameter values.

17.3 A Model of Land Use Conversion with Interaction Effects

We seek a way to explain the pattern of change reflected by the increasing degree of scatteredness of residential development visible in Fig. 17.1. In doing so, we are interested in incorporating the influence of a variety of exogenous features – including accessibility to urban centers, location of public services, and zoning constraints –

as well as the influence of local interaction effects generated by neighboring agents' land use decisions. Each of these landscape features creates its own pattern of spatial heterogeneity, resulting in a complex pattern of spatial variation across the landscape. The monocentric model is clearly not appropriate for this purpose. Because of the particular manner in which space is defined in those models, the only tractable way in which additional spatial variation (e.g., neighborhood effects) can be incorporated is by making these variables a function of distance to the CBD.

Our approach is closer in spirit to the models of by Schelling (1971) and Krugman (1995, 1996b). By abstracting from the standard notion of land rents driven by an accessibility measure to a central district, space can be defined in terms of the relative distance between a land parcel and other features of the landscape – including other land parcels. In this way, the influence of a variety of exogenous features, as well as the spillover effects of neighboring land uses, can be considered. Because we make no attempt to characterize regional land use patterns by solving for an equilibrium solution to the model, this approach requires an alternative means of demonstrating whether the interactions hypothesis is robust to explaining observed exurban patterns of residential development. Elsewhere, we show via simulation of a cellular automata model of land use conversion that sufficiently strong negative interaction effects among neighboring agents result in "repelling" effects that generate a scattered pattern of regional development (Irwin and Bockstael, 1999).[2] In what follows, we develop a simple micro-economic model of land use conversion in which these exogenous features and endogenous interactions are considered. This model is then estimated and the hypothesis regarding whether repelling effects actually exist among residential developed land uses is tested by determining whether the estimated interaction parameter, which captures the spillover effects from neighboring development, is negative and significantly different from zero. The resulting parameter estimates are used in a simulation of future land use pattern in order to investigate the robustness of the model in explaining the qualitative features of observed changes in residential land use pattern – namely the pervasive scatteredness of development and increasing fragmentation of the landscape over time.

The underlying decision model is developed in Irwin and Bockstael (1999), and so we provide only a brief sketch of it here. In developing a model of land use conversion, we start from the viewpoint of a profit-maximizing agent[3] who owns an

[2] This paper applies endogenous interactions theory to a model of agents' land use conversion decisions and investigates whether this hypothesis is consistent with large-scale scattered residential patterns observed in U.S. suburban and urban-rural fringe areas. Interactions among neighboring agents are incorporated into a cellular automaton model of land use conversion by borrowing from a model of interacting particle systems from statistical physics.

[3] The term agent is used to refer to a landowner who could either keep land in an undeveloped use or convert it to residential use. The conversion process may involve selling the undeveloped parcel to another agent – a developer – for conversion. In what follows, we use the term agent throughout, even though, in a strict sense, there may be a distinction between landowner and developer.

undeveloped[4] land parcel and makes a discrete choice in every period regarding the subdivision of the parcel for residential use. The undeveloped parcel is treated as the unit of observation and, therefore, the decision that is modeled is the agent's decision to subdivide the parcel into multiple residential lots or to keep the parcel in an undeveloped use. We do not explicitly deal with commercial development, which makes up a relatively small proportion of developed uses in exurban areas and typically follows residential development in these areas. Conditional on the parcel being undeveloped in the present period, the agent's decision is simplified to a binary choice of converting her parcel to residential use or keeping her parcel in an undeveloped use, such that the present discounted sum of all future expected returns from the land is maximized. Once converted, the land is supplied as residential lots to households, who make location decisions by choosing a bundle of attributes associated with a particular location to maximize utility. Therefore, the agent faces a dynamic optimization problem in which she will choose to convert the parcel to residential use when the expected present discounted value of the parcel in residential use net of conversion costs and opportunity costs is maximized over an infinite time horizon.

The problem is characterized as one of optimal timing of development, since growth pressures are sufficiently strong in this area for most landowners to expect that conversion will be optimal at some time in the future. This amounts to an implicit assumption by market participants of continuous exogenous growth pressures. The criteria for development in time period, t, is a simple one. The expected returns from selling the converted parcel on the residential market, net of the costs of conversion, must exceed the opportunity cost. But the opportunity cost is a complex concept. At minimum, this is the present value of the infinite stream of returns from the land in its current, undeveloped, state. However, given our expectations about growth pressures, this is certainly a lower bound. In reality, the opportunity cost – or foregone returns from initiating development and selling into the residential market today is the present value of taking the same action some time in the future plus the present value of the returns from the undeveloped use until that time. This is an important consideration, since even if development is profitable today, waiting may be more profitable.

Before stating the optimality criteria for conversion, we first define the expected value of a parcel in each of the two possible land use states, where the subscript $s(i,t) = 0$ denotes parcel i in an undeveloped state in period t and $s(i,t) = 1$ denotes parcel i in a residential state in period t. The expected value of parcel i if put in residential use at time t, net of conversion costs, is represented by $\pi_{s=1}(i,t)$. This variable will be a function of:

(a) time-invariant exogenous features that are specific to parcel i, $H_1(i)$, e.g., size, soil type, slope, services available to the parcel, and time-invariant, exogenous location features, e.g., distances to cities, markets, and recreational destinations.

[4] Here, undeveloped uses include agricultural and other resource production-oriented uses of the land, e.g., commercial forestry, as well as land in natural states.

These exogenous variables are constant over time, but spatially correlated over parcels;

(b) the net in uence of spatial externalities generated from the land use of parcels located within a deˇ ned neighborhood of parcel i, that may either increase or decrease the value of parcel i in the developed use in period t, $N_1(i,t)$. This is measured as the relative amount of development within a given neighborhood of each parcel; and

(c) the expected real change in residential prices due to growth pressures in period t relative to $t-1$, designated $\gamma_1(t)$.

Taken together, these imply that the agent's expected value of land in residential use is:

$$\pi_1(i,t) = V\left[H_1(i), N_1(i,t), \gamma_1(t)\right]. \tag{17.1}$$

The net present value of keeping the land in its undeveloped use, $\pi_{s=0}(i,t)$ can be expressed as:

$$\begin{aligned} \pi_0(i,t) &= \sum_{\tau=0}^{\infty} A(i,t+\tau)\delta^{\tau} \\ &= \sum_{\tau=0}^{\infty} A\left[H_0(i), N_0(i,t+\tau), \gamma_0(t+\tau)\right]\delta^{\tau}, \end{aligned} \tag{17.2}$$

where δ is the discount rate and $A(i,t)$ is the proˇ t from the alternative, undeveloped use of parcel i in time t, and is also a function of characteristics of the parcel. Again there will be exogenous and approximately time-invariant characteristics, like soil quality, which we denote, H_0; there will be rates of change in real prices of the marketed goods produced by the undeveloped use of the land, γ_0; and there will be effects on proˇ tability that are driven by surrounding land uses, N_0.

In Irwin and Bockstael (1999), we show that under some plausible assumptions about the expected time paths of the various elements, the optimality rule reduces to a double criteria that must be met for conversion to take place in the current time period, t. This period will be the optimal one for development if, in t, the expected returns from conversion exceed the present value of all foregone income from the alternative, undeveloped use:

$$\begin{aligned} &\pi_1(i,t) > \pi_0(i,t) \rightarrow \\ &V\left[H_1(i), N_1(i,t), \gamma_1(t)\right] > \sum_{\tau=0}^{\infty} A\left[H_0(i), N_0(i,t+\tau), \gamma_0(t+\tau)\right]\delta^{\tau}, \end{aligned} \tag{17.3}$$

and if, in t, the net expected returns from conversion in this period exceed the discounted value of waiting for potentially greater gains next period:

$$\begin{aligned} &\pi_1(i,t) - \pi_0(i,t) > \pi_1(i,t+1) - \pi_0(i,t+1) \rightarrow \\ &V\left[H_1(i), N_1(i,t), \gamma_1(t)\right] - \delta V\left[H_1(i), N_1(i,t+1), \gamma_1(t+1)\right] \\ &\qquad - A\left[H_0(i), N_0(i,t), \gamma_0(t)\right] > 0. \end{aligned} \tag{17.4}$$

In specifying the expected value functions, the distinction is made between the in uence of landscape features that are treated as exogenous characteristics of the landscape, $H_s(i)$, and spatial externalities, $N_s(i,t)$, that are generated by the surrounding pattern of land uses within a deˇ ned neighborhood of parcel i. Because these externalities are generated by the neighboring land use pattern, these effects are clearly endogenous to the land use conversion process. This interaction is captured by assuming that the neighborhood variable is a function of the relative amount of development within the neighborhood at the beginning of that period:

$$N_1(i,t) = \sum_j [D(s(j,t-1)) \cdot M(j,t-1)] / \sum_j M(j,t-1) \text{ for } j \in \Omega_i, \quad (17.5)$$

where $D(.)$ is an indicator variable such that $D(.) = 1$ when $s(j,t-1) = 1$ and $D(.) = 0$ otherwise, $M(j,t-1)$ is the land area of parcel j, and i denotes the set of parcels located within a given neighborhood of parcel i.

Implicit in (17.4) is the notion that exogenous growth pressures in the region may generate potential gains from waiting to develop. to capture this effect, the net present value of waiting till period $t+1$ to develop is rewritten as $\delta\gamma V_1(i,t)$, where δ is the discount factor, $\gamma = 1+r$, and r re ects agents' homogeneous expectations over the rate of increase in the real value of developed land as a result of growth pressures in the region.

Assuming that $V_1(i,t)$ is a linear combination of $N_1(i,t)$ and $H_1(i)$, the agent's optimal conversion rule stated in 17.4 can be rewritten as:

$$\lambda [N(i,t) - \delta\gamma N^e(i,t+1)] + (1-\delta\gamma)H(i) - A(i,t) > 0, \quad (17.6)$$

where λ is the interaction parameter of interest. $N^e(i,t+1)$ represents the agent's expectations over the conˇ guration of neighborhood land use parcels at the beginning of period $t+1$. If agents can not anticipate changes in the neighborhood, their expectations on the conˇ guration of the landscape for period $t+1$ will be identical to their observation of that neighborhood in period t.

Recognizing that we can not measure all the factors that affect the net present value in the developed and undeveloped uses, we add a random error term before restating the problem in a form suitable for estimation. Given this random term $\omega(i,t)$, which captures omitted variables (unknown to the researcher, but not to the decision maker), the model to be estimated can be expressed in probabilistic terms as:

$$\mathrm{P}(i,t) = \mathrm{Prob}[\lambda(1-\delta\gamma)(N(i,t)) + (1-\delta\gamma)H(i) - A(i,t) > \omega(i,t)], \quad (17.7)$$

conditional on (17.3), which states that the proˇ ts from development exceeding the net present value of the parcel in the alternative use.

17.4 Estimation of the Empirical Model

Based on the intertemporal formulation of the agent's conversion decision, it is clear that time plays a crucial role in determining the conversion probability. In fact, given the sort of assumptions made in the previous section regarding continuous growth pressures, the relevant question is not whether conversion occurs, but rather when it occurs. Duration models, in which the timing aspect of a qualitative change from one state to another is explicitly treated, offer a suitable framework for modeling this problem.

Varying assumptions are possible regarding the distribution of durations. Fully parametric models, including the exponential, Weibull, log-normal, log-logistic, and complementary log-log models, can be speciˇ ed. In addition, a semi-parametric approach, commonly referred to as the proportional hazards model or Cox regression model, is also possible. We use this approach to estimate the land use conversion model.[5] In the context of land use, the hazard rate is deˇ ned as the conditional probability that a parcel is developed in period t, given that it has remained in an undeveloped state until time t. Based on the underlying land use decision problem speciˇ ed in (17.7), individual i's contribution to the likelihood function[6] takes the form:

$$\frac{e^{\theta_{it}}}{\sum_{i=1}^{n} Y_{ij} e^{\theta_{jt}}}$$

where,

$$\theta_{it} = \lambda(1-\delta_t\gamma_t)N(i,t-1) + (1-\delta_t\gamma_t)H(x(i)'\beta) - A(z(i)'\phi). \qquad (17.8)$$

$Y_{ij} = 1$ if $t_j > t_i$, where t_i is individual i's duration length until the event occurs, and $Y_{ij} = 0$ otherwise. This mechanism causes only those individuals who have not already exited by a given period to be considered in the portion of the likelihood function associated with that period. The vector of parameters to be estimated is $\lambda,, \beta,, \phi$. The vector $x(i)$ is a time-invariant vector of explanatory variables associated with parcel i and affecting the present value of returns to residential conversions. The vector $z(i)$ is a time-invariant vector of attributes that in uence the expected returns to the parcel in an undeveloped use. $N(i,t-1)$ is a time varying measure of the relative amount of development within parcel i's neighborhood in period $t-1$. δ_t is the time varying discount rate in period t, and γ_t is the time varying growth rate in period t.

17.4.1 The Identiˇ cation Problem

The omitted variables imbedded in the stochastic term in (17.7) arise due to unobserved heterogeneity associated with either the individual agent and/or individual

[5] Because the magnitude and signiˇ cance of the explanatory variables is of greatest concern, the proportional hazards model is useful since it does not require speciˇ cation of the distribution function for the duration length. In addition to this advantage, the method can incorporate time varying covariates, something that is not possible with fully parametric models.

[6] Strictly speaking, the likelihood function is a partial likelihood function, since the form of the likelihood estimated is a ratio of hazards and a common baseline hazard term cancels out.

land parcel that influence conversion decisions. Examples of the former are idiosyncratic factors such as the age and family and financial circumstances of the landowner. These are likely to be distributed somewhat randomly over the landscape. In contrast, the unobserved heterogeneity associated with individual land parcels is likely to be strongly correlated in space as attributes associated with nearby locations are almost certain to be more positively correlated than those associated with locations that are further apart. The presence of unobserved but spatially correlated, heterogeneous features that influence the conversion decision complicates the identification of endogenous interaction effects. If omitted variables are invariant over time and spatially correlated over space, then it will be difficult to distinguish between the influence of such unobserved spatial effects and those of true spatial externalities from surrounding land uses. Even in the absence of true spatial externalities, a positive interaction effect among neighboring parcels will appear to exist.

This version of the identification problem has arisen in a number of contexts outside the land use modeling literature, e.g., the social interactions literature, which is concerned with identifying peer pressure effects from unobserved heterogeneity (Manski, 1993, 1995; Brock and Durlauf, 1998). This same problem arises in the literature on own-state dependence over time, which seeks to separate "true" temporal state dependence (e.g., habitual effects) from "spurious" state dependence (Heckman, 1978, 1981). For example, if an individual's past unemployed state causes a greater probability of current unemployment, then this is "true" state dependence. In contrast, "spurious" state dependence may arise from unobserved heterogeneity across individuals (e.g., differences in education and ability), which, if constant across time, will lead to serial correlation of the errors. The correlation of errors across time creates correlation between the error and the own-state dependence term, represented by the individual's past states, which results in a biased estimate of the own-state dependence parameter.

The identification problem in the land use conversion model is very similar in some respects to the models discussed in these literatures. It is most similar to the social interaction models, both in terms of the source of endogenous effects (i.e., associated with neighboring agents' choices) and the correlation of exogenous variables over space. Analogous to the correlated effects among individuals described by Manski, heterogeneous landscape characteristics that vary over space may generate spatial correlation among neighboring land use decisions. If unobserved, these effects will make decisions appear related, even if they are not, and therefore complicate our ability to discern true state dependence. Although the nature of the development process implies a temporally lagged spatial interaction effect, the presence of time-invariant unobserved heterogeneity creates the same identification problem due to correlated unobservables as that which arises in the simultaneous social interaction models.

Unfortunately, the identification problem in the land use conversion model is further complicated in ways that prevent ready adoption of most of the identification strategies discussed in the literature.[7] As a consequence, exact identiˇ cation of the interaction parameter is not possible. However, it is possible to adapt a strategy of bounding the interaction effect. This approach is suggested by Heckman and Singer (1985), who illustrate the conditions under which the sign of the endogenous effect is identiˇ ed in duration models, where the endogenous term in this case is the duration dependence variable. The question of interest is whether the hazard rate[8] is either increasing (positive duration dependence) or decreasing (negative duration dependence) in the time length of the spell and whether this duration dependence, if it exists, can be distinguished from unobserved heterogeneity across individuals that may cause "spurious" dependence. Due to a data censoring problem that arises naturally in the estimation of these models, the resulting duration dependence parameter will be biased towards negative duration dependence. Heckman and Singer (1985) prove the negative duration bias that results for a model that is estimated without controlling for the effects of unobserved heterogeneity across individuals. They conclude that the direction of the duration dependence is identiˇ ed under these conditions only if the estimated duration dependence parameter is positive. In this case, given the direction of the bias, the true duration dependence effect must also be positive.

In most of the cases considered in the own-state dependence and social interaction literatures, the direction of the bias caused by the unobserved correlation and the true interaction effect are the same. For example, it is usually the case in economic models that the true duration dependence is expected to be negative, e.g., the conditional probability of exit from an unemployed state decreases as the length of the spell increases. Since the bias is also negative, it is impossible to test for the existence of the true duration dependence. In the social interactions case, the correlation among neighboring agents due to unobservables is usually positive (e.g., students perform similarly because they have the same teacher) and the hypothesized interaction effect is always positive. In these cases, the above identiˇ cation condition outlined by Heckman and Singer (1985) is not met and alternative strategies are required to break the correlation between the unobservables and the endogenous term in order to test for the presence of a true interaction effect.

In the land use conversion case, however, this is a feasible approach for identifying the direction of the interaction effect, or more accurately for testing for the existence of a negative interaction effect. Given that the error term and endogenously lagged neighborhood variable are positively correlated, the resulting empirical estimate of the interaction effect will be biased in the positive direction, as shown in the previous section. This implies that the estimated interaction effect bounds the true interaction effect from above. If the estimated effect is negative, then it must hold that the "true" interaction effect is negative for at least some range of the sample and

---

[7] See Irwin and Bockstael (1999) for a fuller discussion of these identiˇ cation strategies and why they are not applicable in this case.

[8] The hazard rate is deˇ ned as the conditional probability of exit from a spell in period t, given that the spell has extended to period t.

over some interval of time. If the estimated interaction effect is positive, however, we can not test for the existence of a true interaction effect.

17.4.2 Specification and Data

Data used to estimate the land use conversion model include spatially defined, micro-level data on land parcels from the Maryland Office of Planning's geo-coded tax assessment data base. The construction of this data set required merging data from several tax assessment data sources, some of which are not geo-coded, in order to compile a 6 year history of "convertible" parcels within a two county study area located in exurban areas of Washington D.C.: Calvert and Charles counties of Maryland. The data set is comprised of all parcels that, as of January 1991, were large enough to accommodate a subdivision of at least five houses given current zoning and could have been converted to residential use. The year of conversion for those that were converted during the period 1991 through 1996 is also included. The data set contains variables that describe the individual parcel, including lot size and land use. Because the centroids of the parcels are geo-coded, it was also possible to locate the parcels in space and, using a Geographic Information System (GIS), to generate a variety of additional spatial attributes associated with the individual parcels. These variables include zoning and distance via the road network to urban centers, such as Washington D.C.. Annual data on the prime rate and on median housing prices from 1991–96 were used to construct estimates of the annual discount factor and growth rates respectively.[9]

The dataset also contains information about the land uses and conversion times of neighboring parcels. The neighborhood variable is constructed as the percent of the neighboring land in a developed (vs. an undeveloped) use in the year prior to the conversion decision. So, for example, conversions that occurred in 1992 were modeled as a function of the percent of neighboring land that was developed as of 1991. Development is defined as all commercial, industrial, and residential uses for which a structure exists on the land parcel, excluding very low density residential development (defined by a lot size of five acres or more). Since this variable changes over time, it was updated for every year the conversion decision was modeled. The extent of the relevant neighborhood around a parcel of interest is essentially an empirical question. It is possible that the direction of the interaction parameter may change with distance, given that different spatial externalities may have different rates of decay. To allow for this possibility, we defined two non-overlapping neighborhoods. Table 17.1 gives the definitions and areas of the various neighborhood indices.

[9] The discount factor was calculated as $\delta = 1/(1+i(t))$, where $i(t)$ = prime rate in period t. The growth rate was calculated as $\gamma = 1 + r(t)$ where $r(t) = [p(t) - p(t-1)]/p(t-1)$ and $p(t)$ is the median housing price for the Washington D.C. metro area. Data on the prime rate came from http://www.hsh.com/indices/prime.html ARM Indexes. Median housing price data for the Washington D.C. metro area from the U.S. Census Bureau, Statistical Abstract of the U.S.

Table 17.1. Extent and Area of Neighborhood Indices

Neighborhood Index	Inner Radius (meters)	Outer Radius (meters)	Area (acres)
N_1	400	930	545
N_2	930	1609	1330

17.4.3 Model Specification

In order to judge the degree to which omission of spatially heterogeneous attributes influences the estimate of the interaction parameter, a series of three nested models was defined. Each of the three models includes the vector of exogenous, time-invariant covariates, $x(i)$, a time-invariant proxy for opportunity costs, $z(i)$, and the neighborhood specification outlined above. Holding the neighborhood specification constant, the nested models differ only in the exogenous features included in $x(i)$ and, consequently, in the amount of unobserved spatial heterogeneity that is relegated to the error term.

The specification of the three nested models is summarized in Table 17.2. Each specification includes a proxy for agricultural profitability (specifically a soil quality indicator) and a nonlinear function of accessibility to Washington D.C.; both are time-invariant over this period. The soil quality variable equals 1 for all natural soils groups designated by the Soil Conservation Service as prime farm land (Maryland Department of State Planning, 1973) and 0 otherwise.[10] Also included in all models are two neighborhood variables, N_1 and N_2, measured as linear functions of the amount of development within each of two neighborhoods of the parcel as defined in Table 17.1. In the first model, Model A, all other sources of spatial heterogeneity are purposefully left in the error term.

Model B incorporates zoning considerations into this specification. The minimum lot size regulation determines the number of lots that can be developed on any given sized parcel and therefore is expected to be significant in the likelihood of conversion. To capture potentially non-linear effects, it is specified as a quadratic term.

Model C incorporates additional observed spatial heterogeneity of the landscape, including: (1) a dummy variable for proximity to a local road, which is coded as 1 if the parcel centroid falls within a $1/4$ mile of local road and 0 otherwise; (2) a dummy variable for steep slope, which would indicate higher costs of conversion, that is coded 1 if the parcel has a steep slope and 0 otherwise; and (3) a dummy variable for public sewer provision, which is coded 1 if the parcel is on public sewer and 0 otherwise. The availability of roads and public sewer are expected to have a positive effect on the hazard rates, since they reduce costs of conversion. The existence of steep terrain is expected to have a negative effect.

[10] Unfortunately this is the only variable that was available to reflect agricultural profitability, since agricultural data at a micro level is not available due to confidentiality restrictions.

Table 17.2. Model Speciˇ cations

Variable	Model A	Model B	Model C
Intercept	X	X	X
Ln(Distance to DC)	X	X	X
Proxy for Opportunity Costs (Soil Dummy)	X	X	X
Inner Neighborhood Land Use Index	X	X	X
Outer Neighborhood Land Use Index	X	X	X
Min Lot Size		X	X
(Min Lot Size)^2		X	X
Proximity to Local Road (Dummy)			X
Steep Slope (Dummy)			X
Public Sewer (Dummy)			X

17.4.4 Empirical Results

The proportional hazards model is set up to estimate the hazard (and not survival) of a parcel's conversion. A positive coefˇ cient indicates that the conversion probability increases in the associated variable. Put another way, a higher value of the variable makes conversion likely to occur sooner rather than later. Tables 17.3–17.4 report the results for each of the model speciˇ cations under two different methods of dealing with tied data, the exact method and Efron's approximation.[11]

In all three nested speciˇ cations of the model, the outer neighborhood measure is negative and signiˇ cantly different from zero. An increase in the amount of neighboring development within the outer neighborhood area causes a decrease in the hazard rate of conversion. The estimated interaction effect becomes more negative with the addition of the minimum lot variable, but then remains essentially constant with the inclusion of additional exogenous features. In all three speciˇ cations, the inner neighborhood variable is not signiˇ cantly different from zero.

Other parameter estimates are consistent with intuition. The hazard rate of a parcel shifts downward with an increase in the opportunity costs associated with conversion, proxied here with the soil quality indicator. The base hazard rate shifts downward at a decreasing rate with the distance to Washington D.C. This accessibility measure is signiˇ cant in the ˇ rst speciˇ cation, Modd A, and becomes highly signiˇ cant in Model B with the inclusion of minimum lot size. This is evidence of the importance of the monocentric model's contention that residential location is a function of accessibility to the central business district. The difference in the signiˇ cance level in the distance parameter between Models A and B suggests that this variable is positively correlated with minimum lot size across parcels. Model B illustrates the highly signiˇ cant in uence of minimum lot size on the hazard rate, which increases at a decreasing rate with increases in the minimum lot size up to about 3.8 acres. Lastly, Model C shows the signiˇ cance of other exogenous features.

[11] For discussion of these methods see, for example, Lee (1992).

Table 17.3. Results from the Proportional Hazards Duration Models of Land Use Conversion, Models A and B

MODEL A	Proportional Hazards Model	
	Ties = Efron	Ties = Exact
	Parameter Estimate	Parameter Estimate
	(Pr $> \chi^2$)	(Pr $> \chi^2$)
Intercept	none	none
Ln(Dist DC)	−10.7962	−10.7965
	(0.0348)	(0.0348)
Opp Costs	−0.5364	−0.5363
	(0.0090)	(0.0090)
%Dev Neighborhood (inner)	12.3127	12.3128
	(0.2227)	(0.2227)
%Dev Neighborhood (outer)	−27.7724	−27.7729
	(0.0375)	(0.0375)
LR Test Statistic*	16.042	16.043
	$p = 0.0030$	$p = 0.0030$
MODEL B	Proportional Hazards Model	
	Ties = Efron	Ties = Exact
	Parameter Estimate	Parameter Estimate
	(Pr $> \chi^2$)	(Pr $> \chi^2$)
Intercept	none	none
Ln(Dist DC)	−30.3661	−30.3697
	(0.0001)	(0.0001)
Opp Costs	−0.5565	−0.5565
	(0.0067)	(0.0067)
Min Lot	80.9075	80.9169
	(0.0001)	(0.0001)
(Min Lot)^2	−21.5798	−21.5825
	(0.0001)	(0.0001)
%Dev Neighborhood (inner)	11.8224	11.8218
	(0.2639)	(0.2640)
%Dev Neighborhood (outer)	−40.0019	−40.0061
	(0.0059)	(0.0059)
LR Test Statistic*	50.034	50.037
	$p = 0.0001$	$p = 0.0001$

* Restricted Likelihood includes intercept term only

Table 17.4. Results from the Proportional Hazards Duration Models of Land Use Conversion, Models C

MODEL C	Proportional Hazards Model	
	Ties = Efron	Ties = Exact
	Parameter Estimate	Parameter Estimate
	(Pr > χ^2)	(Pr > χ^2)
Intercept	none	none
Ln(Dist DC)	−28.2403	−28.2451
	(0.0001)	(0.0001)
Opp Costs	−0.6498	−0.6498
	(0.0020)	(0.0020)
Min Lot	79.7918	79.8035
	(0.0001)	(0.0001)
(Min Lot)^2	−21.1583	−21.1615
	(0.0001)	(0.0001)
Sewer	20.0879	20.0889
	(0.0554)	(0.0554)
Slope	−8.2116	−8.2122
	(0.0381)	(0.0381)
Close to Road	−4.7829	−4.7836
	(0.1276)	(0.1275)
%Dev Neighborhd (inner)	12.4029	12.4024
	(0.2419)	(0.2420)
%Dev Neighborhd (outer)	−39.5839	−39.5896
	(0.0066)	(0.0066)
LR Test Statistic*	59.322	59.326
	$p = 0.0001$	$p = 0.0001$

* Restricted Likelihood includes intercept term only

Public sewer provision and steepness of slope are shown to increase and decrease the parcel's hazard rate respectively, as expected, whereas the coefˇ cient associated with the dummy variable that measures proximity to a local road is not signiˇ cantly different from zero.

Several different explanations for the difference between the negative and significant result on the outer neighborhood parameter and the insigniˇ cant result on the inner neighborhood parameter are possible: (1) there are competing interaction effects at shorter distances that offset each other, (2) there are economies of scale that cause clustering of development on a smaller spatial scale, (3) an own-subdivision effect occurs because of imprecise measurement of the neighborhood, or (4) the correlated, unobserved spatial variation, which is expected to be stronger at shorter distances, masks the interaction effects at these distances.

The first explanation, that of competing interaction effects at shorter distances between parcels, seems plausible. For example, even though negative effects from neighboring development may also exist at closer distances, they may be overcome by positive spillovers from development that are relatively stronger between parcels that are within close proximity. Such positive effects could be generated by an individual's desire to have neighbors or other such "community" effects. If these positive effects decay at a faster rate than the negative spillovers, then this would explain why evidence of negative interaction results for parcels located at an intermediate distance from each other, but not for parcels located in closer proximity to one another.

The second possibility, that economies of scale may lead to the development of adjacent parcels, also seems plausible. For example, due to machinery inputs, the per acre costs of clearing and grading the land and supplying roads is likely to be decreasing in the total number of acres that are developed. Alternatively, securing permits for a group of adjacent parcels may be more expedient than negotiating the permitting process for each developable parcel individually. If such cost considerations encouraged the development of adjacent parcels, then this may explain why there is evidence for negative effects only between parcels located at further distances apart.

Third, the imprecise way in which the neighborhoods are defined, due to unavailable parcel boundaries, may confound the effect. For large parcels, the area that is defined as the inner neighborhood may in reality be part of the parcel under consideration. In this case, the neighborhood effect would be confounded by an own-subdivision effect, which would bias the estimate in a positive direction.

Lastly, given that there is likely spatial autocorrelation left in the error term, the resulting bias of the estimated interaction parameter could wash out any offsetting negative interaction effect. Because spatial autocorrelation among parcels is believed to decrease as the distance between parcels increases, this bias will likely be stronger for the inner neighborhood measure. Empirically, it is not possible to distinguish among these four competing hypotheses for the case of the inner neighborhood measure. Therefore the question of whether interaction effects are present within an adjacent neighborhood area, as defined by the inner neighborhood variable, remains unanswered based on these empirical results.

Despite the uncertainty regarding the direction and magnitude of the inner neighborhood parameter, the case for negative interaction effects associated with the outer neighborhood area is strong. In the absence of any competing hypotheses, we conclude that the negative estimate of the outer neighborhood parameter is evidence of an interaction effect among neighboring agents.

17.5 Predicted Patterns of Development

In order to gauge the robustness of a land use conversion model that accounts for both exogenous and endogenous effects, we performed the following simulation exercise. Future changes in the 1990 land use configuration of Calvert County were simulated using the "Model C" parameter estimates for two different cases: (1) a restricted case, in which the interaction effect was set to zero, and (2) an unrestricted case, in which the estimated interaction effect was included. For each parcel that was "developable" in 1990, the time-invariant exogenous attributes, as well as the time-varing neighborhood land use variable, were calculated. The estimated parameters from Model C were then used to calculate each parcel's likelihood of conversion. In order to translate probabilistic measures of conversion into actual conversion, the effects of exogenous growth pressures were simpliˇ ed by assuming a constant regional demand for new housing. Development rounds were deˇ ned such that one new conversion occurs in each round of development. Given this, the parcel with the highest probability of conversion in each time period was assumed to be the parcel chosen for conversion. Once converted, the probability of a parcel's re-conversion to an undeveloped state was assumed to be very close to zero. The model was simulated for 100 rounds of development for both the restricted and unrestricted cases.

Figure 17.2a shows the comparison of these predicted patterns with observed changes in land use pattern between January 1991 and October 1993, where each point corresponds to a parcel's centroid. In comparing the two predicted patterns (Figs. 17.2b and 17.2c), the pattern simulated with only the exogenous effects appears to have a higher degree of clustered development. Consistent with predictions generated by simulations of the cellular automata model in Irwin and Bockstael (1999), the inclusion of the negative interaction effect generates a pattern that is somewhat more scattered. Interestingly, the actual pattern of residential development between 1991–93 (Fig. 17.2c) appears even somewhat more scattered than either of the predicted patterns.

In order to quantify the differences among the three patterns, a nearest neighbor count statistic was used to summarize each pattern of n points. To calculate these statistics, we adopt the methods outlined in the spatial statistics literature (e.g., Cressie, 1993; Diggle, 1984). The nearest neighbor statistic is a count variable that tallies the number of nearest neighbor points whose inter-point distance falls within each of successively increasing distance ranges. The distance interval d ranges from 0 to the extent of the region so that for $d = d_{max}$, all pairs of points are counted in the count statistic, which is normalized so that it ranges from 0 to 1. In order to gauge the degree of difference among predicted and actual patterns, a quantile-quantile plot is used to compare the statistics from the actual and predicted patterns, which are each calculated for the same values of d. The axes are measured in terms of the proportion of nearest neighbor points that fall within successive distance intervals, d. The degree of difference between the actual pattern and the predicted patterns is evidenced by the degree to which the plot of the statistic measuring the predicted pattern differs from the 45° line.[12]

[12] Diggle (1984) and Cressie (1993) outline a way to use the quantile-quantile plot to statistically test the null hypothesis that the point pattern is generated by a completely spatially random point process (CSR). To do so, the empirical distribution function generated from the observed point pattern is plotted along with upper and lower envelopes from multiple simulations generated under the CSR assumption. This is akin to establishing a conˇ dence

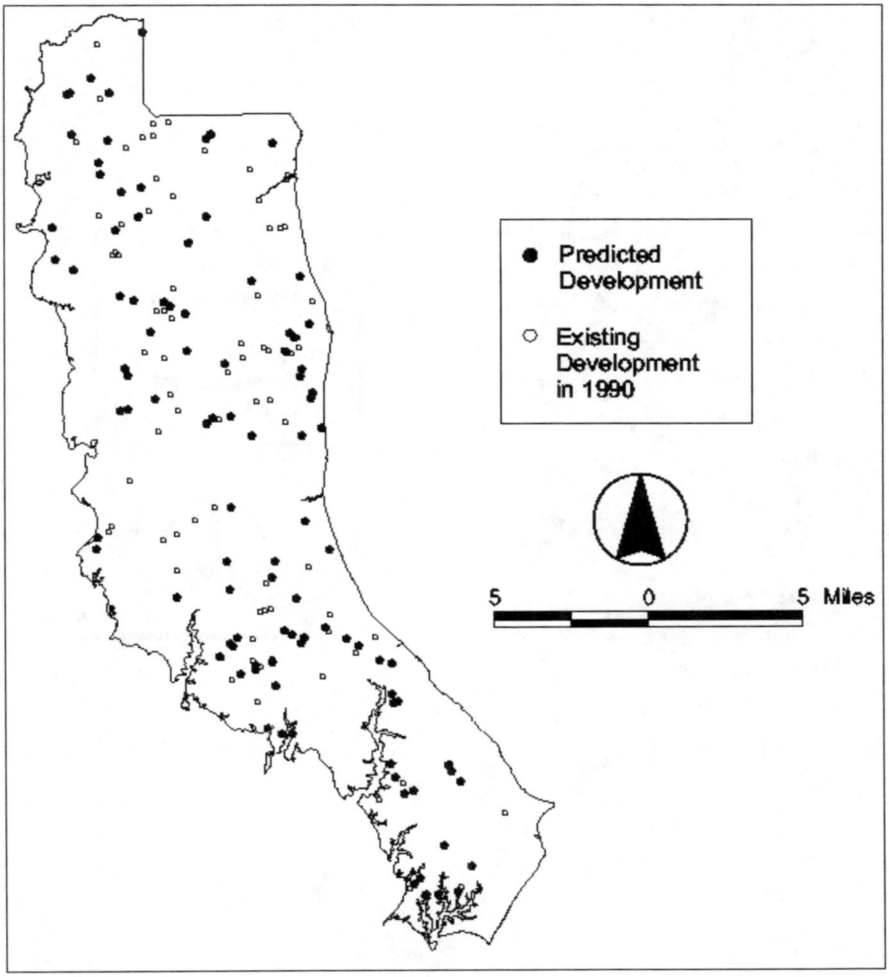

Fig. 17.2a. Observed pattern of residential development between 1991–93

Figure 17.3 shows the comparison between the actual pattern of residential development and the two predicted patterns. The plot of the statistic that corresponds to the pattern simulated with both the interaction and exogenous effects lies relatively close to the 45° line and is therefore qualitatively similar to the actual pattern. In contrast, the plot of the statistic corresponding to the pattern generated with only exogenous effects lies further above the diagonal, suggesting that this pattern has

interval. In our case, we do not use the quantile-quantile plot to perform signiˇ cance tests, but rather as a simple means of comparing the degree of negative and positive spatial correlation exhibited by the actual vs. predicted point patterns.

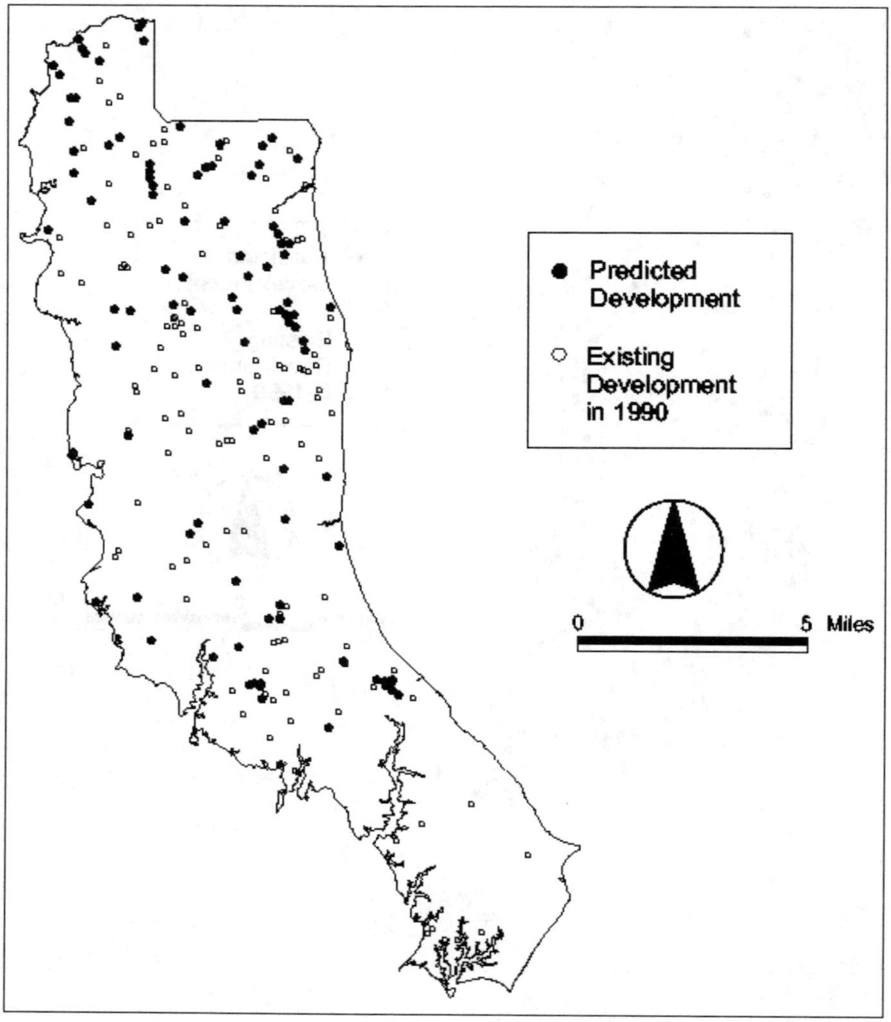

Fig. 17.2b. Simulated pattern of residential development with endogenous and exogenous effects

a higher degree of positive spatial correlation than both the actual pattern and the pattern simulated with the inclusion of the interaction effect. These observations provide further support for the model that incorporates for both exogenous and endogenous effects.

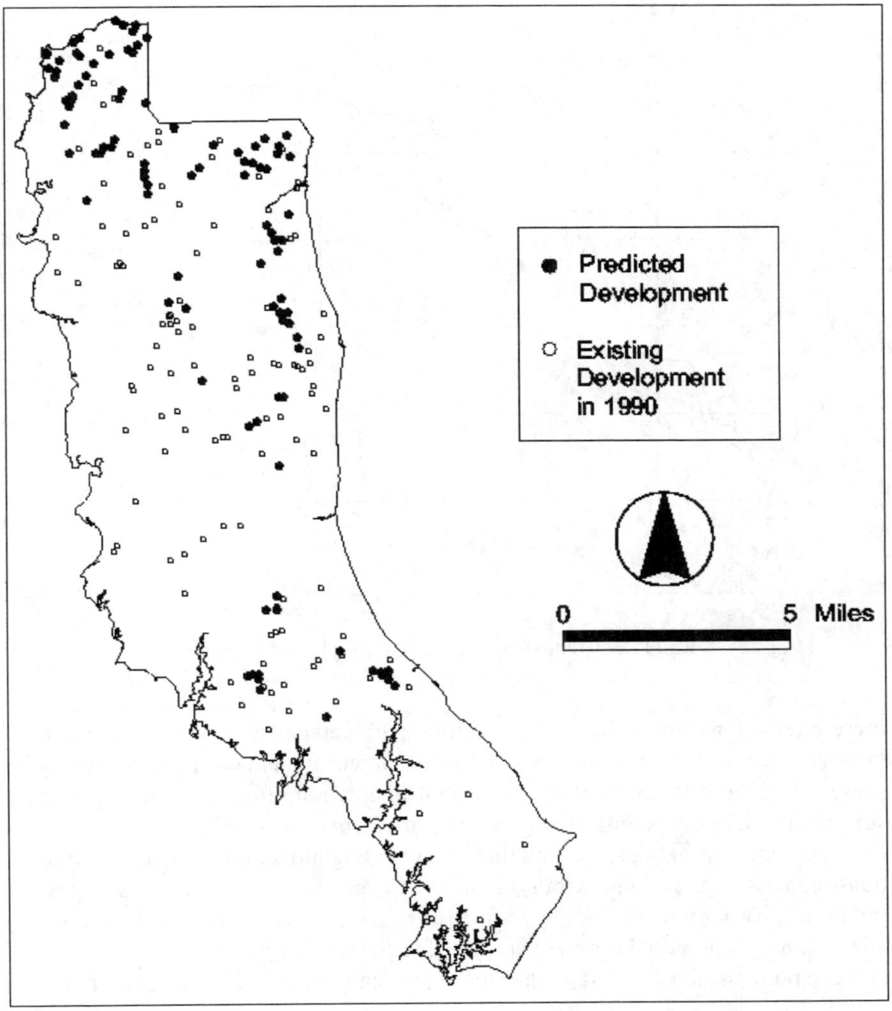

Fig. 17.2c. Simulated pattern of residential development with exogenous effects only

17.6 Conclusions

The presence of negative interaction effects combined with growth pressures in a region may generate very different types of spatial land use patterns than those predicted by the monocentric models or even from a model in which a variety of exogenous landscape features are considered. Depending on the relative magnitudes of the interaction effects, changes in land use pattern may be characterized by various degrees of clustering, scatteredness, and fragmentation. In addition, the evolution of land use pattern over time is potentially much more complex in the presence of

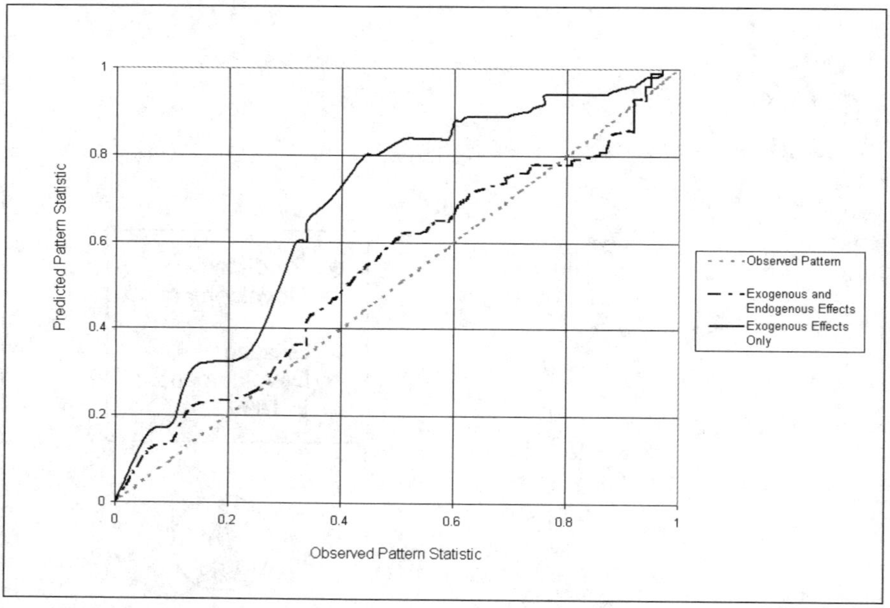

Fig. 17.3. Comparison of Nearest Neighbor Statistics

these interaction effects, due to the resulting path dependency. Past and current decisions in uence future decisions and future changes in land use patterns due to the presence of these temporally lagged spatial interaction effects. In the presence of suf˜ ciently strong repelling effects from positive open space spillovers and negative development externalities, we ˇ nd that the offsetting attracting in uences of exogenous features, e.g., from proximity to the central business district and the supply of public infrastructure, may not be suf˜ cient to mitigate scattered development patterns. In this case, policies aimed at offsetting the repelling effects are necessary to achieve policymakers' stated goal of more concentrated development patterns.

Part V

Trade and Economic Growth

18 Does Trade Liberalization Cause a Race-to-the-Bottom in Environmental Policies? A Spatial Econometric Analysis

Paavo Eliste[1] and Per G. Fredriksson[2]

[1] The World Bank
[2] Southern Methodist University

18.1 Introduction

This chapter explores the impact of openness to trade, and the size of trade flows, on the determination of environmental regulations. Some authors argue that as a result of global trade liberalization countries are likely to relax domestic environmental policy standards in order to increase (or maintain) "competitiveness" (see Esty, 1994; Dua and Esty, 1997; Esty and Geradin, 1997). This could potentially lead to a "race to the bottom," where countries continually undercut the competitors' regulations, or refrain from enacting new environmental policies altogether, a "regulatory chill." Fredriksson (1999) shows in a political economy model that the effect of trade liberalization on politically determined pollution taxes depends on the size of the relative shifts in political power of producer and environmental lobby groups that occur as a result of the liberalization (see also Bommer and Schulze, 1999). Others argue that "ecological dumping" may occur, where environmental policies are set at suboptimally lax levels for strategic reasons (Barrett, 1994; Kennedy, 1994; Rauscher, 1994). Industry and union interests join the environmentalists in their fear that trade liberalization will create "pollution havens" with low stringency of environmental regulations and a comparative advantage in polluting sectors. These fears have given rise to calls for harmonization of environmental policies in regional free trade areas, e.g., across the EU or NAFTA members (Esty and Geradin, 1997).

There is also a growing literature on the effects of environmental regulations on the pattern of trade. Economic theory predicts that more stringent environmental regulations will result in lower exports and greater imports in polluting sectors (see, e.g., Merrifield, 1988; Copeland and Taylor, 1994). However, the literature has found small or insignificant effects of environmental regulations on trade and firm location (see, e.g., Kalt, 1988; Tobey, 1990; van Beers and van den Bergh, 1997).[1]

We focus entirely on the agricultural sector. The objectives are threefold. First, do the agricultural sector environmental regulations in a country's trade partners influence the policies enacted by the country itself? If so, what is the direction of this influence? Second, do countries located in the same geographical area have similar environmental regulations? Finally, we seek to determine whether a country's

[1] See Jaffe *et al.* (1995) for an extensive survey.

openness to trade in uences the stringency of the environmental regulations set in the agricultural sector. To our knowledge spatial econometric techniques have not previously been applied to these issues. Neither has the impact of international trade on the stringency of environmental regulations been explicitly analyzed empirically. Only indirect evidence exists on this issue.[2] The main contribution of this chapter is the application of adequate spatial techniques to the analysis of the relationship between trade and environmental policies.

We model the spatial interdependence between countries by hypothesizing that the stringency of environmental regulations in a given country is (partially) a function of the weighted average of its trade partners' stringency of environmental regulations. We use bilateral export shares as weights. If domestic environmental policies are determined partly by what trade partners do, we expect countries that trade relatively intensively with each other to affect each other relatively more. We also hypothesize that countries which are geographically close, and which therefore may trade more with each other, have similar environmental policies.

We ˇ nd that countries with close trade relations tend to have similar environmental policies. We also provide initial evidence that global trade liberalization may induce countries with relatively lax environmental regulations to upgrade their policies towards the levels of their trade partners with relatively stringent regulations. However, our results do not rule out entirely that a race-to-the-bottom takes place at the regional level. We demonstrate that countries with open trade policies place a considerable weight on the environmental policies implemented by their trade partners which also have open trade regimes, and that this impact is positive. Other ˇ ndings include a positive impact of per capita income and environmental pressures on the stringency of environmental regulations, whereas the producer lobby has a negative effect (see also Eliste and Fredriksson, 1999).

Our ˇ ndings provide some evidence of the impact of trade regimes and economic integration on the political determination of environmental regulations in the agricultural sector. Similar relationships may exist in other sectors. Grossman and Krueger (1993) report that SO_2 levels are signiˇ cantly lower in cities located in countries with relatively more open trade. To the extent that trade openness has similar effects on environmental regulations in both rural and urban areas, we believe that our ˇ ndings provide a partial explanation for their result. However, it should be noted that agricultural production is immobile, and therefore different results may be obtained with data from sectors with a more footloose capital stock.

[2] Fredriksson and Gaston (1999) study the impact of trade openness on the speed of ratiˇ -cation of the 1992 United Nations Climate Change Convention, a test of the "regulatory chill" hypothesis. A few studies have focused on strategic behavior among countries when signing international environmental agreements (IEAs) (Beron *et al.*, 1996), and others study voluntary and non-voluntary control of sulfur and nitrogen emissions in adherence to signed IAEs (Murdoch *et al.*, 1997). Others have analyzed the spatial interaction among US states in the determination of public expenditures (Case *et al.*, 1993), the allocation of local public goods based on the median-voter model (Murdoch *et al.*, 1993), and property tax competition among local governments in the Boston metropolitan area (Brueckner, 1998).

The chapter is organized as follows. Section 18.2 specifies the empirical model. Section 18.3 describes the data and provides a hypothesis specification. Section 18.4 discusses the results, and Sect. 18.5 gives a conclusion and discusses the implications.

18.2 Model Specification

18.2.1 Specification of Spatial Weight Matrixes

The agricultural sector is resource based, i.e., a large part of the capital stock (land) is immobile. Lower environmental regulations may therefore not induce capital movements, thus lowering the incentives for strategic behavior. However, we believe that market participants have incentives to act strategically and politically based on what competing foreign producers and governments do. [3] We model the interdependence between countries by taking into account the stringency of environmental regulations in each country's trade partners. The relative interdependence is determined by the bilateral export flows and the geographical distance between two countries.

We expect the interdependence between two countries to increase with the intensity of trade. This, in turn, should have an impact on a country's environmental regulations, to the extent that a nation cares the most about the environmental regulations set by its closest trading partners. Thus the stringency of each trade partner's environmental regulations is weighed accordingly. Beron *et al.* (1996) argue that asymmetric trade flows give nations different political and economic power over each. This may determine countries' willingness to, e.g., sign an international environmental agreement (IEA). For example, suppose nation A is a country with low stringency of environmental regulations. Assume also that A exports a relatively large share of its total exports to nation B which has highly stringent regulations. In this situation, we may expect nation B to have a relatively large economic power over nation A because nation B's government or consumers can restrict A's access to its markets. Thus, nation A may be forced to increase the stringency of its environmental regulations on the demand from nation B's producers or consumers. By constructing the weights matrix using bilateral export flows we should capture (at least part of) the structure of the economic power among countries.

The first weights matrix is thus defined based on the value of the total agricultural export flows from country j to country i, W^{EXP}.[4] The off-diagonal elements of

---

[3] Another feature of the agricultural sector is that various trade restrictions and price support policies exacerbate market failures in many countries. The environmental effects of commodity programs are relatively well known (see Just and Bockstael, 1991).

[4] We define agricultural exports as food exports (S001) plus non-food agricultural exports (S002). The following countries did not report the value of their agricultural exports for 1990: Bulgaria, Czechoslovakia, Dominican Republic, Mozambique, Nigeria, South Africa, Tanzania, and Zambia. We therefore assume that the value of agricultural exports from these countries equals the value of agricultural imports from these same countries into their trade partners.

the matrix, w_{ij}, denote the share of nation j's total agricultural exports shipped to nation i:

$$w_{ij} = \frac{EXP_{ij}}{\sum_j EXP_{ij}},$$

where EXP_{ij} denotes the value of bilateral agricultural export trade ows between nations i and j, $i \neq j$.

However, a deˇ ciency of this weights matrix is that the aggregated bilateral trade ows may not fully capture the interdependence between countries. For example, two countries located in the same geographical region that produce identical commodities may not export to each other (since they produce identical goods), but to a third country. The weights matrix based on aggregated trade ows does not incorporate this indirect form of interdependence.[5] Countries located in the same geographical area (region) may also have similar environmental regulations due to regional trade agreements that incorporate environmental considerations, e.g., through harmonization.

To account for this deˇ ciency we deˇ ne three different spatial weights matrices based on geographical location. The ˇ rst is a simple contiguity scheme where countries are deˇ ned as neighbors if they share a common border. The resulting general contiguity matrix is deˇ ned as W^{CONT}. The elements of the contiguity matrix are deˇ ned as:

$$w_{ij} = \frac{c_{ij}}{\sum_j c_{ij}},$$

where $c_{ij} = 1$ when country i and j share a common border, and $c_{ij} = 0$ otherwise. There are 50 countries that are connected to some extent in our sample. For some island countries which do not have physical border with their neighbors, we use a speciˇ cation of neighbors based on their geo-economic ties as discussed by Vamvakidis (1998).

The second measure of geographic proximity is based on the shortest great circle distance between each country. The resulting weights matrix is denoted W^{DIST}. The elements of the distance weights matrix are deˇ ned as:

$$w_{ij} = \frac{1/d_{ij}}{1/\sum_j d_{ij}},$$

where d_{ij} is the great circle distance between the geographical centroids of countries i and j. The advantage of the distance matrix is that it enables the weights to capture the geographical proximity of the "island" countries.

The third distance based weights matrix is speciˇ ed as a general contiguity matrix where two countries are deˇ ned as neighbors if the distance between the centroids is less than a predetermined critical value:

$$w_{ij} = 1 \text{ if } DIST_{ij} < DIST_c; \text{ else } w_{ij} = 0,$$

[5] Another problem is that only the aggregated ows are considered, the patterns explored may be different in some commodity groups.

where $DIST_c$ is the critical distance. To capture the possible impact of trade ows at the regional level on environmental regulations we interact this distance based contiguity matrix with the bilateral agricultural exports weights matrix. The resulting matrix is called W_{DIST}^{EXP}, with off-diagonal elements deˇ ned as:

$$w_{ij} = \frac{EXP_{ij}}{\sum_j EXP_{ij}} \text{ if } DIST_{ij} < DIST_c; \text{ else } w_{ij} = 0,$$

where $DIST_c$ is the critical distance.

18.2.2 Econometric Speciˇ cation

The econometric speciˇ cation models a country's stringency of environmental regulations as a function of a weighted average of all other countries' stringency of environmental regulations and the country-speciˇ c explanatory variables:

$$s = \rho W s + X\beta + \varepsilon, \tag{18.1}$$

where s is an n by 1 vector of the values of the stringency of environmental regulations, ρ is a spatial autoregressive parameter, W is a n by n spatial weights matrix, X is a n by k matrix of the exogenous variables, β is a k by 1 vector of regression coef˜ cients, and ε is an n by 1 vector of independent and identically distributed error terms, $\varepsilon \sim N(0, \sigma^2 I)$. The elements of the weights matrix ρW specify the strength of interdependence between each pair of countries, where the stringency of environmental regulations is weighted more heavily the larger the relative trade shares or closer the countries are located to each other.

The simultaneous determination of the stringency of environmental regulations implies that the term Ws is correlated with the error term, ε. Moreover, the multidimensional nature of the dependence implies that Ws in equation (18.1) is correlated with the vector ε, which means that OLS is biased and inconsistent. We can remove the bias by solving (18.1) for the s vector (Anselin, 1988b):

$$s = (I - \rho W)^{-1} X\beta + (I - \rho W)^{-1}\varepsilon. \tag{18.2}$$

Since expression (18.2) is now non-linear in parameters it can be estimated consistently using Maximum Likelihood (ML) techniques. Multiplying X by $(I - \rho W)^{-1}$ implies that the stringency of environmental regulations in a given country depends on the country speciˇ c characteristics (the direct effect) and the characteristics of all other countries it interacts with (the indirect effect).

The spatial dependence may also enter into the regression through the error term. Given $s = X\beta + \varepsilon$ the error structure takes the following form:

$$\varepsilon = \lambda W \varepsilon + \eta, \tag{18.3}$$

where η is the well-behaved error vector, $\eta \sim N(0, \sigma^2 I)$, $W\varepsilon$ is the spatially lagged error term, and λ is the spatial autoregressive coef˜ cient. The consequence of ignoring the spatial error dependence is biased standard errors.

18.3 Data Description and Hypothesis Specification

18.3.1 The Dependent Variable

Our measure of the stringency of environmental regulations is an index based on individual country reports on environmental regulations for the agricultural sector that were compiled for the 1992 United Nations Conference on Environment and Development in Rio (UNCED, 1992). Based on the information gathered, an index (*STRING*) of the stringency of environmental regulations was first developed by Dasgupta *et al.* (1995) for 31 countries. Eliste and Fredriksson (1999) extended the data set to 62 countries using the same methodology as Dasgupta *et al.* (1995). We cannot detect an apparent systematic bias in this index that could drive our results. Below we discuss the independent variables expected to influence *STRING*.

18.3.2 Independent Variables

We define the spatial interaction variables, W_{STRING}, by multiplying each of the two different spatial weights matrices, W (discussed above), with a vector of *STRING*. A negative (positive) sign of the coefficients for the resulting W_{STRING} variables implies that the stringency of environmental regulations decreases (increases) as countries become more integrated via trade, or are located more closely.

The control variables include a trade openness dummy (*OPENdummy*) (Sachs and Warner, 1995), which takes a value of 1 for open and 0 for closed countries. Assuming that environmental quality is a normal good, the demand for environmental quality, and thus the stringency of these regulations, increases with per capita GDP (*GDPpc*). The producers' marginal cost of environmental regulations depends on the size of the agricultural sector. We expect a larger share of total value added from agriculture (*AGDPsh*) to lower *STRING*. Moreover, agricultural pollution has a greater negative impact on welfare if the population density (*POPdensity*) is high, which should result in a greater stringency of environmental regulations. The environmental pressure variables include, first, the share of agricultural land to total land area (*AGLANDsh*), and second, per hectare fertilizer use (*FERTph*). We expect both to have positive impacts on *STRING* (see Just and Antle, 1991; Goklany, 1996). *DEMOCRACYdummy*, which takes a value of 1 when the country is free (democratic) and zero otherwise, controls for institutional factors (Freedom House, 1991). We expect this dummy to have a positive sign.

18.4 Empirical Results

18.4.1 Exploratory Measures of Spatial Interdependence

First, we carry out a test for spatial autocorrelation using the Moran's I statistic (Anselin, 1999). For a row standardized spatial weight matrix, the Moran's I statistic is the ratio of the spatial cross product to the variance:

$$I = \frac{\sum_i \sum_j w_{ij} x_i x_j}{\sum_i x_i^2},$$

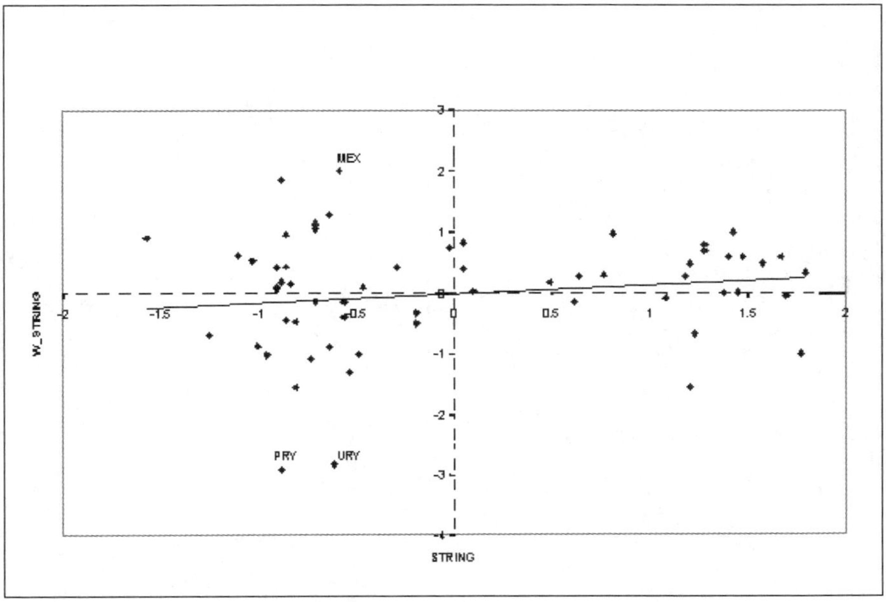

Fig. 18.1a. Stringency of environmental regulations (W^{EXP})

where the x_i's denote the stringency of environmental regulations in country i measured as the deviation from the mean, and w_{ij} are the matching elements of a spatial weight matrix W.[6]

Figures 18.1a–c visualize the structure of spatial autocorrelation between the countries in the sample using the Moran scatterplot suggested by Anselin (1999). We plot the stringency of environmental regulation in country i ($STRING_i$) against its spatially lagged values (W_{STRING_i}) in a standardized form. Deviations larger than two can be considered as outliers. Observations in the upper right hand quadrant of the ˇ gures indicate a positive spatial autocorrelation between high values of $STRING$ and W_{STRING}, and observations in the lower left hand quadrants indicate a positive spatial autocorrelation among low values (i.e., a spatial clustering of countries with similar level of environmental stringency). A negative spatial association is shown in the upper left hand quadrant and the lower right-hand quadrant (i.e. a clustering of dissimilar values).

Figure 18.1a indicates a weak spatial autocorrelation between $STRING$ and the agricultural exports based weights matrix, W^{EXP} (Moran's I statistic equals 0.04 at $p < 0.19$). Out of the sample of 62 countries only 18 fall into the upper right-hand quadrant. This shows a clustering of high stringency countries, i.e., countries with high measures of $STRING$ export to other high $STRING$ countries. At the same time, 19 low $STRING$ countries trade equally often with high $STRING$ coun-

[6] All spatial weights matrixes used in the analysis are row standardized.

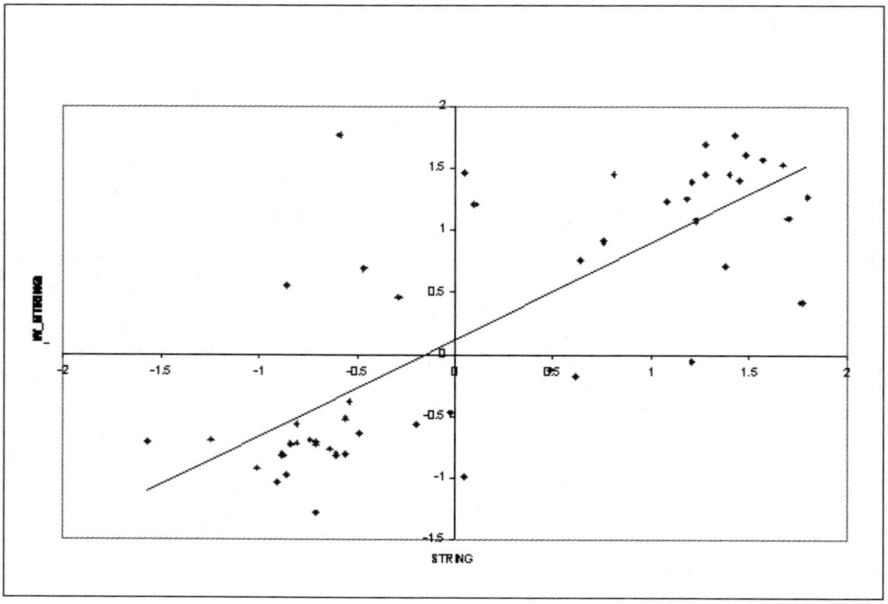

Fig. 18.1b. Stringency of environmental regulations (W^{CONT})

tries (Mexico being one of the most extreme observations), and other low *STRING* countries (Uruguay and Paraguay being outliers). The lower right-hand quadrant of Fig. 18.1a shows high *STRING* countries that trade predominantly with low *STRING* countries. We anticipate that if race-to-the-bottom takes place, it may occur among countries located in the bottom left and bottom right quadrants of Fig 18.1a.

Figures 18.1b and c plot the stringency of environmental regulations against the spatially lagged values using the geographical spatial weights matrixes *WCONT* and *WDIST*. Moran's I statistics are now 0.85 and 0.24 for the contiguity and great circle distance based weighting schemes, respectively, which are signiˇ cant at the 1 percent level. This indicates that geographical location may have an important role in determining environmental regulations. Countries located in the same geographical region tend to have similar high or low values of *STRING*.

Figure 18.2 shows the mean values of the unweighed and agricultural-export-weighed stringency of environmental regulations for OECD (high-income) and non-OECD (low-income) countries. For OECD countries both bars are about the same height, implying that countries with stringent environmental regulations export predominantly to each other. The opposite is the case for non-OECD countries where

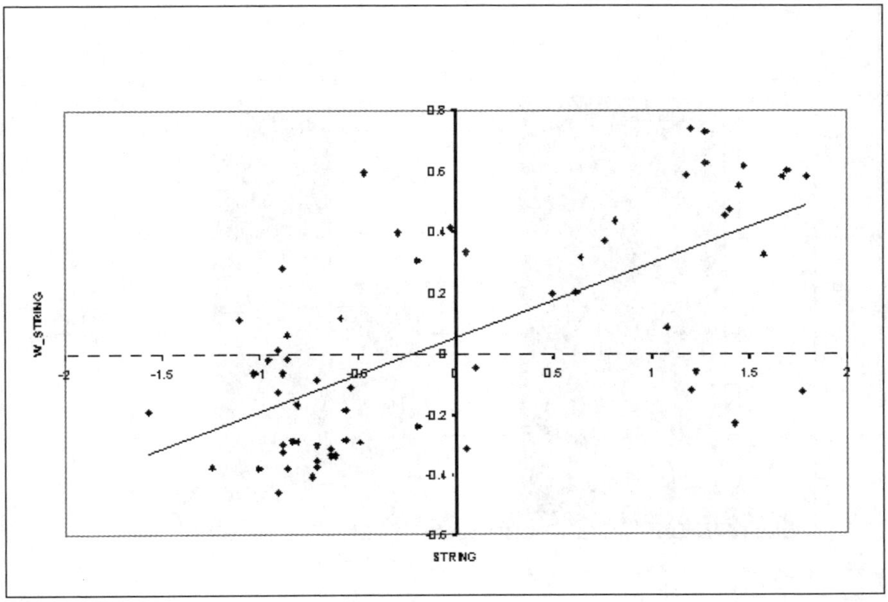

Fig. 18.1c. Stringency of environmental regulations (W^{DIST})

the trade weighted average *STRING* is signiˇ cantly greater than their own stringency of regulations.[7]

It should be noticed that our ˇ ndings are here based on global measures of spatial dependence. Thus, this does not rule out the possibility that various races to the bottom take place regionally. For example, Fig. 18.1a indicates that there are a number of low *STRING* countries that export primarily to other low stringency countries. However, the main focus of the regression analysis below is to investigate the global pattern of spatial dependence.

18.4.2 Regression Analysis

The econometric model estimated here is speciˇ ed as:

$$\begin{aligned} STRING_i = {} & \alpha + \rho WEIGHTstring_i^k + \beta_1 GDPpc_i + \beta_2 GDPsh_i \\ & + \beta_3 POPdensity_i + \beta_4 AGLANDsh_i + \beta_5 FERTph_i \\ & + \beta_6 DEMOCRACYdummy_i + \beta_7 OPENdummy_i + \varepsilon_i, \qquad (18.4) \end{aligned}$$

where k designates the weights matrices, and ε_i is the well-behaved regression residual.

[7] The pattern is consistent with the ˇ ndings of Aten (1997) who ˇ nds that high-income countries trade predominantly with other high-income countries, and low-income countries trade with high-income countries.

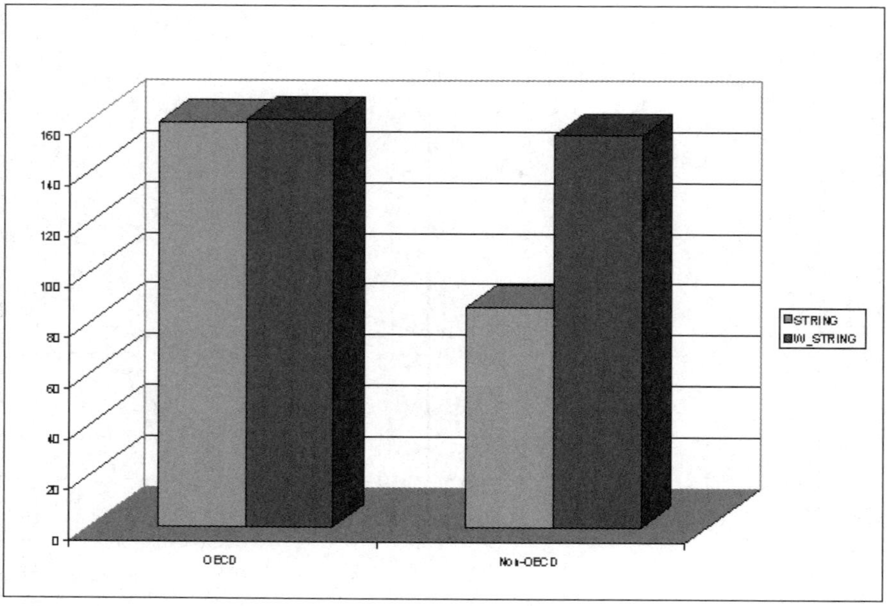

Fig. 18.2. Stringency of environmental regulations (W^{EXP})

Table 18.1 presents the results of the cross-country estimation of the stringency of environmental regulations. The regressions were run using SpaceStat software, using a linear functional form (Anselin, 1992). Regression 1 provides OLS estimates for comparison.

Regressions 2 to 5 in Table 18.1 present Maximum Likelihood estimates for the spatial lag model.[8] The coefˇ cient for the spatial lag term W_{STRING} is positive but insigniˇ cant when using the agricultural export spatial weights matrix, W^{EXP}. The positive coefˇ cient indicates that the likelihood of a given country adopting more stringent environmental regulations is higher if its trade partners have adopted stricter regulations. However, the impact is small. A 1 percent increase in the stringency of a trade partner's environmental regulations increases a country's stringency of environmental regulations by only 0.1 percent.

The positive coefˇ cient for the spatial lag variable based on the geographical weighting scheme (regressions 3 and 4) indicate that countries located in the same geographical region tend to set similar environmental policies. The spatial lag coefˇ cient (W_{STRING}) with the general contiguity based weights matrix (W^{CONT}) is signiˇ cant at the 1 percent level. It may also re ect the fact that neighboring countries tend to have similar agro-climatic conditions and therefore similar production structures. Therefore, they may have similar environmental problems as well as poli-

[8] Alternative model speciˇ cations included semilog, log-log, and linear-log functional forms. The best model ˇ t was achieved with the linear functional form.

Table 18.1. The Impact of Spatially Weighted Stringency of Environmental Regulations on Domestic Environmental Regulations (*STRING*)[1,2]

	Regression					
	1	2	3	4	5	6
W_{STRING}	–	0.057	0.081◇	0.213	0.081◇	–
		(0.374)	(2.458)	(1.148)	(2.458)	
INTERCEPT	76.809*	68.353*	70.832*	55.429◇	70.831*	77.387*
	(8.876)	(2.894)	(8.686)	(2.696)	(8.686)	(9.019)
GDPpc	0.003*	0.003*	0.003*	0.003*	0.003*	0.003*
	(7.362)	(7.870)	(7.562)	(7.285)	(7.562)	(7.230)
AGDPsh	−0.418	−0.422°	−0.312	−0.384°	−0.313	−0.426°
	(−1.652)	(−1.790)	(−1.360)	(−1.626)	(−1.360)	(−1.697)
POPdensity	−0.002	−0.002	−0.001	−0.002	−0.001	−0.002
	(−0.749)	(−0.821)	(−0.464)	(0.889)	(−0.463)	(−0.740)
AGLANDsh	0.223◇	0.221°	0.182°	0.193°	0.182°	0.219◇
	(2.049)	(2.171)	(1.853)	(1.859)	(1.853)	(2.009)
FERTph	0.030°	0.030°	0.026°	0.024	0.028°	0.030°
	(1.775)	(1.877)	(1.681)	(1.405)	(1.681)	(1.761)
DEMOCRACYdummy	0.312	0.111	0.056	0.648	0.054	0.021
	(0.050)	(0.019)	(0.010)	(0.114)	(0.010)	(0.003)
OPENdummy	13.600◇	13.652◇	15.407*	12.986◇	15.408*	–
	(2.065)	(2.214)	(2.605)	(2.135)	(2.605)	
$W_{STRING} * OPENdummy$	–	–	–	–	–	0.088◇
						(2.070)
Spatial weights matrix	–	W^{EXP}	W^{CONT}	W^{DIST}	$W^{EXP}_{DIST1100}$	–
Akaike IC	528.9	530.8	525.2	529.6	525.2	526.6
Spatial BP – test[3]	10.62	10.66	8.95	11.59	8.95	10.29
	(0.16)	(0.15)	(0.26)	(0.11)	(0.26)	(0.11)
LR-test (spatiallag)[3]	–	0.13	5.72	1.28	5.72	–
	(0.72)	(0.02)	(0.26)	(0.02)		
LR-test (spatialerror)[3]	–	1.30	0.06	0.90	0.06	–
	(0.25)	(0.80)	(0.34)	(0.80)		
Number of Observations	62	62	62	62	62	62

[1] Asymptotic z-values in parenthesis

[2] * $p < 0.01$, ◇ $p < 0.05$, ° $p < 0.10$

[3] Probability values in parenthesis

cies. However, the magnitude of the coefficient for the spatial lag variable using the great circle distance (W^{DIST}) is more than twice as great as the coefficient for the contiguity based matrix. A given country's stringency of regulations increases by 0.2 percent when the neighboring countries' stringency increases by 1 percent. However, the coefficient is statistically insignificant.

Next we estimated the spatial lag model using agricultural export flows only among neighboring countries (W^{EXP}_{DIST}). We examined 11 different distance bands ranging from 500 to 1,500 km. The robust Lagrange Multiplier (LM) test was used to search for the appropriate model specification. The best fitting models resulted at critical distance values of 1,000 and 1,100 km. The robust LM tests for 1,000 km and 1,100 km equal 4.73 and 4.99, respectively, which both imply $p < 0.03$. Regression 5 in Table 18.1 presents the results using a critical distance value at 1,100 km ($W^{EXP}_{DIST1,100}$). The coefficient for W_{STRING} is significant at 5 percent level. This suggests that regional trade arrangements may have a strong impact on the determination of environmental regulations. For example, regional trade may lead to harmonization of environmental regulations among close trading partners.

According to the Akaike Information Criterion (AIC) the best fit was obtained with the general contiguity matrix (W^{CONT}) and the matrix using export flows among countries located within the same geographical area ($W^{EXP}_{DIST1,100}$). A spatial Breusch-Pagan (BP) test does not indicate the presence of heteroscedasticity in any of the models. The Likelihood Ratio (LR) test for spatial lag dependence confirms the appropriateness of the spatial lag specification. Moreover, the test for spatial error dependence does not indicate the presence of non-spherical errors, suggesting a good model specification.

The results for the control variables are largely as expected. *GDPpc* is significant at the 1 percent level in all models. *AGDPsh* has the expected negative sign, and is significant at the 10 percent level in models 2 and 4. *POPdensity* is negative but insignificant. Turning to the environmental pressure variables, both *AGLANDsh* and *FERTph* have the expected positive signs and are statistically significant at least at the 10 percent level in most models. *DEMOCRACYdummy* is positive but insignificant, whereas *OPENdummy* is significant at least at the 5 percent level. We interpret the latter result as follows. Countries with more open trade regimes tend to have more stringent environmental regulations. First, more open economies grow faster and thus create a greater surplus that can be used for environmental protection. Second, greater openness may also give a greater exposure to novel ideas such as the benefits of environmental policies. Third, reputational concerns may also play a role, in particular for exporters. Consumer groups in foreign countries may, for instance, demand products with lower pesticide residue. Both exporters and import competitors have an interest in reduced costs, however, and this would tend to put downward pressure on environmental policies. The latter effect does not appear to be important in our data, relatively seen. Finally, greater trade openness induces diffusion of new production technologies, which in addition to higher technical efficiency also may be environmentally more efficient.[9]

Next we turn to the question whether more open countries take into account the environmental regulations of their trade partners in a different way than do relatively closed countries. We interacted *OPENdummy* with the trade weighted (W^{EXP}) spatial lag variable of the stringency of environmental regulations, W_{STRING}. If countries with open trade policies are more receptive to the level of their trade partners' environmental regulations than relatively closed countries, then trade liberalization by high *STRING* countries may have a positive impact on environmental policies and quality. Since W_{STRING} is correlated with the error term we estimated the model using a Two-Stage-Least-Squares Estimator (2SLS).[10]

Regression 6 in Table 18.1 shows that the interaction term $W_{STRING} * OPEN$ is positive and significant at the 5 percent level. Countries with open trade policies take their trade partners environmental policies into account to a greater degree when determining the scope and level of their environmental polices. The environmental regulations set in an open country then partially depend on the regulations set by the open country's trade partners. If these are open this effect should tend to encourage more strict regulations since our earlier regressions show that open countries have greater levels of *STRING*. On the other hand, an open country that opts to reduce the strictness of its regulations may induce a race to the bottom with other countries with liberal trade regimes.

In sum, we find evidence that trade openness has a positive impact on the level of environmental regulations in the agricultural sector. The intuition behind this finding may be that countries that trade with high *STRING* countries may gain a better understanding of the benefits of environmental regulations. This effect is stronger the closer are the trade partners geographically. Moreover, the asymmetric trade interdependence between countries implies that high *STRING* countries may be able to (directly or indirectly) force countries with lax regulations to increase the stringency of their standards. For example, importing nations may impose food safety standards and sanitary rules on agricultural exporters, and thereby influence exporters to adopt stricter regulations. Eco-labeling schemes are also used for, e.g., coffee, bananas, and sugarcane, although this phenomena was small in the year 1990 when our data was collected.

18.5 Conclusions

This chapter tested the hypothesis that trade liberalization induces a race to the bottom in the political determination of environmental regulations in the agricultural sector. Moreover, we explored the hypothesis that neighbors and trade partners influence environmental policies. A novel contribution of the study is the finding that countries do not set their environmental regulations independently. The results of the

[9] Reinhard *et al.* (1997) report a positive relationship between technical efficiency and environmental efficiency in Dutch dairy farms.

[10] The instruments used are all right-hand side variables in model 6.

spatial lag model suggest that countries set more stringent environmental regulations if their close trading partners have relatively strict regulations. There also appears to be a positive relationship between the stringency of environmental regulations and trade openness. We interpret this as the effect of trade liberalization creating greater economic growth and therefore a greater economic surplus available to use for environmental protection. Moreover, reputational effects, increased technology transfers, and a greater exchange of ideas about environmental regulations may play a role.

These findings have policy implications. We cannot find support for the claim that global trade liberalization must halt because environmental policies will suffer. Developing countries that trade relatively heavily with countries with strict regulations have themselves stricter policies. Instead there may be an additional reason for OECD countries to increase the stringency of their environmental regulations, and to trade *more* with developing countries, rather than *less*. Moreover, the next round of trade liberalization talks may set of a chain of events in the area of environmental policy.

More research is needed on these questions, however. The agricultural sector may be a special case because of particularly heavy government intervention distorting prices. Moreover, stricter food safety standards and sanitary measures may be induced by strong consumer demand for less polluting products and production. Moreover, this sector is resource based with immobile capital (land). Thus, lower environmental regulations do not induce firm relocation and capital movements, although the pattern of production and trade should be affected in the long run. The incentive to lower environmental standards may be much greater if this induces an inflow of new and additional capital investments. At the same time, there are NIMBY (Not in My Back Yard) considerations, where environmental policies may be used to discourage local investments in polluting sectors. Spatial econometric techniques would be even more appropriate for the analysis of sectors with mobile capital stocks.

It should also be emphasized that we are not able to infer whether the level and scope of the regulations observed are optimal. We can only explain the variation between countries. Moreover, our results do not imply that environmental quality necessarily must improve with more open trade since scale, composition, and technique effects are also present (see Grossman and Krueger, 1993).

Acknowledgments

We thank Luc Anselin, Dale Colyer, Daniel Esty, David Schorr, David Wheeler, an anonymous referee, and the participants at presentations at West Virginia University and the *Trade and Environment: Preparing for the XXI Century* conference in San Jose, Costa Rica, for helpful comments and discussions. Funding from the Swedish International Development Cooperation Agency (Sida) and the Costa Rica Ministry of Foreign Trade is gratefully acknowledged. The opinions expressed are those of the authors and not those of Sida or the World Bank. The usual disclaimers apply.

19 Regional Economic Growth and Convergence: Insights from a Spatial Econometric Perspective

Bernard Fingleton

University of Cambridge

19.1 Introduction

Economists, economic geographers and regional scientists have suggested different and contrasting explanations of why regions grow at different rates, and what kind of convergence, if any, one might expect from a system of interacting regions. Despite significant differences of approach, there are nevertheless common themes arising from the literature which bring an element of cohesion to a diverse subject matter, namely the relevance for understanding of returns to scale, externalities and catch up mechanisms, and the role of exogenous shocks in real-world turbulence. The chapter first reviews the growth literature, emphasising the importance of these themes, and sets the modelling approach adopted in the chapter in the context of the wider literature. It then gives new expressions for the equilibrium implied by various related models, and an iterative approach is developed to accommodate turbulence leading to "stochastic equilibrium." As an illustration of the potential of the general methodology, the chapter finally focuses on a preferred single equation spatial econometric model (Anselin, 1988b; Anselin and Florax, 1995b). This model leads to substantive empirical evidence regarding causes of productivity growth variations, and the parameter estimates are used to calculate steady-states and stochastic equilibrium for manufacturing productivity ratios for 178 regions of the European Union (EU) (Armstrong, 1995; Cheshire and Carbonaro, 1995).

19.2 Growth Theory: Overview

Neoclassical growth theory, as described by Solow (1956), is a natural starting point for an overview, since most of the theory underpinning regional growth analysis is an adaptation of, or reaction to, the assumptions of basic neoclassical theory. We can appreciate the need for adaptation or change by briefly considering the main tenets of the theory. A fundamental assumption is constant returns to scale (or diminishing returns to capital) and a spatially common technology. Assume that regions are for some reason in disequilibrium, with a misallocation of resources so that capital-labor ratios not at their equilibrium values. As a result of diminishing returns to capital, regions off their equilibrium path with a smaller capital-labor ratio compared with the steady-state value will see faster productivity growth. Regions with a high capital-labor ratio will grow more slowly, so that catch-up occurs until regions move to a common steady-state. Hence in its simplest form, neoclassical growth theory implies the elimination of differences between capital-labor ratios and productivity levels as regions converge to a single equilibrium. At equilibrium, productivity in each region's economy grows at the same rate, which is equal to the exogenously given rate of technical progress but which is unexplained. Needless to say, the simplest form of neoclassical theory has little to offer the regional economic analyst faced with the empirical reality of persistent differences between levels of economic development and varying rates of productivity growth.

Since the empirical evidence for convergence to a single steady-state position is rather mixed, notably for the world economy as a whole and for the regions of the European Union, basic neoclassical theory has been reformulated in an attempt to reconcile theory with empirical reality. Barro and Sala-i-Martin (1992, 1995), and Barro (1991, 1997) develop single equation reduced forms which retains the neoclassical diminishing returns to capital convergence mechanism, but which also introduce dispersed steady-states in place of a single equilibrium point (see also Mankiw *et al.*, 1992; Levine and Revelt, 1992). These so-called Barro-style regressions show that growth is faster the lower the initial GDP per capita level, as predicted by diminishing returns, is faster if Government consumption as a share of GDP is low, and is also faster if the initial level of human capital is higher. They are also characterised by a range of ad-hoc institutional variables such as indices of the rule of law or democracy, plus ancillary variables such as the terms of trade and in ation rate. The additional covariates have the effect that regions converge to different steady-state levels, and therefore allow persistent differences in levels of development observed in the real world, while at the same time enabling estimates of rate of convergence (so called beta convergence) to country- or region-speciˇ c steady-states. Note however that Lee *et al.* (1997) show bias in the estimator of the convergence rate and argue that tests of signiˇ cance using t-statistics are not valid.

Constant returns to scale are a fundamental assumption of neoclassical theory, but there are both theoretical and empirical reasons why increasing returns are preferable for regional analysis, and it has become "almost an article of faith of regional economists that production is characterised by substantial internal and external (agglomeration) economies of scale" (Fingleton and McCombie, 1998). Thus, while diminishing returns to reproducible factors might be inferred from the neoclassical reduced form, the negative association between initial productivity level and subsequent growth can also readily be attributed to other factors. For example regional policies in the EU and technological diffusion may also boost productivity growth in low productivity regions causing regional levels of productivity to converge and possibly offsetting increasing returns which would otherwise lead to divergent behavior. Even in the context of the United States, which is associated with a high level of capital and labor mobility and minimal regulation to hinder the operation of the free market, empirical analysis has supported the hypothesis of increasing returns to scale (McCombie and de Ridder, 1984; Bernat, 1996). Interestingly, even within the more mainstream analysis using (modiˇ ed) Barro style regressions, it is common to ˇ nd the presence of increasing returns to human capital, a finding consistent with endogenous growth theory (Romer, 1986; Lucas, 1988) and theories of technological diffusion (Nelson and Phelps, 1966).

New economic geography (Krugman, 1991a,b; Krugman and Venables, 1995; Ottaviano and Puga, 1998; Puga and Venables, 1997, 1999) also gives a prominent place to increasing returns. From the perspective of economic theory, new tools are now available to accommodate the analytical complexities that are introduced by the presence of imperfect competition resulting from increasing returns. Thus some economic theorists have developed new theory explaining uneven geographical development which is closer to mainstream economics than is the earlier work by economists such as Kaldor. None the less, Kaldor's influence lies behind the new theorising, as Krugman (1991a), states, "we live in an economy closer to Kaldor's vision of a dynamic world driven by cumulative processes than to the standard constant returns model." Despite the progress made, some disadvantages of new economic geography are clearly evident. As presently formulated there does not appear to be much scope for theory testing since the theory is typically abstract and not very amenable to econometric analysis, is at a high level of generality and has little to say about specific places, and is based on some dubious assumptions. Kaldor's vision is still highly pertinent, especially if econometric testing rather than simply deductive analysis is given high priority.

Central to Kaldor's vision is the dynamic Verdoorn Law (Verdoorn, 1949), which came to prominence in regional economic analysis by virtue of its implementation (Kaldor, 1957, 1970) and subsequent formalisation as part of a structural model of cumulative causation explaining why regional productivity growth rates differ (Dixon and Thirlwall, 1975a,b). The basic single equation specification of the dynamic Verdoorn Law holds that there is a linear relationship between the exponential growth rates of labor productivity (p) and output (q) and at its most simple, the Law is tested as a single equation with estimation via OLS. This in itself raises a number of issues, such as the omission of the effect of capital stock growth (k) on productivity growth, and there is the question of whether we need to account for endogeneity either in model structure or in estimation. In addition, the appropriateness of the loglinear production function as the underlying static model has been questioned. We consider these issues in more detail below. Despite its vintage, it still possesses contemporary significance because it embraces endogenous technical progress and is thus clearly a forerunner to new growth theory. Moreover, by also embracing increasing returns to scale, it has attributes that, as we have seen, are very much in vogue as part of the "new economic geography." Most importantly from the empirical regional analysis perspective, it is easily enhanced to incorporate spatial effects or externalities leading directly to spatial econometric models. The model used in the empirical analysis here also incorporates a catch-up term, which at face value is not unlike the way neoclassical convergence is modelled, although the implied mechanism is non-neoclassical. Nevertheless the effect is similar, the catch up element in the model cancels the tendency to diverge and leads inexorably to equilibrium.

Equilibrium implies stability and an absence of change. In a neoclassical world equilibrium comes about because constant returns to scale and a spatially uniform technology have the consequence that per capita income or productivity growth tends to a steady-state value equal to the rate of exogenous technical progress (λ). At steady-state, productivity growth is equal to λ across regions and steady-state levels of productivity (which in realistic neoclassical models will be regionally differentiated) evolve without any change in regional order. Differences in steady-state output growth are determined by each region's employment growth, since output growth depends on employment growth and λ. In the non-neoclassical tradition of Kaldor, cumulative causation models such as in Dixon and Thirlwall (1975a) can also result in a form of equilibrium, depending on the parameterisation. The most realistic outcome (McCombie and Thirlwall, 1994) given reasonable parameter values, is that each region converges on a different constant output growth rate rather than on a divergent growth path.

Apart from the well-known forces leading to equilibrium, there are additional forces due to spatial or regional interaction, although these are conspicuous by their absence from most of the literature. None the less, ideally models of dynamic processes should acknowledge that changes in one region spill over to other regions, so that dynamics and steady-states reflect interregional interdependencies. Additionally, it seems realistic to take account of the role of stochastic, random events, meaning events exogenous to the equation system, such as abrupt changes in policy, government or the social or natural environment that are unpredictable and take on the appearance of random shocks to productivity growth. In a sense these are the antithesis to equilibrium, since a stream of random shocks means continuous change not steady-state. However, it seems important that random events are somehow built into our notion of equilibrium. Theory that ignores the unknown and unknowable and attempts the all-seeing eye of God is irrational.

Quah (1993) also made the point that the notion of a steady-state seems out of tune with the reality of empirical dynamics, criticising the Barro-style methodology because of its assumption of a steady-state growth path which is well approximated by a stable time trend, and pointing out that data do not conform to this implicit assumption and that growth trends in actual economies do not appear to be particularly stable and smooth. He therefore advocated modelling dynamics using Markov chain models, an analytical tool also used by Magrini (1995) and by Fingleton (1997). Markov chains evolve to stochastic equilibrium, which is the stable vector of probabilities of regions attaining different levels of wealth or productivity. This provides a more realistic conception of equilibrium, since it is the probabilities that are fixed, not the regions or countries, which float through the state-space at equilibrium under stochastic influence.

There are however limitations to the Markov chain approach. First, it is not region-specific, but is only concerned with a system of regions in stochastic equilibrium. Secondly, it is somewhat vague in terms of theoretical provenance. Thirdly, it ignores the role of spatial interaction or externalities in regional dynamics. While the notion of stochastic equilibrium provided by the Markov chain literature has partly inspired the approach adopted here, because of these limitations Markov chains *per se* do not provide the analytical cutting edge needed for a deeper understanding of why regional differences occur and persist. Instead, the analysis that follows is based on the dynamic Verdoorn Law with spatial externalities, catch up, and other effects including stochastic turbulence. This is considered the most realistic way to attempt to capture the main empirical "facts" of regional growth dynamics and equilibrium using a model that has an acceptable theoretical basis.

19.3 The Single Equation Approach to the Verdoorn Law

Let us commence with the dynamic Verdoorn Law, as defined in equation (19.1), in order explain productivity growth variations between regions. The coefficient m_0 is the autonomous rate of productivity growth and m_1 is the Verdoorn coefficient, for which a value of about 0.5 is usually found when this specification is fitted to data on manufacturing productivity growth and output growth. This implies that a one percentage point increase in output growth induces an increase in the growth of employment of about one-half of one percentage point and an equivalent increase in the growth of productivity. The assumption is that a proportion of technical change is induced by the growth of output (see Thirlwall, 1983), so that technical change is not an exogenous factor but the outcome of the process of output growth itself. Consequently m_1 reflects the rate of disembodied technical progress (learning by doing) induced by output growth, the effect of the growth of output on capital accumulation and the extent to which this embodies technical progress (Dixon and Thirlwall, 1975a,b). The error term μ reflects the other effects on p which in this initial specification are assumed to behave as random shocks:

$$p = m_0 + m_1 q + \mu. \tag{19.1}$$

We have already suggested that ideally the Verdoorn Law should be augmented, by adding the growth of capital (k) to equation (19.1). In order to see this, consider the conventional Cobb-Douglas production function:

$$Q = A_0 \exp(\lambda t) K^{\alpha} E^{\beta},$$

in which λ is the growth of total factor productivity, Q, K and E are output, capital and employment levels respectively, and α and β are elasticities. On taking (natural) logs, differentiating with respect to time, and rearranging, we obtain a productivity growth (p) as a linear function of output growth (q) and of capital stock growth (k), hence:

$$p = \lambda/\beta + ((\beta - 1)/\beta)q + (\alpha/\beta)k + \mu.$$

Unfortunately, data on capital stock growth *per se* are commonly unavailable at the level of regions or even countries. It therefore has become standard practice to proxy k by the average share of real (gross) equipment investment in GDP, although McCombie and Thirlwall (1994) argue that only by using the net investment-output ratio and assuming that the capital-output ratio is constant across countries, does a credible approximation to the growth of capital occur. However, even the net investment-output ratio may be unavailable. Therefore, the Verdoorn coef˜ cient could possibly be biased as a result of omitting k, unless q and k are orthogonal. One way to resolve the problem (see McCombie and Thirlwall, 1994) is to place a further restriction on the model. Assuming that capital stock growth is equal to output growth (i.e., the capital – output ratio is constant), causes k to drop out of the equation and we are left with a speci˜ cation:

$$p = \lambda/\beta + ((\alpha+\beta-1)/\beta)q + \mu,$$

and if $m_1 = ((\alpha+\beta-1)) > 0$, then $(\alpha+\beta) > 1$ and we have increasing returns. The empirical basis for omitting k is the "stylized fact" that capital stock growth and output growth are approximately the same in most developed economies, an assumption that gains plausibility by conforming closely to the results of empirical tests. For example McCombie and Thirlwall (1994) found that regressing k on q for a sample of developed countries produced a regression slope coef˜ cient that was not signi˜ cantly different from unity.

Note that the interpretation of m_1 should recognize that if the growth of total factor productivity depends on output growth, so that:

$$\lambda = \lambda' + \varphi q + \mu.$$

It then follows that the estimated m_1 combines static returns to scale (equivalent to the Cobb-Douglas coef˜ cients summing to more than 1) and dynamic returns to scale (the change φ in total factor productivity growth per percentage point of output growth), since:

$$p = \lambda'/\beta + ((\varphi+\alpha+\beta-1)/\beta)q + \mu.$$

In fact there is some debate concerning the underlying static model, (McCombie, 1982; Harris and Lau, 1998), and it is unclear whether the loglinear Cobb-Douglas production function is appropriate. Consequently we have further augmented the Verdoorn Law by including variables which are not derived explicitly from this production function. This facilitates the introduction of other variables (let us call them $Z_1, ..., Z_v$) which are also likely to have an effect on productivity growth rates, notably regional policy instruments. The introduction of other regressors provides the opportunity to accommodate a persistent empirical regularity, regional convergence. We have already noted that in a neoclassical world characterized by diminishing returns to reproducible capital and spatially uniform technology, the regional convergence mechanism is due to initially less well-endowed regions having a higher marginal product of capital, so that regressing growth on initial level produces a negative coef˜ cient. In a non-neoclassical world of increasing returns and non-uniform technology, the diffusion of innovations from high to low technology regions is possible, so that technologically de˜ cient regions grow faster because productivity growth is boosted by access to better technology. Fingleton and McCombie (1998) include in their model a variable, essentially the (log) level of productivity in the base year, as a measure of initial level of technology, in order to capture the effect of diffusion of innovations from relatively technologically advanced to more backward regions of the EU.

While it is feasible to represent technology level by productivity level, the productivity level also provides a convenient indirect measure of the strength of regional policy, which is useful since exact data on the diverse forms of regional spending are unavailable. This is based on the assumption that the lower the level of output per worker, the higher the net spending by the EU and national governments on regional assistance. Hence the actual levels of productivity are taken to re ect the combined effect of technology diffusion and regional policy. Rather than base year productivity *per se*, we work with a measure of the productivity gap deˇ ned as $G_0 = 1 - (P_0/P_0^*)$, where P_0 is the region's level of (manufacturing) productivity per (manufacturing) worker in the base year and P_0^* is the highest level of productivity per worker at time 0, analogous to the measure used by Amable (1993), and Fingleton (1998). Combining this and a token ancillary variable Z_1, we develop a nomenclature and refer to (19.2) as the augmented non-spatial effects Verdoorn Law:

$$p = m_0 + m_1 q + m_2 Z_1 + m_3 G_0 + \mu, \tag{19.2}$$

and assume that $\mu \sim \mathrm{N}(0, \sigma^2)$. The term non-spatial effects indicates that model (19.2) fails to take into account the impact of cross-regional externalities, which was one of the themes engendered by the literature mentioned in the introduction. Invoking spatial effects leads inexorably to the two classic alternatives, the spatial lag model and the spatial error model, and therefore the augmented spatial lag Verdoorn Law (19.3) and the augmented spatial error Verdoorn Law (19.4), referred to collectively as augmented spatial effects Verdoorn Laws. In (19.3), Wp denotes spatially lagged productivity growth, equal to the matrix product of W, a standardised matrix deˇ ning inter-regional interaction and the vector p. Hence cell i of vector Wp is the weighted average of productivity growth in regions "surrounding" region i as deˇ ned by W:

$$p = m_0 + m_1 q + m_2 Z_1 + m_3 G_0 + m_4 W p + \mu, \tag{19.3}$$

and $\mu \sim \mathrm{N}(0, \sigma^2)$, an explicit assumption required for consistent estimation via ML. The assumption that, *ceteris paribus*, productivity growth in a region will be boosted by faster productivity growth in surrounding regions seems reasonable if we consider externalities involving technical progress. Following the earlier assumption that capital accumulation embodies technical progress, an assumption that also features in recent endogenous growth theory literature (e.g., Lucas, 1988; Barro and Sala-i-Martin, 1995), let us represent capital accumulation by capital per worker growth (in other words productivity growth given the $k = q$ assumption). Assume that technical change occurring within a particular region is not fully internalized and so spills over to other ˇ rms and individuals. Since at the boundaries between EU (NUTS2) regions are weak barriers and physically separated regions are often well connected, such externalities will not be contained within regions of origin. The result is that ˇ rms and individuals capture externalities generated by productivity growth (in other words capital accumulation) which occurs perhaps in neighboring regions, or in important regions elsewhere. Fingleton and McCombie (1998), Fingleton (1999a), and Puga and Venables (1997) also discuss externalities and regional integration from different perspectives.

The augmented spatial error Verdoorn Law has autoregressive interaction involving the error term, hence:

$$p = m_0 + m_1 q + m_2 Z_1 + m_3 G_0 + \mu, \tag{19.4}$$

and, reverting for convenience to matrix notation,

$$\mu = \rho W \mu + \xi,$$

where $\mu \sim \mathrm{N}(0, \sigma^2 V)$ in which V is the variance-covariance matrix for the autocorrelated errors and ξ is assumed to be distributed as $\xi \sim \mathrm{N}(0, \sigma^2 I)$. This equates to a restricted version of an augmented spatial lag model with both endogenous and exogenous lags, which we call the full unrestricted spatial effects Verdoorn Law:

$$\begin{aligned} p = {} & m_0 + m_1 q + m_2 Z_1 + m_3 G_0 + \rho W p + m_4 W q \\ & + m_5 W Z_1 + m_6 W G_0 + m_7 W \iota + \xi, \end{aligned}$$

in which ι denotes a unit vector. More generally, this is known as the spatial Durbin model.[1]

To show this, we rearrange to obtain:

$$\begin{aligned} (I - \rho W)\, p = {} & m_0 \left(1 + (m_7/m_0) W\right) \iota + m_1 \left(1 + (m_4/m_1) W\right) q \\ & + m_2 \left(1 + (m_5/m_2) W\right) Z_1 + m_3 \left(1 + (m_6/m_3) W\right) G_0 + \xi, \end{aligned}$$

and as a result of introducing the restrictions $\rho = -m_7/m_0 = -m_4/m_1 = -m_5/m_2 = -m_6/m_3$, this reduces to,

$$p = m_0 + m_1 q + m_2 Z_1 + m_3 G_0 + (I - \rho W)^{-1} \xi.$$

The restrictions are usually referred to as the "common factor hypothesis" (Burridge, 1981).

An economic interpretation of the augmented spatial error Verdoorn Law is that regional spillover occurs as a technology externality as hypothesized under (19.3) (Wp) and, because weak regional boundaries merge intra- and extra-regional effects, spillover also involves the direct effects on p. The remote effects of q, Z_1 and W, are represented by Wq, WZ_1 and WG_0 in the extended formulation above. Finally, assume that $\rho = 0$ in the full unrestricted model so that there are only exogenous lags, hence:

$$\begin{aligned} p = {} & m_0 + m_1 q + m_2 Z_1 + m_3 G_0 + m_4 W Q \\ & + m_5 W Z_1 + m_6 W G_0 + m_7 W \iota + \xi, \end{aligned}$$

[1] When we estimate this model (see Table 19.5), the constant is $c = m_0 + m_7$.

which we refer to as the reduced unrestricted spatial effects Verdoorn Law. This model assumes that although there are direct cross-region effects from variables q, Z_1 and W, there is no spatial externality involving technology spillover from capital accumulation in other regions.

For purposes of estimation, we assume that Z, q and G_0 are exogenous, although there exists the possibility that q and Z are also endogenous variables (since G_0 uses base year data, it cannot respond to productivity growth and is treated as exogenous). In the next section details are given of such a specification which involves endogeneity, so that for example q depends on p and *vice versa*.

19.4 A Simultaneous Equation Approach: Problems and Issues

In this section we consider one way of addressing the issue of endogeneity by looking at an example of a simultaneous equation model in which the Verdoorn Law is embedded. The discussion highlights various problems and issues raised by this mode of analysis, which provides an alternative to the single equation approach considered hitherto. Endogeneity is a consequence of the fact that while output growth is hypothesized to determine productivity growth (the Verdoorn Law), the reverse is also a possibility so that fast productivity growth feeds back to stimulate output growth. The reason for this feedback effect is the benefit to competitiveness that ensues from fast productivity growth (or the lack of competitiveness that comes with slow productivity growth). In other words, we assume that the outcome of faster output growth is not simply higher wages or profits, but that costs grow more slowly and this enhances competitiveness (possibly including non-price competitiveness, which involves factors such as quality and reliability). Enhanced competitiveness stimulates demand and hence output growth, which further benefits productivity, competitiveness, demand and output in a virtuous circle, which may also involve faster capital accumulation as a result of faster output growth. Such considerations lead us to the various structural models of cumulative causation embodying the Verdoorn Law (Kaldor, 1970; Myrdal, 1958; Dixon and Thirlwall, 1975a,b; McCombie and Thirlwall, 1994; Targetti and Foti, 1997; Fingleton, 1998).

While structural models take account of simultaneity, they demand theoretical justification that may be hard to find. In the example outlinedbelow, some rationale is provided for the set of equations (19.5a–19.5e) that together comprise the structural model discussed in this Chapter, together with supporting references. There is some scope for avoiding some of these issues in the time series context, as illustrated by Harris and Lau (1998) who avoid the simultaneity problem associated with the single equation Verdoorn Law by estimating long-run cointegration vectors for UK regional time series. This has the advantage of not having to impose strong a priori structural relations and exogeneity assumptions. However as yet there is no comparable error-correction methodology for spatial data (see Fingleton, 1999c) and the Harris and Lau time series method takes no account of spatial effects nor provides much clear insight as to spatial mechanisms.

We set out the simultaneous equation approach in more detail by focussing on a speciˇ c model (reasonably typical of the genre) relating to international variations in total productivity growth (Fingleton, 1998) that includes the Verdoorn Law and which attempts to model spatial effects using a spatially lagged variable. Equations (19.5a–19.5e) summarize the model:

$$p = f(G,\ k, q,)\ ;\ p = a_0 + a_1 G + a_2 k + a_3 q + \mu_1, \tag{19.5a}$$

$$k = f(D)\ ;\ k = b_0 + b_1 D + \mu_2, \tag{19.5b}$$

$$D = f(L,\ WG)\ ;\ D = c_0 + c_1 L + c_2 WG + \mu_3, \tag{19.5c}$$

$$L = f(S,\ G)\ ;\ L = d_0 + d_1 S + d_2 G + \mu_4, \tag{19.5d}$$

$$q = f(p,\ WG)\ ;\ q = e_0 + e_1 p + e_2 WG + \mu_5. \tag{19.5e}$$

In the system, the Verdoorn Law is represented in equation (19.5a) in which productivity growth (p) depends on output growth (q), and q's endogeneity is evident from (19.5e). Equations (19.5b,19.5c,19.5d) show that the model also has endogenously determined capital growth (k), research and development activity (D), measured by the Verspagen (1991) patent registration variable, namely `the sum of the per capita number of patent grants for inhabitants from the country in the US over the period 1962–85', and labor force quality (L), which is proxied by a measure of the extent of secondary schooling. The exogenous variables are primary schooling (S), which is assumed to be culturally determined, the initial productivity gap (G) vis-a-vis the country with the highest productivity level, and spatially lagged productivity gap (WG). In sum, the data come from the Penn World tables (Summers and Heston, 1991) and paper-speciˇ c series as in DeLong and Summers (1991), and Verspagen (1991). While the detailed rationale, data sources and estimation for this speciˇ cation, given in Fingleton (1998), is beyond the scope of this Chapter, an appreciation of the reasoning leading to the structural equations is provided by the following synopsis.

Thus, in addition to the embedded Verdoorn Law linking productivity growth (p) to output growth (q), equation (19.5a) also contains the GDP per worker gap (G) to capture the effect of "aspatial" technology diffusion. We use the deˇ nition $G = (P^* - P)/P^*$ in which P is a country's 1960 productivity level and P^* is the 1960 level for the USA. The assumption is that technological diffusion to countries with initially lower technology levels causes faster growth and hence "catch-up" by "poorer" countries as a result of higher potential beneˇ ts from imitating technology. We assume that the diffusion mechanism is via (global) trade in products and services embodying foreign knowledge and otherwise costly to acquire information (see Grossman and Helpman, 1991a, 1994, who initially introduced trade into endogenous growth models). Thus we assume that "global" trade (meaning primarily non-localized "North-South" trade) gives access to a "technological world standard" but that adoption rates and impacts vary with initial level of technology country-by-country. Different, "spatial" technology spillovers involving more or less neighboring countries, which are attributed to regional trade, are explained below. Unlike the "aspatial" technology diffusion, which depends on conditions internal to the adopting country, spatial diffusion depends on "who your neighbors are." Equation (19.5a) also includes the growth of capital (k) as an explicit determinant of productivity growth.

Equations (19.5b to 19.5d) are based on Amable's (1993) suggestion that capital growth depends on research and development activity (D), and that this is determined by labor force quality (L), a measure of the ability of a country to develop better technologies. Labor force quality (secondary schooling) depends on primary schooling (S) and the initial level of productivity (G). The reason for a G effect is the assumption that the technology level controls educational standards by determining the amount of expenditure and effort individuals devote to education. If there is no market for the skills developed by secondary education, and if the supply of qualified teachers and educational infrastructure is poor, as one might anticipate in a low technology country, then education standards are likely to be lower. Alternatively, high technology countries will demand skills developed by education and possess the requisite infrastructure and personnel.

Also included in (19.5c, 19.5e) is the spatially lagged productivity gap (WG), a term which is included in order to capture spatial spillover. Spillover affects research and development activity (19.5c) by means of the "spatial" innovation diffusion mentioned above. We assume that the level of productivity (qua technology) in surrounding countries (WG) influences research and development activity (19.5c) because, *ceteris paribus*, neighbors will tend to trade more as a result of lower transport costs and because of the existence of regional trading blocks. We assume that firms in countries with low technology regional trading partners will have different private costs and benefits of investing in new innovations compared with firms trading with high technology neighbors. If regional trade involves low technology neighbors, imports will contain fewer innovations and less of a threat to domestic companies facing competition from importers. If it involves high technology neighbors, the threat and opportunity to copy will be high and domestic innovation rates and adoption will tend to be enhanced. Hence we assume that innovativeness is a form of defence mechanism against imports embodying innovations (Baldwin, 1992; Wood, 1998). Additionally, technology-rich neighbors may induce firms to be more innovate in order to penetrate cross-border markets, whereas firms with technology-poor neighbors may be able to capture cross-border markets with lower quality industrial research.

Equation (19.5e), specifying the dependence of q on p and WG, is in fact a reduced form. Output growth is assumed to depend on export growth, which in turn depends on the growth of export prices represented by productivity growth, the growth of prices in neighboring regions (ie competitor prices) and the growth of demand in neighboring regions. Writing neighboring price and demand growth in terms of the (spatially lagged) exogenous variables and simplifying, gives the specification in equation (19.5e).

While theory and the literature may suggest particular simultaneous equations, there are other influences on the specification of the final model. One is that the model should be capable of being estimated. There should be sufficient exogenous variables in the system to ensure that equations are either just or over-identified and estimation does not fail because either the rank or order condition is not satisfied. An equation is just identified if there are exactly enough exogenous variables in the system which are excluded from the equation to act as instrumental variables for the endogenous variables in the equation. Over-identification is when there are more than enough.

Often it is a matter of compromise reaching a final specification, for the constraints of theory, which determine a particular specification for the system *a priori* and anticipates particular parameter estimates (for example a Verdoorn coefficient of about 0.5), may conflict with the requirements for identification. What may be a respectable model from the theoretical perspective may be nonsense from the point of view of estimation, and vice versa. In addition, there are problems posed by choice of estimation method. The results of 2SLS, 3SLS, Limited Information Instrumental Variables and Full Information Maximum Likelihood (FIML) each give different Verdoorn coefficient (a_3) estimates for the model (19.5) fitted to a sample of 60 countries, with values of 0.70078, 0.67425, 0.84118 and 0.56458 respectively! Fortunately, all are reasonably acceptable estimates, but the FIML estimate corresponds most closely to the value of about 0.5 typical of previous work. FIML is at least as efficient asymptotically as any other estimator, but as with the other systems method (3SLS) it is less robust than single equation methods (2SLS) against system misspecification since structural parameter estimates depend on the whole system. From the spatial analysis perspective, IV (2SLS) residual spatial autocorrelation can now be tested using the method of Anselin and Kelejian (1997).

The equilibrium implied by the system of equations (19.5) depends solely on the exogenous variables S, G and WG. In fact we can avoid estimating the system altogether if interest is focussed solely on the equilibrium, and one very simple way to estimate the reduced form is to estimate the OLS regression involving p and the exogenous variables as regressors. This method provides unbiased estimates, although if, as is the case, the structural equations are over-identified, then using the structural parameter estimates to obtain derived reduced form parameters is more efficient. The derived reduced form estimates are however based on an assumption that the over-identifying restrictions are actually true.

In this example, the reduced form for p is:

$$p = r_0 + r_1 S + r_2 WG + r_3 G + \mu_6, \tag{19.6}$$

where,

$$\begin{aligned} r_0 &= (a_3 e_0 + a_2 b_1 c_1 d_0) / (1 - a_3 e_1), \\ r_1 &= (a_2 b_1 c_1 d_1) / (1 - a_3 e_1), \\ r_2 &= (a_2 b_1 c_1 + a_3 e_3)(1 - a_3 e_1), \\ r_3 &= (a_1 + a_2 b_1 c_1 d_2)(1 - a_3 e_1). \end{aligned}$$

Not only are these derived estimates asymptotically more efficient, they are also consistent and have desirable small sample properties. They are however biased

(Kennedy, 1996). We consider expressions giving the steady-state for the reduced form, together with comparable expressions for the earlier single equation models, in the following section.

19.5 Convergence Theory and Methodology

The chapter up to this point has described various single equation models, and a typical structural model, each embodying increasing returns to scale as represented by the Verdoorn Law, innovation diffusion and consequently catch up, and spatial externalities. This section is dedicated to what the models imply for regional equilibrium, and contains some newly derived expressions deˇ ning the steady-states implied by various spatial econometric models. The focus is on the simplest of the models, the augmented non-spatial effects Verdoorn Law, the augmented spatial lag and spatial errors Verdoorn Laws, and the reduced form from the structural equation. Although the previous sections have covered the different models separately, when it comes to considering equilibrium, there are parallels that mean that we can treat them together. Hence, steady-states from single equation augmented (spatial and non-spatial effects) Verdoorn Laws, and reduced forms from the simultaneous equation model, are linked since each steady-state derives from the connection between productivity growth (p) and the level of productivity (P). The expressions in this section, and the equivalent iterative solutions, relate to these different models, the intention being to present, for completeness, some results of wider generality than the actual empirical analysis described later in the chapter, which eventually focuses on one preferred speciˇ cation for the EU regional data set.

19.5.1 Deterministic Equilibrium

Augmented Non-Spatial Effects Verdoorn Laws and Reduced Form This subsection focuses on the deterministic equilibrium for the reduced form of the simultaneous equation system, and for the augmented non-spatial effects Verdoorn Law, which are entirely equivalent in this context. The variables that they have in common are p and $G = 1-(P/P^*)$, so these are denoted as such below. We label the other variables X_1, X_2 so that we can use a common notation for both models: for the augmented non-spatial effects Verdoorn Law $X_1 = q$ and $X_2 = Z_1$, and for the reduced form $X_1 = S$ and $X_2 = WG$.

Ignoring shocks, we deˇ ne convergence as the process by which each region moves from a disequilibrium position to an equilibrium or steady-state position. At equilibrium the productivity growth rate is the same across regions, so that even though productivity levels may increase, the productivity level gap between regions remains stable. With the coefˇ cient $m_3 > 0$, as can be reasonably assumed, the model summarised as equation (19.2) is consistent with such an equilibrium (in theory there are other possible dynamics for $m_3 < 0$, as illustrated by Amable, 1993). The steady-state gap can easily be obtained analytically, since:

$$\mathrm{E}(p) = m_0\mathrm{U} + m_1X_1 + m_2X_2 + m_3G, \tag{19.7}$$

or,

$$p = m_0\mathrm{U} + m_1X_1 + m_2X_2 + m_3G + \mu,$$

and,

$$\mathrm{E}(p^*) = m_0\mathrm{U} + m_1X_1^* + m_2X_2^*, \tag{19.8}$$

where p is the vector of productivity growth by country or region, U is the unit vector, p^* denotes the productivity growth of the region/country with the highest level of productivity in the base period, X_1^* is the productivity leader's value of the ˇ rst exogenous variable and X_2^* the leader's value for the second exogenous variable (in the case of $X_2^* = WG^*$, this equals the matrix product of W^* and G^*, where W^* denotes a matrix with each row identical to the productivity leader's row of W). Hence:

$$\mathrm{E}(p-p^*) = m_1(X_1 - X_1^*) + m_2(X_2 - X_2^*) + m_3G,$$
$$\mathrm{E}(p-p^*) = (m_3 + m_1(X_1 - X_1^*) + m_2(X_2 - X_2^*)) - m_3(\mathrm{U} - G),$$
$$\mathrm{E}(p-p^*) = c - m_3(\mathrm{U} - G).$$

Also $G = \mathrm{U} - (P/P^*) = \mathrm{U} - R$, and therefore the proportional rate of growth of R is:

$$R^\bullet/R = \mathrm{E}(p-p^*) = c - m_3R, \tag{19.9}$$

so that the rate of change of R over time is,

$$R^\bullet = Rc - m_3\mathrm{R}^2,$$

and equilibrium occurs when the rate of change equals zero, hence,

$$R^\bullet = R^e c - m_3 R^{e\,2} = 0,$$

so that,

$$R^e = c/m_3, \tag{19.10}$$

with the other root equal to zero.

This is illustrated in Figs. 19.1 and 19.2 that show typical dynamics of three regions. In Fig. 19.1, two regions have equilibria close to 0.5, but the third has a steady-state near to the level of productivity of the assumed productivity leader. Figure 19.2 shows the precisely the same steady-states obtained by iteration (we label this iteration A). Commencing with the $t = 0$ level of productivity gaps, we see converge to the approximate steady-states. In order to obtain the results for Fig. 19.2, the substitution of $R_t = R_0$ into $R_tc - m_3R_t^2$ gives $R_t^\bullet$, and $R_{t+1} = R_t + R_t^\bullet$ is then used to calculate $R_{t+1}^\bullet$ and hence R_{t+2}. The iterations stop after k iterations when $R_{t+k}^\bullet \approx 0$.

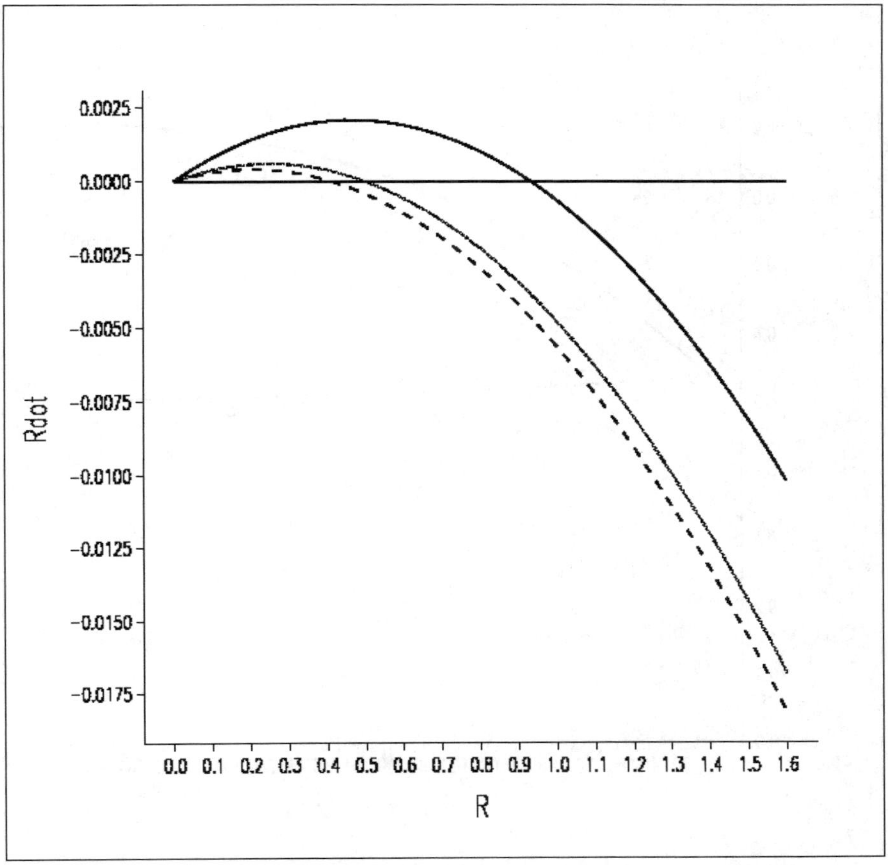

Fig. 19.1. Dynamics for 3 regions

Full Equilibrium Analysis for the Reduced Form In the case of equation (19.6), the reduced form, the method outlined above is a partial equilibrium analysis, because $X_2 = WG$ is kept ˇ xed so that we are assuming that the adjustment process in any one country occurs without corresponding changes in neighboring areas. In order to model the process occurring simultaneously across countries, what is required is a full equilibrium analysis that acknowledges the fact that productivity levels in the surrounding countries are also subject to change. In this case WG is no longer part of c, so that:

$$c = m_3\mathrm{U} + m_1(X_1 - X_1^*),$$
$$R^{\bullet}/R = \mathrm{E}(p - p^*) = c - m_3R + m_2(WG - WG^*),$$
$$R^{\bullet}/R = c - m_3R + m_2W(\mathrm{U} - R) - m_2W^*(\mathrm{U} - R).$$

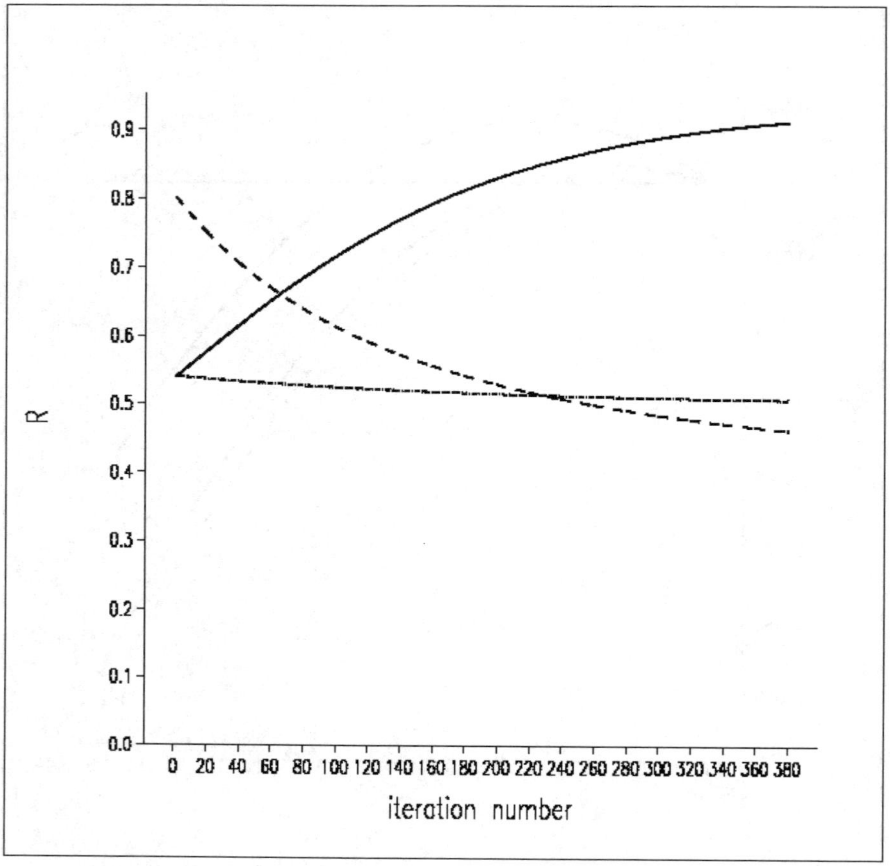

Fig. 19.2. Iterative solution for 3 regions

And if $W^{\#} = (W - W^{*})$, then:

$$R^{\bullet}/R = c - m_3R - m_2W^{\#}R + m_2W^{\#}\mathrm{U}. \tag{19.11}$$

Equilibrium occurs when $R^{\bullet} = 0$, in other words:

$$R^e = \left(m_3I + m_2W^{\#}\right)^{-1}\left(c + m_2W^{\#}\mathrm{U}\right), \tag{19.12}$$

in which I is an identity matrix. Exactly the same result is obtained by iteration. Commencing with $R_t = \mathrm{U} - G$ $(t = 0)$, we calculate:

$$R_t^{\bullet}/R_t = c - m_3R_t - m_2W^{\#}R_t + m_2W^{\#}\mathrm{U}, \tag{19.13a}$$

and hence $R_t^{\bullet}$ and subsequently,

$$R_{t+1} = R_t + r_t^{\bullet}, \tag{19.13b}$$

which replaces R_t in (19.13a) and this gives $R^{\bullet}_{t+1}$ and substituting into (19.13b) we obtain R_{t+2}, so on until $R^{\bullet}_{t+k} \approx 0$. The iterative method has the advantage that it provides the option of switching leadership, a feature that is particularly useful when the stochastic element is introduced as explained below.

Augmented Spatial Lag and Spatial Errors Verdoorn Laws The equilibrium for the augmented spatial error Verdoorn Law (19.4) is given by equation (19.10), since in equation (19.4), $\mathrm{E}(\mu) = 0$, and this gives expectations as in equations (19.7) and (19.8). However, the equilibrium for the augmented spatial lag Verdoorn Law (19.3) requires further analysis. To obtain the spatial lag steady-state, we ˇ rst create a matrix A, the columns of which are the values in the columns of matrix X minus the value for the productivity leader (there is no column of ones corresponding to the constant term). Hence, for the ith row (region) and jth column (variable), $A_{ij} = X_{ij} - X^*_j$. Denote the corresponding vector of coef˜ cients by m (again there is no coef˜ cient for the constant), and it then follows that:

$$R^{\bullet}/R = \left(I - \rho W^{\#}\right)^{-1} Am. \tag{19.14a}$$

The term for G (and thus R) can be separated out so that:

$$R^{\bullet}/R = \left(I - \rho W^{\#}\right)^{-1} A^{\diamondsuit} m^{\diamondsuit} + \left(I - \rho W^{\#}\right)^{-1} G m'',$$

in which $A^{\diamondsuit}$ is A with column G omitted (observe that $G_i = G_i - G^*$), and $m^{\diamondsuit}$ is m with the coef˜ cient for G (denoted m'') omitted. Equilibrium is when $R^{\bullet}/R = 0$, hence:

$$G = (-A^{\diamondsuit} m^{\diamondsuit})(m'')^{-1},$$

and,

$$R^e = \mathrm{U} + (A^{\diamondsuit} m^{\diamondsuit})(m'')^{-1}. \tag{19.14b}$$

Observe that in (19.14b) R^e does not depend on $\rho W p$, apparently the steady-state is not in uenced by spatial interaction! It is the case that without the catch-up term and with only the Verdoorn effect in operation, in equilibrium regions would grow at constant but different rates with the effect that the fastest growing region would tend to a value of 1 and all other regions would tend to 0. The added presence of the catch-up term in the model means that at equilibrium productivity grows at the same rate in all regions. The consequence of this is that the autoregressive spatial lag term is also constant across regions, since the weighted averages of equal productivity growth rates will be constant, and this explains its absence from (19.14b). This apparent lack of a spatial impact is however illusory. First, the transition dynamics leading to equilibrium differ, as can be shown by simulations with $\rho = 0$ and $\rho \neq 0$. Second, omitting Wp biases m so the steady-states will also be biased.

In order to obtain the vector R^e by iteration, we commence with:

$$R^{\bullet}_t/R_t = \left(I - \rho W^{\#}\right)^{-1} A^{\diamondsuit} m^{\diamondsuit} + \left(I - \rho W^{\#}\right)^{-1} G_t m''. \tag{19.15a}$$

In which $t = 0$, and next calculate $R_t^\bullet$ (using $R_0 = \mathrm{U} - G_0$) and thus:

$$R_{t+1} = R_t + R_t^\bullet,$$

and,

$$G_{t+1} = \mathrm{U} - R_{t+1}, \tag{19.15b}$$

which replaces G_t in (19.15a). This gives $R_{t+1}^\bullet$ and hence R_{t+2}, and so on for k iterations until $R_{t+k}^\bullet \approx 0$.

19.5.2 Stochastic Equilibrium

Full Equilibrium Analysis of the Reduced Form The equilibrium (19.12) from the full equilibrium analysis for the reduced form ignores the impact of random shocks. A stochastic term can be introduced into the equivalent iterations (19.14a) and (19.14b), so that (19.14a) becomes:

$$R_t^\bullet / R_t = c - m_3 R_t - m_2 W^\# R_t + m_2 W^\# \mathrm{U} + (\mu - \mu^*),$$

in which $\mu \sim \mathrm{N}(0, \sigma_i^2 I)$, and we calculate $R_t^\bullet$ and subsequently,

$$R_{t+1} = R_t + R_t^\bullet.$$

This iteration can be used to obtain single realisations of the full equilibrium convergence process talking account of the effects of random shocks, or to obtain a "cloud" of realisations as a result of repeating the iteration with different random number streams.

Augmented Spatial Lag and Spatial Errors Verdoorn Laws We ˇ rst commence by considering the equilibrium (19.1) given for non-spatial effects models before ultimately focusing on the augmented spatial effects Verdoorn Laws. For non-spatial effects, an exact equivalent to the iteration A involving R_t and $R_t^\bullet$ to obtain the steady-state given by (19.10) is to calculate $\mathrm{E}(p_t)$ and $\mathrm{E}(p_t^*)$ using $G_t = 1 - (P_t/P_t^*) = 1 - R_t$, and hence $P_{t+1} = P_t(\exp(\mathrm{E}(p_t)))$ and $P_{t+1}^* = P_t^*(\exp(\mathrm{E}(p_t^*)))$ thus giving G_{t+1} and $\mathrm{R}(p_{t+1})$ and $\mathrm{E}(p_{t+1}^*)$, iterating to steady-state at iteration k when $\mathrm{E}(p_{t+k})$ is constant. We label this iteration B. It is convenient to write this in matrix terms using the n by v matrix of "exogenous" variables X_t, hence:

$$\mathrm{E}(p_t) = X_t m, \tag{19.16a}$$
$$P_{t+1} = P_t \left[\exp\left(\mathrm{E}(p_t)\right)\right], \tag{19.16b}$$
$$P_{t+1}^* = P_t^* \left[\exp\left(\mathrm{E}(p_t^*)\right)\right], \tag{19.16c}$$
$$G_{t+1} = 1 - \left(P_{t+1}/P_{t+1}^*\right), \tag{19.16d}$$
$$X_{t+1,v} = G_{t+1}, \tag{19.16e}$$
$$\mathrm{E}(p_{t+1}) = X_{t+1} m. \tag{19.16f}$$

Staying with the non-spatial effects models for the moment, in order to introduce the stochastic element, we replace $\mathrm{E}(p_t)$ by p_t and $\mathrm{E}(p_t^*)$ by p_t^*, by randomly drawing from the assumed error distribution, $\mu_t \sim \mathrm{N}(0, \sigma_i^2)$. The effect is to introduce randomness to P_t, P_t^* and G_t, so that different replications of the "same" iterative process produce different "equilibrium" outcomes.

We now apply the same approach to our augmented spatial effects Verdoorn Law models, thus combining stochastic equilibrium with either spatial errors or with a spatial lag. The approach involves iteration (19.17a19.17b) to (19.17f), starting with $t = 0$ and ending at $t + k$, where k is a suitably large number. Hence, we ˇ rst draw randomly from:

$$\xi_t \sim \mathrm{N}\left(0, \sigma_i^2 I\right),$$

and then calculate either,

$$p_t = (I - \rho W)^{-1} (X_t m + \xi_t), \tag{19.17a}$$

for spatial lag, or,

$$p_t = X_t m + (I - \rho W)^{-1} \xi_t, \tag{19.17b}$$

for spatial error. Given p_t it is then possible to update the level of productivity:

$$P_{t+1} = P_t \exp(p_t), \tag{19.17c}$$

$$P_{t+1}^* = P_t^* \exp(p_t^*), \tag{19.17d}$$

and update the matrix of regressors using,

$$G_{t+1} = 1 - \left(P_{t+1} / P_{t+1}^*\right), \tag{19.17e}$$

$$X_{t+1,v} = G_{t+1}. \tag{19.17f}$$

Assuming that W is also a function of output levels as follows:

$$W_{ij,t+1}^{\diamondsuit} = Q_{i,t+1}^{\alpha} Q_{j,t+1}^{\beta} / d_{ij}^{\gamma}, \tag{19.18a}$$

$$W_{ij,t+1} = W_{ij,t+1}^{\diamondsuit} / \sum_j W_{ij,t+1}^{\diamondsuit}. \tag{19.18b}$$

Clearly there is an element of subjectivity in the design of the W matrix.[2] We know that W matrix design can affect model parameter sampling distributions and diagnostics (Grifˇ th, 1996; Anselin and Florax, 1995b), but now we see that the choice

[2] For instance the value 2 assigned to the coefˇ cients α, β and γ is assumed, since it makes sense empirically. A more ambitious approach would involve a formal search through the parameter space involving maximising the likelihood of the spatial lag or spatial error model, but because of the number of parameters involved, such an undertaking would be computationally complex and is rarely if ever attempted. Moreover, W could be speciˇ ed differently in terms of both functional form and the variables entering into the W equation. For example, in place of output level (Q) in equation (19.18a), one might consider using productivity level (P). The reason for the function adopted is that it seems reasonable as a ˇ rst approximation to treat inter-regional interaction as an increasing function of a measure of economic activity and as a decreasing function of inter-regional distance.

of the "wrong" W matrix could produce a misleading steady-state since this also depends on the parameter estimates. Equations (19.18a,19.18b) give W a time subscript, but this does not itself imply that W changes over time, even though $W^{\diamond}$ does. If q is endogenous and depends on p, then so is W since $Q_{t+1} = Q_t \exp(q_t)$. The result of this feedback will be dynamics that are different (although the steady-state will be the same) from time constant W dynamics. At steady-state with endogenous q, since p is constant then so is q, in other words output in each region grows at the same rate, with the consequence that W is constant. In general, the fact that constant q equates to time constant W matrix provides a useful simpliˇ -cation that we adopt in the empirical simulation that follows. Eliminating the random error ξ_t from the spatial lag model (19.17a) produces exactly the same vector R^e as does (19.14b) or iterations (19.15a) to (19.15b). The same restriction on the spatial error model (19.17b) gives the iteration (19.16a) to (19.16f) and the steady-state (19.10). The empirical work described in the remaining part of the chapter largely involves a speciˇ c application to the EU regional dataset of the iteration (19.17a,19.17c,19.17d,19.17e,19.17f) and the corresponding deterministic steady-states given by (19.15a,19.15b).

19.6 Empirical Convergence Analysis

The chapter has described two approaches leading to steady-state analysis, single equation augmented Verdoorn Laws, and simultaneous equation models embodying the Verdoorn Law as part of the system of equations. In this section the main emphasis is on the augmented spatial lag Verdoorn Law, which turns out to be a better model for the EU data set than the spatial error model or the full and reduced unrestricted spatial effects Verdoorn Laws. Regarding the simultaneous equation model ˇ tted to data for 60 countries, the full equilibrium analysis for the reduced form (19.6) has been published elsewhere, and therefore only a brief summary of the main conclusions is provided here.

19.6.1 The Simultaneous Equation Model

Fingleton (1998) gives full details of the estimates and diagnostics of the simultaneous equations (19.5a–19.5e) and shows that these support the hypotheses and arguments presented in section 19.4. Each of the FIML parameter estimates is appropriately signed, and on referring parameter-standard error ratios to the standard normal distribution it is shown that they are signiˇ cantly different from zero using conventionally accepted Type 1 error rates. The full equilibrium analysis (section 19.5) of the resulting reduced form (equation 19.6) shows that under the model countries converge to widely differing positions vis-a-vis the USA, ranging from complete catch up to extreme poverty (Ethiopia, Mali). A salient feature of these results, which is evidently partly the due to the model's regional effects, is the similar levels of GDP per worker attained by countries in similar locations. For example,

Europe and much of South America catches up the USA, countries in Central America and the Far East tend to an intermediate position, while on the whole African productivity gaps remain large.

19.6.2 The Augmented Spatial Effects Verdoorn Laws

The data used to fit the spatial effects Verdoorn Laws are manufacturing productivity and output for 178 NUTS regions of the European Union (EU) over a period of 20 years (1975–95) (see the Appendix). In the preferred specification, annual average manufacturing productivity growth is a function of manufacturing output growth (GGVA), the basic Verdoorn Law, which is augmented by the catch-up variable (G75) which is a function of manufacturing productivity levels in 1975, as defined in Section 19.3. It is also augmented by the inclusion of an urbanization dummy (URBAN) and a measure of peripherality (LUXDIJ) with respect to the approximate economic centre of gravity of the EU. While a rationale has been given for the other variables, these additional exogenous (Z) variables also require explanation. The assumption is that urban agglomerations possess larger human capital stocks than rural regions, boosting innovation creation and adoption and hence technical progress and manufacturing productivity growth. We therefore anticipate positive coefficient on the urbanization dummy. Regions are categorized as urban if the population density is above 500 inhabitants per square km, a measure suggested as appropriate to Pinelli *et al.* (1998) by the European Commission. A similar human capital gradient is also assumed with distance from the EU centre of gravity. Peripheral regions are generally assumed to be sparsely populated and have lower human capital stocks, perhaps because they tend to be culturally distinct and more dominated by agriculture and services, and thus also have slower technical progress. We therefore anticipate a negative coefficient on the peripherality indicator. Peripherality is measured simply by using the great circle distance of each regional centre from Luxembourg. Pinelli *et al.* (1998) consider a more sophisticated measure that distinguishes between peripheral regions that have different levels of accessibility, but the simpler measure used also here is just as successful in accounting for productivity variations.

Table 19.1 gives the result of fitting the augmented non-spatial effects Verdoorn Law. Table A1 in the Appendix gives IV estimates for the same model, supporting the thesis that the Verdoorn coefficient is minimally affected by any simultaneous equation bias that may result from possible endogeneity involving output growth. The Hausman exogeneity test (omitted variables version), which adds an instrument (see Appendix) to the original model to see if there is a significant improvement, returns a p-value in the relevant F distribution of 0.1905. This is sufficiently large (compared with 0.05) to suggest that there is little correlation between output growth and the errors which would cause simultaneous equation bias.

The estimates in Table 19.1 are very much as anticipated. Evidently the Verdoorn coefficient is a remarkably robust empirical relationship, in this case once again taking a value close to 0.5 suggestive of increasing returns. There is a very

Table 19.1. OLS Estimates of the augmented non-spatial effects Verdoorn Law

Variable	Coeff.	S.D.	t-value	Prob.
CONSTANT	−0.00235384	0.00380086	−0.619292	0.536539
GGVA	0.55545	0.0660132	8.414232	<0.000001
G75	0.0640658	0.00930204	6.887286	<0.000001
LUXDIJ	−1.03223E−05	2.83126E−06	−3.645840	0.000352
URBAN	0.00820547	0.00334538	2.452777	0.015168

Summary	
Regions	178
Variables	5
R^2	0.4650
R^2 adjusted	0.4526
Likelihood	509.194
AIC	−1008.39
SC	−992.478
RSS	0.034137
F-test	37.5847
F-prob	<0.00001
σ^2	0.000197324
σ^2(ML)	0.000191781

signiˇ cant catch-up effect (G75), and productivity growth is slowed by a peripheral location (LUXDIJ) and boosted in urban locations (URBAN). One minor problem with the speciˇ cation is the spurious correlation resulting from the fact that $p = q - e$, where e is the growth of employment, leading an in ated value for R^2. To avoid this it is necessary to regress e on q in order to obtain the correct R^2. This has the result that the regression coefˇ cients are altered, but they can be recovered since, in the context of (19.1), $e = -m_0 + (1 - m_1)q + \mu$. This is shown in practice by the results in Table A2, which show that the correct R^2 is about 35%.

In order to allow diagnostic analysis and subsequent modeling, use is made of the same gravity-based W matrix deˇ ned by (19.18) with t constant and with coefˇ -cients α, β, and γ set equal to the value 2.0. The thesis is that interregional interaction between regions i and j is diminished by increasing (great circle) distance d_{ij} but this effect is attenuated by the levels of manufacturing output Q.

Table 19.2 gives the diagnostics, indicating the presence of heteroscedasticity and residual autocorrelation and, since LM lag exceeds LM error, points to a spatial lag model as the appropriate speciˇ cation.[3] The results of ˇ tting the corresponding

[3] Note that many of the speciˇ cation tests are based on normality of the errors. This is however, rejected by the Kiefer-Salmon test. None the less, the correlation between the normal order statistics and raw residuals equals 0.9893 with an almost perfectly linear plot. Because of the large sample size, the test is very powerful, detecting signiˇ cant deviations from normality that have little practical signiˇ cance.

Table 19.2. Diagnostics for the augmented non-spatial effects Verdoorn Law

Test	Degrees of Freedom	Value	Prob.
Normality Of Errors			
Kiefer-Salmon	2	15.871073	0.000358
Heteroskedasticity			
Koenker-Bassett	4	22.312009	0.000174
Speciˇ cation Robust			
White	13	67.965864	<0.000001
Spatial Dependence			
Moran's I (error)		5.891542	<0.000001
Lagrange Multiplier (error)	1	20.092498	0.000007
Robust LM (error)	1	2.459753	0.116797
Kelejian-Robinson (error)	5	3.343055	0.647257
Lagrange Multiplier (lag)	1	35.035707	<0.000001
Robust LM (lag)	1	17.402962	0.000030
Lagrange Multiplier (SARMA)	2	37.495460	<0.000001
	Condition no.		
Multicollinearity	10.749115		

spatial error model are summarized in the Appendix. The misspeciˇ cation evident in the augmented non-spatial effects Verdoorn Law might also be suggested by the results of the RESET test. Addition of the squared and cubed ˇ tted values to the basic model gives an F ratio of 6.49 with p-value 0.002. However, it should be noted that Florax and Folmer (1992) show that the RESET test has no power against common spatial misspeciˇ cations (an example of this lack of power is provided by Fingleton and McCombie (1998)).

Table 19.3 again provides strong support for increasing returns, with the Verdoorn coefˇ cient close to 0.5, and for catch-up, peripherality and urbanization effects with appropriate signs and levels of signiˇ cance. It also indicates the presence of a very signiˇ cant spatial lag suggesting the existence of cross-region externality effects.

The same model is chosen if we start from most complex model, which is the full unrestricted spatial effects Verdoorn Law (Table A5), and then attempt to impose restrictions. Inference assumes that for two models, one nested within the other, twice the difference in the log-likelihoods approximates an appropriate chi-squared distribution under the null hypothesis that the nested model is true (for examples see

Table 19.3. OLS Estimates of the augmented non-spatial effects Verdoorn Law

Variable	Coeff.	S.D.	z-value	Prob.
W_GGVAPW	0.642234	0.0893122	7.190886	<0.000001
CONSTANT	−0.0192694	0.00397344	−4.849564	0.000001
GGVA	0.495972	0.0601994	8.238824	<0.000001
G75	0.0642047	0.00844908	7.599010	<0.000001
LUXDIJ	−1.41947E−05	2.59694E−06	−5.465922	<0.000001
URBAN	0.00834755	0.00303875	2.747032	0.006014
		Summary		
Regions				178
Variables				6
R^2				0.4619
Sq. corr.				0.5164
Likelihood				522.212
AIC				−1032.42
SC				−1013.33
σ^2				0.000162792

Upton and Fingleton, 1985, 1989; Fingleton, 1999b).[4] Comparing the full speciˇ cation with the spatial error model (Table A3), the null hypothesis is rejected, which amounts to rejection of the common factor hypothesis. Note that the spatial errors diagnostics (see Table A4) also reject the common factor hypothesis.

Comparing the full and reduced spatial effects Verdoorn Laws, the test statistic equals 4.528, identical of course to the likelihood ratio test on the endogenous autoregressive term (W_GGVAPW) in the full model (see Table A6). The reduction from full to reduced model is unacceptable.[5] The implication is that in spite of the presence of the exogenous spatial lags, there is still a signiˇ cant spatial externality. Comparison of the full unrestricted spatial effects model, which we have thus far been unable to simplify, and the spatial lag model, tests the signiˇ cance of the exogenous spatial lags (W_GGVA, W_LUXDIJ, W_URBAN, and WG75), conditional on the presence of the endogenous spatial lag. Twice the difference in the log likelihoods equals 7.9 which has a p-value of about 0.1 when referred to the chi-squared distribution with 4 degrees of freedom, indicating insigniˇ cant exogenous spatial lags as also suggested by the individual t-ratios. The restrictions imposed under the spatial lag speciˇ cation are accepted using the evidence from the joint test. Overall, the insigniˇ cance of the exogenous spatial lags and the persistence of the spatial lag

[4] The use of Maximum Likelihood estimation excludes the possibility of a unit root or unstable process for the spatial lag model. This and related issues are considered by Fingleton (1999c,a).

[5] The are also implications due to the likely presence of multicollinearity, which is inferred from the large condition number associated with the reduced unrestricted model, and which would tend to in ate sampling variances and destabilise parameter estimates.

Table 19.4. Diagnostics for the augmented spatial lag Verdoorn Law

Test	Degrees of Freedom	Value	Prob.
Heteroskedasticity			
Breusch-Pagan	1	62.897068	<0.000001
Spatial B-P test	1	62.941813	<0.000001
Breusch-Pagan	4	69.360529	<0.000001
Spatial B-P test	4	69.416944	<0.000001
Spatial Dependence			
Likelihood Ratio (Lag)	1	26.036358	<0.000001
Lagrange Multiplier (error)	1	0.024267	0.876207

effect in the competing models, suggests a real spatial externality effect that is not simply a by-product of the omission of other spatial lags from the model. Florax and Folmer (1992) provide more details of specification strategies and misspecification testing, and Florax *et al.* (2003) also make explicit the assumptions and relative merits of top down versus bottom up approaches to model selection.

There is one further point to consider, however, which suggests further elaboration of the spatial lag model. The diagnostics in Table 19.4 indicate (random coefficients) heteroscedasticity although no residual spatial dependence, which is reassuring given the large sample size that will make the test very sensitive. Heteroscedasticity is tested by using a linear specification comprising a constant and the single variable LUXDIJ and also by using all four regressors together. The diagnostics for the other models (see Appendix) also point to heteroscedasticity. In fact the very significant heterogeneity according to distance from Luxembourg may possibly be a region-size effect, since the typical size of NUTS 2 regions varies by country.

Table 19.5 gives the ML estimates and summarizes the results of fitting the spatial lag model with groupwise heteroscedasticity.[6] The usual likelihood comparison involving both spatial lag models gives 8.82, which is identical to the value in Table 19.5, indicating a significant improvement as a result of modeling heteroscedasticity. Clearly the spatial externality also remains highly significant even having allowed for spatial heterogeneity. To facilitate this analysis, the groups are formed by dichotomizing the variable LUXDIJ according to whether regions are less than 500 km from Luxembourg, or at least 500 km away. D_LUX_1 denotes core regions, which are seen to have a significantly smaller error variance (0.000101419) than peripheral regions (0.000205032).

[6] An alternative method of accommodating heterogeneity could be to fit heteroscedastic conditional autonormal model (Cressie, 1993), but this involves somewhat different assumptions rather than being a straightforward continuation of the simultaneous autoregressive models considered thus far. Fingleton (1999b) gives an example of the heteroscedastic conditional autonormal model in the context of a critique of the neoclassical reduced form.

Table 19.5. Augmented spatial lag Verdoorn Law: groupwise heteroscedasticity

VARIABLE	Coeff.	S.D.	*z*-value	Prob.
W_GGVAPW	0.650123	0.128873	5.044684	<0.000001
CONSTANT	−0.019942	0.00419611	−4.752487	0.000002
GGVA	0.444352	0.0614415	7.232119	<0.000001
G75	0.06645	0.00767723	8.655467	<0.000001
LUXDIJ	−1.4311E−05	2.48029E−06	−5.769887	<0.000001
URBAN	0.00929659	0.00279149	3.330328	0.000867
GROUP VARIANCES D_LUX_0	0.000205032	2.80647E−05	7.305692	<0.000001
D_LUX_1	0.000101419	1.70634E−05	5.943683	<0.000001
HETEROSCEDASTICITY TEST				
	D.F.	Value	Prob.	
Lik.Ratio	1	8.820509	00.002979	
	Summary			
Regions				178
variables				6
R^2				0.4947
Sq. corr.				0.5447
Likelihood				526.622
AIC				−1041.24
SC				−1022.15
Iterations to convergence				4

The spatial lag estimates in 19.5 were used in the iteration (19.15) to obtain the deterministic equilibrium vector R^e for the augmented spatial lag Verdoorn Law (in practice the same result is provided by iteration (19.17) without the stochastic term). The iterative approach is preferred to the analytical solution (19.14b) since this allows productivity leadership changes. Figure 19.3 shows the results, each region commencing at its base year manufacturing productivity ratio R_0. In order to obtain these dynamics, the exogenous variables LUXDIJ and URBAN are set to their historical values. It seems unreasonable however to assume that historic output growth (GGVA) differences between regions will be maintained at equilibrium. The move to a single European economy suggests a lowering of barriers and enhanced market penetration, so that high demand in faster growing regions will be satisfied by output in other regions, and regions with lagging domestic demand will benefit from higher demand elsewhere. We therefore set the exogenous variable output growth in each region equal to the EU annual average calculated over the period 1975–95 (equal to 0.01731) in the simulation exercise. This is also a convenient assumption because of the implications for the W matrix, as discussed in Sect. 19.5.

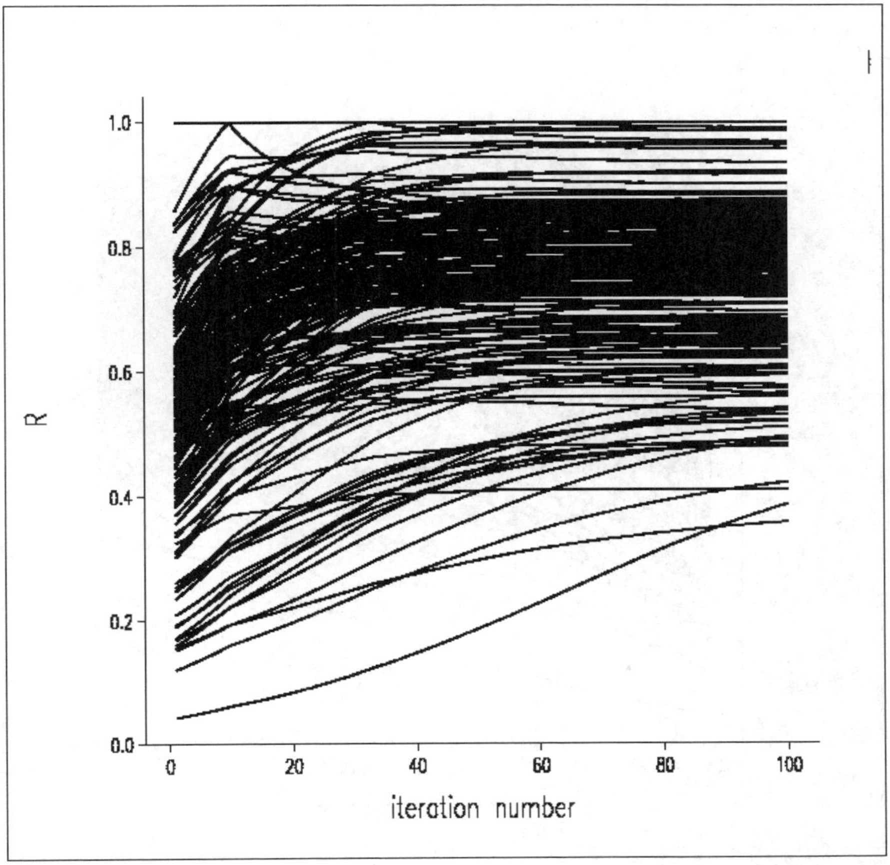

Fig. 19.3. Deterministic solution (178 EU regions)

In Fig. 19.3 we see the deterministic equilibria converge after about 75 iterations to a stable steady-state. The regions do not converge to the same steady-state, but to their own equilibrium positions. Of course the paths traced by the individual economies are idealised, showing none of the turbulence of the real world. Turbulence is injected by iteration (19.17), in which we assume $\sigma_i^2 = 0.0001$ when region i is within 500 km of Luxembourg, and $\sigma_i^2 = 0.0002$ otherwise. Figure 19.4 is the result of a single simulation. It is noticeable in this that the paths traced by the individual economies are not independent. Simultaneous reactions occur when the leadership changes, since then the fastest growing region replaces a slower growing region as productivity leader, and the other regions show a commensurate fall. This shows up as peaks and valleys in Figs. 19.3 and 19.4. This can also occur if there is no leadership change, if a large shock favors the leader's growth at the expense of the other regions, or causes a comparatively large slowdown. Moreover, because

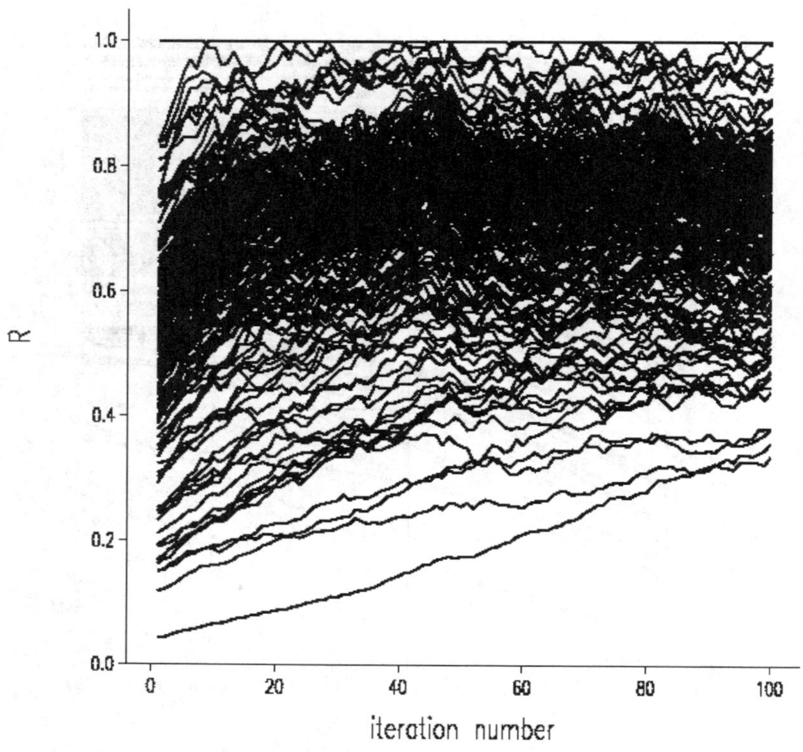

Fig. 19.4. Stochastic solution (178 EU regions)

regions in disequilibrium interact spatially, we evidently have turbulence cycles of faster and slower growth as the net outcome of shocks simultaneously transmitted across regions. Rerunning the simulation using different random number streams produces the same phenomena, although the peaks and valleys are in different positions.

The stochastic simulation in Fig. 19.4 is simply a one-off run that depends on the random numbers that happen to occur. In order to see the average outcome, we run iteration (19.17) 100 times using 100 different random number streams. The results are summarized by the G mean ($G = 1 - R$) and variance for each region. The overall distribution of the 178 mean values of G is presented in Fig. 19.5, which shows that equilibrium involves convergence in manufacturing productivity gaps compared with the gaps that have existed historically in the EU. The mean gap in 1975 was equal to 0.494, and this fell to 0.453 in 1995. In contrast the estimated mean (of the G means) at equilibrium is 0.2801. Part of the reason for large historical gaps is the existence of a few high-performing regions, for example, notwithstanding its peripheral and comparatively rural economy, we have seen a remarkable increase in manufacturing productivity in the Ireland in recent years. In order to minimize the impact of these speci˘ c anomalies, Fig. 19.5 also includes the distribution of the average gap for the period 1975–95. The much more convergent pattern in equilibrium re ects the fact that the special anomalous circumstances are assumed not to exist at equilibrium. Converge re ects the effect of spillovers between regions (the spatial lag or externality effect) and the diffusion of technology and regional policy (the catch-up effect). However, we see that convergence is not to a single pan-European steady-state as is predicted by simple neoclassical theory, but to a dispersed distribution in which differences between manufacturing productivity are maintained by differences in peripherality and urbanization levels. If these differences are to be reduced, policy should therefore focus on increasing the rate of technical progress in peripheral and rural regions.

19.7 Conclusions

The spatial analysis of economic growth in much mainstream economics literature can be criticised because it fails to utilise developments in spatial econometrics, is tied to neoclassical economic theory and the unrealistic assumption of diminishing returns, and fails to take account of turbulence as observed in the real world. New economic geography attempts to introduce theoretical coherence assuming increasing returns, and Verdoorn Law and Markov chain analysis have also been used to address some of these issues. This chapter represents an attempt to combine aspects of these approaches to further develop a new line of analysis.

The empirical analysis in the chapter is based on a single equation augmented spatial lag Verdoorn Law, in which manufacturing productivity growth depends on the exogenous variables manufacturing output growth, the initial level of productivity gap, peripherality and urbanisation, plus an autoregressive spatial lag effect. The estimates obtained reaf˜ rm the now well-established empirical observation of a Verdoorn coef˜ cient close to 0.5, consistent with increasing returns. The estimates provide empirical support for the hypotheses that there are signi˘ cant cross-region externalities that are attributed to the spillover of technical change resulting from capital accumulation. The catch up effect is attributed to innovation diffusion enhanced by policy instruments, and the faster productivity growth in urban and non-peripheral regions is attributed to larger human capital stocks inducing faster technical progress.

The model estimates are used to obtain new expressions for deterministic and stochastic equilibrium in which regional dynamics are treated as a co-ordinated ensemble, with the movement of one region simultaneously in uencing, and being in uence by, the movement of other regions. These are used to obtain equilibrium distributions of manufacturing productivity ratios and gaps for 178 NUTS 2 regions of the EU. These show that, in the absence of stronger policy intervention, differ-

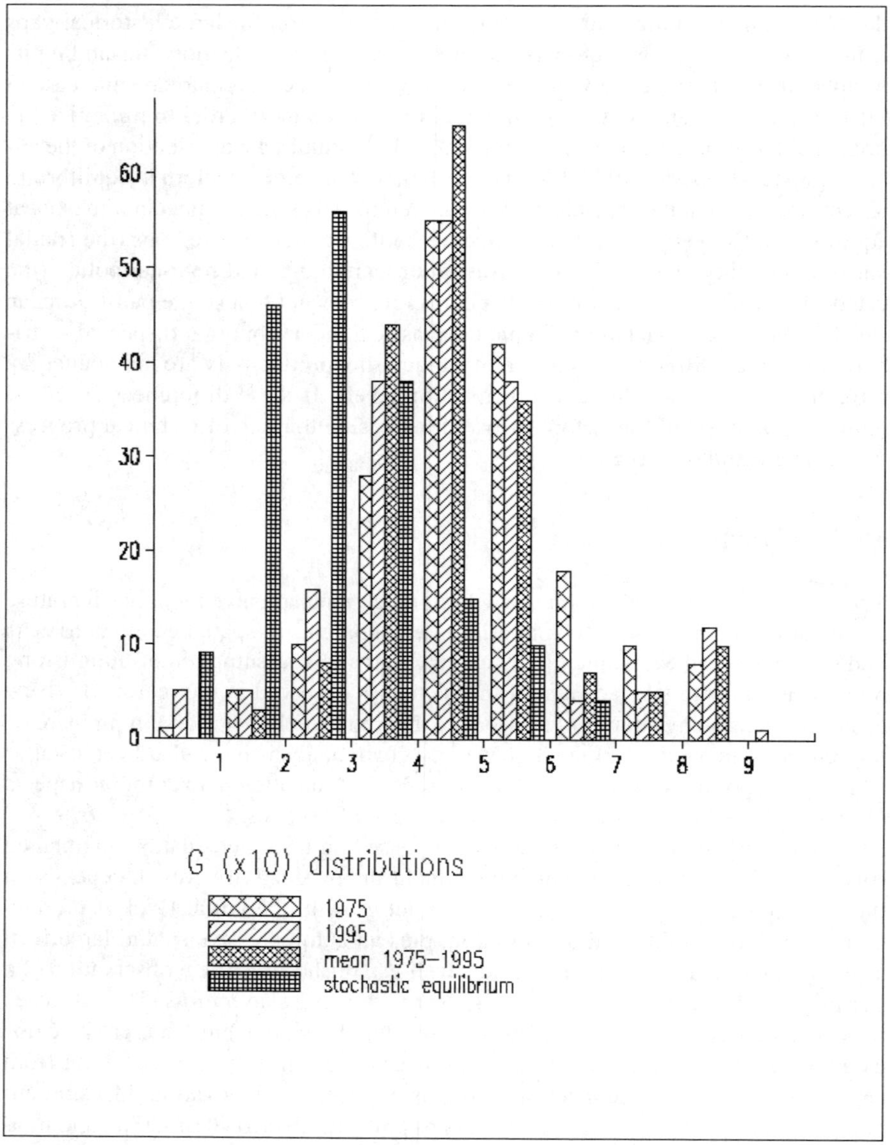

Fig. 19.5. Empirical and simulated *G* distributions

ences between productivity levels are likely to be maintained even if output growth is equalised. However, these should be set in the context of turbulent regional dynamics rather than a smooth progression to dispersed steady-states.

Table A1. IV(2SLS) estimates of the augmented non-spatial effects Verdoorn Law

Parameter	Estimate	*t* (corrected for 2SLS)
Constant	−0.00377	−0.9754
IGGVA	0.5107	7.0908
G75	0.06868	7.4110
LUXDIJ	−0.00001032	−3.6382
URBAN	0.00813	2.3977

Appendix: Description of Data

The data, consisting of measures for *p*, *q*, *e* and *G*, covering 178 NUTS 2 regions of the EU, are taken, with permission, from Cambridge Econometrics' European Regional Databank. While there are problems of measurement associated with regional (and national!) economic data, these series are undoubtedly the most consistent and accurate available for the EU as a whole. Productivity growth (*p* or GGVAPW) is represented by the average annual (exponential) growth of manufacturing (and energy) gross value added per worker over the period 1975–95. Similarly, output growth (*q* or GGVA) is the average annual growth of manufacturing gross value added over the period. Gross value added is measured in constant (1985) ecus. Since Groningen and Flevoland have anomalous manufacturing GVA and GVA per worker values (largely due to uctuations in gas production in Groningen and possibly also due to commuting), in these cases the Dutch national averages were used. In the case of Hamburg, it is apparent that commuting may also be a distorting in uence because of the (NUTS 1) region's small spatial extent. Hence a more appropriate deˇ nition of the city was used, the travel to work area (ROR05) which comprises the NUTS 1 region of Hamburg and the surrounding Kreise that qualify as part of the functional urban area. For example, in 1990, the Hamburg TTWA had a population of 2.9 m people, compared with 1.6 m people for the NUTS 1 region. This provides a more realistic per worker GVA.

IV or 2SLS estimation has frequently been carried out in cross-sectional growth analysis by treating lagged variables as predetermined, for example Barro and Sala-i-Martin (1995) use the average investment ratio for 1960–64 as an instrument for the average for 1965–75, although it is not always certain that using a lagged variable as an instrumental variable will solve the problem by being independent of the error term (Barro and Sala-i-Martin argue that lag values are reasonable candidates as instruments because the correlation of the residuals in the growth regressions between their two periods is insubstantial). In the current analysis there is an absence of earlier data, hence we use an alternative approach which is a variant of instrumental variables estimation that has been suggested in the context of the errors in variables problem (Kennedy, 1992). We adapt the method suggested by Durbin's 1954 which uses as an instrumental variable rank orders (1,2,3 etc.... denoting the highest, second, third etc. value) in place of an endogenous variable. Evidently this produces consistent estimates under fairly general conditions (Johnston, 1984, see)

Table A2. The augmented non-spatial effects Verdoorn Law with manufacturing employment growth as the dependent variable

Variable	Coeff.	S.D.	*t*-value	Prob.
CONSTANT	0.00235384	0.00380086	0.619292	0.536539
GGVA	0.44455	0.0660132	6.734254	<0.000001
G75	−0.0640658	0.00930204	−6.887286	<0.000001
LUXDIJ	1.03223E−05	2.83126E−06	3.645840	0.000352
URBAN	−0.00820547	0.00334538	−2.452777	0.015168
		Summary		
regions				178
variables				5
R^2				0.3468
R^2 adjusted				0.3317
likelihood				509.194
AIC				−1008.39
SC				−992.478
RSS				0.034137
F-test				22.9620
F-prob				<0.000001
σ^2				0.000197324
σ^2 (ML)				0.000191781

although Maddala (1992) warns that if the errors are large, the ranks will be correlated with the errors and the estimators inconsistent. In this case the ranks were used as the ˇ rst stage of two stage least squares to provide the instrumental variable IGGVA.

Table A3. Maximum likelihood estimates of the augmented spatial error Verdoorn Law

Variable	Coeff.	S.D.	z-value	Prob.
CONSTANT	−0.00506377	0.00749376	−0.675732	0.499211
GGVA	0.518839	0.0622928	8.329036	<0.000001
G75	0.0608311	0.00854873	7.115801	<0.000001
LUXDIJ	−1.3371E−05	2.95198E−06	−4.529498	<0.000001
URBAN	0.00656228	0.00282258	2.324923	0.020076
LAMBDA	0.786602	0.0647262	12.152762	<0.000001
		Summary		
regions				178
variables				5
R^2				0.3960
Sq corrn.				0.4550
R2(Buse)				0.4536
likelihood				520.991
AIC				−1031.98
SC				−1016.07
σ^2				0.000163081

Table A4. Augmented spatial error Verdoorn Law : diagnostics

Test	Degrees of freedom	Value	Prob.
HETEROSKEDASTICITY			
Breusch-Pagan	4	65.082447	<0.000001
Spatial B-P test	4	65.153157	<0.000001
SPATIAL ERROR DEPENDENCE			
Likelihood Ratio	1	23.594794	0.000001
COMMON FACTOR HYPOTHESIS			
Likelihood Ratio	4	10.341621	0.035050
Wald	4	12.321395	0.015115
SPATIAL LAG DEPENDENCE			
Lagrange Multiplier	1	0.706906	0.400473

Table A5. The full unrestricted spatial effects Verdoorn Law

Variable	Coeff.	S.D.	z-value	Prob.
CONSTANT	−0.0215826	0.0122882	−1.756363	0.079026
W_GGVAPC	0.485011	0.122601	3.956005	0.000076
GGVA	0.504768	0.0623567	8.094844	<0.00001
G75	0.0625927	0.00850855	7.356448	<0.00001
LUXDIJ	−1.85978E−05	3.86386E−06	−4.813263	0.000001
URBAN	0.00743201	0.0030283	2.454181	0.014121
W_GGVA	−0.218508	0.303754	−0.719360	0.471919
W_G75	−0.000619914	0.043697	−0.014187	0.988681
W_LUXDIJ	2.83958E−05	1.7207E−05	1.650245	0.098893
W_URBAN	0.00261002	0.0072566	0.359675	0.719090
		Summary		
regions				178
variables				10
R^2				0.5529
Sq corrn.				0.5619
likelihood				526.162
AIC				−1032.32
SC				−1000.51
σ^2				0.000157033

Table A6. Diagnostics: the full unrestricted spatial effects Verdoorn Law

Test	Degrees of freedom	Value	Prob.
HETEROSKEDASTICITY			
Breusch-Pagan	1	56.392445	<0.000001
Spatial B-P test	1	56.422106	<0.000001
SPATIAL DEPENDENCE			
Likelihood Ratio (Lag)	1	4.528454	0.033336
Lagrange Multiplier (error)	1	0.275748	0.599502

Table A7. The reduced unrestricted spatial effects Verdoorn Law

Variable	Coeff.	S.D.	*t*-value	Prob.
CONSTANT	−0.035937	0.0123217	−2.916549	0.004020
GGVA	0.50871	0.0651122	7.812818	<0.000001
G75	0.0617617	0.00888259	6.953118	<0.000001
LUXDIJ	−1.96417E−05	4.02486E−06	−4.880108	0.000002
URBAN	0.00779819	0.00315842	2.469018	0.014543
W_GGVA	0.379528	0.310311	1.223056	0.223011
W_G75	0.0487325	0.0431399	1.129638	0.260230
W_LUXDIJ	2.44087E−05	1.79525E−05	1.359627	0.175759
W_URBAN	0.0152581	0.00736639	2.071317	0.039849
		Summary		
regions				178
variables				9
R^2				0.5464
R^2 adjusted				0.5250
likelihood				523.898
AIC				−1029.80
SC				−1001.16
RSS				0.0289385
F-test				25.4507
F-prob				<0.000001
σ^2				0.000171234
σ^2 (ML)				0.000162576

Table A8. Diagnostics: the reduced unrestricted spatial effects Verdoorn Law

Test	Degrees of freedom	Value	Prob.
NORMALITY OF ERRORS			
Jarque-Bera	2	41.534396	<0.000001
HETEROSKEDASTICITY			
Koenker-Bassett	1	22.891738	0.000002
SPATIAL DEPENDENCE			
Moran's *I* (error)		2.276043	0.022843
Lagrange Multiplier (error)	1	1.054496	0.304474
Robust LM (error)	1	1.967359	0.160728
Kelejian-Robinson (error)	9	6.852399	0.652484
Lagrange Multiplier (lag)	1	2.357202	0.124706
Robust LM (lag)	1	3.270065	0.070555
Lagrange Multiplier (SARMA)		4.324561	0.115062
	Condition no.		
multicollinearity	60.249863		

20 Growth and Externalities Across Economies: An Empirical Analysis Using Spatial Econometrics

Esther Vayá, Enrique López-Bazo, Rosina Moreno, and Jordi Suriñach

University of Barcelona

20.1 Introduction

Recent theoretical models of economic growth emphasize the importance of external effects for the accumulation of factors of production (Romer, 1986, 1990; Lucas, 1988). Externalities imply that an increase in the stock of reproducible factors leads to an improvement in the level of technology that cannot be fully appropriated by the agent making the investment. As a result, the aggregate or social return on the investment is greater than the private return obtained by the individual agent. A crucial assumption is usually that externalities spread over the entire economy, affecting the level of technology of each individual firm.

We start from the recognition that external effects are relevant with respect to economic growth. We also expect externalities to spill over the barriers of economies, in line with the idea of across-economy interactions outlined in Lucas (1993). When we assume across-economy spillovers in human capital, the theoretical implication is that all economies converge to the same steady state, regardless of differences in initial conditions. This prediction, however, seems to be at odds with empirical evidence. We therefore use a somewhat stricter set of restrictions, and assume that there are spatial limits to the spread of externalities. In addition, we postulate that the diffusion of innovations will always be easier within groups of closely related economies ("clubs"). Essentially, we agree with Durlauf and Quah (1999), who state that "[i]t is easy to see that if we allow[ed] natural groupings of economies to form, so that economies within a group interact more with each other than with those outside, then the "average" H [human capital] that they converge to will, in general, vary across groups." In the case of regions integrated in a particular area, the economies can be thought of as interacting strongly with each other. When these relationships have an influence on economic growth, the models built to explain such growth patterns must explicitly include some measure of the linkages across economies.

This chapter therefore discusses the importance of across-region relationships in growth, and the dynamics towards the steady state. Specifically, we start with the presentation of a simple growth model in which the diffusion of knowledge due to investments in capital is not confined to the limits of the economy in which the innovation is generated, but knowledge spills over into neighboring economies. Subsequently, we derive the growth equation for this simple model. We use spatial econometrics techniques to test for the existence, and estimate the magnitude, of these externalities.

The remainder of this chapter is organized as follows. Section 20.2 presents a theoretical and empirical justification for the consideration of regional spillovers in economic growth. In Sect. 20.3, we introduce a simple growth model including across-region externalities, and we derive the corresponding growth equation. Section 20.4 describes the empirical models in detail, and shows how to use statistical tests and estimation procedures to identify the existence of spillovers, and estimate their magnitude. In Sect. 20.5, we discuss various pivotal spatial econometric topics related to the testing and estimation procedures used in the analysis. Section 20.6 presents empirical evidence for the case of Spanish and European regions, and we conclude in Sect. 20.7.

20.2 Do Spatial Externalities Matter?

20.2.1 Theoretical Issues

Recent studies show that linkages across economies are likely to be important in explaining economic growth. A well-known source generating such linkages is investments in R&D. The knowledge generated by foreign R&D investments may spill over among trade partners (Coe and Helpman, 1995; Park, 1995; Helpman, 1997). Most analyses consider R&D investments in foreign countries embodied in traded goods as the main channel for technological diffusion. However, this is not necessarily the only relevant channel. Keller (1998) points out that multiple channels for technological diffusion are likely to exist, and he shows that import-weighted R&D investments in foreign countries do not achieve a higher statistical significance than random combinations.

Technological diffusion between national economies is of considerable relevance, and the diffusion of technology is often assumed to be even stronger between regions of the same economy. This may be true for the trade channel as well as for other channels. Although oftentimes laboratories and R&D centers are located in a limited number of regions, this does not preclude firms located outside those regions from using the extra-regionally generated knowledge. Especially in the case of *public* R&D centers, governments will be concerned to ensure that the results of R&D spread throughout the entire territory (López-Bazo *et al.*, 1998).

The diffusion of technology is likely to be higher among regions that are geographically close to each other as compared to economies that are more distant. This may be due to the fact that the relative amount of traded goods is relatively high, but local social conditions can also be relevant.[1] Rodriguez-Pose (1999) shows that local social conditions play an important role in the way economies incorporate and adapt ongoing innovations. It is obvious that the probability of neighboring economies sharing similar local conditions is relatively high, and the transfer of technology between neighboring regions is therefore also likely to be more intense.

[1] When analyzing the role of distance in spatial diffusion of technology, both contagious and hierarchical spread of knowledge is of interest (see Cliff and Ord, 1981; Morrill *et al.*, 1988, for a discussion).

Vayá *et al.* (1998) suggest that the transfer of technology in integrated regional economies is facilitated by common markets for skilled labor and final goods, access to similar types of capital throughout the entire area, and managerial talent.

Pecuniary externalities can also evoke a concentration of firms in macro-areas spanning several regions (Glaeser *et al.*, 1992; Venables, 1996). In this case, externalities at the firm level are passed on to the aggregate regional level. Once centrifugal forces (costs of production in a specific location relative to those of other locations) surpass the effects of agglomeration economies in a region, it is plausible that firms start looking for locations in contiguous regions. In these nearby regions, production costs are lower, while at the same time firms can still take advantage of the external economies, because of the short distances involved in interaction. This hypothesis is in line with the process of progressive industrialization in the periphery, as proposed in Puga and Venables (1996). If the distance between economies is an important location factor, agglomeration economies operating at a supra-regional level give rise to an external regional effect.

Finally, Kubo (1995) presents a theoretical model in which production inputs of a particular region are relevant for the output of another region. This model shows how the likelihood of even or uneven development depends on the magnitude of the internal returns to scale of both regions, as well as on the value of the externality.

20.2.2 Empirical Evidence

Despite the apparent significance of regional spillovers, most of the empirical literature focuses on the analysis of externalities at the firm or industry level (Caballero and Lyons, 1990; Raut, 1995; Burnside, 1996; Ravallion and Jalan, 1996). Costello (1993) shows how total factor productivity growth is more strongly correlated across industries within one country than across countries within one industry. However, Kollmann (1995) observes that correlations across industries within a region are weaker than across regions within an industry. He also reports that productivity growth is more strongly correlated across the regions of the U.S. than across the G7 countries. Quah (1996) presents additional evidence for the relevance of regional spillovers. He shows that the distribution of the per capita product in regions of the E.U., conditioned on the level in neighboring regions, is more strongly concentrated than the real distribution. Vayá (1998), López-Bazo *et al.* (1999) and Rey and Montouri (1999) detect strong spatial dependence in the distribution of per capita product or productivity for different geographical areas.

In all the above cases, the spatial distribution of economic growth or wealth appears to be far from random or equal. This may be caused by spatial autocorrelation in investment rates and in the average level of technology of each economy. In the former case, similarities in saving rates and other preference parameters may largely explain the autocorrelation. In the latter case, the greater intensity of knowledge diffusion across neighboring economies and the presence of agglomeration economies that surpass regional barriers are likely to be influential.

Theoretically as well as empirically, the above argument underlines the importance of external economies that cross the weak, and sometimes artificial, regional boundaries. However, to the best of our knowledge, only a limited number of papers consider the performance of other economies in explaining growth. Barro and Sala-i-Martin (1995), following the proposal in Chua (1993), include a weighted average of per capita income of a country's immediate geographical neighbors as a regressor in the growth equation. Their ˇ ndings provide support for an external effect across countries, although the effect is small. Ciccone (1996) observes how a large fraction of growth in total factor productivity spreads out to neighbors for a large sample of 98 countries. Moreno and Trehan (1997), using distance and other measures of proximity, show that the neighbors' growth affects a country's growth, while Ades and Chua (1997), considering political instability in nearby countries rather than economic growth, observe a signiˇ cant spillover effect. Fingleton and McCombie (1998) provide evidence for the regional case. They ˇ nd a signiˇ cant externality effect in labor productivity for a sample of E.U. regions.

Studies of this kind are, however, few in number, and the existing studies apply a rather ad hoc approach to modeling spillovers across economies. In the following sections, we address the theory behind modeling spatial spillover effects as well as the issue of how externalities can be measured and empirically assessed.

20.3 A Simple Growth Model With Spillovers Across Regions

We start-off by describing a simple model of economic growth, initially proposed in López-Bazo *et al.* (1998). In this model, externalities arising from an increase in the level of technology in neighboring regions are explicitly considered.[2] Subsequently, we follow Vayá *et al.* (1998) in deriving a growth equation in the presence of externalities.

We consider a simple economy in which labor productivity in region i and period t, y_{it}, is a function of a vector of reproducible factors per worker synthesized in k_{it} (for instance, physical or human capital), and the state of the technology, A_{it}:

$$y_{it} = A_{it} k_{it}^{\alpha}, \tag{20.1}$$

with decreasing returns in factor accumulation ($\alpha < 1$).

We introduce two key assumptions. First, we allow for externalities due to the accumulation of capital within a regional economy. Consequently, following the reasoning in Romer (1986); Lucas (1988), the aggregate level of technology is a function of the aggregate level of k.[3] Second, we assume the existence of externalities related to the aggregate level of technology of the neighbors which, in turn, are linked to their capital stock. This means that innovations or ideas, linked to investments in k, can ow across economies. Therefore:

$$A_{it} = \Delta k_{it}^{\delta} k_{\rho it}^{\gamma}, \tag{20.2}$$

[2] We use the concept of neighborhood in a broad sense, and do not strictly conˇ ne it to physical contiguity.

[3] We consider the aggregate level of technology as a function of capital intensity rather than capital stock, in order to avoid scale effects.

where Δ is an exogenous component, for the sake of simplicity assumed to be constant,[4] δ is the measure of the degree of external returns within the region, and the subscript ρi denotes the set of regional neighbors of region i. As a result, $k_{\rho i}$ is the amount of capital per worker at the neighbors of region i. The parameter γ measures the regional spillover effect, which is assumed positive: when $k_{\rho it}$ increases by one percent (causing an increase in technology of those regions), technology in region i increases by γ percent.

Clearly, when $\delta = \gamma = 0$ and $\alpha < 1$ we are dealing with the traditional Solow-Swan production speciˇ cation, whereas the Romer-Lucas speciˇ cation with (general) external effects is captured by $\delta > 0$ and $\gamma = 0$.

Substituting (20.2) in (20.1) leads to:

$$y_{it} = \Delta k_{it}^{\tau} k_{\rho it}^{\gamma}, \tag{20.3}$$

where $\tau = \alpha + \delta$. When a regional economy increases its stock of reproducible factors, it obtains a return of τ. If its neighbors simultaneously increase their stocks as well, there is a spillover effect that raises the returns in region i to $\tau + \gamma$. Productivity in region i also increases with $k_{\rho i}$ even in the case of no further investment in k_i. This is because of the diffusion of technology from the neighbors making the stock of capital in region i more productive.

The growth rate of k_i is:

$$\frac{\boldsymbol{k}_i}{k_i} = s\Delta k_i^{-(1-\tau)} k_{\rho i}^{\gamma} - (d+n), \tag{20.4}$$

where s is the saving rate (considered exogenous for the sake of simplicity), and $(d + n)$ the effective rate of depreciation (the temporal subscript is omitted to simplify notation). The rate of investment in k_i is a decreasing function of its stock in the case of decreasing returns within the region ($\tau < 1$), while it is an increasing function of the stock at the neighbors. This means that investments in reproducible factors is greater in those regions located in areas with high stocks of these factors, because externalities across regions within the area increase the returns on these investments. In contrast, incentives to invest will be lower in a region surrounded by others with low capital intensity.

Under the assumption of similar capital intensity in the steady state in all regional economies, $k_i^* = k_{\rho i}^* = k^*$, the growth of the economy in the equilibrium is:[5]

$$g_k = \frac{\boldsymbol{k}}{k} = s\Delta k^{-(1-(\tau+\gamma))} - (d+n). \tag{20.5}$$

4 The assumption of a common exogenous effect across economies is relaxed in the empirical analysis, because we allow for differences across groups of economies.

5 This will be the case when there are no boundaries to externalities across economies. When technological diffusion and/or agglomeration economies are limited by distance, what follows is true only for the economies within each geographical club.

In the long-run, g_k is zero, so if $\tau+\gamma<1$ the economy converges to the following steady state capital intensity:

$$k^* = \left(\frac{s\Delta}{n+d}\right)^{\frac{1}{1-(\tau+\gamma)}}, \tag{20.6}$$

or, in terms of productivity,

$$y^* = \Delta^{\frac{1}{1-(\tau+\gamma)}} \left(\frac{s}{n+d}\right)^{\frac{\tau+\gamma}{1-(\tau+\gamma)}}. \tag{20.7}$$

The steady state depends on the usual technological and preference parameters and on the strength of the regional externalities. A stronger regional interdependence induces a higher stock of capital per worker. In this case, all regions share a common steady state because returns to investment in a group of neighbors are globally a decreasing function of the average intensity in this group. Therefore, productivity equalizes within groups and across groups in equilibrium. However, if for any group of regions, externalities are strong enough to cause $\tau+\gamma \geq 1$, we face endogenous growth for that group despite decreasing returns to reproducible factors at the regional level. In this case, the initial gap across regions continues to exist, or the gap can even increase, preventing convergence to steady state levels in the long run.

20.3.1 A Growth Equation in the Presence of Regional Spillovers

Under the assumption of decreasing returns to scale, it is possible to derive the dynamic path to the equilibrium associated with the growth model above. After log-linearization, using a ˇ rst order Taylor expansion of (20.4) around the steady state, we obtain:

$$(\ln k_{it} - \ln k_{i0}) = (1-e^{-\beta t})(\ln k^* - \ln k_{i0}), \tag{20.8}$$

where $\beta = (1-\tau)(n+d)$ is the usual speed of convergence. In addition, taking into account that:

$$\ln k_{it} = \frac{\ln y_{it} - \ln \Delta - \gamma \ln k_{\rho it}}{\tau} \text{ and } \ln k^* = \frac{\ln y^* - \ln \Delta}{\tau+\gamma}, \tag{20.9}$$

we can obtain (20.8) in terms of labor productivity as,

$$\begin{aligned}(\ln y_{it} - \ln y_{i0}) = \xi - (1-e^{-\beta t})\ln y_{i0} + \gamma(\ln k_{\rho it} - \ln k_{\rho i0}) \\ +\gamma(1-e^{-\beta y})\ln k_{\rho it},\end{aligned} \tag{20.10}$$

where ξ is a constant that measures the level of y in the long-run equilibrium as,

$$\begin{aligned}\xi = (1-e^{-\beta t})\left[\frac{1-\gamma}{1-(\tau-\gamma)}\ln \Delta + \frac{\tau}{1-(\tau+\gamma)}\ln s \right.\\ \left. - \frac{\tau}{1-(\tau+\gamma)}\ln (n+d)\right].\end{aligned} \tag{20.11}$$

From the above expressions three conclusions can be drawn. First, the consideration of regional externalities does not affect the speed of convergence, as β is a function of the usual parameters τ, n and d. Second, two new elements appear in the growth equation: the growth rate of capital per worker at the neighbors, and their initial levels. In the presence of positive interregional spillovers ($\gamma > 0$), both variables increase the growth of productivity in region i. Finally, the assumption of growth diffusion across regions has a positive effect on the steady state of both capital intensity and labor productivity.

20.4 Empirical Specifications

In the preceding section, we describe the main characteristics of a growth model that includes the effects of economic activity in nearby regions. Although we derive different expressions for the steady state levels and the growth rates, empirical counterparts of these expressions are necessary. We note that the variables corresponding to the spillover effect are included in the empirical specifications on the basis of the theoretical model, and hence not as an empirically determined add-on.

20.4.1 An Empirical Production Function with Regional Spillovers

We aim for a specification that allows us to estimate the magnitude of regional spillovers as well as to distinguish between internal and external returns within the region. Following Mankiw *et al.* (1992), we include both physical and human capital in the production function:

$$y_{it} = A_{it} k_{it}^{\theta_k} h_{it}^{\theta_h}, \tag{20.12}$$

where y_{it} is the average level of output per worker in region i during period t, k_{it} and h_{it} the average levels of physical and human capital per worker, respectively, and $\theta_l (l = k,\ h)$ measures the average internal returns at the firm level.[6] As stated above, A_{it} is partially endogenous in this model, reflecting both an externality within region i to investments in k and h, and the technological interdependence across neighboring regional economies. So,

$$A_{it} = \Delta k_{it}^{\delta_k} h_{it}^{\delta_h} A_{\rho it}^{\gamma}, \tag{20.13}$$

where Δ represents the exogenous level of technology, and $\delta_l (l = k,\ h)$ is the measure of external returns to physical and human capital within the region, as caused by the effects of the accumulation of these factors in each region. $A_{\rho it}$ is total factor productivity of the neighbors of region i, collecting the process of diffusion of ideas

[6] If $Y_{ji} = A_{ji} k_{ji}^{\theta_k} h_{ji}^{\theta_h}$ is the production technology for firm j in region i during any time-period, (20.12) can be obtained in the usual manner by averaging across firms in region i, although under the rather strong assumption of homogeneity across firms.

and innovations across close regions, γ being the intensity of these interdependencies. From (20.12) we can rewrite $A_{\rho it}$ as:

$$A_{\rho it} = \frac{y_{\rho it}}{k_{\rho it}^{\theta_k} h_{\rho it}^{\theta_h}}. \tag{20.14}$$

As usual, in order to simplify the final specification, we assume the same value for internal and external returns for all regions, as well as the same intensity for the across-regions spillovers, γ.[7] Then, after substituting (20.13) and (20.14) in (20.12) and log-linearization, we obtain the final expression for the production function as:[8]

$$\begin{aligned} \ln y_{it} &= \ln \Delta + (\theta_k + \delta_k) \ln k_{it} + (\theta_h + \delta_k) \ln h_{it} \\ &\quad + \gamma(\ln y_{\rho it} - \theta_k \ln k_{\rho it} - \theta_h \ln h_{\rho it}). \end{aligned} \tag{20.15}$$

Equation (20.15) shows that $(\theta_k + \delta_k)$ and $(\theta_h + \delta_h)$ reflect the strength of total returns (i.e., internal plus external within the region) to physical and human capital associated with the stock of domestic factors. We can, however, actually obtain an estimate of the internal returns (θ_l) from the parameters associated with factors at the neighbors. In this way, we will be able to obtain an estimate of internal returns to physical and human capital, *social* returns within the region (or externalities within the regional economy), δ_l, and the parameter measuring the externalities across regions, γ. As noted by Ciccone (1996), all of these can be obtained using aggregate regional or national data.

20.4.2 An Empirical Specification for the Growth Equation with Regional Spillovers

When homogeneous data for the stock of capital of a sample of economies are available, estimation of the parameters in (20.10) is straightforward. However, it is commonly the case that the only data available are figures for output per inhabitant or per worker (for instance, for the U.S. states, regions in Europe, or large samples of countries). As a result, (20.10) cannot be estimated in these cases. Nevertheless, based on the assumption that any one region alone is large enough to exert a significant effect on the productivity in the group of neighboring regions as a whole, and that all regions share the same intra-regional return parameters (α and δ),

$$\ln k_{\rho it} = \frac{\ln y_{\rho it} - \ln \Delta}{\tau}. \tag{20.16}$$

[7] Coe and Helpman (1995); Kubo (1995) suggest considering differences in the strength of the externalities depending on the characteristics of each region.

[8] This specification has much in common with the specification in Ciccone (1996). However, Ciccone uses a moving average specification, whereas we use a spatial autoregressive representation.

We obtain the expression for the growth equation in terms of initial productivity and growth at the neighbors by substituting (20.16) in (20.10):

$$(\ln y_{it} - \ln y_{i0}) = \xi' - (1 - e^{-\beta t})\ln y_{i0} + \frac{\gamma}{\tau}(\ln y_{\rho it} - \ln y_{\rho i0}) + \frac{\gamma}{\tau}(1 - e^{-\beta t})\ln y_{\rho i0}, \tag{20.17}$$

with,

$$\xi' = (1 - e^{-\beta t})\left[\frac{\tau^2 - \gamma^2 + \gamma^2(\tau+\gamma)}{\tau(\tau+\gamma)\,[1-(\tau+\gamma)]}\ln\Delta + \frac{\tau}{1-(\tau+\gamma)}\ln s - \frac{\tau}{1-(\tau+\gamma)}\ln(n+d)\right]. \tag{20.18}$$

Note that the coef˜ cients affecting growth and initial levels at the neighbors now depend on the ratio between the external effect and the returns within the region ($\varphi = \gamma/\tau$). It is therefore not feasible to estimate the externality parameter for this model. We can only assess its importance when compared with the returns to reproducible factors within the region.

20.5 The Spatial Econometrics of Considering Externalities Across Economies

The expressions for the production function and the growth equation with externalities, (20.15) and (20.17), respectively, have much in common with spatial process models commonly used in spatial econometrics. In fact, empirically, externalities across economies translate into dependence across the units of analysis.

The production function, with pooled data for n regions and T time-periods, given in (20.15), can be rewritten as:

$$\ln y = \ln\Delta + (\theta_k + \delta_k)\ln k + (\theta_h + \delta_h)\ln h + \gamma(W_1\ln y - \theta_k W_1 \ln k - \theta_h W_1 \ln h) + v,\ v \sim \mathrm{N}(0, \sigma^2 I), \tag{20.19}$$

with a well-behaved error term and variables represented as vectors containing information for each region ($i = 1, ..., n$) and time-period ($t = 1, ..., T$). The vector Δ collects any potential differences in the exogenous level of technology across regional economies and over time.[9] The variables $W_1 \ln y$, $W_1 \ln k$ and $W_1 \ln h$ are the spatial lagged variables for labor productivity, and physical and human capital per worker, respectively, and contain a weighted average of the respective values at the

[9] Following Mankiw *et al.* (1992) and Islam (1995), $\ln\Delta$ can differ for each region since it may not only re ect initial differences in technology, but also in differences in resource endowments, climate, and institutional conditions.

neighbors. The matrix I is the nT by nT identity matrix. Finally, W_1 has the same dimension, and can be expressed as:

$$W_1 = \begin{pmatrix} C^{11} & O & O & . & O \\ O & C^{22} & O & . & O \\ O & O & C^{33} & . & O \\ . & . & . & . & . \\ O & O & O & . & C^{TT} \end{pmatrix}, \tag{20.20}$$

with O being a n by n matrix of zeros and C^{tt} a n by n matrix of spatial weights, where each of its elements, c_{ij}^{tt}, re ects the interaction between region i and region j in period t.

In a similar way, the growth equation in (20.17) can be rewritten as:

$$\begin{aligned} g_y &= a - (1 - e^{-\beta})\ln y + \varphi W_1 g_y \\ &+ \varphi(1 - e^{-\beta}) W_1 \ln y \\ &+ v, \; v \sim \mathrm{N}(0, \sigma^2 I), \end{aligned} \tag{20.21}$$

where for yearly data, g_y denotes annual growth rates, and variables are represented as $(n$ by $(T-1))$ by 1 vectors containing information for each region and time period t (with $t = 2, ..., T$ for g_y, and $t = 1, ..., T-1$ for $\ln y$), and I is the $(n$ by $(T-1))$ by $(n$ by $(T-1))$ identity matrix. As above, $W_1 g_y$ and $W_1 \ln y$ are the spatially lagged variables for the growth rates and the initial level of income, respectively. Finally, a collects any difference in the steady states across economies. It may be composed of variables that mimic the factors included in ξ' in (20.18).

It is important to note that both empirical speciˇ cations differ from the spatial autoregressive (AR) error model. In the case of the growth equation, the spatial AR error model can be expressed as:

$$\begin{cases} g_y = & a - (1 - e^{-\beta})\ln y + u \\ u = & \lambda W_1 u + v, \; v \sim \mathrm{N}(0, \sigma^2 I) \end{cases}, \tag{20.22}$$

which in the so-called common factor hypothesis speciˇ cation (COMFAC) reads as:

$$\begin{aligned} g_y &= (I - \lambda W_1)a - (1 - e^{-\beta})\ln y \\ &+ \lambda W_1 g_y + \lambda(1 - e^{-\beta}) W_1 \ln y \\ &+ v, \; v \sim \mathrm{N}(0, \sigma^2 I). \end{aligned} \tag{20.23}$$

The two formulations are identical if the nonlinear constraints on the parameters are met (see Anselin, 1988b, for details). The restrictions on the parameters involving growth rates and the initial conditions match those in our speciˇ cations, but in the COMFAC representation the spatial lag of the variables affecting the steady state, summarized by a in the empirical speciˇ cation, in uences the growth rates as well. If the COMFAC model is correct, transitional dynamics for an economy do not only depend on the distance to the economy's own steady state, but also on the distance of the neighbors to their steady state. In our model, given in (20.21), the latter distance does not exert any direct in uence.

The same reasoning applies to the production function. In this case, our assumptions imply that the exogenous level of technology (ln Δ) at the neighbors does not directly in uence labor productivity in the economy. It should, however, be noticed that the exogenous level of technology in the whole system affects labor productivity in a given economy through the effect of the lag of the endogenous variable. The same applies to the steady state level in the growth equation. The COMFAC speciˇ cation violates such an assumption. Alternatively, (20.19) and (20.21) can be seen as a mixed regressive-spatial regressive model in the terminology of Anselin (1988b), and Florax and Folmer (1992). In the production function, only the spatial lags of ln *k* and ln *h* are included among the regressors, and in the growth equation the spatial lag of initial income, although with the theoretical restrictions on the parameters.

Rewriting the two empirical models in terms of the speciˇ cations used in spatial econometrics has three major advantages. First, as mentioned above, different hypotheses as to the sources of the externalities can be deˇ ned by means of the speciˇ cation of the weights matrix. Second, the statistical signiˇ cance of regional externalities can be checked. Third, the intensity of the across-region externalities can be quantiˇ ed consistently.

20.5.1 Deˇ nition of the Weights Matrices

The assumption of technological dependence across neighboring regions leads to the appearance of spatial lags in both the production function and the growth equation. By means of precisely deˇ ning the concept of "neighborhood," several hypotheses about the process of technology diffusion can be considered. First, we can identify two possibilities, depending on the timing of the absorption of the external effects: the externality is generated in an economy and incorporated by other economies either within one period of time, or over several time-periods. In the former case, considering contemporaneous spatial dependence will sufˇ ce to account for external effects. This is the assumption in the empirical exercises in which we deal with long-run relationships. In this case, the matrix of weights will be deˇ ned as W_1 in (20.20), and the empirical speciˇ cations are given by (20.19) and (20.21).

The other possibility requires the inclusion of further lag terms in (20.19) and (20.21), referring to the effect on current productivity or growth rates of spatial interactions in previous time-periods. For instance, for a ˇ rst order autoregressive process characterizing the spatial dependence across economies, we can write the growth equation as:

$$\begin{aligned} g_y = {} & a-(1-e^{-\beta})\ln y+\varphi_1 W_1 g_y+\varphi_2 W_2 g_y \\ & +\varphi_1(1-e^{-\beta})W_1 \ln y+\varphi_2(1-e^{-\beta})W_2 \ln y+v, \end{aligned} \tag{20.24}$$

where $v \sim \mathrm{N}(0, \sigma^2 I)$, W_1 is deˇ ned as in (20.20), and W_2 as:

$$W_2 = \begin{pmatrix} O & O & O & . & O \\ C^{21} & O & O & . & O \\ O & C^{32} & O & . & O \\ . & . & . & . & . \\ O & O & . & C^{T(T-1)} & O \end{pmatrix}, \tag{20.25}$$

with $C^{t(t-1)}$ as a matrix of weights where the elements $c_{ij}^{t(t-1)}$ re ect the interaction between region i in period t and region j in period $t-1$. It should be noted that for the parameters of the model to be identiˇ ed, the error term should not show an AR spatial process unless the weight matrices for the lag dependence and the error dependence are different (Anselin, 1988b; Anselin and Florax, 1995c). In (20.24) this should be the case for contemporaneous spatial dependence as well as for spatial dependence with a one-period lag. Consequently, the assumption of non-contemporaneous spatial dependence requires stronger independence assumptions with respect to the error term for the model to be identiˇ ed.

With respect to contemporaneous dependence, the next step is to deˇ ne the matrices C^{tt} on the diagonal of W_1. There are at least three possible speciˇ cations:

1. Adjacent regions take direct advantage of the diffusion of technology. In this case, $W_1 + I_T \otimes C$, where I_T is a T by T identity matrix, and each element c_{ij} of C is deˇ ned as $S_{ij}/\sum_{j=1}^{n} S_{ij}$, where S_{ij} is a contiguity factor that equals 1 when i and j are neighbors and 0 otherwise. The matrix C does not change over time, unless the in uence of each of the contiguous regions is deˇ ned by, for instance, their population or production. In that case, the matrices on the diagonal of W_1 are different for different time-periods.[10]
2. In the context of spatial diffusion, it is logical to think of technology spreading over space, with the in uence regions declining with increasing distance. This is easily incorporated in the weights matrix deˇ ning each element of C as the squared inverse distance between the geographical centers of each region.
3. In line with the literature on technological diffusion among trade partners, we can think of economies exchanging intermediate goods (embodying innovations) as taking advantage of technological improvements and pecuniary externalities in other economies. A possible deˇ nition for the weights of C_{tt} in (20.20) is, for instance, the percentage of goods that region i buys from region j as a share of the total volume of goods imported by the former in each period. This speciˇ cation is in accordance with a number of applied spatial econometric studies, in which different economic priors are used (see, for instance, Case *et al.*, 1993; Molho, 1995; Aten, 1997).

Regardless of the deˇ nition of C, our models hypothesize that changes in the exogenous variables spread throughout the spatial system, because of the presence of the spatial lag of the endogenous variable among the regressors.

---

[10] As noted by the reviewers, there is no a priori reason to assume that the spatial parameters are equal over time, especially when the weights matrix is allowed to change.

The above specifications of the weights matrix satisfy the necessary regularity conditions required to guarantee the desirable properties of estimators and tests. Specifically, the weights are nonnegative and remain finite (see Anselin, 1988b, for further details), they are row-standardized, and in the case of the inverse distance weights matrix the choice of the metric does not have an important bearing on the results.

20.5.2 Testing for the Existence of Externalities Across Economies

Empirical studies that estimate returns to scale and the rate of convergence have dealt with expressions such as (20.19) and (20.21), but they exclude spatially lagged variables. However, erroneous omission of spatially lagged variables, when significant, will cause problems of spatial dependence. This, in turn, evokes misleading statistical inferences for the traditional specifications. Diagnostic checking for spatial dependence in models of this kind is therefore advisable.

Presuming the data fit the empirical models described above, we can expect the Lagrange Multiplier (LM) tests for spatial dependence (Anselin and Florax, 1995c) to reject the null hypotheses of no spatial dependence. This is independent of the nonlinearity in the parameters of the models. The nonlinearity in the parameters does not affect the asymptotic properties of the tests, because under the null hypotheses of no spatial dependence the empirical models are linear.

In the case of the growth equation, the absence of spillovers implies $\varphi = 0$ in (20.21). The model under the null is the traditional growth equation:

$$\begin{aligned} g_y &= a - (1 - e^{-\beta})\ln y + v \\ &= X\theta + v, \;\; v \sim \mathrm{N}(0, \sigma^2 I). \end{aligned} \tag{20.26}$$

The LM test for the null hypothesis $\varphi = 0$ is given by:

$$\mathrm{LM} - \mathrm{EXT} = \left(\frac{\hat{v}' W_1 [g_y + (1 - e^{-\beta})\ln y]}{\hat{\sigma}^2} \right) [T_1 + G]^{-1}, \tag{20.27}$$

where $\hat{v}$ are the residuals of (20.26), $T_1 = \mathrm{tr}(W_1' W_1 + W_1^2)$, and:

$$G = \frac{1}{\hat{\sigma}^2} \left[\left(W_1 [X\hat{\theta} + (1 - e^{-\beta})\ln y] \right)' M \left(W_1 [X\hat{\theta} + (1 - e^{-\beta})\ln y] \right) \right] \tag{20.28}$$

with $M = I - X(X'X)^{-1}X'$. The expression for the test follows immediately, noting that:

$$\frac{\partial \lfloor g_y - h(\cdot) \rfloor}{\partial \varphi} = -W_1 [g_y + (1 - e^{-\beta})\ln y], \tag{20.29}$$

where $h(\cdot)$ is the nonlinear function of the parameters on the right hand side of the empirical model.[11] Under the null hypothesis of no externalities across economies, the LM-EXT test is distributed as χ^2 with one degree of freedom.[12]

In the context of the production function (20.19) the LM-EXT can be used to test for the existence of externalities, although for the statistic to be "identiˇ ed," the external within the economy effects, $\delta_l (l = k, h)$, needs to be zero. This is a consequence of the fact that the parameter for externalities within the economy can only be determined under the alternative of the existence of externalities across economies. Otherwise, we cannot distinguish internal (θ_l) from external intraregional (δ_l) returns.

20.5.3 Estimating the Magnitude of Externalities Across Economies

The inclusion of the spatial lag of the endogenous variable requires estimation methods that guarantee consistency. Some additional subtleties also need to be discussed. Speciˇ cally, the model implies nonlinear restrictions on the parameters. From (20.19) it is clear how the internal to the ˇ rm returns (θ_l) affect the stock in both region i and its neighboring regions. Additionally, the parameter measuring the magnitude of the externalities affects the level of technology at the neighbors, and hence interacts with the θ_l parameters. Similarly, this applies to (20.21), in which the parameter re ecting the magnitude of the externality interacts with the term that involves the rate of convergence measuring the role of the initial level of income at the neighbors. In both cases, the parameter affecting the spatial lag of the endogenous variable is involved in the restrictions. The estimation procedure will have to account for this fact.

A Maximum Likelihood (ML) estimator provides a consistent estimator in the presence of a spatially lagged endogenous variable. In this speciˇ c case, the Likelihood function to be maximized needs to include the restrictions on the parameters. For example, in the case of the growth equation,

$$\begin{aligned} L(\beta,\varphi,\sigma^2) = {} & \sum_i \ln\,(1-\varphi\upsilon_i) - \frac{N\times(T-1)}{2}\ln\,(2\pi) - \frac{N\times(T-1)}{2}\ln\,(\sigma^2) \\ & - \frac{1}{2\sigma^2}[g_y - \varphi W_1 g_y - a + (1-e^{-\beta})\ln y - \varphi(1-e^{-\beta})W_1 \ln y]', \\ & [g_y - \varphi W_1 g_y - a + (1-e^{-\beta})\ln y - \varphi(1-e^{-\beta})W_1 \ln y] \qquad (20.30) \end{aligned}$$

with $\upsilon(i = 1,\ldots,n)$ by $(T-1)$ as the eigenvalues of the weights matrix W_1.

The maximization of (20.30) provides ML estimates of the parameters when the usual conditions are satisˇ ed. The assumption of a different steady state for each economy, operationalized by means of unobservable ˇ xed regional effects, is not feasible because it causes an incidental parameter problem.[13] This is not the

[11] Details about the derivation are available from the authors upon request.

[12] As in the LM-ERR and LM-LAG tests, the degrees of freedom for the test should be T instead of 1 when the spatial parameter is allowed to change over time.

[13] We would like to thank the editors for pointing out this problem.

case for a random effects model. We do not consider a random effects speciˇ cation because, as is well known from the econometrics literature, the unobservable effects are correlated with the regressors, causing inconsistency in the estimation of the parameters. This is obviously the case for both the production function and the growth equation. Ideally, we would want to include variables approaching differences across economies in the exogenous level of technology and the steady state. Unfortunately, this is not feasible because of lack of data. We therefore adopt an intermediate solution, and approximate regional differences in the level of technology and the steady state by means of dummy variables for different *types* of regions.

Our empirical speciˇ cation is identiˇ ed because the columns of the matrix of the pseudo-regressors are linearly independent, both in the case of the growth equation and the production function. Hence, parameter estimates under the constraints implied by the functional form can be obtained from the data. As before, a problem arises in the production function when $\gamma = 0$, preventing the identiˇ cation of internal and external returns within the economy.

With this in mind, the maximization of (20.30) can be achieved by a standard optimization routine. The estimation process can be simpliˇ ed slightly by adapting the procedure for a spatial endogenous lag model suggested in Anselin (1988b). Given that the spatial parameter, φ in (20.21), is involved in the restriction, the estimation process proceeds along the following steps:

1. For values of $0 \leq \varphi \leq 1$,[14] compute the matrix of pseudo-regressors to be used in step 2 for the nonlinear speciˇ cation:

$$X_0 = \frac{\partial f(\cdot)}{\partial \theta'}, \tag{20.31}$$

   where θ is the vector of parameters, and $f(\cdot) = a - (1 - e^{-\beta})\ln y + \varphi(1 - e^{-\beta})W_1 \ln y$, as can be seen using the fact that the term in brackets in (20.30) can be written as $Ag_y - f(\cdot)$, with $A = I - \varphi W_1$.
2. For each value of φ carry out ordinary least square (OLS) of X_0 on g_y, and of X_0 on Wg_y. This yields $\hat{\theta}_0$ and $\hat{\theta}_L$.
3. Compute the residuals $\hat{v}_0$ and $\hat{v}_L$, and evaluate the concentrated likelihood for each value of φ as:

$$\begin{aligned} L_C = {} & \sum_i \ln\,(1 - \varphi\upsilon_i) - \frac{N \times (T-1)}{2} \ln\,(2\pi) \\ & - \frac{N \times (T-1)}{2} \ln \left[\frac{N \times (T-1)}{1} (\hat{v}_0 - \varphi\hat{v}_L)'\,(\hat{v}_0 - \varphi\hat{v}_L) \right] \\ & - \frac{N \times (T-1)}{2} \end{aligned} \tag{20.32}$$

[14] We constrain φ in order to exclude the possibility of negative spillovers, and returns to external effects which are higher than returns within the economy. Negative values for φ considerably reduce the likelihood in all cases.

4. Select the value of φ maximizing L_C, and compute the associated estimate of θ.

The same procedure can be applied with respect to the production function including external effects.

20.6 Empirical Evidence

In this section, we report results on the relevance of regional spillovers in the production function and in the growth equation, for two different samples of regional economies. In the case of the production function, we present results for Spanish regions, and we analyze European regions using the growth equation. The unavailability of homogeneous data for physical and human capital in the E.U. precludes the estimation of the production function for the set of all European regions. Results in this section are obtained using code in Gauss v.3.2.8 (available upon request).

20.6.1 Externalities Across Economies in the Production Function: Results for Spanish Regions

We estimate the production function for data on the Spanish regions at the NUTSII Eurostat level for the period 1964–93. Data are obtained from three sources. Gross value added at constant 1990 prices and employment figures are taken from *Renta Nacional de España y su distribución provincial*, published by Banco Bilbao-Vizcaya. The net stock of privately held physical capital, measured in constant prices, is taken from Fundación BBV (1995). Human capital is operationalized as the fraction of employed that at least started secondary schooling, taken from Pérez and Serrano (1998). Because the first source publishes data every two years except for 1964 and 1967, we have information for 15 periods. The panel data set therefore contains 255 observations, consisting of 17 regions and 15 time-periods.

We define two different weights matrices, one measures physical contiguity and another, trade flows between regions.[15] In the latter matrix the factor S_{ij} is operationalized as the total volume of transported goods by road and train with origin in region j and destination in region i. These data are available from the Eurostat REGIO database.[16]

Although our empirical model incorporates regional externalities on the basis of theoretical considerations, we follow the reasoning in Sect. 20.5 and start by estimating the production function without spatial spillovers in order to determine whether their omission causes spatial dependence in the residuals. As mentioned above, we introduce a regional dummy to capture spatial heterogeneity in the exogenous level of technology. It takes the value of one for regions with higher than average productivity over the entire period. We also include a time-trend to pick

15 The scale differences of Spanish regions and their small number prevent the square inverse matrix from picking up the real proximity across regions in this case.

16 We use the average volume from 1988 to 1992 as homogeneous trade, since data prior to 1988 are not available. The submatrices on the diagonal of W_1 are the same for each year.

Table 20.1. Results for the production function without externalities across economies for the Spanish regions (OLS)

		OLS
Intraregional returns to physical capital $(\theta_k + \delta_k)$		0.309
		(0.029)
Intraregional returns to human capital $(\theta_h + \delta_h)$		0.486
		(0.024)
$\ln L$		316.265
Spatial statistics	First order contiguity	Trade flows
LM-ERR	11.473	3.568
	[0.001]	[0.059]
LM-LAG	7.564	5.353
	[0.006]	[0.021]
LM-EXT$(H_0: \gamma = 0)$	11.393	4.158
	[0.001]	[0.041]

Note: Robust standard errors in parentheses, and probability values in square brackets. A time-trend and the regional dummy described in the main text are included in the estimation. The dataset contains 17 regions and 15 time-periods.

up exogenous technical progress. The spatial dummy and the time-trend are significant in all estimation runs, with the regional dummy indicating a higher level of exogenous technology in the more productive regions.

Results for the OLS estimation are summarized in Table 20.1. We mentioned above that this specification does not allow the separation of intraregional internal and external returns to physical and human capital. In the case of physical capital we obtain $\theta_k + \delta_k = 0.31$, which approximately corresponds to the share of physical capital in national income in developed economies. For human capital, intraregional returns are equal to 0.48. Both results are highly significant. With respect to spatial autocorrelation, the results of the traditional LM tests (LM-ERR and LM-LAG) clearly indicate that the null hypothesis of no spatial autocorrelation should be rejected. This applies to the first order contiguity matrix as well as the trade flows matrix. The LM-EXT statistic defined in (20.27) specifically tests for the existence of externalities. The results clearly support our hypothesis that for Spanish regions diffusion of technology is relevant, between contiguous regions as well as between trade partners.

The ML estimation of (20.19) is summarized in Table 20.2. The Likelihood Ratio (LR) test for the joint significance of the spatial lags, computed using the two spatial matrices, confirms the appropriateness of the empirical model. Concerning the contiguity matrix, we obtain $\theta_k + \delta_k = 0.28$ and $\theta_k = 0.22$, indicating the existence of a positive intraregional externality to physical capital, δ_k. In the case of human capital, we find $\theta_h + \delta_h = 0.49$ and $\theta_h = 0.46$. Consequently, over 20 percent

Table 20.2. Results for the production function with externalities across economies for the Spanish regions (ML)

	ML	
	First order contiguity	Trade ows
Intraregional returns to physical capital $(\theta_k + \delta_k)$	0.284	0.283
	(0.034)	(0.032)
Intraregional returns to human capital $(\theta_h + \delta_h)$	0.489	0.498
	(0.025)	(0.025)
Externality across regions γ	0.276	0.223
	(0.077)	(0.096)
Internal returns to physical capital (θ_k)	0.216	0.096
	(0.202)	(0.281)
Internal returns to human capital (θ_h)	0.456	0.374
	(0.176)	(0.321)
ln L	322.863	319.619
LR Test[1]	13.195	6.708
	[0.004]	[0.082]

See the note to Table 20.1.

[1] Likelihood Ratio Test for the global signiˇ cance of the spatial regressors.

of the returns to accumulation in physical capital is due to externalities related to the production factor within the economy. This implies that the larger the aggregate stock of physical capital in an economy, the higher the total effect of new individual investments in this factor. This is in accordance with the results obtained by Ciccone (1996), for a wider sample of countries. Our estimates reveal a much smaller intraregional effect to externalities related to human capital investments (approximately 6 percent). Although smaller in magnitude, this effect supports the reasoning in Lucas (1988).

The estimate for γ equals 0.28 for the intensity of regional externalities, and is signiˇ cantly different from zero. This implies that the hypothesis of the absence of interregional spillovers is strongly rejected, and signals that a 10 percent increase in the level of total factor productivity of the neighbors, raises the level of technology in the region by almost 3 percent. This result supports the idea of strong technological spillovers or interdependencies across neighboring regions. The estimated value is lower than the estimate obtained by Ciccone (1996) for a wide sample of countries,[17] although it is in line with the results for European regions obtained by Fingleton and McCombie (1998).

Table 20.2 also shows the results for the speciˇ cation using the trade ows weights matrix. The results are rather similar, which may be due to trade mainly

[17] Ciccone obtains a value of 0.58 for international technology spillovers using the Mankiw *et al.* (1992) sample of 98 countries.

occurring between contiguous economies. What is surprising, however, are the high values for the intraregional externalities, especially for physical capital, although caution is required because of the large estimated standard errors for θ_k and θ_h.

We also estimate the model, for both specifications of the weight matrix, without imposing restrictions on the parameters. The test results reveal that the restrictions are strongly supported by the data, and indicate that the results for the Spanish regions support our hypothesis of the relevance of externalities across regional economies in the production process.

20.6.2 Externalities Across Economies in the Growth Equation: Results for European Regions

We examine 108 regions in the E.U. over the time-period 1975–92 to test for the presence of across-region spillovers in the growth equation. The main data source is the Eurostat REGIO database, complemented with other sources at the national level. The variable of interest is labor productivity, divided by the average value for the E.U. Lack of homogeneous data for capital stocks prevents the estimation of (20.10), and we therefore estimate (20.17) instead.

Table 20.3 summarizes the results for the growth equation prior to including the spatial externalities. The estimation is carried out for a cross-section dataset (where the endogenous variable is the labor productivity growth rate over the period 1975–92, and the initial income level is measured in 1975) and for a dataset with pooled yearly data. In order to reflect differences in the steady state for groups of European regions, we include dummy variables for three categories of regions: those with less than 80 percent of the average labor productivity in the E.U., between 80 and 110 percent, and more than 110 percent. The dummy variables are significant in all estimation runs, indicating the a priori expected differences in steady states. For the cross-section dataset, we observe significant, though slow convergence within each group of regions. The rate of convergence is slightly higher than the usual 2 percent per year (2.3 percent). The convergence rate equals 3.3 percent for the pooled dataset.

The spatial dimensions are included using two alternative specifications of the weights matrix, the physical first order contiguity matrix and the squared inverse distance matrix, both row-standardized. Unfortunately, data for trade flows between E.U. regions are not available. However, if trade flows decrease with distance we may interpret the results for the inverse distance matrix as a good proxy for externalities related to traded goods. The significance of the LM-ERR and the LM-LAG tests in Table 20.3 show that the omission of interregional externalities causes spatial dependence in the residuals of the traditional growth equation. The LM-EXT statistic, testing for the significance of the externality with the proposed empirical model as the alternative hypothesis, clearly rejects the null hypothesis of no externalities across economies. This is the case for both the cross-section and the pooled sample. Subsequently, following the same reasoning as in the Spanish case, the model in (20.21) is estimated.

Table 20.3. Results for the growth equation without externalities across economies for the European regions (OLS)

	OLS			
	Cross-section data		Pooled data	
Rate of convergence (β)	0.023		0.033	
	(0.005)		(0.005)	
ln L	67.608		2756.178	
Spatial statistics	First order contiguity	Squared inverse distance	First order contiguity	Squared inverse distance
LM-ERR	38.727	58.470	1170.772	1872.528
	[0.000]	[0.000]	[0.000]	[0.000]
LM-LAG	34.493	43.874	1162.923	1838.166
	[0.000]	[0.000]	[0.000]	[0.000]
LM-EXT	22.520	22.336	1125.199	1714.875
($H_0 : \varphi = 0$)	[0.000]	[0.000]	[0.000]	[0.000]

Note: Robust standard errors in parentheses, and probability values in square brackets. Dummies for three categories of regions as described in the main text are included in the estimation. The dataset contains 108 regions and 17 time-periods.

The results for the ML estimation are summarized in Table 20.4. For the cross-section dataset and the ˇ rst order contiguity matrix, the estimate of φ indicates that the effects of the interregional spillovers amount to approximately two-thirds of the returns within the region. This implies that, under the assumption of the share of physical capital in total income being approximately 0.3, the externalities across economies to capital accumulation will be close to 0.2. This result is quite similar to the value observed in the preceding analysis for the Spanish regions. The estimates also reveal that a one percentage-point increase in labor productivity in neighboring regions causes an increase of 0.011 in the regional growth rate. At the same time, the effect of the acceleration in growth rates of the neighbors is of crucial importance: a one percentage-point increase in the weighted growth rate of neighboring regions is associated with an increase of 0.63 in the regional growth rate. Following the reasoning in Sect. 20.2, we can think of this process as the income level of the neighbors affecting economic growth of a region because of technological and/or pecuniary spillovers. This effect is concerned with supply-side externalities. A high value for the implied parameter affecting $W_1 g_y$ shows that a large proportion of regional growth is due to a contagion effect (growth rates are greater when neighboring economies are also growing at high rates, and smaller when neighboring economies are stagnating or growing slowly). This effect can be considered a demand-side externality, following from, for instance, demand from neighbors for ˇ nal goods or inputs produced in the region under consideration. Following Caballero and Lyons' seminal paper 1990, this type of externality is frequently consid-

Table 20.4. Results for the growth equation without externalities across economies for the European regions (ML)

	ML			
	Cross-section data		Pooled data	
	First order contiguity	Squared inverse distance	First order contiguity	Squared inverse distance
Rate of convergence (β)	0.022	0.024	0.032	0.035
	(0.005)	(0.005)	(0.006)	(0.006)
Relative externality (φ)	0.630	0.840	0.686	0.919
	(0.063)	(0.069)	(0.016)	(0.011)
Implied $\varphi(1-e^{-\beta})$[1]	0.011	0.016	0.021	0.031
	(0.003)	(0.004)	(0.004)	(0.005)
ln L	88.912	89.020	3297.762	3309.17
LR Test[2]	42.608	42.824	1083.168	1111.984
	[0.000]	[0.000]	[0.000]	[0.000]

See the note to Table 20.3.

[1] In the cross-section model the implied parameter is given by $\varphi(1-e^{-\beta T})/T$.

[2] Likelihood Ratio Test for the global signiˇ cance of the spatial regressors.

ered for the case of industries. It should be noted that the rate of convergence is not affected by the omission of across-region externalities. Finally, the use of a squared inverse distance matrix results in higher estimates for the supply and demand-side externalities.

Results for the pooled data are qualitatively similar. For the ˇ rst order contiguity matrix, the supply and demand-side externalities are highly signiˇ cant. In this estimation, a one-point increase in the growth rate of the neighbors leads to an increase of approximately 0.7 points in the regional growth rate. This estimate is higher than the corresponding estimate for the cross-section data. The difference may be due to the demand-side externality also capturing a pro-cyclical relationship in growth among neighboring economies, since we are considering annual growth rates. The results also show that a one-point increase in labor productivity of the neighbors implies a 0.02 point increase in the regional growth rate. The rate of convergence merely shows minor changes as compared to the estimation omitting the externalities. Finally, for the speciˇ cation using the squared inverse distance matrix, the estimated parameters are slightly higher, but the conclusions are very similar to those obtained for the speciˇ cation with the contiguity matrix.

20.7 Conclusions

We consider the role of externalities across regional economies in explaining economic growth, in this chapter. A simple growth model, in which the regional level of technology depends on the level of technology in neighboring regions, is presented. The model presumes that technology is a function of capital stock and of the ow of innovations and ideas across neighboring regions. From this simple model, we deduct that the regional growth rate is a (positive) function of the capital stock of neighboring regions. Such a relationship may counteract the neoclassical tendency of decreasing returns to capital in a regional setting. Subsequently, we derive a growth equation approximating the transitional dynamics to the steady state. It shows how growth rates are positively affected by both investments and the existing stock of neighboring economies, leaving aside the fact that interregional externalities contribute to raising the steady state level. It should be noted that the parameter measuring the speed of convergence is not affected by the spillover effect.

Spatial econometric techniques provide a natural framework to test for the occurrence of interregional externalities, and to estimate their magnitude. We present and discuss operational empirical models for both the production function and the growth equation. These models necessitate slight and subtle changes to the tests for spatial effects and the spatial Maximum Likelihood estimators. By means of differentiating the operationalization of the weights matrix, we are able to distinguish differing hypotheses regarding the interregional interaction through externalities. In the empirical part of the chapter, we establish the relevance of these externalities for regions of Spain as well as for regions in Europe.

The empirical results presented in this chapter have several implications for regional policy. First, policies aimed at stimulating economic growth in lesser developed regions should account for the spillover effects with neighboring regions. Our results suggest, that a region surrounded by prosperous economies can achieve higher growth rates. The reverse is also true. Similarly, coordinated investments in groups of lagging regions is likely to be more successful than isolated investment-inducing actions. This ˇ nding supports the desirability of creating supra-regional agencies promoting regional investment, because supra-regional agencies are prone to internalizing the spatial externalities. Second, the prevalence of interregional externalities can create a "poverty trap" based on geographical location. The fundamental question is of course, how can an economy escape the poverty trap? Clearly, the required efforts will be less when neighbors simultaneously invest resources. If not, individual efforts may be sub-optimal.

In future research, we will perform a more exhaustive analysis of the channels through which interregional spread of externalities may take place. We assume stability in the parameters of the model, particularly the spatial dependence parameters, in the current analyses. Relaxing this assumption may have non-negligible effects on the outcome of the analysis. The same holds for our current lack of specifying appropriate dynamics in the process of interregional diffusion of externalities. It would also be interesting to apply panel data techniques different from the ones used in this chapter. At the same time, continuing the development and adaptation of spatial econometric tools to handle space-time data sets will be conducive to a continued and intensiˇ ed in-depth treatment of various issues in applied economic growth research.

Acknowledgements

We would like to thank the editors for helpful comments and suggestions. Part of this research was ˇ nanced by the *Plan Nacional de I+D*, Spanish Ministry of Education, Project 2FD97-1004-C03-01.

References

Abbot A. 1997. Of time and space: The contemporary relevance of the Chicago school, *Social Forces*, 75: 1149–1182.

Acs Z., Anselin L. and Varga A. 2002. Patents and innovation counts as measures of regional production of new knowledge, *Research Policy*, 31: 1069–1085.

Ades A. and Chua H. 1997. The neighbor's curse: Regional instability and economic growth, *Journal of Economic Growth*, 2: 279–304.

Advisory Commission on Intergovernmental Relations 1985. *Cigarette Tax Evasion: A Second Look*, ACIR, Washington, DC.

Agnihotri S., Palmer-Jones R. and Parikh A. 2002. Missing women in Indian districts: A quantitative analysis, *Structural Change and Economic Dynamics*, 13: 285–314.

Ahn H. and Powell J.L. 1993. Semiparametric estimation of censored selection models with a nonparametric selection mechanism, *Journal of Econometrics*, 58: 3–29.

Aizer A. and Currie J. 2002. *Networks or Neighborhoods? Correlations in the Use of Publicly-Funded Maternity Care in California*, Working Paper 9209, NBER, Cambridge, MA.

Akerlof G.A. 1997. Social distance and social decisions, *Econometrica*, 65: 1005–1027.

Albert J.H. and Chib S. 1993. Bayesian analysis of binary and polychotomous response data, *Journal of the American Statistical Association*, 88: 669–679.

Amable B. 1993. Catch-up and convergence: A model of cumulative growth, *International Review of Applied Economics*, 7: 1–25.

Amemiya T. 1971. The estimation of the variances in a variance components model, *International Economic Review*, 12: 1–13.

Amemiya T. 1985. *Advanced Econometrics*, Harvard University Press, Cambridge, MA.

Anas A. 1981. The estimation of multinomial logit models of joint location and travel mode choice from aggregated data, *Journal of Regional Science*, 21: 223–243.

Anas A. 1983. Discrete choice theory, information theory, and the multinomial logit and gravity models, *Transportation Research B*, 17: 13–23.

Anselin L. 1980. Estimation methods for spatial autoregressive structures, Ph.D. thesis, Cornell University.

Anselin L. 1981. Small sample properties of estimators for the linear model with a spatial autoregressive structure in the disturbance, *Modeling and Simulation*, 12: 899–904.

Anselin L. 1986. Some further notes on spatial models and regional science, *Journal of Regional Science*, 26: 799–802.

Anselin L. 1988a. Lagrange Multiplier test diagnostics for spatial dependence and spatial heterogeneity, *Geographical Analysis*, 20: 1–17.

Anselin L. 1988b. *Spatial Econometrics, Methods and Models*, Kluwer Academic, Boston, MA.

Anselin L. 1990. Some robust approaches to testing and estimation in spatial econometrics, *Regional Science and Urban Economics*, 20: 141–163.

Anselin L. 1992. *SpaceStat, a Software Program for Analysis of Spatial Data*, National Center for Geographic Information and Analysis (NCGIA), University of California, Santa Barbara, CA.

Anselin L. 1993. Discrete space autoregressive models, in Goodchild M.F., Parks B.O. and Steyaert T. (eds.) *Environmental Modeling With GIS*, Oxford University Press, Oxford, UK, 454–469.

Anselin L. 1996. The Moran scatterplot as an exploratory spatial data analysis tool to assess local instability in spatial association, in Fischer M.M., Scholten H.J. and Unwin D. (eds.) *Spatial analytical perspectives on GIS*, Taylor and Francis, London, UK, 111–125.

Anselin L. 1999. Interactive techniques and explanatory spatial data analysis, in Longley P.A., Goodchild M.F., Maguire D.J. and Rhind D.W. (eds.) *Geographical Information Systems: Principles, Techniques, Management and Applications*, John Wiley and Sons, New York, NY, 251–264.

Anselin L. 2000. Computing environments for spatial data analysis, *Journal of Geographical Systems*, 2: 201–220.

Anselin L. 2001a. Rao's score test in spatial econometrics, *Journal of Statistical Planning and Inference*, 97: 113–139.

Anselin L. 2001b. Spatial econometrics, in Baltagi B. (ed.) *A Companion to Theoretical Econometrics*, Basil Blackwell, Oxford, UK, 310–330.

Anselin L. 2001c. Spatial effects in econometric practice in environmental and resource economics, *American Journal of Agricultural Economics*, 83: 705–710.

Anselin L. 2002. Under the hood: Issues in the speciˇ cation and interpretation of spatial regression models, *Agricultural Economics*, 17: 247–267.

Anselin L. 2003a. *GeoDa 0.9 User's Guide*, Spatial Analysis Laboratory (SAL). Department of Agricultural and Consumer Economics, University of Illinois, Urbana-Champaign, IL.

Anselin L. 2003b. Spatial externalities, *International Regional Science Review*, 26: 147–152.

Anselin L. 2003c. Spatial externalities, spatial multipliers and spatial econometrics, *International Regional Science Review*, 26: 153–166.

Anselin L. and Bera A.K. 1998. Spatial dependence in linear regression models with an introduction to spatial econometrics, in Ullah A. and Giles D. (eds.) *Handbook of Applied Economic Statistics*, Marcel Dekker, New York, NY, 237–289.

Anselin L. and Florax R.J.G.M. 1995a. Introduction, in Anselin L. and Florax R.J.G.M. (eds.) *New Directions in Spatial Econometrics*, Springer-Verlag, Berlin, Germany, 3–18.

Anselin L. and Florax R.J.G.M. 1995b. *New Directions in Spatial Econometrics*, Springer-Verlag, Berlin, Germany.

Anselin L. and Florax R.J.G.M. 1995c. Small sample properties of tests for spatial dependence in regression models: Some further results, in Anselin L. and Florax

R.J.G.M. (eds.) *New Directions in Spatial Econometrics*, Springer-Verlag, Berlin, Germany, 21–74.

Anselin L. and Le Gallo J. 2003. *Panel Data Spatial Econometrics with PySpace*, Spatial Analysis Laboratory (SAL). Department of Agricultural and Consumer Economics, University of Illinois, Urbana-Champaign, IL.

Anselin L. and Grif˜ th D. 1988. Do spatial effects really matter in regression analysis, *Papers, Regional Science Association*, 65: 11–34.

Anselin L. and Kelejian H.H. 1997. Testing for spatial error autocorrelation in the presence of endogenous regressors, *International Regional Science Review*, 20: 153–180.

Anselin L. and Moreno R. 2003. Properties of tests for spatial error components, *Regional Science and Urban Economics*, 33: 595–618.

Anselin L. and Rey S.J. 1991. Properties of tests for spatial dependence in linear regression models, *Geographical Analysis*, 23: 110–131.

Anselin L. and Rey S.J. 1997. Introduction to the special issue on spatial econometrics, *International Regional Science Review*, 20: 1–7.

Anselin L. and Rey S.J. 2002. *New Tools for Spatial Data Analysis: Proceedings of a Workshop*, Center for Spatially Integrated Social Science, University of California, Santa Barbara, CA (CD-ROM).

Anselin L., Bera A.K., Florax R.J.G.M. and Yoon M.J. 1996. Simple diagnostic tests for spatial dependence, *Regional Science and Urban Economics*, 26: 77–104.

Anselin L., Varga A. and Acs Z. 1997. Local geographic spillovers between university research and high technology innovations, *Journal of Urban Economics*, 42: 422–448.

Anselin L., Rey S.J. and Talen E. 2000. The expanded and revised IRSR subject and author index, *International Regional Science Review*, 23: 345–349.

Armstrong H.W. 1995. Convergence among regions of the European Union 1950–1990, *Papers in Regional Science*, 74: 143–152.

Arrow K.J. 1962. The economic implications of learning by doing, *Review of Economic Studies*, 29: 155–173.

Aschauer D. 1989. Is public expenditure productive, *Journal of Monetary Economics*, 23: 177–200.

Ashenfelter O., Harmon C. and Oosterbeek H. 1999. A review of estimates of the schooling/earnings relationship, with tests for publication bias, *Labour Economics*, 6: 453–470.

Aten B.H. 1997. Does space matter? International comparisons of the prices of tradables and nontradables, *International Regional Science Review*, 20: 35–52.

Atkinson S. and Crocker T. 1987. Bayesian approach to assessing the robustness of hedonic property value studies, *Journal of Applied Econometrics*, 1: 27–45.

Avery R.B., Hansen L.P. and Holtz V.J. 1983. Multiperiod probit models and orthogonality condition estimation, *International Economic Review*, 24: 21–35.

Azariadis C. and Drazen A. 1990. Threshold externalities in economic development, *Quarterly Journal of Economics*, 105: 501–526.

Baillie R. and Baltagi B. 1998. Prediction from the regression model with one-way error components, in Pesaran H., Lahiri K., Hsiao C. and Lee L.F. (eds.) *Analysis*

of Panel Data and Limited Dependent Variable Models, Cambridge University Press, Cambridge, UK.
Baldwin R. 1992. Measurable dynamic gains from trade, *Journal of Political Economy*, 100: 162–174.
Baller R.D. and Richardson K. 2002. Social integration, imitation and the geographic patterning of suicide, *American Sociological Review*, 67: 873–888.
Baller R.D., Anselin L., Messner S.F., Deane G. and Hawkins D. 2001. Structural covariates of U.S. county homicide rates: Incorporating spatial effects, *Criminology*, 39: 561–590.
Baltagi B. 1995. *Econometric Analysis of Panel Data*, John Wiley and Sons, Chichester, UK.
Baltagi B. 2001. *Econometric Analysis of Panel Data (Second Edition)*, John Wiley and Sons, Chichester, UK.
Baltagi B. and Levin D. 1986. Estimating dynamic demand for cigarettes using panel data, *Review of Economics and Statistics*, 48: 148–155.
Baltagi B. and Li D. 2001a. Double length arti˘ cial regressions for testing spatial dependence, *Econometric Reviews*, 20: 31–40.
Baltagi B. and Li D. 2001b. LM tests for functional form and spatial error correlation, *International Regional Science Review*, 24: 194–225.
Baltagi B., Song S.H. and Koh W. 2003. Testing panel data regression models with spatial error correlation, *Journal of Econometrics*, in press.
Bao S., Barkley D.L. and Henry M.S. 1995. RAS-an integrated regional analysis system with ARC/INFO, *Computers, Environment, and Urban Systems*, 19: 37–56.
Barkley D.L., Henry M.S. and Bao S. 1994. Metropolitan growth: Boon or bane to nearby rural areas, *Choices*, 14–19.
Barrett S. 1994. Strategic environmental policy and international trade, *Journal of Public Economics*, 54: 325–338.
Barro R. 1991. Economic growth in a cross section of countries, *Quarterly Journal of Economics*, 106: 407–443.
Barro R. 1997. *Determinants of Economic Growth: A Cross-Country Empirical Study*, MIT Press, Cambridge, MA.
Barro R. and Sala-i-Martin X. 1992. Convergence, *Journal of Political Economy*, 100: 223–251.
Barro R. and Sala-i-Martin X. 1995. *Economic Growth*, McGraw Hill Inc, New York, NY.
Barry R.P. and Pace R.K. 1999. Monte Carlo estimates of the log determinant of large sparse matrices, *Linear Algebra and its Applications*, 289: 41–54.
Bartels C. and Hordijk L. 1977. On the power of the generalized Moran contiguity coef˘ cient in testing for spatial autocorrelation among regression disturbances, *Regional Science and Urban Economics*, 7: 83–101.
Bartelsman E., Caballero R. and Lyons R. 1994. Customer- and supplier-driven externalities, *American Economic Review*, 84: 1075–1085.
Bartik T. 1987. The estimation of demand parameters in hedonic models, *Journal of Political Economy*, 95: 81–88.

Bastian C.T., McLeod D.M., Germino M.J., Reiners W.A. and Blasko B.J. 2002. Environmental amenities and agricultural land values: A hedonic model using geographic information systems data, *Ecological Economics*, 40: 337–349.

Bavaud F. 1998. Models for spatial weights: A systematic look, *Geographical Analysis*, 30: 153–171.

Baybeck B. and Huckfeldt R. 2002. Spatially dispersed ties among interdependent citizens: Connecting individuals and aggregates, *Political Analysis*, 10: 261–275.

Becker G., Grossman M. and Murphy K. 1994. An empirical analysis of cigarette addiction, *American Economic Review*, 84: 396–418.

Becker R.A., Chambers J.M. and Wilks A.R. 1998. *The New S Language*, Chapman and Hall, London, UK.

Bell K.P. and Bockstael N.E. 2000. Applying the generalized moments estimation approach to spatial problems involving microlevel data, *The Review of Economics and Statistics*, 82: 72–82.

Belsley D.A., Kuh E. and Welsch R.E. 1980. *Regression Diagnostics: Identifying In uential Data and Source of Collinearity*, John Wiley and Sons, New York, NY.

Benhabib J. and Spiegel M. 1994. The role of human capital in economic development: Evidence for aggregate cross-country rate, *Journal of Monetary Economics*, 34: 143–173.

Bera A.K. and Ullah A. 1991. Rao's score test in econometrics, *Journal of Quantitative Economics*, 7: 189–220.

Bera A.K. and Yoon M.J. 1993. Speciˇ cation testing with locally misspeciˇ ed alternatives, *Econometric Theory*, 9: 649–658.

Bernat G. 1996. Does manufacturing matter? A spatial econometric view of Kaldor's laws, *Journal of Regional Science*, 36: 463–477.

Berndt E. and Hanson B. 1992. Measuring the contribution of public infrastructure capital in Sweden, *Scandinavian Journal of Economics*, 94: 151–172.

Berndt E.R. 1991. *The Practice of Econometrics: Classic and Contemporary*, Addison-Wesley, Cambridge, MA.

Beron K.J. and Vijverberg W.P. 2003. Probit in a spatial context: A Monte Carlo aproach, in Anselin L., Florax R.J.G.M. and Rey S.J. (eds.) *Advances in Spatial Econometrics*, Springer-Verlag, Heidelberg.

Beron K.J., Murdoch J.C. and Vijverberg W.P. 1996. *Why Cooperate? An Independent Probit Model of Network Correlations*, Working paper, School of Social Sciences, University of Texas at Dallas, Richardson, TX 75083.

Beron K.J., Murdoch J.C. and Vijverberg W.P. 2003. Why cooperate? public goods, economic power, and the Montreal Protocol, *The Review of Economics and Statistics*, 85: 286–297.

Besag J.P., Green D.H. and Mengersen K. 1995. Bayesian computation and stochastic systems, *Statistical Science*, 10: 3–66.

Best N.G., Arnold R.A., Thomas A., Waller L.A. and Conlon E.M. 1999. Bayesian models for spatially correlated disease and exposure data, in Bernardo J., Berger J., Dawid A. and Smith F. (eds.) *Bayesian Statistics 6*, Oxford University Press, New York, NY, 131–156.

Bijmolt T. and Pieters R. 2001. Meta-analysis in marketing when studies contain multiple measurements, *Marketing Letters*, 12: 157–169.

Bivand R.S. 2001. More on spatial data analysis, *R News*, 1: 13–17.

Bivand R.S. 2002a. Implementing spatial data analysis software tools in R, in Anselin L. and Rey S.J. (eds.) *New Tools for Spatial Data Analysis: Proceedings of a Workshop*, Center for Spatially Integrated Social Science, University of California, Santa Barbara, CA (CD-ROM).

Bivand R.S. 2002b. Spatial econometrics functions in R: Classes and methods, *Journal of Geographical Systems*, 4: 405–421.

Bivand R.S. and Gebhardt A. 2000. Implementing functions for spatial statistical analysis using the R language, *Journal of Geographical Systems*, 2: 307–317.

Bivand R.S. and Szymanski S. 1997. Spatial dependence through local yardstick competition: Theory and testing, *Economics Letters*, 55: 257–265.

Blommestein H.J. and Koper N.A. 1998. The in uence of sample size on the degree of redundancy in spatial lag operators, *Journal of Econometrics*, 82: 317–333.

Boarnet M.G. 1992. Intra-metropolitan growth patterns: The nature and causes of population and employment changes within an urban area, Ph.D. thesis, Princeton University.

Boarnet M.G. 1994a. An empirical model of intrametropolitan population and employment, *Papers in Regional Science*, 73: 135–153.

Boarnet M.G. 1994b. The monocentric model and employment location, *Journal of Urban Economics*, 36: 79–97.

Boarnet M.G. 1998. Spillovers and the locational effects of public infrastructure, *Journal of Regional Science*, 38: 381–400.

Boarnet M.G. and Glazer A. 2002. Federal grants and yardstick competition, *Journal of Urban Economics*, 52: 53–64.

Bockstael N.E. 1996. Modeling economics and ecology: The importance of a spatial perpective, *American Journal of Agricultural Economics*, 78: 1168–1180.

Bockstael N.E. and Geoghegan J. 1999. *The Supply of Sprawl*, Working paper, Dept. of Agricultural and Resource Economics, University of Maryland, College Park, MD.

Bogue D. 1953. *Population Growth in Standard Metropolitan Areas 1900–1950*, Scripts Foundation in Research in Population Problems, Oxford, OH.

Bolduc D., Fortin B. and Fournier M.A. 1996. The effect of incentive policies on the practice location of doctors: A multinomial probit analysis, *Journal of Labor Economics*, 14: 703–732.

Bolduc D., Fortin B. and Gordon S. 1997. Multinomial probit estimation of spatially interdependent choices: An empirical comparison of two new techniques, *International Regional Science Review*, 20: 77–101.

Bommer R. and Schulze G. 1999. Environmental improvement with trade liberalization, *European Journal of Political Economy*, 15: 639–661.

Börsch-Supan A. and Hajivassiliou V.A. 1993. Smooth unbiased multivariate probability simulators for Maximum Likelihood estimation of limited dependent variable models, *Journal of Econometrics*, 58: 347–368.

Box G. and Cox D.R. 1964. An analysis of transformations, *Journal of the Royal Statistical Society, B*, 26: 211–243.

Box G. and Jenkins G. 1976. *Time Series Analysis: Forcasting and Control*, Holden-Day, San Francisco.

Brandsma A. and Ketellapper R. 1979. Further evidence on alternative procedures for testing of spatial autocorrelation among regression disturbances, in Bartels C. and Ketellapper R. (eds.) *Exploratory and Explanatory Statistical Analysis of Spatial Data*, Martinus Nijhoff, Boston, MA, 111–136.

Brett C. and Pinkse J. 1997. Those taxes all over the map! A test for spatial independence of municipal tax rates in British Columbia, *International Regional Science Review*, 20: 131–151.

Breusch T. and Pagan A. 1980. The Lagrange Multiplier tests and its applications to model speciˇ cation in econometrics, *Review of Economic Studies*, 47: 239–253.

Brock W.A. and Durlauf S.N. 1998. *Discrete Choice with Social Interactions I: Theory*, Working Paper 9521, Social System Research Institute, University of Wisconsin, Madison, WI.

Brock W.A. and Durlauf S.N. 2001. Discrete choice with social interactions, *Review of Economic Studies*, 68: 235–260.

Brown J. and Rosen H. 1982. On the estimation of structural hedonic price models, *Econometrica*, 50: 765–768.

Brueckner J.K. 1998. Testing for strategic interaction among local governments: The case of growth controls, *Journal of Urban Economics*, 44: 438–467.

Brueckner J.K. 2003. Strategic interaction among governments: An overview of empirical studies, *International Regional Science Review*, 26: 175–188.

Brueckner J.K. and Saavedra L.A. 2001. Do local governments engage in strategic tax competition? *National Tax Journal*, 54: 203–229.

Brunsdon C., Fotheringham A. and Charlton M. 1996. Geographically Weighted Regression: A method for exploring spatial nonstationarity, *Geographical Analysis*, 28: 281–298.

Buettner T. 2003. Tax base effects and ˇ scal externalities of local capital taxation: Evidence from a panel of German jurisdictions, *Journal of Urban Economics*, 54: 110–128.

Burbidge J.B., Magee L. and Robb A.L. 1988. Alternative transformations to handle extreme values of the dependent variable, *Journal of the American Statistical Association*, 83: 123–127.

Burnside C. 1996. Production function regressions, returns to scale and externalities, *Journal of Monetary Economics*, 37: 177–201.

Burridge P. 1980. On the Cliff-Ord test for spatial autocorrelation, *Journal of the Royal Statistical Society, B*, 42: 107–108.

Burridge P. 1981. Testing for a common factor in a spatial autoregression model, *Environment and Planning A*, 13: 795–800.

Caballero R. and Lyons T. 1989. *The Role of External Economies in US Manufacturing*, Working Paper 3033, NBER, Cambridge, MA.

Caballero R. and Lyons T. 1990. Internal versus external economies in European industry, *European Economic Review*, 34: 805–830.

Caballero R. and Lyons T. 1992. External effects in US procyclical productivity, *Journal of Monetary Economics*, 29: 209–225.

Can A. 1992. Speciˇ cation and estimation of hedonic housing models, *Regional Science and Urban Economics*, 22: 453–474.

Can A. 1998. Geographic information systems (GIS) in housing and mortgage ˇ - nance, Editor's introduction, *Journal of Housing Research*, 9: 1–4.

Can A. and Megbolugbe I. 1997. Spatial dependence and house price index construction, *Journal of Real Estate Finance and Economics*, 14: 203–222.

Cano-Guervós R., Chica-Olmo J. and Hermoso-Gutiérrez J.A. 2003. A geostatistical method to deˇ ne districts within a city, *Journal of Real Estate Finance and Economics*, 27: 61–85.

Card D. and Krueger A.B. 1995. Time-series minimum-wage studies: A meta-analysis, *American Economic Review*, 85: 238–243.

Carlino G. and Mills E. 1986. The role of agglomeration potential in population and employment growth, Working Paper No. 86–13. Federal Reserve Bank of Philadelphia, PA.

Carlino G. and Mills E. 1987. The determinants of county growth, *Journal of Regional Science*, 27: 39–54.

Case A.C. 1991. Spatial patterns in household demand, *Econometrica*, 59: 953–966.

Case A.C. 1992. Neighborhood in uence and technological change, *Regional Science and Urban Economics*, 22: 491–508.

Case A.C., Rosen H. and Hines J.R. 1993. Budget spillovers and ˇ scal policy interdependence: Evidence from the states, *Journal of Public Economics*, 52: 285–307.

Casella G. and George E. 1992. Explaining the Gibbs sampler, *American Statistician*, 46: 167–174.

Casetti E. 1972. Generating models by the expansion method: Applications to geographical research, *Geographical Analysis*, 4: 81–91.

Casetti E. 1992. Bayesian regression and the expansion method, *Geographical Analysis*, 24: 58–74.

Cassell E. and Mendelsohn R. 1985. The choice of functional forms for hedonic price equations, *Journal of Urban Economics*, 18: 135–142.

Chambers J.M. 1998. *Programming with Data*, Springer-Verlag, New York, NY.

Chambers J.M. and Hastie T.J. 1992. *Statistical Models in S*, Chapman and Hall, London, UK.

Chambers R. 1988. *Applied Production Analysis*, Cambridge University Press, Cambridge, UK.

Chang W. 1981. Production externalities, variable returns to scale, and theory of trade, *International Economic Review*, 22: 511–525.

Chen X. and Conley T.G. 2001. A new semiparametric spatial model for panel time series, *Journal of Econometrics*, 105: 59–83.

Cheshire P. and Carbonaro G. 1995. Convergence-divergence in regional growth rates: An empty black box? in Armstrong H.W. and Vickerman R.W. (eds.) *Convergence and Divergence Among European Regions*, European Research in Regional Science, Pion, London, UK.

Chib S. 1992. Bayes inference in the tobit censored regression model, *Journal of Econometrics*, 51: 79–99.

Cho W.K.T. 2003. Contagion effects and ethnic contribution networks, *American Journal of Political Science*, 47: 368–387.

Chua H. 1993. Regional spillovers and economic growth, Ph.D. thesis, Harvard University.

Ciccone A. 1996. *Externalities and Interdependent Growth: Theory and Evidence*, Working paper, Department of Economics, University of California at Berkeley and University Pompeu Fabra, Barcelona, Spain.

Clapp J.M., Kim H. and Gelfand A.E. 2002. Predicting spatial patterns of house prices using LPR and Bayesian smoothing, *Real Estate Economics*, 30: 505–532.

Clayton D.G. 1991. A Monte Carlo method for Bayesian inference in frailty models, *Biometrics*, 47: 467–85.

Cleveland W. and Devlin S. 1988. Locally weighted regression: An approach to regression analysis by local ˇ tting, *Journal of the American Statistical Association*, 82: 596–610.

Cliff A. and Ord J.K. 1971. Evaluating the percentage points of a spatial autocorrlation coefˇ cient, *Geographical Analysis*, 3: 51–62.

Cliff A. and Ord J.K. 1972. Testing for spatial autocorrelation among regression residuals, *Geographical Analysis*, 4: 267–284.

Cliff A. and Ord J.K. 1973. *Spatial Autocorrelation*, Pion, London, UK.

Cliff A. and Ord J.K. 1975. The choice of a test for spatial autocorrelation, in Davis J. and McCullagh M. (eds.) *Display and Analysis of Spatial Data*, John Wiley and Sons, Chichester, UK.

Cliff A. and Ord J.K. 1981. *Spatial Processes: Models and Applications*, Pion, London, UK.

Coe D. and Helpman E. 1995. International R&D spillovers, *European Economic Review*, 39: 859–887.

Cohen J. and Tita G. 1999. Editors' introduction, *Journal of Quantitative Criminology*, 15: 373–378.

Conley T.G. 1999. GMM estimation with cross sectional dependence, *Journal of Econometrics*, 92: 1–45.

Conley T.G. and Ligon E. 2002. Economic distance, spillovers and cross country comparisons, *Journal of Economic Growth*, 7: 157–187.

Conley T.G. and Topa G. 2002. Socio-economic distance and spatial patterns in unemployment, *Journal of Applied Econometrics*, 17: 303–327.

Copeland B. and Taylor M. 1994. North-South trade and the environment, *Quarterly Journal of Economics*, 755–787.

Costello D. 1993. A cross-country, cross-industry comparison of productivity growth, *Journal of Political Economy*, 101: 207–222.

Cox D.R. 1970. *Analysis of Binary Data*, Methuen, London, UK.

Cox D.R. and Snell E.J. 1968. A general deˇ nition of residuals, *Journal of the Royal Statistical Society Series B*, 39: 248–275.

Cressie N. 1993. *Statistics for Spatial Data*, John Wiley and Sons, New York, NY.

Cressie N. and Read T.R.C. 1985. Do sudden infant deaths come in clusters? *Statistics and Decisions*, 2: 333–349.

Cropper M.L., Deck L.B. and McConnell K. 1988. On the choice of functional forms for hedonic price functions, *The Review of Economics and Statistics*, 70: 668–675.

Das D., Kelejian H.H. and Prucha I.R. 2003. Finite sample properties of estimators of spatial autoregressive models with autoregressive disturbances, *Papers in Regional Science*, 82: 1–27.

Dasgupta S., Mody A., Roy S. and Wheeler D. 1995. *Environmental Regulation and Development. A Cross-Country Empirical Analysis*, Working paper, No. 1448, World Bank, Washington, DC.

Davidson J. 1994. *Stochastic Limit Theory*, Oxford University Press, Oxford, UK.

Davidson R. and MacKinnon J.G. 1993. *Estimation and Inference in Econometrics*, Oxford University Press, Oxford, UK.

de Boor C. 1978. *A Practical Guide to Splines*, Springer-Verlag, Berlin, Germany.

de Boor C. 1999. *Matlab Spline Toolbox User Guide Version 2.0.1*, Mathworks, Natick, MA.

de Frutos R.F. and Pereira A.M. 1993. *Public Capital and Aggregate Growth in the United States: Is Public Capital Productive?*, Working paper, Department of Economics, University of California, San Diego, CA.

de Graaff T., Florax R.J.G.M., Nijkamp P. and Reggiani A. 2001. A general misspeciˇ cation test for spatial regression models: Dependence, heterogeneity, and nonlinearity, *Journal of Regional Science*, 41: 255–276.

de la Fuente A. 1996. *Infraestructuras y productividad: Un panorama y algunos resultados para las regiones españolas*, Working Paper 52.96, Instituto de Análisis Económico, Universidad Autónoma de Barcelona, Barcelona, Spain.

Deitz R. 1993. A joint model of residential and urban employment location, *Journal of Urban Economics*, 44: 197–215.

DeLong J. and Summers L. 1991. Equipment investment and economic growth, *The Quarterly Journal of Economics*, 106: 445–502.

Dempster A.P., Laird N.M. and Rubin D. 1977. Maximum Likelihood from incomplete data via the EM algorithm, *Journal of the Royal Statistical Society B*, 39: 1–38.

Diamond J. 1982. Aggregate demand management in search equilibrium, *Journal of Political Economy*, 90: 881–894.

Dietz R.D. 2002. The estimation of neighborhood effects in the social sciences: An interdisciplinary approach, *Social Science Research*, 31: 539–575.

Diggle P. 1984. *Statistical Analysis of Spatial Point Patterns*, Academic Press, London, UK.

Dixit A. and Stiglitz J.E. 1977. Monopolistic competition and optimum product diversity, *American Economic Review*, 67: 297–308.

Dixon R. and Thirlwall A. 1975a. A model of regional growth rate differences on Kaldorian lines, *Oxford Economic Papers*, 27: 201–214.

Dixon R. and Thirlwall A. 1975b. *Regional Growth and Unemployment in the United Kingdom*, Macmillan, London, UK.

Dobkins L.H. and Ioannides Y.M. 1998. Spatial interactions among US cities, presented at the 1998 North American Meeting of the Econometric Society, Chicago, IL.

Dowd M.R. and LeSage J.P. 1997. Analysis of spatial contiguity in uences on state price level formation, *International Journal of Forecasting*, 13: 245–253.

Driscoll J.C. and Kraay A.C. 1998. Consistent covariance matrix estimation with spatially dependent panel data, *The Review of Economics and Statistics*, 80: 549–560.

Dua A. and Esty D. 1997. *Sustaining the Asia Paciˇ c Miracle*, Institute for International Economics, Washington, DC.

Duan N. 1983. Smearing estimate: A nonparametric retransformation method, *Journal of the American Statistical Association*, 78: 605–610.

Dubin R. 1988. Estimation of regression coefˇ cients in the presence of spatially autocorrelated errors, *The Review of Economics and Statistics*, 70: 466–474.

Dubin R. 1992. Spatial autocorrelation and neighborhood quality, *Regional Science and Urban Economics*, 22: 433–452.

Dubin R. 2003. Robustness of spatial autocorrelation speciˇ cation, some Monte Carlo evidence, *Journal of Regional Science*, 43: 221–248.

Dubin R., Pace R.K. and Thibodeau T.G. 1999. Spatial autoregression techniques for real estate data, *Journal of Real Estate Literature*, 7: 79–95.

Duffy-Deno K. and Eberts R. 1991. Public infrastruture and regional economic development, *Journal of Urban Economics*, 30: 329–343.

Durbin J. 1954. Errors in variables, *Review of the International Statistical Institute*, 22: 23–32.

Durlauf S.N. 1991. *Nonergodic Economic Growth*, Working Paper 3719, NBER, Cambridge, MA.

Durlauf S.N. and Quah D.T. 1999. The new empirics of economic growth, in Taylor J. and Woodford M. (eds.) *Handbook of Macroeconomics*, North Holland Elsevier Science, Amsterdam, The Netherlands, 231–304.

Eaton J. and Eckstein Z. 1997. Cities and growth: Theory and evidence from France and Japan, *Regional Science and Urban Economics*, 27: 443–474.

Eckert J.K. 1990. *Property Appraisals and Assessment Administration*, International Association of Assessing Ofˇ cers, Chicago, IL.

Eckert J.K. and O' Connor P.M. 1992. Computer-assisted review assurance (CARA): A California case study, *Property Tax Journal*, 11: 59–80.

Efron B. and Tibshirani R. 1986. Bootstrap methods for standard errors, conˇ dence intervals, and other measures of statistical accuracy, *Statistical Science*, 1: 57–77.

Eilers P.H. and Marx B.D. 1996. Flexible smoothing with B-splines and penalties, *Statistical Science*, 11: 89–121.

Elhorst J.P. 2001. Dynamic models in space and time, *Geographical Analysis*, 33: 119–140.

Elhorst J.P. 2003. Speciˇ cation and estimation of spatial panel data models, *International Regional Science Review*, 26: 244–268.

Eliste P. and Fredriksson P.G. 1999. The political economy of environmental regulations, government assistance, and foreign trade, in Fredriksson P.G. (ed.) *Trade, Global Policy, and the Environment*, The World Bank, Washington, DC.

Ellison G. and Glaeser E.L. 1997. Geographic concentration in US manufacturing industries: A dartboard approach, *Journal of Political Economy*, 105: 889–927.

Engle R., Hendry D.F. and Richard J.F. 1983. Exogeneity, *Econometrica*, 51: 277–304.

Epple D. 1987. Hedonic prices and implicit markets: Estimating demand and supply functions for differentiated products, *Journal of Political Economy*, 95: 59–80.

ESRI 1992. *ARC/INFO Geographic Information System (revision 6.0)*, Environmental Systems Research Institute (ESRI), Redlands, CA.

Esty D. 1994. *Greening the GATT: Trade, Environment, and the Future*, Institute for International Economics, Washington, DC.

Esty D. and Geradin D. 1997. Market access, competitiveness, and harmonization: Environmental protection in regional trade agreements, *The Harvard Environmental Law Review*, 21: 265–336.

Fingleton B. 1997. Speciˇ cation and testing of Markov chain models: An application to convergence in the European union, *Oxford Bulletin of Economics and Statistics*, 59: 385–403.

Fingleton B. 1998. *International Economic Growth: Simultaneous Equation Models Incorporating Regional Effects*, Working paper, Department of Land Economy, University of Cambridge, Cambridge, UK.

Fingleton B. 1999a. *Economic Geography with Spatial Econometrics: A Third Way to Analyse Economic Development and Equlibrium, with Application to the EU Regions*, Working paper, Department of Economics, European University Institute, Florence, Italy.

Fingleton B. 1999b. Estimates of time to economic convergence: An analysis of regions of the European union, *International Regional Science Review*, 22: 5–35.

Fingleton B. 1999c. Spurious spatial regression: Some Monte Carlo results with a spatial unit root and spatial cointegration, *Journal of Regional Science*, 39: 1–19.

Fingleton B. and McCombie J.S.L. 1998. Increasing returns and economic growth: Some evidence for manufacturing from the European Union regions, *Oxford Economic Papers*, 50: 89–105.

Florax R.J.G.M. 1992. *The University: A Regional Booster? Economic Impacts of Academic Knowledge Infrastructure*, Avebury, Aldershot, UK.

Florax R.J.G.M. 2002. Methodological pitfalls in meta-analysis: Publication bias, in Florax R.J.G.M., Nijkamp P. and Willis K. (eds.) *Comparative Environmental Economic Assessment*, Edward Elgar, Cheltenham, UK, 177–207.

Florax R.J.G.M. and Folmer H. 1992. Speciˇ cation and estimation of spatial linear regression models: Monte Carlo evaluation of pre-test estimators, *Regional Science and Urban Economics*, 22: 405–432.

Florax R.J.G.M. and Rey S.J. 1995. The impact of misspeciˇ ed spatial structure in linear regression models, in Anselin L. and Florax R.J.G.M. (eds.) *New Directions in Spatial Econometrics*, Springer-Verlag, Berlin, Germany, 111–135.

Florax R.J.G.M. and van der Vlist A. 2003. Spatial econometric data analysis: Moving beyond traditional models, *International Regional Science Review*, 26: 223–243.

Florax R.J.G.M., Folmer H. and Rey S.J. 1998. *The Relevance of Hendry's Methodology: Experimental Simulation Results for Linear Spatial Models*, Working Paper 98-125/4, Tinbergen Institute, Amsterdam, The Netherlands.

Florax R.J.G.M., de Groot H.L.F. and de Mooij R. 2002a. *Meta-Analysis: A Tool for Upgrading Inputs of Macroeconomic Policy Models*, Working Paper 2002-041/3, Tinbergen Institute, Amsterdam, The Netherlands.

Florax R.J.G.M., de Groot H.L.F. and Heijungs R. 2002b. *The Empirical Economic Growth Literature: Robustness, Signiˇcance and Size*, Working Paper 2002-040/3, Tinbergen Institute, Amsterdam, The Netherlands.

Florax R.J.G.M., Folmer H. and Rey S.J. 2003. Speciˇcation searches in spatial econometrics: The relevance of Hendry's methodology, *Regional Science and Urban Economics*, 33: 557–579.

Follain J.R. and Jimenez E. 1985. Estimating the demand for housing characteristics: Survey and critique, *Regional Science and Urban Economics*, 15: 77–107.

Fotheringham A. 1991. Migration and spatial structure: The development of the competing destinations model, in Stillwell J. and Congdon P. (eds.) *Migration Models: Macro and Micro Approaches*, Bellhaven, London, UK, 57–72.

Fotheringham A., Brunsdon C. and Charlton M. 1998. Geographically Weighted Regression: A natural evolution of the Expansion Method of spatial data analysis, *Environment and Planning A*, 30: 1905–1927.

Fotheringham A., Brunsdon C. and Charlton M. 2002. *Geographically Weighted Regression*, John Wiley and Sons, Chichester, UK.

Fox K.A. 1974. *Social Indicators and Social Theory: Elements of an Operational System*, John Wiley and Sons, New York, NY.

Fredriksson P.G. 1999. The political economy of trade liberalization and environmental policy, *Southern Economic Journal*, 65: 513–525.

Fredriksson P.G. and Gaston N. 1999. The importance of trade for the ratiˇcation of the 1992 climate change convention, in Fredriksson P.G. (ed.) *Trade, Global Policy, and the Environment*, The World Bank, Washington, DC, chapter 12.

Freedom House 1991. *Freedom in the World: Political Rights and Civil Liberties*, Freedom House, New York, NY.

Freeman III A. 1974. On estimating air pollution control beneˇts from land value studies, *Journal of Environmental Economics and Management*, 1: 74–83.

Freeman III A. 1979. *The Beneˇts of Environmental Improvement*, Resources for the Future, Johns Hopkins Press, Baltimore, MD.

Freund J. and Walpole R. 1980. *Mathematical Statistics, 3rd Edition*, Prentice-Hall, Upper Saddle River, NJ.

Fujita M. and Krugman P. 2004. The new economic geography: where now, and to where, *Papers in Regional Science*, 83: 139–164.

Fujita M. and Mori T. 1997. Structural stability and evolution of urban systems, *Regional Science and Urban Economics*, 27: 399–442.

Fundación BBV 1995. *El stock de capital en la economia española*, Banco Bilbao Vizcaya, Bilbao, Spain.

Gaile G. 1980. The spread-backwash concept, *Regional Studies*, 14: 15–25.

Gamerman D., Moreira A.R. and Rue H. 2003. Space-varying regression models: Speciˇcations and simulation, *Computational Statistics and Data Analysis*, 42: 513–533.

Garc´a-Mila T. and McGuire T. 1992. The contribution of publicly provided inputs to states' economies, *Regional Science and Urban Economics*, 22: 229–241.

Garc´a-Mila T., McGuire T. and Porter R. 1996. The effect of public capital in state-level production functions reconsidered, *The Review of Economics and Statistics*, 78: 177–180.

Gelfand A.E. 1998. Spatio-temporal modeling of residential sales data, *Journal of Business and Economic Statistics*, 16: 312–321.

Gelfand A.E. and Smith A.F. 1990. Sampling-based approaches to calculating marginal densities, *Journal of the American Statistical Association*, 85: 398–409.

Gelfand A.E., Hills S.E., Racine-Poon A. and Smith A.F. 1990. Illustration of Bayesian inference in normal data models using Gibbs sampling, *Journal of the American Statistical Association*, 85: 972–985.

Gelfand A.E., Ghosh S.K., Knight J.R. and Sirmans C. 1998. Spatio-temporal modeling of residential sales data, *Journal of Business and Economic Statistics*, 16: 312–321.

Gelfand A.E., Kim H.J., Sirmans C. and Banerjee S. 2003. Spatial modeling with spatially varying coef˜ cient processes, *Journal of the American Statistical Association*, 98: 387–396.

Gelman A., Carlin J.B., Stern H.S. and Rubin D. 1995. *Bayesian Data Analysis*, Chapman and Hall, London, UK.

Geman S. and Geman D. 1984. Stochastic relaxation, Gibbs distributions, and the Bayesian restoration of images, *IEEE Transactions on Pattern Analysis and Machine Intelligence*, 6: 721–741.

Geoghegan J., Wainger L. and Bockstael N.E. 1997. Spatial landscape indices in a hedonic framework: an ecological economics analysis using GIS, *Ecological Economics*, 23: 251–264.

Getis A. and Grif˜ th D. 2002. Comparative spatial ˜ ltering in regression analysis, *Geographical Analysis*, 34: 130–140.

Getis A. and Ord J.K. 1992. The analysis of spatial association by distance statistics, *Geographical Analysis*, 24: 189–206.

Geweke J. 1986. Exact inference in the inequality constrained normal linear regression model, *Journal of Applied Econometrics*, 1: 127–141.

Geweke J. 1989. Bayesian inference in econometric models using Monte Carlo integration, *Econometrica*, 57: 1317–1340.

Geweke J. 1993. Bayesian treatment of the independent Student-t linear model, *Journal of Applied Econometrics*, 8: 19–40.

Giacomini R. and Granger C.W. 2003. Aggregation of space-time processes, *Journal of Econometrics*, in press.

Gilks W., Richardson S. and Spiegelhalter D. 1996. *Markov Chain Monte Carlo in Practice*, Chapman and Hall, London, UK.

Gillen K., Thibodeau T.G. and Wachter S. 2001. Anisotropic autocorrelation in house prices, *Journal of Real Estate Finance and Economics*, 23: 5–30.

Gilley O. and Pace R.K. 1995. Improving hedonic estimation with an inequality restricted estimator, *The Review of Economics and Statistics*, 77: 609–621.

Gimpel J.G. 1999. *Separate Destinations: Migration, Immigration and the Politics of Places*, University of Michigan Press, Ann Arbor, MI.

Gimpel J.G. and Schuknecht J.E. 2003. Political participation and the accessibility of the ballot box, *Political Geography*, 22: 471–488.

Glaeser E.L., Kallal H., Scheinkman J. and Schleifer A. 1992. Growth in cities, *Journal of Political Economy*, 100: 1126–1152.

Glaeser E.L., Sacerdote B.I. and Scheinkman J. 1996. Crime and social interactions, *Quarterly Journal of Economics*, 111: 507–548.

Glaeser E.L., Sacerdote B.I. and Scheinkman J. 2002. *The Social Multiplier*, Working Paper 9153, NBER, Cambridge, MA.

Gleditsch K.S. and Ward M.D. 2000. War and peace in space and time: The role of democratization, *International Studies Quarterly*, 44: 1–29.

Godfrey L. 1988. *Misspeciˇ cation Tests in Econometrics*, Cambridge University Press, Cambridge, UK.

Goklany I. 1996. Factors affecting environmental impacts: The effect of technology on long-term trends in cropland, air pollution, and water related diseases, *Ambio*, 25: 497–503.

Golany G. 1982. Selecting sites for new settlements in arid lands: Negev case study, *Energy and Building*, 4: 23–41.

Goldberger A.S. 1962. Best linear unbiased prediction in the generalized linear regression model, *Journal of the American Statistical Association*, 57: 369–375.

Golub G. and van Loan C. 1989. *Matrix Computations, 2nd Edition*, John Hopkins University Press, Baltimore, MD.

Goodchild M.F., Anselin L., Appelbaum R. and Harthorn B. 2000. Toward spatially integrated social science, *International Regional Science Review*, 23: 139–159.

Goodman A.C. and Thibodeau T.G. 1995. Age-related heteroskedasticity in hedonic house price equations, *Journal of Housing Research*, 6: 25–42.

Graves P., Murdoch J.C., Thayer M. and Waldman D. 1988. The robustness of hedonic price estimation: Urban air quality, *Land Economics*, 64: 220–233.

Greene W.H. 1997. *Econometric Analysis, 3rd Edition*, Prentice-Hall, Upper Saddle River, NJ, third.

Grifˇ th D. 1981. Modelling urban population density in a multicentered city, *Journal of Urban Economics*, 9: 298–310.

Grifˇ th D. 1985. An evaluation of correction techniques for boundary effects in spatial statistical analysis: Contemporary methods, *Geographical Analysis*, 17: 81–88.

Grifˇ th D. 1987. *Spatial Autocorrelation: A Primer*, Association of American Geographers, Washington, DC.

Grifˇ th D. 1988. *Advanced Spatial Statistics*, Kluwer, Dordrecht, The Netherlands.

Grifˇ th D. 1995. An evaluation of correction techniques for boundary effects and missing value techniques, *Geographical Analysis*, 17: 81–88.

Grifˇ th D. 1996. Some guidelines for specifying the geographic weights matrix contained in spatial statistical models, in Arlinghaus S. and Grifˇ th D.A. (eds.) *Practical Handbook of Spatial Statistics*, CRC Press, Boca Raton, FL.

Grifˇ th D. and Amrhein C. 1982. *Discriminating Between Solutions to the Boundary Value Problem in Spatial Statistical Analysis*, Working paper, Paper presented in the AAG Middle States Meeting, October 22-23, Montclair State College, Upper Monclair, NJ.

Grif˜ th D. and Amrhein C. 1983. An evaluation of correction techniques for boundary effects in spatial statistical analysis: Traditional methods, *Geographical Analysis*, 15: 352–60.

Grif˜ th D., Paelinck J. and van Gastel R. 1998. The Box-Cox transformation: Computational and interpretation features of the parameters, in Grif˜ th D., Amrhein C. and Huriot J.M. (eds.) *Econometric Advances in Spatial Modelling and Methodology*, Kluwer Academic, Dordrecht, The Netherlands, 46–56.

Grossman G. and Helpman E. 1991a. *Innovation and Growth in the World Economy*, MIT Press, Cambridge, MA.

Grossman G. and Helpman E. 1991b. Trade, knowledge spillovers and growth, *European Economic Review*, 35: 517–526.

Grossman G. and Helpman E. 1994. Endogenous innovation in the theory of growth, *Journal of Economic Perspectives*, 8: 23–44.

Grossman G. and Krueger A.B. 1993. Environmental impacts of a North American free trade agreement, in Garber P. (ed.) *The US-Mexico Free Trade Agreement*, MIT Press, Cambridge, MA, 13–56.

Guttorp P. 2000. Environmental statistics, *Journal of the American Statistical Association*, 95: 289–292.

Haining R. 1977. Model speci˜ cation in stationary random ˜ elds, *Geographical Analysis*, 9: 107–129.

Haining R. 1978. The moving average model for spatial interaction, *Transactions and Papers, Institute of British Geographers*, 202–225.

Haining R., Grif˜ th D. and Bennett R. 1983. Simulating two-dimensional autocorrelated surfaces, *Geographical Analysis*, 15: 247–253.

Hajivassiliou V.A. 1990. *Smooth Estimation of Panel Data LDV Models*, Working paper, Department of Economics, Yale University, New Haven, CT.

Hajivassiliou V.A. 1993. Simulation estimation methods for limited dependent variable models, in Maddala G., Rao C. and Vinod H. (eds.) *Handbook of Statistics*, North-Holland, Amsterdam, The Netherlands, 519–542.

Hajivassiliou V.A., McFadden D. and Ruud P.A. 1996. Simulation of multivariate normal rectangular probabilities: Theoretical computational results, *Journal of Econometrics*, 72: 85–134.

Halvorsen R. and Pollakowski H. 1981. Choice of functional form for hedonic price equation, *Journal of Urban Economics*, 10: 37–49.

Hansen L.P. 1982. Large sample properties of generalized method of moments estimators, *Econometrica*, 50: 1029–1054.

Hanson G. 1998. *Market Potential, Increasing Returns and Geographic Concentration*, Working paper, Department of Economics, University of Michigan, Ann Arbor, MI.

Harris R. and Lau E. 1998. Verdoorn's law and increasing returns to scale in the UK regions 1968–1991: Some new estimates based on the cointegration approach, *Oxford Economic Papers*, 50: 201–219.

Harrison D. and Rubinfeld D.L. 1978. Hedonic housing prices and the demand for clean air, *Journal of Environmental Economics and Management*, 5: 81–102.

Hartshorne R. 1939. *The Nature of Geography*, Association of American Geographers, Lancaster, PA.

Hastie T.J. and Tibshirani R. 1990. *Generalized Additive Models*, Chapman and Hall, London, UK.

Hastings W.K. 1970. Monte Carlo sampling methods using Markov chains and their applications, *Biometrika*, 57: 97–109.

Hausman J. 1978. Speciˇ cation tests in econometrics, *Econometrica*, 46: 1251–1271.

Hautsch N. and Klotz S. 2003. Estimating the neighborhood in uence on decision makers: Theory and an application on the analysis of innovation decisions, *Journal of Economic Behavior and Organization*, 52: 97–113.

Heckman J. 1978. Simple statistical models for discrete panel data developed and applied to tests of the hypothesis of true state dependence against the hypothesis of spurious state dependence, *Annales de L'INSEE*, 30-31: 227–269.

Heckman J. 1981. Statistical models for discrete panel data, in Manski C.F. and McFadden D. (eds.) *Structural Analysis of Discrete Data with Econometric Applications*, MIT Press, Cambridge, MA, 114–180.

Heckman J. 2001. Micro-data, heterogeneity, and the evaluation of public policy: Nobel lecture, *Journal of Political Economy*, 109: 673–748.

Heckman J. and Singer B. 1985. Social science duration analysis, in Heckman J. and Singer B. (eds.) *Longitudinal Analysis of Labor Market Data*, Cambridge University Press, Cambridge, UK, 39–110.

Hedges L. 1997. The promise of replication in labour economics, *Labour Economics*, 4: 111–114.

Hedges L. and Olkin I. 1985. *Statistical Methods for Meta-Analysis*, Academic Press, New York, NY.

Helpman E. 1997. *R&D and Productivity: The International Connection*, Working Paper 6101, NBER, Cambridge, MA.

Helpman E. 1998. The size of regions, in Pines D., Sadka E. and Zilcha I. (eds.) *Topics in Public Economics*, Cambridge University Press, Cambridge, UK, 33–54.

Henderson J.V. 1974. The types and size of cities, *American Economic Review*, 64: 640–656.

Henderson J.V. 1988. *Urban Development: Theory, Facts and Illusion*, Oxford University Press, Oxford, UK.

Henderson J.V. 1992. Where does an industry locate, *Journal of Urban Economics*, 35: 83–104.

Hendry D.F. 1979. Predictive failure and econometric modelling in macroeconomics: The transactions demand for money, in Ormerod P. (ed.) *Economic Modelling*, Heinemann, London, UK.

Hendry D.F. 1984. Monte Carlo experimentation in econometrics, in Griliches Z. and Intriligator M. (eds.) *Handbook of Econometrics*, vol. II, North-Holland, Amsterdam, The Netherlands, 937–976.

Hendry D.F., Pagan A. and Sargan D. 1984. Dynamic speciˇ cation, in Griliches Z. and Intrilligator M. (eds.) *Handbook of Econometrics*, North Holland, Amsterdam, The Netherlands, 1025–1102.

Henry M.S., Barkley D.L., Bao S. and Brooks K. 1994. Estimates of subcounty linkages in selected southern FEAs, Paper Presented at the Regional Science Association, International, Niagara Falls, Ontario.

Henry M.S., Barkley D.L., Bao S. and Brooks K. 1997. The hinterland's stake in metropolitan growth: Evidence from selected southern regions, *Journal of Regional Science*, 37: 479–501.

Herberg H., Kemp M. and Tawada M. 1982. Further implications of variable returns to scale, *Journal of International Economics*, 13: 65–84.

Hoff P.D., Raftery A.E. and Handcock M.S. 2002. Latent space approaches to social network analysis, *Journal of the American Statistical Association*, 97: 1090–1098.

Holloway G., Shankar B. and Rahman S. 2002. Bayesian spatial probit estimation: A primer and an application to HYV rice adoption, *Agricultural Economics*, 27: 383–402.

Holtz-Eakin D. 1994. Public-sector capital and the productivity puzzle, *The Review of Economics and Statistics*, 76: 12–21.

Holtz-Eakin D. and Lovely M. 1996. Scale economies, returns to variety, and the productivity of public infrastructure, *Regional Science and Urban Economics*, 26: 105–123.

Holtz-Eakin D. and Schwartz A. 1995. Spatial productivity spillovers from public infrastructure: Evidence from state highways, *International Tax and Public Finance*, 2: 459–468.

Horowitz J. and Härdle W. 1996. Direct semiparametric estimation of single-index models with discrete covariates, *Journal of the American Statistical Association*, 91: 1632–1640.

Hsiao C. 1986. *Analysis of Panel Data*, Cambridge University Press, Cambridge, UK.

Hughes D. and Holland D. 1994. Core-periphery economic linkage: A measure of spread and possible backwash effects for the Washington economy, *Land Economics*, 70: 364–377.

ICBS 1999. *Statistical Abstract of Israel (Annual)*, Israel Central Bureau of Statistics, Jerusalem, Israel.

Ihaka R. and Gentleman R. 1996. R: A language for data analysis and graphics, *Journal of Computational and Graphical Statistics*, 5: 299–314.

Ioannides Y.M. 1994. Product differentiation and economic growth in a system of cities, *Regional Science and Urban Economics*, 24: 461–484.

Irwin E.G. 2002. The effects of open space on residential property values, *Land Economics*, 78: 465–480.

Irwin E.G. and Bockstael N.E. 1999. *Interacting Agents, Spatial Externalities, and the Endogenous Evolution of Land Use Pattern*, Working paper, Department of Agricultural and Resource Economics, University of Maryland, College Park, MD.

Irwin E.G. and Bockstael N.E. 2001. The problem of identifying land use spillovers: Measuring the effects of open space on residential property values, *American Journal of Agricultural Economics*, 83: 698–704.

Irwin E.G. and Bockstael N.E. 2002. Interacting agents, spatial externalities and the evolution of residential land use patterns, *Journal of Economic Geography*, 2: 31–54.

Islam N. 1995. Growth empirics: A panel data approach, *Quarterly Journal of Economics*, 110: 1127–1170.

Jaffe A., Peterson S., Portney P. and Stavins R.N. 1995. Environmental regulation and the competitiveness of U.S. manufacturing: What does the evidence tell us? *Journal of Economic Literature*, 33: 132–163.

Johnston J. 1984. *Econometric Methods*, McGraw-Hill, New York, NY.

Johnston J. and DiNardo J. 1997. *Econometric Methods, 4th Edition*, McGraw-Hill, New York, NY.

Jones J.P. and Casetti E. 1992. *Applications of the Expansion Method*, Routledge, New York, NY.

Jovanovic B., Lach S. and Lary V. 1992. *Growth, and Human Capital's Role as an Investment in Cost Reduction*, Mimeo.

Judge G.G., Grif˜ ths W.E., Hill R.C. and Lee T.S. 1985. *The Theory and Practice of Econometrics*, John Wiley and Sons, New York, NY.

Just R. and Antle J. 1991. Effects of commodity program structure on resource use and the environment, in Just R.E. and Bockstael N.E. (eds.) *Commodity and Resource Policies in Agricultural Systems*, Springer-Verlag, Berlin, Germany.

Just R. and Bockstael N.E. 1991. *Commodity and Resource Policies in Agricultural Systems*, Springer-Verlag, Berlin, Germany.

Kahn S. and Lang K. 1988. Ef˜ cient estimation of structural hedonic systems, *International Economic Review*, 29: 157–166.

Kaldor N. 1957. A model of economic growth, *Economic Journal*, 67: 591–624.

Kaldor N. 1970. The case for regional policies, *Scottish Journal of Political Economy*, 17: 37–48.

Kalnins A. 2003. Hamburger prices and spatial econometrics, *Journal of Economics and Management Strategy*, 12, in press.

Kalt J. 1988. The impact of domestic environmental regulatory policies on U.S. international competitiveness, in Spence A. and Hazard H. (eds.) *International Competitiveness*, Harper and Row, Ballinger, Cambridge, MA.

Kaluzny S., Vega S., Cardoso T. and Shelly A. 1997. *S+SpatialStats User's Manual*, Springer-Verlag, New York, NY.

Kanemoto Y. 1980. *Theories of Urban Externalities*, Elsevier, North Holland, Amsterdam, The Netherlands.

Keane M.P. 1993. Simulation estimation for panel data models with limited dependent variables, in Maddala G., Rao C. and Vinod H. (eds.) *HandBook of Statistics*, North-Holland, Amsterdam, The Netherlands, 545–571.

Keane M.P. 1994. A computationally practical simulator estimator for panel data, *Econometrica*, 62: 95–116.

Kelejian H.H. and Oates W.E. 1989. *Introduction to Econometrics*, Harper and Row, New York, NY.

Kelejian H.H. and Prucha I.R. 1997. Estimation of spatial regression models with autoregressive errors by two stage least squares procedures: A serious problem, *International Regional Science Review*, 20: 103–111.

Kelejian H.H. and Prucha I.R. 1998. A generalized spatial two stage least squares procedure for estimating a spatial autoregressive model with autoregressive disturbances, *Journal of Real Estate Finance and Economics*, 17: 99–121.
Kelejian H.H. and Prucha I.R. 1999. A generalized moments estimator for the autoregressive parameter in a spatial model, *International Economic Review*, 40: 509–533.
Kelejian H.H. and Prucha I.R. 2001. On the asymptotic distribution of the Moran I test statistic with applications, *Journal of Econometrics*, 104: 219–257.
Kelejian H.H. and Prucha I.R. 2002. 2SLS and OLS in a spatial autoregressive model with equal spatial weights, *Regional Science and Urban Economics*, 32: 691–707.
Kelejian H.H. and Prucha I.R. 2003. Estimation of simultaneous systems of spatially interrelated cross sectional equations, *Journal of Econometrics*, in press.
Kelejian H.H. and Robinson D.P. 1992. Spatial autocorrelation: A new computationally simple test with an application to per capita county policy expenditures, *Regional Science and Urban Economics*, 22: 317–331.
Kelejian H.H. and Robinson D.P. 1993. A suggested method of estimation method for spatial interdependent models with autocorrelated errors, and an application to a county expenditure model, *Papers in Regional Science*, 72: 297–312.
Kelejian H.H. and Robinson D.P. 1995. Spatial correlation: A suggested alternative to the autoregressive model, in Anselin L. and Florax R. (eds.) *New Directions in Spatial Econometrics*, Springer-Verlag, Berlin, Germany, 75–95.
Kelejian H.H. and Robinson D.P. 1997. Infrastructure productivity estimation and its underlying econometric speciˇ cations: A sensitivity analysis, *Papers in Regional Science*, 76: 115–131.
Kelejian H.H. and Robinson D.P. 1998. A suggested test for autocorrelation and/or heteroskedasticity and corresponding Monte Carlo results, *Regional Science and Urban Economics*, 28: 389–417.
Kelejian H.H. and Yuzefovich Y. 2001. *Properties of Tests for Spatial Error Components: A Further Analysis*, Working paper, Working Paper.
Keller W. 1997. *Trade and the Transmission of Technology*, Working paper, University of Wisconsin and NBER, Madison, WI.
Keller W. 1998. Are international R&D spillovers trade-related? Analyzing spillovers among randomly matched trade partners, *European Economic Review*, 42: 1469–1481.
Kennedy P. 1992. *A Guide to Econometrics*, MIT Press, Cambridge, MA.
Kennedy P. 1996. *A Guide to Econometrics*, MIT Press, Cambridge, MA.
Kennedy P.W. 1994. Equilibrium pollution taxes in open economies with imperfect competition, *Journal of Environmental Economics and Management*, 27: 49–63.
Kim C.W., Phipps T.T. and Anselin L. 2003a. Measuring the beneˇ ts of air quality improvement: A spatial hedonic approach, *Journal of Environmental Economics and Management*, 45: 24–39.
Kim J., Elliott E. and Wang D.M. 2003b. A spatial analysis of county-level outcomes in US presidential elections: 1988–2000, *Electoral Studies*, in press.
King M. 1981. A small sample property of the Cliff-Ord test for spatial correlation, *Journal of the Royal Statistical Association B*, 43: 263–264.

Kiriacou G. 1991. *Level and Growth Effects of Human Capital: A Cross-Country Study of the Convergence Hypothesis*, Working Paper 91-26, New York University, New York, NY.

Klepper S. and Leamer E. 1984. Consistent sets of regression estimates with errors in all variables, *Econometrica*, 51: 153–183.

Knight J.R., Sirmans C. and Turnbull G. 1994. List price signaling and buyer behavior in the housing market, *Journal of Real Estate Finance and Economics*, 9: 177–192.

Knox P. 1994. *Urbanization: An Introduction to Urban Geography*, Prentice-Hall, Englewood Cliffs, NJ.

Kollmann K. 1995. The correlation of productivity growth across regions and industries in the United States, *Economics Letters*, 47: 229–250.

Krakover S. 1987. Clusters of cities versus city region in regional planning, *Environment and Planning A*, 19: 1375–1386.

Krugman P. 1991a. *Geography and Trade*, Leuven University Press and MIT Press, Leuven, Belgium and Cambridge, MA.

Krugman P. 1991b. Increasing returns and economic geography, *Journal of Political Economy*, 99: 438–499.

Krugman P. 1992. *A Dynamic Spatial Model*, Working Paper 4219, NBER, Cambridge, MA.

Krugman P. 1995. *Development, Geography, and Economic Theory*, MIT Press, Cambridge, MA.

Krugman P. 1996a. Confronting the mystery of urban hierarchy, *Journal of the Japanese and International Economies*, 10: 399–418.

Krugman P. 1996b. *The Self-Organizing Economy*, Blackwell Publishers, Cambridge, MA.

Krugman P. 1998. Space: The ˇ nal frontier, *Journal of Economic Perspectives*, 12: 161–174.

Krugman P. and Venables A.J. 1995. Globalization and the inequality of nations, *Quarterly Journal of Economics*, 110: 857–880.

Kubo Y. 1995. Scale economies, regional externalities and the possibility of uneven regional development, *Journal of Regional Science*, 35: 318–328.

Lahatte A. 2003. Restrictions on the autoregressive parameters of share systems with spatial dependence, *Economics Letters*, 78: 225–229.

Lahiri S.N. 1996. On the inconsistency of estimators under inˇ ll asymptotics for spatial data, *Sankhya A*, 58: 403–417.

Lay D. 1997. *Linear Algebra and its Applications*, Addison-Wesley, Reading, MA.

Leamer E. 1983a. Model choice and speciˇ cation analysis, in Griliches Z. and Intriligator M.D. (eds.) *Handbook of Econometrics I*, North-Holland, Amsterdam, The Netherlands, 286–330.

Leamer E. 1983b. Reporting the fragility of regression estimates, *The Review of Economics and Statistics*, 64: 306–317.

Lee E. 1992. *Statistical Methods for Survival Data Analysis, 2nd Ed.*, John Wiley and Sons, New York, NY.

Lee K., Pesaran M. and Smith R. 1997. Growth and convergence in a multi-country empirical stochastic Solow model, *Journal of Applied Econometrics*, 12: 357–392.

Lee L.F. 1998. Simulated Maximum Likelihood estimation of dynamic discrete choice statistical models: Some Monte Carlo results, *Journal of Econometrics*, 82: 1–35.

Lee L.F. 2002. Consistency and ef˜ ciency of least squares estimation for mixed regressive, spatial autoregressive models, *Econometric Theory*, 18: 252–277.

Leenders R.T.A.J. 2002. Modeling social in uence through network autocorrelation: Constructing the weights matrix, *Social Networks*, 24: 21–47.

LeSage J.P. 1997a. Bayesian estimation of spatial autoregressive models, *International Regional Science Review*, 20: 113–129.

LeSage J.P. 1997b. *Bayesian Estimation of Spatial Probit/Tobit Models*, Working paper, Department of Economics, University of Toledo, Toledo, OH.

LeSage J.P. 1999. *Spatial Econometrics*, The Web Book of Regional Science, Regional Research Institute, West Virginia University, Morgantown, WV.

LeSage J.P. 2000. Bayesian estimation of limited dependent variable spatial autoregressive models, *Geographical Analysis*, 32: 19–35.

LeSage J.P. and Krivelyova A. 1999. A spatial prior for Bayesian vector autoregressive models, *Journal of Regional Science*, 39: 297–317.

Levine R. and Revelt D. 1992. A sensitivity analysis of cross-country growth regressions, *American Economic Review*, 82: 942–963.

Li B. 1995. Implementing spatial statistics on parallel computers, in Arlinghaus S. and Grif˜ th D. (eds.) *Practical Handbook of Spatial Statistics*, CRC Press, Boca Raton, FL, 107–148.

Lindley D.V. 1971. The estimation of many parameters, in Godambe V. and Sprott D. (eds.) *Foundations of Statistical Inference*, Holt, Reinhart, and Winston, Toronto, ON, 435–453.

López-Bazo E., Vayá E., Moreno R. and Suriñach J. 1998. *Grow, Neighbor, Grow, Grow ...Neighbor be Good!*, Working paper, Department of Econometrics, Statistics and Spanish Economy, University of Barcelona, Barcelona, Spain.

López-Bazo E., Vayá E., Mora A. and Suriñach J. 1999. Regional economic dynamics and convergence in the European Union, *The Annals of Regional Science*, 22: 1–28.

Lucas R. Jr. 1988. On the mechanics of developement planning, *Journal of Monetary Economics*, 22: 3–42.

Lucas R. Jr. 1993. Making a miracle, *Econometrica*, 61: 251–272.

Lynch L. and Lovell S.J. 2003. Combining spatial and survey data to explain participation in agricultural land preservation programs, *Land Economics*, 79: 259–276.

MacKinnon J.G. 1991. Critical values for cointegration tests, in Engle R. and Granger C. (eds.) *Long-Run Economic Relationships*, Oxford University Press, Oxford, UK, 267–276.

Maddala G.S. 1983. *Limited-Dependent and Qualitative Variables in Econometrics*, Cambridge University Press, Cambridge, UK.

Maddala G.S. 1992. *Introduction to Econometrics*, Macmillan Publishing Company, New York, NY.

Magrini S. 1995. *Economic Convergence in the European Union: A Markov Chain Approach*, Working paper, Urban and Regional Economics, University of Reading, UK.

Mankiw N., Romer D. and Weil D. 1992. A contribution to the empirics of economic growth, *Quarterly Journal of Economics*, 107: 407–437.

Manski C.F. 1975. Maximum score estimation of a stochastic utility model of choice, *Journal of Econometrics*, 3: 205–228.

Manski C.F. 1993. Identiˇ cation of endogenous social effects: The re ection problem, *Review of Economic Studies*, 60: 531–542.

Manski C.F. 1995. *Identiˇ cation Problems in the Social Sciences*, Harvard University Press, Cambridge, MA.

Manski C.F. 2000. Economic analysis of social interactions, *Journal of Economic Perspectives*, 14: 115–136.

Manski C.F. and Thompson S. 1986. Operational characteristics of maximum score estimation, *Journal of Econometrics*, 32: 85–108.

Marshall A. 1920. *Principles of Economics*, MacMillan and Co, London, UK.

Martin T. and Ottaviano G. 1999. Growing locations: Industry location in a model of endogenous growth, *European Economic Review*, 43: 281–302.

Maryland Department of State Planning 1973. *Natural Soil Groups of Maryland*, Technical Series Publication 199.

Mas M., Maudos J., Perez F. and Uriel E. 1996. Infrastructures and productivity in the Spanish regions, *Regional Studies*, 30: 641–649.

Matula D.W. and Sokal R.R. 1980. Properties of Gabriel graphs relevant to geographic variation research and the clustering of points in the plane, *Geographic Analysis*, 12: 205–222.

McCombie J.S.L. 1982. Economic growth, Kaldor's laws and the static-dynamic Verdoorn law paradox, *Applied Economics*, 14: 279–294.

McCombie J.S.L. and de Ridder J. 1984. The Verdoorn law controversy: Some new empirical evidence using US state data, *Oxford Economic Papers*, 36: 268–284.

McCombie J.S.L. and Thirlwall A. 1994. *Economic Growth and the Balance of Payments Constraint*, McMillan, Basingstoke, UK.

McFadden D. 1989. A method of simulated moments for estimation of discrete response models without numerical integration, *Econometrica*, 57: 995–1026.

McMillan J., Ullah A. and Vinod H. 1989. Estimation of the shape of the demand curve by nonparametric Kernel methods, in Raj B. (ed.) *Advances in Econometrics and Modelling*, Kluwer Academic Press, Dordrecht, The Netherlands, 85–92.

McMillen D.P. 1992. Probit with spatial autocorrelation, *Journal of Regional Science*, 32: 335–348.

McMillen D.P. 1995a. Selection bias in spatial econometric models, *Journal of Regional Science*, 35: 417–436.

McMillen D.P. 1995b. Spatial effects in probit models: A Monte Carlo investigation, in Anselin L. and Florax R.J.G.M. (eds.) *New Directions in Spatial Econometrics*, Springer-Verlag, Berlin, Germany, 189–228.

McMillen D.P. 1996. One hundred ˇ fty years of land values in Chicago: A nonparametric approach, *Journal of Urban Economics*, 40: 100–124.

McMillen D.P. and McDonald J.F. 1997. A nonparametric analysis of employment density in a polycentric city, *Journal of Regional Science*, 37: 591–612.
McMillen D.P. and McDonald J.F. 1999. Land use before zoning: The case of 1920's Chicago, *Regional Science and Urban Economics*, 29: 473–489.
Meeker W. and Escobar L. 1995. Teaching about approximate conˇ dence regions based on Maximum Likelihood estimates, *The American Statistician*, 49: 48–53.
Meese R. and Wallace N. 1991. Nonparametric estimation of dynamic hedonic price models and the construction of residential housing price indices, *Journal of the American Real Estate and Urban Economics Association*, 19: 308–332.
Merriˇ eld J. 1988. The impact of selected abatement strategies on transnational pollution, the terms of trade, and factor rewards: A general equlibrium approach, *Journal of Environmental Economics and Management*, 15: 259–284.
Messner S.F. and Anselin L. 2004. Spatial analyses of homicide with areal data, in Goodchild M. and Janelle D. (eds.) *Spatially Integrated Social Science*, Oxford University Press, New York, NY, 127–144.
Metropolis N., Rosenbluth A., Rosenbluth M., Teller A. and Teller E. 1953. Equation of state calculations by fast computing machines, *Journal of Chemical Physics*, 21: 1087–1092.
Mills E. and Price R. 1984. Metropolitan suburbanization and central city problems, *Journal of Urban Economics*, 15: 1–17.
Miyao T. and Kanemoto Y. 1987. *Urban Dynamics and Urban Externalities*, Harwood Academic Publishers, New York, NY.
Molho I. 1995. Spatial autocorrelation in British unemployment, *Journal of Regional Science*, 35: 641–658.
Moran P. 1948. The interpretation of statistical maps, *Journal of the Royal Statistical Society B*, 10: 243–251.
Moran P. 1950a. Notes on continuous stochastic phenomena, *Biometrika*, 37: 17–23.
Moran P. 1950b. A test for the serial independence of residuals, *Biometrika*, 37: 178–181.
Moreno R. 1998. Infraestructuras, externalidades y crecimiento regional: Algunas aportaciones para el caso regional español, Ph.D. thesis, University of Barcelona, Barcelona, Spain.
Moreno R. and Trehan B. 1997. Location and the growth of nations, *Journal of Economic Growth*, 2: 399–418.
Moreno R., Artis M., López-Bazo E. and Suriñach J. 1997. Evidence on the complex link between infrastructure and regional growth, *International Journal of Development Planning Literature*, 12: 81–108.
Moreno R., López-Bazo E. and Artis M. 1998. *Public Capital, Private Capital and Costs of Production: Short and Long Run Effects*, Working paper, Department of Econometrics, Statistics and Spanish Economy, University of Barcelona, Barcelona, Spain.
Morenoff J.D. and Sampson R.J. 1997. Violent crime and the spatial dynamics of neighborhood transition: Chicago, 1970–1990, *Social Forces*, 76: 31–64.
Morenoff J.D., Sampson R.J. and Raudenbush S.W. 2001. Neighborhood inequality, collective efˇ cacy, and the spatial dynamics of urban violence, *Criminology*, 39: 517–559.

Morrill R., Gaile G. and Thrall G. 1988. *Spatial Diffusion*, Sage Publications, Newbury Park, CA.

Morrison C. and Schwartz A. 1996. State infrastructure and productive performance, *The American Economic Review*, 86: 1095–1111.

Moulton B. 1986. Random group effects and the prediction of regression estimates, *Journal of of Econometrics*, 32: 385–397.

Munnell A.H. 1990a. How does infrastructure affect regional economic performance? in Munnell A.H. (ed.) *Is There A Shortfall in Public Capital Investment*, Proceedings Federal Reserve Bank of Boston Conference.

Munnell A.H. 1990b. How does public infrastructure affect regional economic performance? *New England Economic Review*, 11–32.

Mur J. 1999. Testing for spatial autocorrelation: Moving average versus autoregressive processes, *Environment and Planning A*, 31: 137–1382.

Mur J. and Tr´vez F.J. 2003. Unit roots and deterministic trends in spatial econometric models, *International Regional Science Review*, 26: 289–312.

Murdoch J.C., Rahmatian M. and Thayer M. 1993. A spatially autoregressive median voter model of recreational expenditures, *Public Finance Quarterly*, 21: 334–350.

Murdoch J.C., Sandler T. and Sargent K. 1997. A tale of two collectives: Sulphur versus nitrogen oxides emission reduction in Europe, *Economica*, 64: 281–301.

Murdoch J.C., Sandler T. and Vijverberg W.P. 2003. The participation decision versus the level of participation in an environmental strategy: A spatial probit analysis, *Journal of Public Economics*, 87: 337–362.

Myrdal G. 1958. *Economic Theory and Underdeveloped Regions*, Duckworth, London, UK.

Nadiri M.I. and Kim S. 1996. *International R&D Spillovers, Trade and Productivity in Major OECD Countries*, Working paper, NBER, Cambridge, MA.

Nelson G.C. 2002. Introduction to the special issue on spatial analysis, *Agricultural Economics*, 27: 197–200.

Nelson G.C. and Hellerstein D. 1997. Do roads cause deforestation? Using satellite images in econometric analysis of land use, *American Journal of Agricultural Economics*, 79: 80–88.

Nelson G.C., Harris V. and Stone S.W. 2001. Deforestation, land use, and property rights: Empirical evidence from Darién, Panama, *Land Economics*, 77: 187–205.

Nelson R. and Phelps E. 1966. Investment in humans technological diffusion and economic growth, *American Economic Review*, 56: 69–75.

Newey W. and West K. 1987. A simple positive semi-deˇ nite heteroskedasticity and autocorrelation consistent covariance matrix, *Econometrica*, 55: 703–708.

Nychka D. 2000. Challenges in understanding the atmosphere, *Journal of the American Statistical Association*, 95: 972–975.

Ord J.K. 1975. Estimation methods for models of spatial interaction, *Journal of the American Statistical Association*, 70: 120–126.

Ottaviano G. and Puga D. 1998. Agglomeration in the global economy: A survey of the new economic geography, *World Economy*, 21: 707–731.

Pace R.K. 1997. Performing large spatial regressions and autoregressions, *Economics Letters*, 54: 283–291.

Pace R.K. and Barry R.P. 1997a. Fast CARs, *Journal of Statistical Computation and Simulation*, 59: 123–147.
Pace R.K. and Barry R.P. 1997b. Quick computation of spatial autoregressive estimators, *Geographical Analysis*, 29: 232–246.
Pace R.K. and Barry R.P. 1997c. Sparse spatial autoregressions, *Statistics and Probability Letters*, 33: 291–297.
Pace R.K. and Barry R.P. 1998. *Spatial Statistics Toolbox 1.0*, Real Estate Research Institute, Louisiana State University, Baton Rouge, LA.
Pace R.K. and Gilley O. 1997. Using the spatial conˇ gurationof the data to improve estimation, *Journal of Real Estate Finance and Economics*, 14: 333–340.
Pace R.K. and Gilley O. 1998. Generalizing OLS and grid estimators, *Real Estate Economics*, 26: 331–347.
Pace R.K. and LeSage J.P. 2002. Semiparametric Maximum Likelihood estimates of spatial dependence, *Geographical Analysis*, 34: 76–90.
Pace R.K. and LeSage J.P. 2003a. Chebyshev approximation of log-determinants of spatial weights matrices, *Computational Statistics and Data Analysis*, in press.
Pace R.K. and LeSage J.P. 2003b. Likelihood dominance spatial inference, *Geographical Analysis*, 35: 133–147.
Pace R.K. and Zou D. 2000. Closed-form maximum likelihood estimates of nearest-neighbor spatial dependence, *Geographical Analysis*, 32: 154–172.
Pace R.K., Barry R.P., Clapp J.M. and Rodriquez M. 1998a. Spatiotemporal autoregressive models of neighborhood effects, *Journal of Real Estate Finance and Economics*, 17: 15–33.
Pace R.K., Barry R.P. and Sirmans C. 1998b. Spatial statistics and real estate, *Journal of Real Estate Finance and Economics*, 17: 5–13.
Palmquist R. 1984. Estimating demand for the characteristics of housing, *The Review of Economics and Statistics*, 66: 394–404.
Park W. 1995. International R&D spillovers and OECD economic growth, *Economic Inquiry*, 33: 571–591.
Paterson R.W. and Boyle K.J. 2002. Out of sight, out of mind? Using GIS to incorporate visibility in hedonic property value models, *Land Economics*, 78: 417–425.
Pérez F. and Serrano L. 1998. *Capital humano, crecimiento económico y desarrollo regional en españa (1964–1997)*, Working paper, Fundación Bancaja.
Pesaran M. and Smith R. 1995. Estimating long-run relationships from dynamic heterogenous panels, *Journal of Econometrics*, 68: 79–113.
Pinelli D., Giacometti R., Lewney R. and Fingleton B. 1998. *European Regional Competitiveness Indicators*, Working paper, Department of Land Economy, University of Cambridge, Cambridge, UK.
Pinkse J. 1993. On the computation of semiparametric estimates in limited dependent variables models, *Journal of Econometrics*, 58: 185–205.
Pinkse J. 1998. A consistent nonparametric test for serial independence, *Journal of Econometrics*, 84: 205–231.
Pinkse J. 1999. *Asymptotics of the Moran Test and a Test for Spatial Correlation in Probit Models*, Working paper, Department of Economics, University of British Columbia, Vancouver, BC.

Pinkse J. and Slade M.E. 1998. Contracting in space: An application of spatial statistics to discrete-choice models, *Journal of Econometrics*, 85: 125–154.

Pinkse J., Slade M.E. and Brett C. 2002. Spatial price competition: A semiparametric approach, *Econometrica*, 70: 1111–1153.

Pisati M. 2001. Tools for spatial data analysis, *Stata Technical Bulletin*, 60: 21–37.

Plantinga A.J., Lubowski R.N. and Stavins R.N. 2002. The effect of potential land development on agricultural prices, *Journal of Urban Economics*, 52: 561–581.

Poirier D. and Ruud P.A. 1988. Probit with dependent observations, *Review of Economic Studies*, 55: 593–614.

Portnov B.A. and Erell E. 2001. *Urban Clustering: The Bene˘ ts and Drawbacks of Location*, Ashgate, Aldershot, UK.

Powell J.L., Stock J.H. and Stoker T.M. 1989. Semiparametric estimation of index coef˜ cients, *Econometrica*, 57: 1403–1430.

Puga D. 1996. The rise and fall of economic agglomerations, Paper presented at CEPR Workshop, Location and Regional Convergence/Divergence, CORE, Louvain-la-Neuve, Belgium.

Puga D. and Venables A.J. 1996. The spread of industry: Spatial agglomeration in economic development, *Journal of the Japanese and International Economics*, 10: 440–464.

Puga D. and Venables A.J. 1997. Preferential trading arrangements and industrial location, *Journal of International Economics*, 43: 347–68.

Puga D. and Venables A.J. 1999. Agglomeration and economic development: Import substitution vs. trade liberalization, *Economic Journal*, 109: 292–311.

Quah D.T. 1993. Empirical cross-section dynamics in economic growth, *European Economic Review*, 37: 426–434.

Quah D.T. 1996. Regional convergence clusters across Europe, *European Economic Review*, 40: 951–958.

Raftery A.E. 2000. Statistics in sociology, 1950-2000, *Journal of the American Statistical Association*, 95: 654–661.

Ramsey J. 1988. Monotone regression splines in action, *Statistical Science*, 3: 425–441.

Rauscher M. 1994. On ecological dumping, *Oxford Economic Papers*, 46: 822–840.

Raut L. 1995. R&D spillovers and productivity growth: Evidence from Indian private ˘ rms, *Journal of Development Economics*, 48: 1–23.

Ravallion M. and Jalan J. 1996. Growth divergence due to spatial externalities, *Economics Letters*, 53: 227–232.

Reinhard S., Lovell C.K. and Thijssen G. 1997. *Econometric Estimation of Technical and Environmental Ef˜ ciency: An Application of Dutch Dairy farms*, Working paper, Agricultural Economics Research Institute, The Hague, The Netherlands.

Revelli F. 2001. Spatial patterns in local taxation: Tax mimicking or error mimicking? *Applied Economics*, 33: 1101–1107.

Revelli F. 2002a. Local taxes, national politics and spatial interactions in English district election results, *European Journal of Political Economy*, 18: 28–299.

Revelli F. 2002b. Testing the tax mimicking versus expenditure spillover hypotheses using English data, *Applied Economics*, 34: 1723–1731.

Revelli F. 2003. Reaction or interaction? Spatial process identiˇ cation in multi-tiered government structures, *Journal of Urban Economics*, 53: 29–53.

Rey S.J. 2001. Spatial empirics for economic growth and convergence, *Geographical Analysis*, 33: 195–214.

Rey S.J. 2004. Spatial analysis of regional economic growth, inequality and change, in Goodchild M. and Janelle D. (eds.) *Spatially Integrated Social Science*, Oxford University Press, New York, NY, 280–299.

Rey S.J. and Montouri B.D. 1999. U.S. regional income convergence: A spatial econometric perspective, *Regional Studies*, 33: 143–156.

Ridker R. and Henning J. 1967. The determinants of residential property values with special reference to air pollution, *The Review of Economics and Statistics*, 49: 246–257.

Rietveld P. 1995. Infrastructure and spatial economic development, *Annals of Regional Science*, 29: 117–119.

Ripley B.D. 1988. *Statistical Inference for Spatial Processes*, Cambridge University Press, Cambridge, UK.

Rodriguez-Pose A. 1999. Innovation prone and innovation averse societies: Economic performance in Europe, *Growth and Change*, 30: 75–105.

Roe B., Irwin E.G. and Sharp J.S. 2002. Pigs in space: Modeling the spatial structure of hog production in traditional and nontraditional production regions, *American Journal of Agricultural Economics*, 84: 259–278.

Romer P.M. 1986. Increasing returns and long-run growth, *Journal of Political Economy*, 94: 1003–1037.

Romer P.M. 1990. Endogenous technical change, *Journal of Political Economy*, 98: S71–S102.

Rosen S. 1974. Hedonic prices and implicit markets: Product differentiation in pure competition, *Journal of Political Economy*, 82: 34–55.

Rosenblatt M. 1956. A central limit theorem and a strong mixing condition, *Proceedings of the National Academy of Sciences*, 42: 43–47.

Rosenthal R. 1991. *Meta-Analytic Procedures for Social Research*, Sage, London, UK.

Royle J.A. and Berliner L.M. 1999. A hierarchical approach to multivariate spatial modeling and prediction, *Journal of Agricultural, Biological and Environmental Statistics*, 4: 29–56.

Ruud P.A. 1991. Extensions of estimation methods using the EM algorithm, *Journal of Econometrics*, 49: 305–341.

Saavedra L.A. 2000. A model of welfare competition with evidence from AFDC, *Journal of Urban Economics*, 47: 248–279.

Saavedra L.A. 2003. Tests for spatial lag dependence based on method of moments estimation, *Regional Science and Urban Economics*, 33: 27–58.

Sachs J.D. and Warner A. 1995. Economic reform and the process of global integration, *Brookings Papers on Economic Activity*, 26: 1–118.

Sampson R.J., Morenoff J.D. and Earls F. 1999. Beyond social capital: Spatial dynamics of collective efˇ cacy for children, *American Sociological Review*, 64: 633–660.

Sampson R.J., Morenoff J.D. and Gannon-Rowley T. 2002. Assessing "neighborhood effects": Social processes and new directions in research, *Annual Review of Sociology*, 28: 443–478.

Schankerman M. and Nadiri M.I. 1986. A test of static equilibrium models and rates of return to quasi-ˇ xed factors, with an application to the Bell system, *Journal of Econometrics*, 33: 97–118.

Schelling T. 1971. Dynamic models of segregation, *Journal of Mathematical Sociology*, 1: 143–186.

Schmidt P. 1976. *Econometrics*, Marcel Dekker, New York, NY.

Schmitt B. 1996. Advantages comparatifs, dynamique de population et dynamique d'emploi des espaces ruraux, *Revue d'Économie Régionale et Urbaine*, 2: 362–382.

Scitovsky T. 1954. Two concepts of external economies, *Journal of Political Economy*, 62: 143–151.

Seitz H. and Licht G. 1995. The impact of public infrastructure capital on regional manufacturing production cost, *Regional Studies*, 29: 231–240.

Sen A. 1976. Large sample size distribution of statistics used in testing for spatial autocorrelation, *Geographical Analysis*, 9: 175–184.

Shephard R. 1953. *Cost and Production Functions*, Princeton University Press, Princeton, NJ.

Shikin E. and Plis A. 1995. *Handbook on Splines for the User*, CRC Press, Boca Raton, FL.

Shilton L. and Craig S. 1999. Spatial patterns of headquarters, *Journal of Real Estate Research*, 17: 341–364.

Shroder M. 1995. Games the states don't play: Welfare beneˇ ts and the theory of ˇ scal federalism, *Review of Economics and Statistics*, 77: 183–191.

Smirnov O. and Anselin L. 2001. Fast maximum likelihood estimation of very large spatial autoregressive models: A characteristic polynomial approach, *Computational Statistics and Data Analysis*, 35: 301–319.

Smith V. and Huang J. 1993. Hedonic models and air pollution: Twenty-ˇ ve years and counting, *Environmental and Resource Economics*, 3: 381–394.

Smith V. and Huang J. 1995. Can markets value air quality? A meta-analysis of hedonic property value models, *Journal of Political Economy*, 103: 209–227.

Smith V. and Pattanayak K. 2002. Is meta-analysis a Noah's ark for non-market valuation, *Environmental and Resource Economics*, 22: 271–296.

Sneek J. and Rietveld P. 1997. *Estimating Spatial ARMA Models*, Working Paper Discussion paper 97-643/3, Tinbergen Institute, Amsterdam, The Netherlands.

Solow R. 1956. A contribution to the theory of economic growth, *Quarterly Journal of Economics*, 70: 65–94.

Spitzer J. 1984. Variance estimates in models with the Box-Cox transformation: Implications for estimation and hypothesis testing, *The Review of Economics and Statistics*, 66: 645–652.

Stanley T. 2001. Wheat from chaff: Meta-analysis as quantitative literature review, *Journal of Economic Perspectives*, 15: 131–150.

Starr H. 2001. Using Geographic Information Systems to revisit enduring rivalries: The case of Israel, *Geopolitics*, 5: 37–56.

Steinnes D.N. 1977. Causality and intraurban location, *Journal of Urban Economics*, 4: 69–79.

Steinnes D.N. and Fisher W.D. 1974. An econometric model of intraurban location, *Journal of Regional Science*, 14: 65–80.

Stern S. 1997. Simulation-based estimation, *Journal of Economic Literature*, 35: 2006–2039.

Stetzer F. 1982. Specifying weights in spatial forecasting models: The results of some experiments, *Environment and Planning A*, 14: 571–584.

Suárez F. 1992. Economias de escala, poder de mercado y externalidades: Medición de las fuentes del crecimiento español, *Investigaciones Económicas*, 16: 411–441.

Subramanian S. and Carson R.T. 1988. Robust regression in the presence of heteroskedasticity, in Rhodes G. and Fomby T. (eds.) *Advances in Econometrics*, JAI Press, Greenwich, CT, 85–138.

Summers R. and Heston A. 1991. The Penn World Tables (Mark 5): An expanded set of international comparisons, 1950–1988, *Quarterly Journal of Economics*, 106: 327–369.

Sutton A., Abrams K., Jones D., Sheldon T. and Song F. 2001. *Methods for Meta-Analysis in Medical Research*, John Wiley and Sons, Chichester, UK.

Swann G.M., Prevezer M. and Stout D. 1998. *The Dynamic of Industrial Clustering: International Comparisons in Computing and Biotechnology*, Oxford University Press, Oxford, UK.

Tabuchi T. 1998. Urban agglomeration and dispersion: A synthesis of Alonso and Krugman, *Journal of Urban Economics*, 44: 333–351.

Targetti F. and Foti A. 1997. Growth and productivity: A model of cumulative growth and catch-up, *Cambridge Journal of Economics*, 21: 27–43.

Taub A. 1979. Prediction in the context of the variance components model, *Journal of Econometrics*, 10: 103–107.

Terui N. and Kikuchi M. 1994. The size-adjusted critical region of Moran's I test statistics for autocorrelation and its application to geographic areas, *Geographical Analysis*, 26: 213–227.

Theil H. and Goldberger A.S. 1961. On pure and mixed statistical estimation in economics, *International Economic Review*, 2: 65–78.

Thibodeau T.G. 2003. Marking single-family property values to market, *Real Estate Economics*, 31: 1–22.

Thirlwall A. 1983. Symposium on Kaldor's laws, *Journal of Post Keynesian Economics*, 5: 341–429.

Thomas A. 1996. *Increasing Returns, Congestion Costs, and the Geographic Concentration of Firms*, Working paper, International Monetary Fund, Washington, DC.

Thomas D.C. 2000. Some contributions of statistics to environmental epidemiology, *Journal of the American Statistical Association*, 95: 315–319.

Thorsnes P. and McMillen D.P. 1998. Land value and parcel size: A semiparametric analysis, *Journal of Real Estate Finance and Economics*, 17: 233–244.

Thurston L. and Yezer A. 1994. Causality in the suburbanization of population and employment, *Journal of Urban Economics*, 35: 105–118.

Tibshirani R. and Hastie T.J. 1987. Local likelihood estimation, *Journal of the American Statistical Association*, 82: 559–567.

Tiefelsdorf M. 2000. Modelling spatial processes, in *Lecture Notes in Earth Sciences*, Volume 87, Springer-Verlag, Berlin, Germany.

Tiefelsdorf M. 2002. The saddlepoint approximation of Moran's I's and local Moran's I_i's reference distributions and their numerical evaluation, *Geographical Analysis*, 34: 187–206.

Tiefelsdorf M. and Boots B. 1995. The exact distribution of Moran's I, *Environment and Planning A*, 27: 985–999.

Tiefelsdorf M., Grif˜ th D. and Boots B. 1999. A variance-stabilizing coding scheme for spatial link matrices, *Environment and Planning A*, 31: 165–180.

Tobey J. 1990. The effects of domestic environmental policies on patterns of world trade: An empirical test, *Kyklos*, 43: 191–209.

Topa G. 2001. Social interactions, local spillover and unemployment, *Review of Economic Studies*, 68: 261–295.

Ullah A. and Singh R.S. 1989. Estimation of a probability density function with applications to nonparametric inference in econometrics, in Raj B. (ed.) *Advances in Econometrics and Modelling*, Kluwer Academic Press, Dordrecht, The Netherlands.

UNCED 1992. *Nations of the Earth Report, Vol. I–III*, United Nations, Geneva, Switzerland.

Upton G.J. and Fingleton B. 1985. *Spatial Data Analysis by Example I*, John Wiley and Sons, New York, NY.

Upton G.J. and Fingleton B. 1989. *Spatial Data Analysis by Example II*, John Wiley and Sons, Chichester, UK.

USEPA 1993. *National Air Quality and Emissions Trends Report, 1992*, US Government Printing Of˜ ce, Research Triangle Park, NC.

Vamvakidis A. 1998. Regional integration and economic growth, *The World Bank Economic Review*, 12: 251–270.

van Beers C. and van den Bergh J. 1997. An empirical multi-country analysis of the impact of environmental policy on foreign trade ows, *Kyklos*, 50: 29–46.

Vayá E. 1998. Localización, crecimiento y externalidades regionales. Una propuesta basada en la Econometria Espacial, Ph.D. thesis, University of Barcelona, Barcelona, Spain.

Vayá E., López-Bazo E. and Artis M. 1998. *Growth, Convergence and (Why Not?) Regional Externalities*, Working Paper E98/31, Divisió de Ciences Juridiques, Economiques i Socials, Collecció d'Economia, University of Barcelona, Barcelona, Spain.

Velázquez F. 1993. Econom´as de escala y tamaños óptimos en la industria española (1980–1986), *Investigaciones Económicas*, 17: 507–525.

Venables A.J. 1996. Equilibrium locations of vertically linked industries, *International Economic Review*, 37: 341–359.

Verdoorn P. 1949. Fattori che regolano lo sviluppo della produttivita del lavoro, *L'Industria*, 1: 3–10.

Verspagen B. 1991. A new empirical approach to catching up or falling behind, *Structural Change and Economic Dynamics*, 2: 359–380.

Verspagen B. 1997. Estimating international technology spillovers using tech ow matrices, *Weltwirtschaftliches Archiv*, 133: 226–248.

Vijverberg W.P. 1997. Monte Carlo evaluation of multivariate normal probabilities, *Journal of Econometrics*, 76: 281–307.

Vijverberg W.P. 1999. *Rectangular and Wedge-Shaped Multivariate Normal Probabilities*, Working paper, School of Social Sciences, University of Texas at Dallas, Richardson, TX.

Walcott S.M. 1999. High tech in the deep south: biomedical ˇ rm clusters in metropolitan Atlanta, *Growth and Change*, 30: 48–74.

Wall M.M. 2003. A close look at the spatial structure implied by the CAR and SAR models, *Journal of Statistical Planning and Inference*, in press.

Waller L.A., Carlin B., Hong X. and Gelfand A.E. 1997. Hierarchical spatio-temporal mapping of disease rates, *Journal of the American Statistical Association*, 92: 607–617.

Wansbeek T. and Kapteyn A. 1978. The separation of individual variation and systematic change in the analysis of panel data, *Annales de l'INSEE*, 30-31: 659–680.

Wansbeek T. and Kapteyn A. 1983. A note on spectral decomposition and Maximum Likelihood estimation of ANOVA models with balanced data, *Statistics and Probability Letters*, 1: 213–215.

Ward M.D. 2002. The development and application of spatial analysis for political methodology, *Political Geography*, 21: 155–158.

Ward M.D. and O'Loughlin J. 2002. Spatial processes and political methodology: Introduction to the special issue, *Political Analysis*, 10: 211–216.

Werczberger E. 1987. A dynamic model of urban land use with externalities, *Regional Science and Urban Economics*, 17: 391–410.

Wikle C.K., Berliner L.M. and Cressie N. 1998. Hierarchical Bayesian space-time models, *Environmental and Ecological Statistics*, 5: 117–154.

Wolpert R.L. and Ickstadt K. 1998. Poisson/Gamma random ˇ eld models for spatial statistics, *Biometrika*, 85: 251–267.

Wood A. 1998. Globalization and the rise in labour market inequalities, *Economic Journal*, 108: 1463–1482.

Zellner A. 1971. *An Introduction to Bayesian Inference in Econometrics*, John Wiley and Sons, New York, NY.

Author Index

Index

List of Contributors

LUC ANSELIN
Department of Agricultural and Consumer Economics
University of Illinois, Urbana–Champaign
Urbana, IL 61801
USA
anselin@uiuc.edu

MANUEL ARTÍS
Research Group "Analisi Quantitativa Regional" (AQR)
Department of Econometrics, Statistics, and Spanish Economy
University of Barcelona
08034 Barcelona
Spain
artis@eco.ub.es

BADI H. BALTAGI
Department of Economics
Texas A&M University
College Station, TX 77843
USA
badi@econmail.tamu.edu

DAVID BARKLEY
Department of Agricultural and Applied Economics
Clemson University
Clemson, SC 29634
USA
dbrkly@clemson.edu

RONALD BARRY
Department of Mathematical Sciences
University of Alaska
Fairbanks, Alaska 99775
USA
FFRPB@aurora.alaska.edu

KURT J. BERON
School of Social Science
University of Texas at Dallas
Richardson, TX 75083
USA
kberon@utdallas.edu

ROGER S. BIVAND
Department of Economics
Norwegian School of Economics and Business Administration
N-5045 Bergen
Norway
Roger.Bivand@nhh.no

MARLON G. BOARNET
Department of Urban and Regional Planning
University of California, Irvine
Irvine, CA 92717
USA
mgboarne@uci.edu

NANCY BOCKSTAEL
Department of Agricultural and Resource Economics
University of Maryland
College Park, MD 20742
USA
nancyb@arec.umd.edu

THOMAS DE GRAAFF
Department of Spatial Economics
Free University
1081 HV Amsterdam
The Netherlands
tgraaff@feweb.vu.nl

PAAVO ELISTE
Poverty Reduction and Economic Management
East Asia and Paciˇ c Region
The World Bank
Washington, DC 20433
USA
peliste@worldbank.org

BERNARD FINGLETON
Department of Land Economy
University of Cambridge
Cambridge CB2 9EP
UK
bf100@cam.ac.uk

MARK M. FLEMING
Fannie Mae Foundation
Washington, DC 20016
USA
mark_ eming@fanniemae.com

RAYMOND J.G.M. FLORAX
Department of Spatial Economics
Free University
1081 HV Amsterdam
The Netherlands
r orax@feweb.vu.nl

PER G. FREDRIKSSON
Department of Economics
Southern Methodist University
Dallas, TX 75275
USA
pfredrik@mail.smu.edu

MICHAEL F. GOODCHILD
Department of Geography
University of California, Santa Barbara
Santa Barbara, CA 93106
USA
good@ncgia.ucsb.edu

YAW HANSON
Fannie Mae
Washington, DC 20016
USA
yhanson@fanniemae.com

MARK HENRY
Department of Agricultural and Applied Economics
Clemson University
Clemson, SC 29634
USA
mhenry@clemson.edu

YANNIS M. IOANNIDES
Department of Economics
Tufts University
Medford, MA 02155
USA
yioannid@tufts.edu

ELENA IRWIN
Department of Agricultural, Environmental, and Development Economics
The Ohio State University
Columbus, OH 43210
USA
irwin.78@osu.edu

HARRY H. KELEJIAN
Department of Economics
University of Maryland
College Park, MD 20742
USA
kelejian@econ.umd.edu

JAMES P. LESAGE
Department of Economics
University of Toledo
Toledo, OH 43606
USA
jpl@jpl.econ.utoledo.edu

DONG LI
Department of Economics
Kansas State University
Manhattan, KS 66506
USA
dongli@ksu.edu

ENRIQUE LÓPEZ-BAZO
Research Group "Analisi Quantitativa Regional" (AQR)
Department of Econometrics, Statistics, and Spanish Economy
University of Barcelona
08034 Barcelona
Spain
elopez@eco.ub.es

JOHN F. MCDONALD
Department of Economics
University of Illinois at Chicago
Chicago, IL 60607
USA
john.f.mcdonald@uic.edu

DANIEL P. MCMILLEN
Departments of Economics and Finance
University of Illinois at Chicago
Chicago, IL 60607
USA
mcmillen@uic.edu

ROSINA MORENO
Research Group "Analisi Quantitativa Regional" (AQR)
Department of Econometrics, Statistics, and Spanish Economy
University of Barcelona
08034 Barcelona
Spain
rmore@eco.ub.es

JAMES C. MURDOCH
School of Social Science
University of Texas at Dallas
Richardson, TX 75083
USA
murdoch@utdallas.edu

R. KELLEY PACE
E.J. Ourso College of Business Administration
Louisiana State University
Baton Rouge, LA 70803
USA
kelley@pace.am

JORIS PINKSE
Department of Economics
The Pennsylvania State University
University Park, PA 16802
USA
joris@psu.edu

BORIS A. PORTNOV
Desert Architecture and Urban Planning Unit
Jacob Blaustein Institute for Desert Research
Ben-Gurion University of the Negev
Midreshet, Ben-Gurion 84990
Israel
portnov@bgumail.bgu.ac.il

SERGIO J. REY
Department of Geography
San Diego State University
San Diego, CA 92182
USA
rey@typhoon.sdsu.edu

DENNIS P. ROBINSON
Institute for Economic Advancement
University of Arkansas at Little Rock
Little Rock, AR 72204
USA
dprobinson1@ualr.edu

C.F. SIRMANS
Center for Real Estate and Urban Studies
University of Connecticut
Storrs, CT 06269
USA
cf@sba.uconn.edu

V. CARLOS SLAWSON JR.
E.J. Ourso College of Business Administration
Louisiana State University
Baton Rouge, LA 70803
USA
cslawson@lsu.edu

JORDI SURIÑACH
Research Group "Analisi Quantitativa Regional" (AQR)
Department of Econometrics, Statistics, and Spanish Economy
University of Barcelona
08034 Barcelona
Spain
surinach@eco.ub.es

MARK A. THAYER
Department of Economics
San Diego State University
San Diego, CA 92182
USA
mthayer@mail.sdsu.edu

ESTHER VAYÁ
Research Group "Analisi Quantitativa Regional" (AQR)
Department of Econometrics, Statistics, and Spanish Economy
University of Barcelona
08034 Barcelona
Spain
evaya@eco.ub.es

Wim P.M. Vijverberg
School of Social Sciences
University of Texas at Dallas
Richardson, TX 75083
USA
vijver@utdallas.edu